# Bayesian Inference
# Volume II

# The International Library of Critical Writings in Econometrics

*Series Editors:* Mark Blaug
*Professor Emeritus, University of London*
*Consultant Professor, University of Buckingham*
*Visiting Professor, University of Exeter*
Adrian C. Darnell
*Senior Lecturer in Economics, University of Durham*

1. The Econometrics of Panel Data (Volumes I and II)
   *G.S. Maddala*

2. Macroeconometric Modelling (Volumes I and II)
   *Kenneth F. Wallis*

3. Simultaneous Equation Estimation
   *Carl F. Christ*

4. The History of Econometrics (Volumes I and II)
   *Adrian C. Darnell*

5. Time Series (Volumes I and II)
   *Andrew Harvey*

6. The Methodology of Econometrics (Volumes I and II)
   *Dale J. Poirier*

7. Bayesian Inference (Volumes I and II)
   *Nicholas G. Polson and George C. Tiao*

Future titles will include:

General to Specific Modelling
*Neil R. Ericsson*

Selection Bias
*James J. Heckman*

Specification Problems

Nonlinear Models

# Bayesian Inference
# Volume II

*Edited by*

## Nicholas G. Polson and George C. Tiao

*Graduate School of Business,*
*University of Chicago, US*

THE INTERNATIONAL LIBRARY OF CRITICAL WRITINGS IN ECONOMETRICS

An Elgar Reference Collection

Published by
Edward Elgar Publishing Limited
Gower House
Croft Road
Aldershot
Hants GU11 3HR
England

Edward Elgar Publishing Company
Old Post Road
Brookfield
Vermont 05036
USA



**British Library Cataloguing in Publication Data**
Bayesian Inference. – (International
Library of Critical Writings in
Econometrics; Vol. 7)
I. Polson, Nicholas II. Tiao, George C.
III. Series
330.01519542

**Library of Congress Cataloguing in Publication Data**
Bayesian inference / edited by Nicholas G. Polson and George C. Tiao.
p. cm. — (International library of critical writings in
econometrics; 7) (An Elgar reference collection)
Includes bibliographical references and index.
1. Econometrics. 2. Bayesian statistical decision theory.
3. Economics, Mathematical. I. Polson, Nicholas G., 1963– .
II. Tiao, George C., 1933– . III. Series. IV. Series: An Elgar
reference collection.
HB139.B3948 1995
330'.01'519542—dc20 94–40799
CIP

ISBN 1 85278 668 X (2 volume set)

Printed in Great Britain at the University Press, Cambridge

# Contents

# Acknowledgements

The editor and publishers wish to thank the following who have kindly given permission for the use of copyright material.

American Statistical Association for articles: Brule M. Hill (1965), 'Inference about Variance Components in the One-Way Model', *Journal of the American Statistical Association*, **60** (311), 806–25; William H. DuMouchel and Jeffrey E. Harris (1983), 'Bayes Methods for Combining the Results of Cancer Studies in Humans and Other Species', *Journal of the American Statistical Association*, **78** (382), June, 293–315; Luke Tierney and Joseph B. Kadane (1986), 'Accurate Approximations for Posterior Moments and Marginal Densities', *Journal of the American Statistical Association*, **81** (393), March, 82–6; Martin A. Tanner and Wing Hung Wong (1987), 'The Calculation of Posterior Distributions by Data Augmentation (with Discussion', *Journal of the American Statistical Association*, **82** (398), June, 528–50; Alan E. Gelfand and Adrian F.M. Smith (1990), 'Sampling-Based Approaches to Calculating Marginal Densities', *Journal of the American Statistical Association*, **85** (410), June, 398–409;

Basil Blackwell for articles: Seymour Geisser and Jerome Cornfield (1963), 'Posterior Distributions for Multivariate Normal Parameters', *Journal of the Royal Statistical Society*, **B**, **25** (2), 368–76; D.V. Lindley and A.F.M. Smith (1972), 'Bayes Estimates for the Linear Model', *Journal of the Royal Statistical Society*, **B**, **34** (1), 1–41; C.C. Heyde and I.M. Johnstone (1979), 'On Asymptotic Posterior Normality for Stochastic Processes', *Journal of the Royal Statistical Society*, **B**, **41** (2), 184–9.

Econometric Society for articles: Arnold Zellner (1985), 'Bayesian Econometrics', *Econometrica*, **53** (2), March, 253–69; John Geweke (1989), 'Bayesian Inference in Econometric Models Using Monte Carlo Integration', *Econometrica*, **57** (6), November, 1317–39.

Institute of Electrical and Electronics Engineers, Inc. for article: Stuart Geman and Donald Geman (1984), 'Stochastic Relaxation, Gibbs Distributions, and the Bayesian Restoration of Images', *IEEE Transactions on Pattern Analysis and Machine Intelligence*, **6** (6), November, 721–41.

Institute of Mathematical Statistics for article: Dennis V. Lindley (1964), 'The Bayesian Analysis of Contingency Tables', *Annals of Mathematical Statistics*, **35** (4), 1622–43; Thomas S. Ferguson (1974), 'Prior Distributions on Spaces of Probability Measures', *The Annals of Statistics*, **2** (4), July, 615–29.

Joseph B. Kadane and James M. Dickey for their own article: (1980), 'Bayesian Decision Theory and the Simplification of Models', in Jan Kmenta and James B. Ramsey (eds), *Evaluation of Econometric Models*, 245–68.

Elsevier Science Publishers B.V. for excerpt: Jacques H. Drèze (1974), 'Bayesian Theory of Identification in Simultaneous Equation Models', in S.E. Fienberg and A. Zellner (eds), *Studies in Bayesian Econometrics in Honor of Leonard J. Savage*, 159–74.

Every effort has been made to trace all the copyright holders but if any have been inadvertently overlooked the publishers will be pleased to make the necessary arrangement at the first opportunity.

In addition the publishers wish to thank the Library of the London School of Economics and Political Science, and the Photographic Unit of the University of London Library for their assistance in obtaining these articles.

# Part I
# Hierarchical, Multivariate and Non-Parametric Models

# [1]

Article from the *Journal of the Royal Statistical Society*, 1963, **B**, **25** (2), 368–76.



## Posterior Distributions for Multivariate Normal Parameters

By Seymour Geisser and Jerome Cornfield

*National Institutes of Health*

[Received January 1963. Revised May 1963]

Summary

A class of Bayes posterior distributions is obtained for the parameters of the multivariate normal distribution. These are compared with their fiducial and confidence counterparts.

### 1. Introduction

Cornish (1961), following an earlier suggestion of Fisher (1954), has derived a joint fiducial distribution of the means of a $p$-variate normal distribution. The hyper-ellipsoidal fiducial region on the means given by this distribution is not the same as that given by the confidence region based on Hotelling's $T^2$. In most earlier examples of differences between fiducial and confidence intervals, the disagreement has been obscured either by uncertainty as to the existence of a confidence interval based only on sufficient statistics, as in the Behrens–Fisher problem, or by the existence of two or more fiducial solutions, as in the ratio of two normal means. But the present disagreement is unequivocal since the confidence region is well defined and no other fiducial region has yet been produced to confound the issue.

We have thought it of interest to examine this disagreement from the point of view of Bayes, using a prior distribution with an adjustable parameter, $v$. This examination leads to a posterior distribution of the means which reduces to the Fisher–Cornish density (F–C) for $v = 2$ and to what we will call the Hotelling density (H) for $v = p+1$ (Section 2). We have investigated the effect of a choice of $v$ on the marginal distribution of a single mean and on the marginal distribution of the elements of the covariance matrix (Section 3). The main result here is that there is no value of $v$ that will lead to both Student's density as a marginal density for a single mean and Hotelling's density as the joint density for all the means. We also demonstrate that the posterior distribution of the elements of the covariance matrix in the bivariate case is different from the corresponding fiducial distribution for all values of $v$.

Assuming we have $N>p$ observations from a $p$-variate normal distribution with vector mean $\boldsymbol{\mu}' = (\mu_1, \ldots, \mu_p)$ and covariance matrix $\boldsymbol{\Sigma} = \{\sigma_{ij}\}$, we then calculate from the observations the sample vector mean $\bar{\mathbf{x}}' = (\bar{x}_1, \ldots, \bar{x}_p)$ and the sample covariance matrix $\mathbf{S} = \{s_{ij}\}$, where

$$s_{ij} = (N-1)^{-1} \sum_k (x_{ki}-\bar{x}_i)(x_{kj}-\bar{x}_j) \quad \text{and} \quad \bar{x}_i = N^{-1} \sum_k x_{ki}.$$

In this notation, the Fisher–Cornish fiducial density of the means assumes the form

$$\phi(\boldsymbol{\mu}\,|\,\mathbf{S}, \bar{\mathbf{x}})\,d\boldsymbol{\mu} = \left(\frac{N}{(N-1)\pi}\right)^{\frac{1}{2}p} \frac{\Gamma\{(N-1+p)/2\}}{\Gamma\{(N-1)/2\}\,|\mathbf{S}|^{\frac{1}{2}}} \times \left\{1+\frac{N(\bar{\mathbf{x}}-\boldsymbol{\mu})'\mathbf{S}^{-1}(\bar{\mathbf{x}}-\boldsymbol{\mu})}{N-1}\right\}^{\frac{1}{2}(N-1+p)} d\boldsymbol{\mu}, \qquad (1.1)$$

where

$$d\boldsymbol{\mu} = \prod_{i=1}^{p} d\mu_i.$$

Cornish shows that the quantity $T^2 = N(\bar{\mathbf{x}}-\boldsymbol{\mu})'\mathbf{S}^{-1}(\bar{\mathbf{x}}-\boldsymbol{\mu})$, where $\boldsymbol{\mu}$ is variable and $\bar{\mathbf{x}}$ and $\mathbf{S}$ are fixed, is distributed like $pF(p, N-1)$, where $F(\alpha,\beta)$ represents the $F$-distribution with $\alpha$ and $\beta$ degrees of freedom (see equation 2.9 and the following discussion). This differs from the density of $T^2$ as given by Hotelling (where $\bar{\mathbf{x}}$ and $\mathbf{S}$ are variable and $\boldsymbol{\mu}$ is fixed), namely, $\{p(N-1)/(N-p)\}\,F(p, N-p)$. This leads to a hyperellipsoidal confidence region which differs from the fiducial region except for the case $p = 1$.

## 2. A Bayes Procedure

For notational convenience let

$$\boldsymbol{\Sigma}^{-1} = \{\sigma^{ij}\},\quad d\boldsymbol{\Sigma}^{-1} = \prod_{i\geqslant j} d\sigma^{ij},\quad d\mathbf{S} = \prod_{i\geqslant j} ds_{ij},\quad d\boldsymbol{\mu} = \prod_{i} d\mu_i \quad\text{and}\quad d\bar{\mathbf{x}} = \prod_{i} d\bar{x}_i.$$

For $N > p$, $\boldsymbol{\Sigma}$ and $\mathbf{S}$ positive definite, we have

$$f(\mathbf{S}\,|\,\boldsymbol{\Sigma}^{-1})\,d\mathbf{S} = K(p,N)\,|(N-1)\,\mathbf{S}|^{\frac{1}{2}(N-2-p)}\,|\boldsymbol{\Sigma}^{-1}|^{\frac{1}{2}(N-1)} \times \exp\{-\tfrac{1}{2}\operatorname{tr}\boldsymbol{\Sigma}^{-1}(N-1)\,\mathbf{S}\}\,d\mathbf{S}, \tag{2.1}$$

where

$$(N-1)^{\frac{1}{2}p(p+1)}\,K^{-1}(p,N) = 2^{\frac{1}{2}p(N-1)}\,\pi^{\frac{1}{4}p(p-1)}\prod_{i=1}^{p}\Gamma\left(\frac{N-i}{2}\right),$$

and

$$f(\bar{\mathbf{x}}\,|\,\boldsymbol{\mu},\boldsymbol{\Sigma}^{-1})\,d\bar{\mathbf{x}} = \left(\frac{N}{2\pi}\right)^{\frac{1}{2}p}|\boldsymbol{\Sigma}^{-1}|^{\frac{1}{2}}\exp\{-\tfrac{1}{2}\operatorname{tr}\boldsymbol{\Sigma}^{-1}N(\bar{\mathbf{x}}-\boldsymbol{\mu})(\bar{\mathbf{x}}-\boldsymbol{\mu})'\}\,d\bar{\mathbf{x}}. \tag{2.2}$$

Let the prior joint distribution of $\boldsymbol{\Sigma}^{-1}$ and $\boldsymbol{\mu}$ be

$$g_v(\boldsymbol{\mu},\boldsymbol{\Sigma}^{-1})\,d\boldsymbol{\mu}\,d\boldsymbol{\Sigma}^{-1} \propto |\boldsymbol{\Sigma}|^{\frac{1}{2}v}\,d\boldsymbol{\mu}\,d\boldsymbol{\Sigma}^{-1}, \tag{2.3}$$

where $v$ may be any integer positive or negative such that $v \leqslant N$. A justification for this prior distribution can be given in terms of an objective Bayesian criterion which, however, will not be discussed. Here we will merely explore its consequences. Therefore, since $\bar{\mathbf{x}}$ and $\mathbf{S}$ are independent,

$$P_v(\boldsymbol{\Sigma}^{-1},\boldsymbol{\mu}\,|\,\bar{\mathbf{x}},\mathbf{S})\,d\boldsymbol{\Sigma}^{-1}\,d\boldsymbol{\mu} \propto f(\mathbf{S}\,|\,\boldsymbol{\Sigma}^{-1})\,f(\bar{\mathbf{x}}\,|\,\boldsymbol{\mu},\boldsymbol{\Sigma}^{-1})\,g_v(\boldsymbol{\mu},\boldsymbol{\Sigma}^{-1})\,d\boldsymbol{\Sigma}^{-1}\,d\boldsymbol{\mu}. \tag{2.4}$$

Substitution of (2.1) and (2.2) in (2.4) yields

$$P_v(\boldsymbol{\Sigma}^{-1},\boldsymbol{\mu}\,|\,\bar{\mathbf{x}},\mathbf{S})\,d\boldsymbol{\Sigma}^{-1}\,d\boldsymbol{\mu} \propto |\boldsymbol{\Sigma}^{-1}|^{\frac{1}{2}(N-v)} \times \exp[-\tfrac{1}{2}\operatorname{tr}\boldsymbol{\Sigma}^{-1}\{(N-1)\,\mathbf{S}+N(\bar{\mathbf{x}}-\boldsymbol{\mu})(\bar{\mathbf{x}}-\boldsymbol{\mu})'\}]\,d\boldsymbol{\Sigma}^{-1}\,d\boldsymbol{\mu}. \tag{2.5}$$

Integrating out $\boldsymbol{\mu}$ in (2.5) we obtain the posterior distribution of $\boldsymbol{\Sigma}^{-1}$ to be

$$P_v(\boldsymbol{\Sigma}^{-1}\,|\,\mathbf{S})\,d\boldsymbol{\Sigma}^{-1} \propto |\boldsymbol{\Sigma}^{-1}|^{\frac{1}{2}(N-v-1)}\exp\{-\tfrac{1}{2}\operatorname{tr}\boldsymbol{\Sigma}^{-1}(N-1)\,\mathbf{S}\}\,d\boldsymbol{\Sigma}^{-1}. \tag{2.6}$$

Integrating out $\boldsymbol{\Sigma}^{-1}$ in (2.5), using the Wishart–Bartlett density result, we obtain the posterior marginal density of $\boldsymbol{\mu}$ given $\bar{\mathbf{x}}$ and $\mathbf{S}$ to be

$$P_v(\boldsymbol{\mu}\,|\,\bar{\mathbf{x}},\mathbf{S})\,d\boldsymbol{\mu} \propto \frac{|(N-1)\,\mathbf{S}|^{\frac{1}{2}(N+p-v)}\,d\boldsymbol{\mu}}{|(N-1)\,\mathbf{S}+N(\bar{\mathbf{x}}-\boldsymbol{\mu})(\bar{\mathbf{x}}-\boldsymbol{\mu})'|^{\frac{1}{2}(N+p-v+1)}}. \tag{2.7}$$

Since

$$\frac{|(N-1)\mathbf{S}|^{\frac{1}{2}(N+p-v)}}{|(N-1)\mathbf{S}+N(\bar{\mathbf{x}}-\boldsymbol{\mu})(\bar{\mathbf{x}}-\boldsymbol{\mu})'|^{\frac{1}{2}(N+p-v+1)}}$$
$$=\frac{\{1+[N(\bar{\mathbf{x}}-\boldsymbol{\mu})'\mathbf{S}^{-1}(\bar{\mathbf{x}}-\boldsymbol{\mu})]/(N-1)\}^{-\frac{1}{2}(N+p-v+1)}}{|(N-1)\mathbf{S}|^{\frac{1}{2}}}, \tag{2.8}$$

we obtain, after inserting the norming constant,

$$P_v(\boldsymbol{\mu}\,|\,\bar{\mathbf{x}},\mathbf{S})\,d\boldsymbol{\mu}=\left(\frac{N}{\pi(N-1)}\right)^{\frac{1}{2}p}\frac{\Gamma\{(N+p-v+1)/2\}}{\Gamma\{(N-v+1)/2\}|\mathbf{S}|^{\frac{1}{2}}}$$
$$\times\left\{1+\frac{N(\bar{\mathbf{x}}-\boldsymbol{\mu})'\mathbf{S}^{-1}(\bar{\mathbf{x}}-\boldsymbol{\mu})}{N-1}\right\}^{-\frac{1}{2}(N+p-v+1)}d\boldsymbol{\mu}. \tag{2.9}$$

For $v=2$ this reduces to (1.1), the F–C (Fisher–Cornish) posterior fiducial distribution of $\boldsymbol{\mu}$. Further we may find for any integral $v \leqslant N$ the posterior distribution of $T^2 = N(\bar{\mathbf{x}}-\boldsymbol{\mu})'\mathbf{S}^{-1}(\bar{\mathbf{x}}-\boldsymbol{\mu})$ by noting that $T^2$ is constant on ellipsoids, so that the density is immediately found from the expression for the surface area of an ellipsoid, which is in turn easily obtained by an appropriate orthogonal transformation to yield the result that $T^2$ is distributed like $\{p(N-1)/(N-v+1)\}\,F(p,N-v+1)$. Thus, for $v=2$ we have the F–C posterior fiducial density of $T^2$, $pF(p,N-1)$, and for $v=p+1$ we have the usual H (Hotelling) density of $T^2$, $\{p(N-1)/(N-p)\}\,F(p,N-p)$. This also indicates that for any fixed $N$, $\bar{\mathbf{x}}$ and $\mathbf{S}$ and probability $1-\alpha$, the posterior interval on the means given by a $p$-dimensional hyperellipsoid will monotonically decrease as $v$ decreases.

## 3. CONSEQUENCES OF THE POSTERIOR DISTRIBUTIONS

It is of interest to point out some consequences of the posterior distribution of $\boldsymbol{\mu}$ and $\boldsymbol{\Sigma}^{-1}$. First we shall consider the marginal distribution of $\boldsymbol{\mu}$ given by (2.9). Principally for convenience, but also obviously because of the distributions that result, let us denote $v=2$ as the Bayes F–C case and $v=p+1$ as the Bayes H case. We have already remarked on the fact that the distribution of $T^2$ is, for all $p$, invariant only in the H case. By invariant, we mean that the distribution of $T^2$ is the same whether we consider $\bar{\mathbf{x}}$ and $\mathbf{S}$ variable with $\boldsymbol{\mu}$ fixed or the converse, $\boldsymbol{\mu}$ variable with $\bar{\mathbf{x}}$ and $\mathbf{S}$ fixed. We also note that for $p=1$ both cases coincide and the posterior distribution of $s_{11}^{-\frac{1}{2}}(\bar{x}_1-\mu_1)\sqrt{N}$ is exactly "Student's" $t$ with $N-1$ degrees of freedom. Now for $p>1$ we make the transformation $\eta_i = \mathbf{a}_i'(\boldsymbol{\mu}-\bar{\mathbf{x}})$ in (2.9) where

$$\mathbf{a}_i' = (a_{i1},\ldots,a_{ip}).$$

In this transformation we permit the vector $\mathbf{a}_1$ to be arbitrary, subject only to the condition that at least one element is non-zero, and choose the subsequent vectors $\mathbf{a}_2,\ldots,\mathbf{a}_p$ such that $(\mathbf{a}_1'\mathbf{a}_1)^{-\frac{1}{2}}\mathbf{A}=\mathbf{R}$ is orthogonal where the matrix of column vectors $(\mathbf{a}_1,\mathbf{a}_2,\ldots,\mathbf{a}_p)=\mathbf{A}$. Therefore the new vector variable $\boldsymbol{\eta}'=(\boldsymbol{\mu}-\bar{\mathbf{x}})'\mathbf{A}$ has density

$$P_v(\boldsymbol{\eta})\,d\boldsymbol{\eta}\propto\left[1+\frac{N\boldsymbol{\eta}'(\mathbf{R}'\mathbf{S}\mathbf{R})^{-1}\boldsymbol{\eta}}{(N-1)(\mathbf{a}_1'\mathbf{a}_1)}\right]^{-\frac{1}{2}(N+p-v+1)}d\boldsymbol{\eta}. \tag{3.1}$$

Since $\mathbf{R'SR}$ is positive definite symmetric we have $\mathbf{R'SR} = \mathbf{TT'}$, where

$$\mathbf{T} = \begin{bmatrix} t_{11} & . & . & 0 & 0 \\ & . & . & & . \\ & . & & . & . \\ & . & & . & 0 \\ t_{p1} & . & . & . & t_{pp} \end{bmatrix}, \tag{3.2}$$

a lower triangular matrix. It is then easily seen that

$$t_{11}^2 = (\mathbf{a}_1'\,\mathbf{a}_1)^{-1}\,\mathbf{a}_1'\,\mathbf{S}\mathbf{a}_1. \tag{3.3}$$

Letting $\boldsymbol{\eta} = \mathbf{T}\boldsymbol{\alpha}$ and noting that $\alpha_1 = \eta_1\,t_{11}^{-1}$ we get the distribution of $\boldsymbol{\alpha}' = (\alpha_1, \ldots, \alpha_p)$ as

$$P_v(\boldsymbol{\alpha})\,d\boldsymbol{\alpha} \propto B^{-\frac{1}{2}(N+p-v+1)} \left[1 + \frac{N(\alpha_2^2 + \ldots + \alpha_p^2)}{B(\mathbf{a}_1'\,\mathbf{a}_1)(N-1)}\right]^{-\frac{1}{2}(N+p-v+1)} d\boldsymbol{\alpha}, \tag{3.4}$$

where

$$B = 1 + \frac{N\alpha_1^2}{(\mathbf{a}_1'\,\mathbf{a}_1)(N-1)}. \tag{3.5}$$

Since the factor in (3.4) which involves $\alpha_2, \ldots, \alpha_p$ is basically of the form (2.13), we may easily integrate out $\alpha_2, \ldots, \alpha_p$. This results in

$$P_v(\alpha_1)\,d\alpha_1 \propto B^{-\frac{1}{2}(N-v+2)}\,d\alpha_1. \tag{3.6}$$

Hence

$$\frac{\sqrt{(N)}\,\alpha_1}{(\mathbf{a}_1'\,\mathbf{a}_1)^{\frac{1}{2}}} = \frac{\sqrt{(N)}\,\mathbf{a}_1'(\boldsymbol{\mu} - \bar{\mathbf{x}})}{(\mathbf{a}_1'\,\mathbf{S}\mathbf{a}_1)^{\frac{1}{2}}} \tag{3.7}$$

is distributed like $(N-1)^{\frac{1}{2}}(N+1-v)^{-\frac{1}{2}}\,t$ where $t$ has the "Student" distribution with $N+1-v$ degrees of freedom. We note then that this quantity is exactly $t$ with $N-1$ degrees of freedom only in the F–C case.

Special cases of particular interest are: (1) the elements of $a_1$ are all zero except for a 1 in the $j$th position; (2) the elements of $a_1$ are all zero except for a 1 in the $j$th position and a $-1$ in the $i$th position. The former case yields

$$s_{jj}^{-\frac{1}{2}}(\mu_j - \bar{x}_j)\sqrt{N} \quad \text{as} \quad (N-1)^{\frac{1}{2}}(N+1-v)^{-\frac{1}{2}}\,t$$

with $t$ having $N+1-v$ degrees of freedom; the latter case yields the same distribution for the quantity

$$(s_{ii} + s_{jj} - 2s_{ij})^{-\frac{1}{2}}(\mu_i - \mu_j + \bar{x}_j - \bar{x}_i)\sqrt{N}.$$

Hence we contrast the invariance of Hotelling's $T^2$ which obtains in the H case with the invariance of "Student's" $t$ which obtains in the F–C case.

We now turn to the marginal posterior distribution which involves the elements of $\boldsymbol{\Sigma}^{-1}$, i.e. equation (2.6). For our purpose it is sufficient to focus our attention on the bivariate case $p = 2$. Transforming from the elements of $\boldsymbol{\Sigma}^{-1}$ to $\sigma_1$, $\sigma_2$ and $\rho$ we can easily obtain the posterior distribution of the latter to be

$$P_v(\sigma_1, \sigma_2, \rho \,|\, s_1, s_2, r) \propto \frac{\exp\{-\frac{1}{2}\operatorname{tr}(N-1)\boldsymbol{\Sigma}^{-1}\mathbf{S}\}}{(1-\rho^2)^{\frac{1}{2}(N-v+5)}\,(\sigma_1\,\sigma_2)^{N+3-v}}. \tag{3.8}$$

If we compare (3.8) with the fiducial distribution of these parameters as given by Fisher in equations 226 through 228 of (1959) we observe that the only possibility of equality, based on the exponents of $\sigma_1$ and $\sigma_2$, is for the case $v = 3$ or, since $p = 2$, the H case. (Note that Fisher's equation 226 follows from his previous equation only if the differential element is

$$\frac{d\sigma_1}{\sigma_1}\frac{d\sigma_2}{\sigma_2}$$

rather than $d\sigma_1 d\sigma_2$ as given.) By consideration of the marginal distribution of $\rho$ we can show that this is not possible since it differs from the fiducial distribution of $\rho$. Using the usual transforms

$$\frac{s_1 s_2}{\sigma_1 \sigma_2'} = \alpha, \quad \frac{s_1 \sigma_2}{\sigma_1 s_2} = e^{\beta}, \tag{3.9}$$

and then integrating out $\alpha$ and $\beta$, we obtain the posterior marginal density of $\rho$ to be

$$P_v(\rho | r) = \pi^{-1}(N+1-v)(1-r^2)^{\frac{1}{2}(N+2-v)}(1-\rho^2)^{\frac{1}{2}(N-v-1)} I_{N+2-v}(\rho r), \tag{3.10}$$

where

$$I_m(\rho r) = \int_0^\infty \frac{d\beta}{(\cosh \beta - \rho r)^m}. \tag{3.11}$$

Hence the posterior marginal distribution of $\rho$ depends only on $r$ and has the same functional form as the distribution of $r$ given $\rho$, with $\rho$ and $r$ interchanged and with $N+1-v$ instead of $N-2$ degrees of freedom. Note that for $v = 3$, the H case, the degrees of freedom are invariant, i.e. the same as for $r$ given $\rho$.

We shall now show that the fiducial density $\phi(\rho | r)$ as defined in equation 3.18 is not equal to $P_v(\rho | r)$. The distribution of $r$ for given $\rho$ based on $N$ observations is

$$f(r | \rho, N) = \pi^{-1}(N-2)(1-r^2)^{\frac{1}{2}(N-4)}(1-\rho^2)^{\frac{1}{2}(N-1)} I_{N-1}(\rho r). \tag{3.12}$$

Now it was shown by Hotelling (1953) and Harley (1957) that

$$h(r) = a \sin^{-1} r + b \tag{3.13}$$

is the only function of $r$ such that

$$E\{h(r)\} = h(\rho). \tag{3.14}$$

Hence

$$\int_{-1}^{1} \cos^{-1} r f(r | \rho)\, dr = \cos^{-1} \rho. \tag{3.15}$$

Integrating (3.15) by parts, and writing

$$F(x | \rho) = \int_{-1}^{x} f(r | \rho)\, dr, \tag{3.16}$$

we obtain

$$\int_{-1}^{1} (1-r^2)^{-\frac{1}{2}} F(r | \rho)\, dr = \cos^{-1} \rho. \tag{3.17}$$

Taking the partial derivative of both sides with respect to $\rho$, and noting that

$$-\frac{\partial F(r | \rho)}{\partial \rho} = \phi(\rho | r), \tag{3.18}$$

we obtain

$$\int_{-1}^{1} (1-r^2)^{-\frac{1}{2}} \phi(\rho|r)\, dr = (1-\rho^2)^{-\frac{1}{2}}. \tag{3.19}$$

We now show that $P_v(\rho|r)$ does not satisfy (3.19). By virtue of (3.12) we have

$$P_v(\rho|r) = (1-r^2)^{\frac{3}{2}} (1-\rho^2)^{-\frac{3}{2}} f(r|\rho, N+3-v), \tag{3.20}$$

where $f(r|\rho, N+3-v)$ is the density of $r$ given $\rho$ based on $N+3-v$ observations. Therefore

$$\int_{-1}^{1} (1-r^2)^{-\frac{1}{2}} P_v(\rho|r)\, dr = \int_{-1}^{1} (1-r^2)(1-\rho^2)^{-\frac{3}{2}}$$

$$\times f(r|\rho, N+3-v)\, dr = (1-\rho^2)^{-\frac{3}{2}} E(1-r^2). \tag{3.21}$$

Hence, if $\phi(\rho|r) = P_v(\rho|r)$ for any $v$, we must have

$$(1-\rho^2)^{-\frac{3}{2}} E(1-r^2) = (1-\rho^2)^{-\frac{1}{2}}, \tag{3.22}$$

or

$$E(1-r^2) = (1-\rho^2). \tag{3.23}$$

However, (3.23) is false since the only function in general which will satisfy (3.14) is $h(r) = a \sin^{-1} r + b$. Hence

$$\phi(\rho|r) \neq P_v(\rho|r). \tag{3.24}$$

It is quite likely that no Bayes approach will lead to the fiducial form as given by Fisher (1959, p. 172). Perhaps a multiparameter extension of Lindley's (1958) argument could be made to cover this case.

One other point of interest involves the derivation of a pseudo-posterior distribution of $\rho$ by multiplying the marginal prior distribution of $\rho$ by $f(r|\rho)$. We denote the pseudo-posterior distribution obtained this way by $P'_v(\rho|r)$ and investigate the circumstances under which $P'_v(\rho|r) = P_v(\rho|r)$. We can transform the prior distribution (2.3) into

$$g_v(\rho, \sigma_1, \sigma_2)\, d\mu\, d\rho\, d\sigma_1\, d\sigma_2 \propto (1-\rho^2)^{\frac{1}{2}v-3} \sigma_1^{v-4} \sigma_2^{v-4}\, d\mu\, d\sigma_1\, d\sigma_2\, d\rho, \tag{3.25}$$

so that $\rho, \sigma_1, \sigma_2$ are independent in the prior distribution. Consider the pseudo-posterior distribution of $\rho, P'_v(\rho|r)$ obtained in the following manner.

$$P'_v(\rho|r) = \frac{f(r|\rho)\, g_v(\rho)}{g(r)}, \tag{3.26}$$

where

$$g_v(\rho) = (1-\rho^2)^{\frac{1}{2}v-3}$$

and

$$g(r) = \int f(r|\rho)\, g_v(\rho)\, d\rho. \tag{3.27}$$

Therefore

$$P'_v(\rho|r) \propto (1-\rho^2)^{\frac{1}{2}(N+v-7)} I_{N-1}(\rho r). \tag{3.28}$$

Hence only for $v = 3$, the H case, is

$$P'_v(\rho|r) = P_v(\rho|r). \tag{3.29}$$

Therefore, for the class of Bayes prior distributions studied here, only the H case is invariant under this type of manipulation.

For convenience we present in Table 1 a summary of the results of this Section, comparing the Bayes H case and the Bayes F–C case with the corresponding fiducial distributions. An entry of $\phi$ in the table means that the distribution is identical with its corresponding fiducial distribution. $\bar{\phi}$ denotes the fact that the fiducial distribution is demonstrably different. $P(\theta | D)$ denotes the posterior distribution of the parameters given the data. The last three entries summarize some of the invariances.

TABLE 1

| $P(\theta \mid D)$ | *Bayes H case* | *Bayes F–C case* |
|---|---|---|
| $P(\sigma_1, \sigma_2, \rho \mid D)$ | $\bar{\phi}$ | $\bar{\phi}$ |
| $P(\sigma_1, \sigma_2 \mid \rho, D)$ | $\phi$ | $\bar{\phi}$ |
| $P(\rho \mid D)$ | $\bar{\phi}$ | $\bar{\phi}$ |
| $P(\boldsymbol{\mu} \mid \Sigma, D)$ | $\phi$ | $\phi$ |
| $P(\boldsymbol{\mu} \mid D)$ | $\bar{\phi}$ | $\phi$ |
| $P(T^2 \mid \bar{\mathbf{x}}, \mathrm{S}) = f(T^2 \mid \boldsymbol{\mu})$ | yes | no |
| $P(t \mid \bar{\mathbf{x}}, \mathrm{S}) = f(t \mid \boldsymbol{\mu})$ | no | yes |
| $P(\rho \mid r) = P'(\rho \mid r)$ | yes | no |

## 4. DISCUSSION

Fisher (1959) has derived the simultaneous fiducial distribution of the five parameters of the bivariate normal distribution with the apparent purpose of refuting previous demonstrations of its non-uniqueness (Mauldon, 1955). Cornish (1961), however, obtains the joint fiducial distribution of means, in agreement with Fisher (1954), from a pivotal quantity and not from the joint fiducial distribution of all the parameters. It is therefore pertinent to ask whether integration of Fisher's (1959) distribution, with respect to $\sigma_1$, $\sigma_2$ and $\rho$, yields the Fisher–Cornish result for $p = 2$. We discussed this point with A. P. Dempster and he devised an elegant proof demonstrating that these results are actually inconsistent for the case $r = 0$ (Dempster, 1963). His proof hinges on the fact that $\phi(\rho | 0) = P_2(\rho | 0)$, which he establishes by an indirect argument. Actually this particular fact may be deduced directly from the definition of

$$\phi(\rho | 0) = \frac{\partial}{\partial \rho} \int_0^1 f(r | \rho)\, dr,$$

where we use the alternative form for

$$f(r | \rho) \propto (1-\rho^2)^{\frac{1}{2}(N-1)} (1-r^2)^{\frac{1}{2}(N-4)} \sum_{j=0}^{\infty} \Gamma^2\left(\frac{N+j-1}{2}\right) \frac{(2\rho r)^j}{j!}, \tag{4.1}$$

so that

$$\int_0^1 f(r | \rho)\, dr \propto (1-\rho^2)^{\frac{1}{2}(N-1)} \sum \frac{\Gamma\{(j+1)/2\}\, \Gamma\{(N+j-1)/2\}}{j!} (2\rho)^j. \tag{4.2}$$

Taking the partial derivative of both sides with respect to $\rho$, we obtain after a bit of algebraic simplification,

$$\phi(\rho | 0) \propto (1-\rho^2)^{\frac{1}{2}(N-3)} \propto P_2(\rho | 0). \tag{4.3}$$

This implies, curiously enough, that

$$\phi(\rho, \sigma_1, \sigma_2 | s_1, s_2, 0) = P_2(\rho | 0) P_3(\sigma_1, \sigma_2 | \rho, s_1, s_2, 0) \tag{4.4}$$

is a product of the factors of the two different Bayes posterior distributions. The result also shows that the confidence limits for $\rho$, given $r = 0$, are symmetric about 0, suggesting that the graphs given in Table 15 of Pearson and Hartley (1954), particularly for $N = 3$, are not entirely accurate.

The product of (4.4) and $\phi(\mu_1, \mu_2 | \rho, \sigma_1, \sigma_2)$ given by Fisher (1959, equation 230) is the joint fiducial density of the five parameters when $r = 0$. It is an easy exercise to show that integration of this density with respect to $\sigma_1$, $\sigma_2$ and $\rho$ does not yield the Fisher–Cornish result. For those who still believe in the consistency of the fiducial argument, unconstrained by restrictions on regions of integration, the demonstration that Fisher's simultaneous fiducial distribution does not lead to his asserted fiducial marginal should be illuminating, since this is by no means an artificial concoction.

It is well known that some confidence intervals are non-Bayesian in the sense that, even for any one confidence coefficient, no prior distribution will lead to a posterior probability interval identical with the confidence interval for all points in the sample space. The most accurate unbiased confidence interval on the ratio of two normal means (Lehmann, 1959, pp. 182–183) is an example of such an interval. (Recall that for each point in the sample space there is an $\alpha > 0$, such that the interval with confidence coefficient $1-\alpha$ is $-\infty$ to $\infty$, whereas the posterior probability for this interval is 1 for all prior distributions.) Similarly, some confidence intervals are Bayesian in the sense that some prior distribution will lead to posterior probability and confidence intervals which are identical for all points in the sample space and all confidence coefficients—as in the case of a normal mean. The present results are a reminder that it is possible to have confidence intervals on two or more functions of a set of parameters which are Bayesian individually, but not collectively. Thus, the confidence interval given by Student's distribution is Bayesian, and so is the region given by Hotelling's $T^2$, but no prior distribution exists which leads to both of them.

It is possible to regard a confidence interval as a prescription for setting odds on the location of the parameter point (Buehler, 1959). That is to say, someone who "believes" the interval should be willing to bet that it contains the parameter point, risking amount $1-\alpha$ to win $\alpha$, or to bet that it does not, risking $\alpha$ to win $1-\alpha$. If, and only if, the confidence set is Bayesian for all values of $\alpha$ will it be impossible for an opposing better to choose his bets in such a way, sometimes betting for and sometimes against the set, as to have positive expectation of gain for every point in a finite parameter space (Cornfield, unpublished). If confidence sets which are individually, but not collectively, Bayesian are used to set simultaneous betting odds, it then follows that it is possible for an opposing better to find such a combination of bets. In this sense the Student and Hotelling confidence intervals are inconsistent with each other, even to those unwilling to assume the existence of prior distributions of unknown constants. (Extension to infinite parameter spaces presents difficulties, but not for the necessity aspect of the theorem, which is involved here.)

One is therefore led to ask how to choose among the family of prior distributions of multivariate normal parameters that has been considered in this paper. This is a larger issue than can be discussed here. It raises both the question of objective versus subjective Bayes' approaches, and of what the criteria should be for choosing among different objective methods of obtaining prior distributions. It is clear, however, because of the existence of confidence intervals which are individually, but not collectively Bayesian, that the hope of finding a "nice" criterion that invariably leads to Bayesian confidence limits when they exist cannot be realized.

*Note added in proof.* In connection with the sentence following equation (3.24), it is easy to show, using a recently published proof of Brillinger, that the joint fiducial distribution of $\rho$, $\sigma_1$ and $\sigma_2$ is not a Bayes posterior distribution for any prior distribution. See Brillinger, D. R. (1963), "Necessary and sufficient conditions for a statistical problem to be invariant under a Lie group", *Ann. math. Statist.*, **34**, 492–500.

## REFERENCES

BUEHLER, R. J. (1959), "Some validity criteria for statistical inference", *Ann. math. Statist.*, **30**, 845–863.

CORNISH, E. A. (1961), "Simultaneous fiducial distribution of location parameters", *Tech. Paper*, no. 8, CSIRO, Melbourne, Australia.

DEMPSTER, A. P. (1963), "Further examples of inconsistencies in the fiducial argument", *Ann. math. Statist.*, **34**, 873–891.

FISHER, R. A. (1954), "Discussion on the symposium on interval estimation", *J. R. statist. Soc.* B, **16**, 212–213.

—— (1959), *Statistical Methods and Scientific Inference*, 2nd ed. Edinburgh: Oliver & Boyd.

HARLEY, B. I. (1957), "Further properties of an angular transformation of the correlation coefficient", *Biometrika*, **44**, 273–275.

HOTELLING, H. (1953), "New light on the correlation coefficient and its transforms", *J. R. statist. Soc.* B, **15**, 193–232.

LEHMANN, E. L. (1959), *Testing Statistical Hypotheses.* New York: Wiley.

LINDLEY, D. V. (1958), "Fiducial distributions and Bayes' theorem", *J. R. statist. Soc.* B, **20**, 102–107.

MAULDON, J. G. (1955), "Pivotal quantities for Wishart's and related distributions, and a paradox in fiducial theory", *J. R. statist. Soc.* B, **17**, 79–85.

PEARSON, E. S. and HARTLEY, H. O. (1954), *Biometrika Tables for Statisticians*, **1**. Cambridge University Press.

# [2]

Article from *Annals of Mathematical Statistics*, 1964, **35** (4), 1622–43.

## THE BAYESIAN ANALYSIS OF CONTINGENCY TABLES

BY DENNIS V. LINDLEY

*University College of Wales and Harvard Business School*

**Summary.** This paper describes how data from a multinomial distribution, and in particular data in the form of a contingency table, may be studied by using a prior distribution of the parameters and expressing the results in the form of a posterior distribution, or some aspects thereof, of the parameters. The analysis used must depend on the prior distribution and the form described here only applies to a certain type of prior knowledge but, for reasons given below, it is believed that this type is of frequent occurrence. The binomial situation is first considered and the results obtained there suggest a general result for the multinomial distribution, which is then established. A few remarks on Bayesian analysis in general enable the result to be applied, first to certain multinomial problems and then, with the aid of another general result, to contingency tables. The method used there has close connections with the Analysis of Variance and these connections are examined, particularly with a view to simplifying the analysis of contingency tables involving three or more factors.

**1. Binomial distributions.** Although it will appear as a special case of results to be established for the general multinomial situation, it is instructive to begin with the binomial distribution which suggested the generalizations. Let $N$ independent trials with constant probability $\theta$ of success result in $n$ successes and $(N - n)$ failures. The likelihood is

$$\theta^n(1 - \theta)^{N-n}. \tag{1.1}$$

As other authors have remarked, it is convenient to take as prior distribution one with density proportional to

$$\theta^a(1 - \theta)^b \tag{1.2}$$

for suitable $a, b > -1$. The limiting case where $a$ and $b$ both tend to $-1$ is also important. The posterior density is then proportional to

$$\theta^{a+n}(1 - \theta)^{b+N-n}. \tag{1.3}$$

The known relation between this beta-distribution and the $F$-distribution enables the last result to be expressed by saying that

$$F = \left(\frac{b + N - n + 1}{a + n + 1}\right)\left(\frac{\theta}{1 - \theta}\right) \tag{1.4}$$

has an $F$-distribution on $[2(a + n + 1), 2(b + N - n + 1)]$ degrees of freedom. When Fisher first introduced the distribution he suggested using the $z$-transfor-

Received 30 December 1963; revised 4 August 1964.

mation of it on the grounds that $z$ was more nearly normally distributed than was $F$.

Since the distribution of the ratio of two independent $\chi^2$-variables is $F$, the use of the logarithmic transformation of $F$ and its approximate normality is intimately related to the same transformation of a $\chi^2$, or gamma, variable and its approximate normality. The latter transformation has been discussed by Bartlett and Kendall [2] who remark that the "transformation may safely be used for $n = 10$ and over, more tentatively from $n = 5$ to $n = 9$, and probably not at all below $n = 5$." (Here $n$ is the degrees of freedom for $\chi^2$.) In our context the degrees of freedom are not less than twice the number of successes (or of failures) so that the transformation can certainly be used provided the smaller of these two numbers is 5 or more and probably when 3 or 4. (The recommendation is similar to that for the $\chi^2$-approximation to Pearson's goodness-of-fit statistic which, as we shall see below, we recommend be replaced by another statistic. We have more to say about this at the end of Section 3.)

Fisher used $z = \frac{1}{2} \ln F$: in this context it is more convenient to omit the half and use the natural logarithm of $F$. Fisher's approximation then amounts to saying that if $F$ has $\nu_1$ and $\nu_2$ degrees of freedom then $\ln F$ is approximately normal with mean $\ln \{(\nu_1 - 1)/(\nu_2 - 1)\}$ and variance $2(\nu_1^{-1} + \nu_2^{-1})$. In both moments the terms omitted are of order $\nu_1^{-2}$, $(\nu_1\nu_2)^{-1}$ and $\nu_2^{-2}$. For $\ln\{\theta/(1 - \theta)\}$ the mean is therefore

$$\ln \{(a + n + \tfrac{1}{2})/(N - n + b + \tfrac{1}{2})\}, \tag{1.5}$$

and the variance

$$(a + n + 1)^{-1} + (b + N - n + 1)^{-1}; \tag{1.6}$$

the approximations being good for large $a + n$ and $b + N - n$, probably above 5.

One is thus led to consider the natural logarithm of the odds, $\theta/(1 - \theta)$, in favour of success. The initial reason for using it here is its convenient approximate posterior distribution, but reflection suggests that it is a convenient quantity to consider for other reasons. If the value of $\theta$ were known, and less than $\frac{1}{2}$, it would be usual to quote its value to a fixed number of decimal places: the argument being that it is proportional changes in $\theta$ that are typically of interest. In these circumstances the logarithm of $\theta$ could be quoted to a fixed number of places and a given change in it would be of equal significance for all $\theta$. But the argument should be symmetric in success and failure and hence the logarithm of the odds, or simply the log-odds, can be used. A further, and more important argument in favour of using the log-odds is the existence of certain additive properties that they have in situations to be discussed below: essentially these arise from the fact that the important property of independence is expressed multiplicatively and therefore, in terms of logarithms, additively. The results of this paper will be expressed entirely in terms of log-odds and their multinomial generalizations.

We next consider the prior distribution (1.2). We propose to consider the limiting case $a = b = -1$ and justify this choice by two main arguments as follows. In many situations the prior knowledge of $\theta$ is small, and there is some advantage in considering an analysis for a standard prior which could be used in most applications where no appreciable prior knowledge is available. If the log-odds is to be used then the original Bayes hypothesis would suggest taking all values of it to be equally likely. This gives $a = b = -1$. That this is sensible can be seen by noting that the results of trials cannot decrease the powers of $\theta$ and $(1 - \theta)$ in the beta-distribution and presumably cannot decrease the amount of information about $\theta$. Consequently as small values as possible for $a$ and $b$ would correspond to the least prior information about $\theta$. In order that the prior distribution be proper (i.e. integrate to 1) it is necessary that $a$ and $b$ both exceed $-1$ so that $a = b = -1$ is a lower bound, though not attained, within the class of proper prior distributions. The resulting distribution is improper but this need not concern us since we will be using approximations where the degrees of freedom for $F$ are large and the posterior distributions are proper.

A second reason for restricting attention to $a = b = -1$ is that this special case can be regarded as a canonical form for all prior distributions belonging to the beta-family (1.2). By this we mean that any beta-distribution can be reduced to the canonical form $a = b = -1$. To see this we remark that if the prior distribution is given by (1.2) it may be regarded as a posterior distribution to data consisting of $(a + 1)$ successes and $(b + 1)$ failures, this data having prior distribution with $a = b = -1$. Consequently if the actual data yield $n$ successes and $(N - n)$ failures we may regard the total knowledge of $\theta$ to consist of $(n + a + 1)$ successes and $(N - n + b + 1)$ failures for which the prior has the canonical form. Consequently by adding $(a + 1)$ and $(b + 1)$ respectively to the numbers of successes and failures actually observed the beta-distribution may be replaced by the canonical form. Of course, the method is not available for prior distributions which are not beta: see remarks in Section 8 below.

In many applications it will probably be true that the values of $a$ and $b$ for the prior distribution will be small. If the sample values, $n$ and $(N - n)$, the numbers of successes and failures, are large, we see from (1.5) and (1.6) that the actual small values of $a$ and $b$ will not be important. It is therefore reasonable to take $a = b = -1$. With $a = b = -1$ the posterior distribution of $\ln \{\theta/(1 - \theta)\}$ is approximately normal with mean

$$(1.7) \qquad \ln \{(n - \tfrac{1}{2})/(N - n - \tfrac{1}{2})\}$$

and variance $n^{-1} + (N - n)^{-1}$. For notational simplicity the $\frac{1}{2}$'s which occur in (1.7) will be omitted, with little loss in accuracy, and we shall write

$$(1.8) \qquad \ln \{\theta/(1 - \theta)\} \sim N[\ln \{n/(N - n)\}, n^{-1} + (N - n)^{-1}].$$

Throughout the paper some improvement in the approximations for the prior with $a = b = -1$ may be effected by reducing the observations by $\frac{1}{2}$ in calculating the posterior means. In words (1.8) says that the true log-odds is approximately normally distributed about the sample log-odds with variance equal to the sum of the reciprocals of the numbers in the two classes, success and failure.

The remainder of the paper is concerned with generalizations of the result in the last sentence, in particular with the meaning of log-odds in the multinomial situation, and with applications.

**2. Multinomial distributions.** In the multinomial situation a slightly different notation is desirable. Let $k$ denote the number of classes: in the binomial case $k = 2$. Let $\theta_1, \theta_2, \cdots, \theta_k$ denote the probabilities for the $k$ classes: necessarily $\theta_i \geqq 0$ for all $i$, and $\sum \theta_i = 1$ (all summations in this section run from 1 to $k$). Let $n_1, n_2, \cdots, n_k$ denote the observed numbers in each of the classes when $N = \sum n_i$ independent trials are made with the above probabilities for the classes.

We use another suggestion of Fisher's. He remarked that if $n_i (i = 1, 2, \cdots k)$ were independent Poisson variables with means $\Psi_i$, then the conditional distribution of them, given $N = \sum n_i$, would be multinomial as just described with $\theta_i = \Psi_i / \sum \Psi_i$. The proof is immediate since $N$ has a Poisson distribution with mean $\sum \Psi_i$ and hence

$$\begin{aligned} p(n_1, n_2, \cdots, n_k \mid N) &= p(n_1, n_2, \cdots, n_k)/p(N) \\ &= \frac{e^{-\Sigma\Psi_i} \prod(\Psi_i^{n_i}/n_i\,!)}{e^{-\Sigma\Psi_i}(\sum\Psi_i)^N/N!} \qquad (2.1) \\ &= N! \prod(\theta_i^{n_i}/n_i\,!), \end{aligned}$$

as required. An alternative way of regarding this result is to remark that the probability distribution of the Poisson $n_i$ factors into the distribution of $N$, which depends only on $\theta = \sum \Psi_i$, and the conditional multinomial distribution of the $n_i$ given $N$, which depends only on $\theta_1, \theta_2, \cdots, \theta_k$. If the prior distribution of the $\Psi_i$ similarly factors into one part which depends only on $\theta$ and another which depends only on the $\theta_i$, the same will be true of the posterior distribution. Consequently, under these conditions on the prior, the posterior distribution of the $\theta_i$ will depend only on the multinomial part of the likelihood. Thus this posterior may be obtained by the Poisson device.

An appropriate prior distribution for the $\Psi_i$ may be obtained by supposing that their logarithms are independent and uniformly distributed over the whole real line. This prior distribution factors in the way described above. The Jacobian of the transformation from $\Psi_1, \Psi_2, \cdots, \Psi_k$ to $\theta, \theta_1, \theta_2, \cdots, \theta_{k-1}$ is easily found to be $\theta^{k-1}$, whence

$$\frac{d\Psi_1 d\Psi_2 \cdots d\Psi_k}{\Psi_1\Psi_2 \cdots \Psi_k} = \theta^{k-1} \frac{d\theta d\theta_1 \cdots d\theta_{k-1}}{\theta^k \theta_1 \cdots \theta_k} = \left(\frac{d\theta}{\theta}\right)\left(\frac{d\theta_1 \cdots d\theta_{k-1}}{\theta_1 \cdots \theta_k}\right) \qquad (2.2)$$

as required. The posterior distribution of the $\theta_i$ is then obtained by multiplying the last factor in (2.2) by the likelihood from (2.1) with the result proportional to

$$\prod \theta_i^{n_i - 1}. \qquad (2.3)$$

We seek an approximation to (2.3). We find this by considering the Poisson distribution. The posterior distribution of the $\Psi_i$, given the $n_i$ (now Poisson variables) is proportional to

$$\prod (e^{-\Psi_i}\Psi_i^{n_i-1}). \tag{2.4}$$

Thus the $\Psi_i$ are independent and each is distributed in a Type III or Gamma distribution. But, as explained above, if a variable has such a distribution, then its logarithm is approximately normally distributed for all but small values of $n_i$. It is easy to calculate the mean and variance with the result that the $\ln \Psi_i$ are independent and approximately normally distributed with means $\ln n_i$ and variances $n_i^{-1}$.

Let $a_1, a_2, \cdots a_k$ be a set of constants with $\sum a_i = 0$. In common with the nomenclature used in the design and analysis of experiments, a linear form in quantities with coefficients which add to zero will be termed a *contrast* in those quantities. Consider a contrast in the $\ln \Psi_i$.

$$\sum a_i \ln \Psi_i = \sum a_i \ln (\theta\theta_i) = \sum a_i \ln \theta_i$$

since $\sum a_i = 0$. Consequently a contrast in the $\ln \Psi_i$ is equally a contrast in the $\ln \theta_i$. But the contrasts in the $\ln \Psi_i$ are approximately normally distributed and any set of them is approximately jointly normally distributed. The same must therefore apply to the $\ln \theta_i$ and, by the arguments given earlier, these distributions of the $\ln \theta_i$ apply to multinomial sampling. Consequently we have the following

THEOREM 1. *If the random variables $n_1, n_2, \cdots, n_k$ have a multinomial distribution with parameters $\theta_1, \theta_2, \cdots, \theta_k$; and if the prior distribution of the $\theta_i$ has density proportional to $(\prod \theta_i)^{-1}$ over the region $\theta_i \geqq 0$, $\sum \theta_i = 1$: then if the constants $a_{pi}$ ($p = 1, 2, \cdots, m$; $i = 1, 2, \cdots, k$; $m < k$) satisfy $\sum_i a_{pi} = 0$, the joint posterior distribution of the contrasts $\sum_i a_{pi} \ln \theta_i$ ($p = 1, 2, \cdots, m$) is approximately normal with means*

$$\sum_i a_{pi} \ln n_i \tag{2.5}$$

*and covariances (variances when $p = q$)*

$$\sum_i a_{pi} a_{qi} n_i^{-1}. \tag{2.6}$$

The expressions for the means and covariances follow from the independence of the $\ln \Psi_i$ and their means and variances.

The means given by (2.5) may be written

$$\sum_i a_{pi} \ln (n_i/N), \tag{2.7}$$

since $\sum_i a_{pi} = 0$. This may be preferred since the individual terms in the sum are then of the same order as those in the corresponding contrast $\sum_i a_{pi} \ln \theta_i$: $\theta_i$ being of the same order as $n_i/N$.

In the binomial case, $k = 2$, the only contrasts are multiples of $\ln \theta_1 - \ln \theta_2 = \ln \{\theta/(1 - \theta)\}$, the log-odds, in the notation of Section 1. Consequently the earlier result for the binomial is a special case of the theorem. In view of the generalization we shall call a contrast in the $\ln \theta_i$ a *log-odds*, despite the fact that the odds do not necessarily refer to a single contrast of one event with another. Thus $\ln \theta_1 - \ln \theta_3$, for example, is the logarithm of genuine conditional odds, $\theta_1/\theta_3$, but we shall use the name for $\ln \theta_1 - 2 \ln \theta_2 + \ln \theta_3$ which has not this property. The theorem can now be expressed as saying that the approximate posterior distribution of the log-odds is as it would be if the $\ln \theta_i$ were approximately $N(\ln n_i, n_i^{-1})$ and independent. (Of course, the $\ln \theta_i$ are not themselves even approximately independent since $\sum \theta_i = 1$.) Consequently if attention is confined to log-odds then the simple normal result may be used: in particular techniques of the Analysis of Variance are available. We shall see below that many parameters of interest in the analysis of multinomial data, particularly in the form of contingency tables, are expressible as log-odds, and that therefore the normal theory can be applied. As in the binomial case the approximation can be improved by subtracting $\frac{1}{2}$ from $n_i$ in (2.5).

**3. Remarks on Bayesian analysis.** Before proceeding to discuss application of Theorem 1, it is necessary to clarify a few points in statistical analyses that overtly use Bayes theorem and a prior distribution. The object of such an analysis is to provide the posterior distribution of the parameters or, if only some of them are of interest, the marginal posterior distribution of those. If the data are subsequently to be used as a basis for decision then the posterior distribution provides the necessary material for the calculation of the best decision. Once the posterior distribution has been obtained the only problem remaining is the descriptive one of how to present it. If we follow classical statistics and its concept of a confidence interval, it is rather natural to summarize the posterior distribution by giving an interval which contains a rather large, say 95%, amount of the posterior probability. This has the disadvantage of only giving information about the tails of the distribution and does not provide, for example, any good idea of the most probable values. The provision of the median and other quantiles in addition to the 95% values would help. For some purposes even an interval may be thought to be an overelaborate summary, particularly in cases where one value of the parameter is of especial interest. In these cases it may be felt adequate to say whether this value is a probable one on the basis of the posterior distribution. Again borrowing an idea from classical statistics, where the confidence set is the set of values that would not be rejected if tested as null hypotheses in a significance test at the level of the confidence set, we can perform a Bayesian significance test by seeing whether the particular value lies within the interval containing proportion $(1 - \alpha)$ of the posterior distribution. If not, then it can be said that this value is significant at level $\alpha$: in the sense that it does not belong to a set of values having reasonably high posterior probability.

Of course there are many intervals containing a given proportion of the posterior distribution. We shall choose that interval (typically unique) which is such that no values within it are less probable than any value without it. This provides the shortest interval in the usual sense. These ideas are discussed in much more detail in [15] but we would point out here that no account has been taken, in formulating this concept of a significance test, in any economic or general decision-theoretic considerations. This seems to be in the same spirit as a classical significance test. Neither have we incorporated strong prior ideas about the value of especial interest. Our prior distributions are "smooth" in the neighbourhood of the special value. This is in marked contrast to Jeffreys' significance tests [10] where there is a concentration of prior probability on the special value.

We now apply these ideas to the posterior distributions of log-odds. The normal posterior distribution is particularly easy to understand for a single parameter. For several parameters the multivariate form is not so simple to comprehend. Let $\phi_1, \phi_2, \cdots, \phi_s$ be $s$ linearly independent log-odds with means $m_i$ and covariances $v_{ij}$, say. The expressions for these means and covariances follow from (2.5) and (2.6). The joint density of the $\phi_i$ is constant on the ellipsoids

$$\sum_{i,j=1}^{s} (\phi_i - m_i)v^{ij}(\phi_j - m_j) = c, \tag{3.1}$$

where $v^{ij}$ are the elements of the inverse of the dispersion matrix, and $c$ is any positive constant Furthermore the left-hand side of (3.1) is distributed as $\chi^2$ on $s$ degrees of freedom. If $\chi^2_\alpha$ is the upper $100\alpha\%$ point of this distribution, the posterior probability that

$$\sum_{i,j=1}^{s} (\phi_i - m_i)v^{ij}(\phi_j - m_j) \leqq \chi^2_\alpha \tag{3.2}$$

is $(1 - \alpha)$. With $\alpha = 0.05$, the posterior probability of (3.2) is 95% but, of course, any value of $\alpha$ may be used. In particular if the values $\phi_i = \phi_i^{(0)}$ are of interest, they would seem unlikely if they did not satisfy (3.2) with $\phi_i = \phi_i^{(0)}$ $(i = 1, 2, \cdots, s)$. This provides a significance test of the hypothesis that $\phi_i = \phi_i^{(0)}$. If $\phi_i^{(0)} = 0$ $(i = 1, 2, \cdots, s)$ the relevant statistic reduces to

$$\sum_{i,j=1}^{s} m_i v^{ij} m_j \tag{3.3}$$

and may be compared with $\chi^2$.

In applications of (3.2) and (3.3) it should be remembered that although the $m_i$ and $v^{ij}$ are statistics and the $\phi_i$ are parameters, it is the $\phi_i$ that are the random variables. In the usual argument, wherein the $m_i$ are random variables, normally distributed about means $\phi_i$ with covariances $v_{ij}$ (supposed known), statements like (3.2) and (3.3) are still correct. In particular the comparison of (3.3) with the $\chi^2$ distribution is basic to the Analysis of Variance, the variance being supposed known. Consequently in analyzing the log-odds by the Bayesian methods just described, we have available the methods of the Analysis of Variance. In the following sections we will continually make use of ideas derived from that classical field.

Before proceeding to the applications of Theorem 1, two remarks need to be made in the light of these comments on Bayesian analyses. The first concerns the invariance of methods based on (3.2) and (3.3) under linear transformations of the log-odds. This is because the quadratic forms are so invariant. A consequence of this remark is the observation that it is possible to replace any particular set of log-odds by linear transformations thereof which may simplify the analysis.

The second remark concerns the approximation to the Type III distribution (Equation (2.4)) that is basic to the methods. There are, of course other, and better, approximations: for example, the Wilson-Hilferty one using $\chi^{\frac{2}{3}}$; but they cannot compare in convenience with the log-odds. One approximation that does merit serious attention is the classical $\chi^2$-approximation based on the statistic $\sum(O - E)^2/E$ in an obvious notation. This can be used in a Bayesian analysis along the same lines as just explained for (3.3): details are given in [15]. It is hoped to compare the results obtained from the $\chi^2$ and log-odds approaches in another paper. Preliminary numerical work suggests that $\chi^2$ is worse than the log-odds in supplying an approximation to the posterior distribution. This is not surprising in view of the fact that $\chi^2$ was not designed to do this. However, if a uniform prior distribution for $\theta$ in the binomial case replaces the uniform for $\ln \theta$ used above, the effect being to increase the $n_i$ by 1 in the log-odds results, then $\chi^2$ seems to be a good approximation to the posterior distribution.

**4. Certain multinomial problems.** In the multinomial situation described in Section 2, let $\theta_i = \theta_i^{(0)}$ $(i = 1, 2, \cdots k)$ be a simple null hypothesis specifying the values of all the parameters. We now proceed to derive a test of this null hypothesis using the above results on the posterior distribution of log-odds and the principle of Bayesian analysis just outlined. The posterior density of the $\ln \Psi_i$ (from the Poisson distribution) is proportional to

$$\exp\{-\tfrac{1}{2}\sum(\ln \Psi_i - \ln n_i)^2 n_i\}, \tag{4.1}$$

using the approximation. This distribution will also apply to the $\ln \theta_i$ provided that only contrasts are considered. Now (4.1) may be rewritten, using the usual breakdown of a sum of squares into a sum of squares about a weighted mean plus a term involving the mean. Let $u_i = \ln \Psi_i - \ln n_i$, normally distributed about zero with variance $n_i^{-1}$; then (4.1) is

$$\exp\{-\tfrac{1}{2}[\sum n_i(u_i - \bar{u})^2 + N\bar{u}^2]\} \tag{4.2}$$

with $\bar{u} = \sum n_i u_i / \sum n_i$. Make an orthogonal linear transformation from $u_i$ to new variables, one of which is a multiple of $\bar{u}$. The remainder will necessarily involve only contrasts of the $u_i$, and hence log-odds, and will be appropriate to the multinomial situation. Consequently the posterior distribution of the log-odds is proportional to $\exp\{-\frac{1}{2}\sum n_i(u_i - \bar{u})^2\}$ and, as in the argument leading to

(3.2), the posterior probability that $\sum n_i(u_i - \bar{u})^2 \leqq \chi_\alpha^2$ is $(1 - \alpha)$, the $\chi^2$ having $(k - 1)$ degrees of freedom. The test of the null hypothesis will be significant at the $\alpha$ level if the null values for $u_i$ do not belong to the confidence set. The null value for $u_i - \bar{u}$ is $(\ln \theta_i^{(0)} - \ln n_i) - \{\sum n_i(\ln \theta_i^{(0)} - \ln n_i)\}/\sum n_i$. If $n_i/N$ replace $n_i$, as they may do without affecting the expression, we have here an expression involving comparisons of the logarithms of the *observed* proportions $n_i/N$, with the logarithms of the *expected* proportions, $\theta_i^{(0)}$. The difference between these is $d_i$, say, and the expression to be compared with $\chi_\alpha^2$ is the weighted sum of squares of the $d_i$ about their weighted mean; the weights being $n_i$.

The test just derived replaces, in the present analysis, the usual test based on Pearson's expression $\sum (O - E)^2/E$: using instead the deviations $(\ln O - \ln E)$. Now the Pearson test may be extended to provide a test of a composite null hypothesis by replacing the expected numbers, $E$, under the simple null by the expected numbers under the maximum likelihood value for the composite null: the $\chi^2$ criterion losing a degree of freedom for each parameter estimated by maximum likelihood. (Any other asymptotically efficient method may be used.) The same extension to $\chi^2$ is similarly available in a Bayesian analysis and details are given in [15]. It is also possible to extend the test just derived to the composite case by replacing $\theta_i^{(0)}$ by any asymptotically efficient estimate, $\hat{\theta}_i$, of $\theta_i$ under the composite null hypothesis. The same test criterion is used with $\hat{\theta}_i$ replacing $\theta_i^{(0)}$ and the degrees of freedom reduced by one for each parameter estimated. We do not give the proof here since it closely parallels that for Pearson's criterion which is rather long. The interested reader with the above reference beside him should have no difficulty in making the necessary alterations to the log-odds case. An example of the use of this method is provided by a problem concerning linkage in genetics. The observational material consists of a multinomial with $k = 4$, under the null hypothesis the four probabilities are respectively $9 + \alpha$, $3 - \alpha$, $3 - \alpha$, and $1 + \alpha$, each divided by 16. There $\alpha$ is the linkage parameter and under the null may be asymptotically estimated by $\hat{\alpha}$, say. The deviations, $\ln\{(9 + \hat{\alpha})/16\} - \ln n_i/N$ etc., between log-expected and log-observed, are then found and the test criterion is the weighted sum of squares of these about their weighted mean. This can be referred to $\chi^2$ on $(k - 1) - 1$, here 2, degrees of freedom. It is only necessary to use this method when the null hypothesis cannot be expressed in terms of log-odds. When it can, simpler methods are available as we shall see below.

One advantage of the present approach over the usual $\chi^2$-analysis is that it provides the whole of the posterior distribution and not just that aspect of it provided by a significance test. Thus (4.2), when integrated with respect to $\bar{u}$, provides the posterior density of all the $(k - 1)$ log-odds. If any log-odds are of special interest then they may considered on their own, integrating out the unwanted log-odds from the joint posterior distribution. A simpler approach is to consider the required log-odds directly. For example, suppose one is interested in the conditional probability of an observation falling in the first class given that it has fallen in the first two; that is, $\theta_1/(\theta_1 + \theta_2)$. Denote this parameter by $\phi$, then $\ln[\phi/(1-\phi)] = \ln\theta_1 - \ln\theta_2$ is a log-odds with variance $n_1^{-1} + n_2^{-1}$. Hence $\ln[\phi/(1-\phi)]$ has a normal posterior distribution with mean $\ln n_1 - \ln n_2$ and variance $n_1^{-1} + n_2^{-1}$. Note that the same result may be obtained by regarding $(n_1 + n_2)$ as fixed and treating the situation as binomial with $\phi$ as the parameter.

This argument does not immediately extend to the situation where the parameter cannot be expressed in terms of log-odds. An example is provided by $\theta_1 + \theta_2$. However the example may be handled by combining the first two classes, when $\theta_1 + \theta_2$ is simply the probability of falling into this new class, and inferences may be made using the results of Section 2. But this method involves a prior distribution over the parameters of the new $(k-1)$ classes that needs to be shown to be equivalent to the original prior over the $k$ classes. Otherwise one is changing the prior to suit the analysis which may be a good approximation but is not correct Bayesian argument. In fact, no approximation is involved and the two priors are equivalent. Thus, in order to make inferences about $\theta_1 + \theta_2$ the original multinomial over $k$ classes can be replaced by a binomial over two. To establish the equivalence of the priors it is convenient to pass to contingency tables, which exhibit the combination of classes in its commonest form. In a contingency table of $r$ rows and $s$ columns, $k = rs$ and if one is interested in the row probabilities, these are the sum of the $s$ probabilities in that row.

**5. Two-way contingency tables.** A two-way contingency table typically arises through each of a number of items being classified into one of $r$ exclusive and exhaustive classes $A_1, A_2, \cdots A_r$, and simultaneously into one of $s$ exclusive and exhaustive classes $B_1, B_2, \cdots B_s$. The probability $\theta_{ij}$ of an item being classified as both $A_i$ and $B_j$ is supposed the same for each item. If $N$ items are independently classified, the numbers $n_{ij}(i = 1, 2, \cdots r; j = 1, 2, \cdots s)$ of items in classes $A_i$ and $B_j$ are multinomially distributed with index $N$ and parameters $\theta_{ij}$. The probability of being classified as $A_i$ is $\sum_{j=1}^{s}\theta_{ij} = \theta_{i\cdot}$, say. $\theta_{\cdot j} = \sum_{i=1}^{r}\theta_{ij}$ is similarly the probability of being classified as $B_j$. In the usual language, the description just given refers to a contingency table with neither margin (only the total, $N$) fixed. We shall have something to say about contingency tables with one or both margins fixed below.

It is often convenient to parameterize the contingency table other than through the $\theta_{ij}$. For example, the probabilities of the classes $A_i(i = 1, 2, \cdots r)$, $\theta_{i\cdot}$, and the conditional probabilities of the classes $B_j$, given the $A_i$, $\theta_{ij}/\theta_{i\cdot}$, may be used instead. Inferences about these cannot be made directly until their prior distribution has been determined. To this end we prove the following

THEOREM 2. *If the prior distribution of $\theta_{ij}(i = 1, 2, \cdots r: j = 1, 2, \cdots s)$ is proportional to $\prod_{ij}\theta_{ij}^{-1}$, then the prior distribution of $\theta_{i\cdot}$ and $\phi_{ij} = \theta_{ij}/\theta_{i\cdot}$ is proportional to*

$$\prod_i \theta_{i\cdot}^{-1} \prod_{i,j} \phi_{ij}^{-1}. \tag{5.1}$$

Consider the change from parameters $\theta_{ij}$, for all $i$ and $j$ except $i = r, j = s$, to $\theta_{i\cdot}(i < r)$ and $\phi_{ij}(j < s)$. Certain values have to be excluded because of the constraints that exist amongst the parameters; namely $\sum_{ij} \theta_{ij} = \sum_i \theta_{i\cdot} = \sum_j \phi_{ij} = 1$. We have

$$\begin{aligned} \theta_{ij} &= \phi_{ij}\theta_{i\cdot} & i < r, j < s, \\ \theta_{rj} &= \phi_{rj}(1 - \sum_{i<r} \theta_{i\cdot}) & j < s, \\ \theta_{is} &= (1 - \sum_{j<s} \phi_{ij})\theta_{i\cdot} & i < r. \end{aligned} \tag{5.2}$$

The Jacobian of the transformation from $\theta_{ij}$ to the new parameters is easily calculated to be

$$\prod_{i=1}^{r} \theta_{i\cdot}^{s-1}. \tag{5.3}$$

Consequently the joint density of $\phi_{ij}$ and $\theta_{i\cdot}$ is proportional to the product of $\prod_{i,j} \theta_{ij}^{-1}$ and (5.3): that is to

$$\prod_{i=1}^{r} \prod_{j=1}^{s} (\theta_{i\cdot}\phi_{ij})^{-1} \prod_{i=1}^{r} \theta_{i\cdot}^{s-1} = \prod \theta_{i\cdot}^{-1} \prod \phi_{ij}^{-1},$$

as required.

The theorem establishes that the prior distribution on the $rs$ classes is consistent with the same prior distribution on the reduced number $r$ of classes. In particular the argument used at the end of the previous section is justified, and generally classes may be amalgamated and the approximation of Theorem 1 applied to the reduced number of parameters. Even more can be said because of properties of the likelihood function.

The multinomial likelihood for the contingency table is proportional to

$$\prod_{ij} \theta_{ij}^{n_{ij}} = \prod_{i} \theta_{i\cdot}^{n_{i\cdot}} \prod_{i,j} \phi_{ij}^{n_{ij}},$$

where $n_{i\cdot} = \sum_j n_{ij}$, a product of a function of the $\theta_{i\cdot}$ and a function of the $\phi_{ij}$. The theorem proved that the prior similarly factors. The same is therefore true of the posterior distribution and inferences concerning $\theta_{i\cdot}$ can be made independently of those concerning $\phi_{ij}$. Inferences about the marginal probabilities $\theta_{i\cdot}$ follow as for the multinomial on $r$ classes. Inferences for $\phi_{ij} (j = 1, 2, \cdots s)$ for any $i$, similarly follow using a multinomial on $s$ classes since the prior distribution $\prod_j \phi_{ij}^{-1}$ is of the usual form. There is a symmetrical result in terms of $\theta_{\cdot j}$ and $\theta_{ij}/\theta_{\cdot j}$ using the other margin based on the $B$-classification.

Another consequence of these results is that inferences about the conditional probabilities, $\phi_{ij}$, can be made irrespective of the distribution of the $n_{i\cdot}$. The reason for this is that the only part of the likelihood used in the inference is the conditional probability of $n_{ij}$ given $n_{i\cdot}$ : the distribution of the $n_{i\cdot}$ is not involved. Contingency tables often arise with the $n_{i\cdot}$ fixed, rather than random variables, as when the number of items in each class $A_i$ is selected deterministically and then classified according to the $B$-classification. The table is often said to have one margin fixed. Contingency tables with one margin fixed may be analyzed in the same way as tables with only $N$ fixed; at least as far as the $\phi_{ij}$ are concerned, inferences about the $\theta_{i\cdot}$ are not possible in that case.

Let us apply the results to the $2 \times 2$ table ($r = s = 2$). A convenient parameterization is through $\theta_{1\cdot} = \theta_{11} + \theta_{12}$, the probability of the class $A_1$, and $\phi_{11}$ and $\phi_{21}$, the probabilities of the classification $B_1$ given $A_1$ and $A_2$ respectively. Combining the results of the two theorems we see that the posterior distributions are approximately as follows:

(a) $\ln\{\theta_{1\cdot}/\theta_{2\cdot}\}$ is normal with mean $\ln\{n_{1\cdot}/n_{2\cdot}\}$ and variance $n_{1\cdot}^{-1} + n_{2\cdot}^{-1}$,
(b) $\ln\{\phi_{11}/\phi_{12}\}$ is normal with mean $\ln\{n_{11}/n_{12}\}$ and variance $n_{11}^{-1} + n_{12}^{-1}$,
(c) $\ln\{\phi_{21}/\phi_{22})$ is normal with mean $\ln\{n_{21}/n_{22}\}$ and variance $n_{21}^{-1} + n_{22}^{-1}$,

and these three distributions are independent. All three results follow because in each case we are dealing with a binomial situation and the three likelihoods and their corresponding priors factorize. Symmetrical results are available with $A$ and $B$ interchanged. Inference (a) is not available if the margin corresponding to the $A$-classification is fixed.

Usually a more interesting parameter is one describing the relationship between the two classifications. Indeed, much of the literature on $2 \times 2$ tables is only concerned with this aspect and even more specifically is devoted to examining whether the two classifications are independent: that is, $\theta_{ij} = \theta_{i\cdot}\theta_{\cdot j}$. Two tests available are Fisher's exact test and the usual $\chi^2$-approximation. The latter has a Bayesian justification: no similar rationalization seems to exist for Fisher's method though progress might be possible on the lines suggested by Bahadur and described below. Measurement of the association between the classifications can be carried out in a number of ways: we seek for one in terms of log-odds. If the classifications are independent $\phi_{11} = p(B_1 \mid A_1) = \theta_{11}/(\theta_{11} + \theta_{12}) = \theta_{21}/(\theta_{21} + \theta_{22}) = p(B_1 \mid A_2) = \phi_{12}$ and this may be written either as

$$\theta_{11}/\theta_{12} = \theta_{21}/\theta_{22} \quad \text{or} \quad \phi_{11}/\phi_{12} = \phi_{21}/\phi_{22},$$

or in words, the odds for the $B$-classification are the same within $A_1$ and $A_2$. Hence a possible parameter to consider is the log-odds

$$\begin{aligned} \phi &= \ln\theta_{11} - \ln\theta_{21} - \ln\theta_{12} + \ln\theta_{22}, \\ &= \ln\phi_{11} - \ln\phi_{21} - \ln\phi_{12} + \ln\phi_{22}. \end{aligned} \tag{5.4}$$

By the main theorem, or by a combination of (b) and (c) above, this parameter is approximately normally distributed with mean

$$\ln n_{11} - \ln n_{21} - \ln n_{12} + \ln n_{22} \tag{5.5}$$

and variance

$$n_{11}^{-1} + n_{21}^{-1} + n_{12}^{-1} + n_{22}^{-1}. \tag{5.6}$$

The null hypothesis of independence is $\phi = 0$ and may be tested by referring

$$\frac{(\ln n_{11} - \ln n_{21} - \ln n_{12} + \ln n_{22})^2}{n_{11}^{-1} + n_{21}^{-1} + n_{12}^{-1} + n_{22}^{-1}} \tag{5.7}$$

to $\chi^2$ on one degree of freedom. The approximation can probably be improved by subtracting $\frac{1}{2}$ from each of the $n_{ij}$ in the numerator before taking logarithms. This result is available for tables with one or no margins fixed. If $\phi$ is used as a measure of association, Bayesian confidence intervals for it may easily be obtained using (5.5) and (5.6). The natural parameters to use with $\phi$ are $\theta_{1\cdot}$ and $\theta_{\cdot 1}$, the marginal probabilities.

It will be helpful in passing to tables larger than $2 \times 2$ to note certain similarities between the analysis and an analysis of variance. If the table (with no margins fixed) is analyzed in terms of $\theta_{1\cdot}$, $\theta_{\cdot 1}$ and $\phi$, the first two correspond to main effects of the $A$- and $B$-classifications separately. The last corresponds to an interaction between the two classifications: indeed, the form of (5.4) is exactly that of an interaction based on the logarithms of the probabilities. However, if $\phi$ is thought of as an interaction a corresponding main effect would be

$$\text{(5.8)} \qquad \ln \theta_{11} - \ln \theta_{21} + \ln \theta_{12} - \ln \theta_{22},$$

which is usually of no interest. What is of interest, and what has been used, is ((a) above)

$$\text{(5.9)} \qquad \ln \{\theta_{1\cdot}/\theta_{2\cdot}\} = \ln (\theta_{11} + \theta_{12}) - \ln (\theta_{21} + \theta_{22}).$$

Even had (5.8) been used, it should be noted that it is not independent of $\phi$ since the $\ln \theta_{ij}$ do not have equal variances. Whilst therefore analysis of variance ideas are conceptually useful in the study of contingency tables the breakdown of the total variation is into parts that are of separate interest and are not necessarily or typically independent: neither do the individual contributions have an additive property that is often demanded of analyses of variance or of $\chi^2$. The interdependence of the posterior distributions of $\theta_{1\cdot}$, $\theta_{\cdot 1}$ and $\phi$ is as follows. $\theta_{1\cdot}$ is independent of $\phi$ ((a)—(c) above) as is $\theta_{\cdot 1}$. But $\theta_{1\cdot}$ is not independent of $\theta_{\cdot 1}$ nor is $\phi$ independent of the pair $(\theta_{1\cdot}, \theta_{\cdot 1})$. No parameterization that is symmetrical in the two classifications and gives independent distributions seems possible.

The $2 \times 2$ contingency table occasionally arises with both margins, $n_{i\cdot}$ and $n_{\cdot j}$, fixed. The classic example is the lady tasting tea. The above ideas do not seem to lead to a Bayesian solution partly because of the difficulty mentioned at the end of the last paragraph. Bahadur, in a private communication, has suggested a parameterization of the conditional distribution, given the two margins, that does lead to a Bayesian analysis. If the two classifications are independent it is easy to calculate this distribution for $n_{11}$ : it is hypergeometric. Denote it by $p(n_{11})$. Then the suggestion is to use

$$\text{(5.10)} \qquad p(n_{11} \mid \theta) = e^{\theta n_{11}} p(n_{11}) / \sum_n e^{\theta n} p(n)$$

for the distribution when the classifications are not independent. The parameter $\theta$ measures the amount of association between the two classifications: $\theta = 0$ corresponding to independence. The likelihood is sufficiently simple for Bayesian analysis to be possible but its value is limited by the difficulty in interpretation of $\theta$. But this is perhaps a criticism of the design of the experiment because in any case it is difficult to see how from any analysis one could infer what the chance was of the lady classifying correctly any future cup of tea presented to her. The distributions (5.10) have been used by Elfving in a Neyman-Pearson study of the problem.

The extension of the analysis of the $2 \times 2$ table to the $r \times 2$ table for general $r$ is straightforward. Inferences about $\theta_{.1}$ or about the $\theta_{i.}(i = 1, 2, \cdots r)$ can be made in the ways described above for the binomial and multinomial respectively. The null hypothesis of no association between the two classifications is that $\theta_{ij}/\theta_{i.} = \phi_{ij}$ does not depend on $i$. In terms of log-odds this is the same as saying

$$\ln \theta_{i1} - \ln \theta_{i2} \tag{5.11}$$

does not depend on $i$, or that all the log-odds in (5.11) are equal. Now these log-odds are independent and hence their posterior density has logarithm proportional to

$$\sum\{(\ln \theta_{i1} - \ln \theta_{i2}) - (\ln n_{i1} - \ln n_{i2})\}^2(n_{i1}^{-1} + n_{i2}^{-1})^{-1}. \tag{5.12}$$

This may be written (compare the passage from (4.1) to (4.2)) in terms of a weighted sum of squares about the mean and a term involving the mean. Consequently a test that all of (5.11) are equal can be made with the former. If $x_i = \ln n_{i1} - \ln n_{i2}$ and $m_i = n_{i1}^{-1} + n_{i2}^{-1}$ the test criterion is

$$\sum (x_i - x.)^2 m_i, \tag{5.13}$$

where $x. = \sum m_i x_i / \sum m_i$, and is referred to $\chi^2$ on $(r - 1)$ degrees of freedom·

This test may be derived another way, which serves again to establish a relationship between these ideas and the analysis of variance. The hypothesis (5.11) is equivalent to saying that there exist constants $a_i$, $b_j$ $(i = 1, 2, \cdots r; j = 1, 2)$ such that

$$\ln \theta_{ij} = a_i + b_j. \tag{5.14}$$

Furthermore, provided only contrasts are considered, the $\ln \theta_{ij}$ are independently normally distributed with known variances $n_{ij}^{-1}$. Consequently the hypothesis (5.14) is the hypothesis of no interaction in a two-way analysis of variance with the usual normal distribution theory, the variances being known but unequal. The roles of parameters and observations are interchanged: it is the parameters that are the random variables. The hypothesis may be tested by the usual method of considering the appropriate residual sum of squares when the model (5.14) is fitted. The details are given in Section 4.4 of [18]. That this method is available in the Bayesian framework is shown in detail in [15], but that it should be so is intuitively obvious from the results given in Section 3 of this paper. It is not difficult to demonstrate that the method of analysis of variance leads to the same result as (5.13). It may be noted that the approach based on (5.13) is simpler than Scheffé's approach when one of the classifications is dichotomous.

New difficulties arise when we pass to the general $r \times s$ table with both $r$ and $s$ greater than 2. It is not possible to find log-odds that are both independent and reflect the hypothesis of interest, namely that the two classifications are independent. It is easy to satisfy the latter requirement but without the former the quadratic form corresponding to (5.12) is no longer a simple sum of squares. Independence cannot be achieved since the variances of the $\ln \theta_{ij}$ are chance quantities. For example, consider a test of no association between the classifications in a $3 \times 3$ table. Four log-odds whose vanishing would be equivalent to the independence of the classifications are

$$\begin{aligned} &\ln \theta_{11} - \ln \theta_{12} - \ln \theta_{21} + \ln \theta_{22}, \\ &\ln \theta_{11} - \ln \theta_{13} - \ln \theta_{21} + \ln \theta_{23}, \\ &\ln \theta_{21} - \ln \theta_{22} - \ln \theta_{31} + \ln \theta_{32}, \\ &\ln \theta_{21} - \ln \theta_{23} - \ln \theta_{31} + \ln \theta_{33}. \end{aligned} \tag{5.15}$$

But these are correlated; for example, the covariance between the first and second is $n_{11}^{-1} + n_{21}^{-1}$. The direct way to proceed from here is to determine the dispersion matrix, $A$, of the log-odds in (5.15) and also the sample values of the same contrasts: namely the expressions that result from replacing $\theta_{ij}$ in (5.15) by $n_{ij}$. If $n$ is the column vector of the sample values then the relevant quadratic form is $n'A^{-1}n$ which may be referred to $\chi^2$ on four degrees of freedom. The same quadratic form would result whatever log-odds were chosen since they must be linear in the log-odds used in (5.15) and, as mentioned in Section 3, the method is invariant under linear transformations. The calculation is not too prohibitive on an electronic computer. If only the test is required, and not confidence limits for the log-odds, it is not even necessary to invert $A$, since if the triangular resolution $A = T'T$ is used, the quadratic form is the sum of squares $m'm$ where $T'm = n$.

Nevertheless the computations involve square matrices of size $(s - 1) \times (t - 1)$, and it may be better to proceed directly to the analysis of variance approach and calculate the criterion as a residual sum of squares. To do this it is necessary to solve linear equations in either $(r - 1)$ or $(s - 1)$ unknowns, whichever is the smaller. The matrices involved are therefore much smaller in size, but the manipulations on them are more complicated and the computations of the individual entries less simple. The direct calculations can be considerably simplified by using methods suggested by Goodman [9]. He shows that the inversion problems can in part be easily solved and that the final matrices that have to be inverted in any numerical case are substantially smaller than described above. For details we refer the reader to Goodman's paper.

**6. Three-way contingency tables.** We next consider the problem of $r \times s \times t$ tables in which individuals are classified in each of three ways: the two already referred to and also the exclusive and exhaustive classes $C_1, C_2, \cdots C_t$. Denote the numbers and probabilities in the classes by $n_{ijk}$ and $\theta_{ijk}$ respectively. As in the two-way table it is convenient to reparameterize the situation and, for example, use

$$(6.1)\quad \theta_{i\cdot\cdot} = \sum_{j,k} \theta_{ijk}, \qquad \phi_{ij\cdot} = \sum_k \theta_{ijk}/\theta_{i\cdot\cdot}, \qquad \Psi_{ijk} = \theta_{ijk}/\sum_k \theta_{ijk}.$$

Then $\theta_{i\cdot\cdot}$ is the probability of $A_i$, $\phi_{ij}$ is the probability of $B_j$, given $A_i$, and $\Psi_{ijk}$ is the probability of $C_k$ given both $A_i$ and $B_j$. Clearly $\theta_{ijk} = \theta_{i\cdot\cdot}\phi_{ij\cdot}\Psi_{ijk}$. Repeated applications of Theorem 2 show that if the prior density is proportional to $\prod \theta_{ijk}^{-1}$ then the posterior distributions of $\{\theta_{i\cdot\cdot}\}$, $\{\phi_{ij\cdot}\}$ and $\{\Psi_{ijk}\}$ will be independent. Furthermore, from the form of the likelihood function, the distribution of $\{\phi_{ij\cdot}\}$ will not depend on the distribution (if any) of $n_{i\cdot\cdot}$, nor will the distribution of $\{\Psi_{ijk}\}$ depend on the distribution (if any) of $n_{ij\cdot}$. These results enable tables in which one or two of the classifications are nonrandom to be analyzed. The methods already discussed enable analyses of $\{\theta_{i\cdot\cdot}\}$ and $\{\phi_{ij\cdot}\}$ to be made: we therefore consider $\{\Psi_{ijk}\}$.

The fact that the methods for two-way tables do extend to larger tables is, we claim, one of the main advantages of them. The classical methods based on the $\chi^2$-statistic present complications when an attempt is made to extend them. One extension to $2 \times 2 \times 2$ tables that is well known is Bartlett's [1] definition of no three-factor interaction. The hypothesis is that

$$(6.2)\qquad \frac{\theta_{111}\theta_{221}}{\theta_{211}\theta_{121}} = \frac{\theta_{112}\theta_{222}}{\theta_{212}\theta_{122}}.$$

This can equivalently be written with $\Psi$ everywhere replacing $\theta$, the suffixes remaining unaltered. If logarithms are taken of both sides of (6.2) we are led to consider

$$(6.3)\qquad \begin{aligned}\Psi = \{\ln \theta_{111} - \ln \theta_{211} - \ln \theta_{121} + \ln \theta_{221}\}\\ - \{\ln \theta_{112} - \ln \theta_{212} - \ln \theta_{122} + \ln \theta_{222}\},\end{aligned}$$

which is a log-odds, and whose vanishing corresponds to no interaction in Bartlett's sense. $\Psi$ is easily interpreted as the difference for classes $C_1$ and $C_2$ of the measure of association (5.4) between the $A$- and $B$-classifications. By the symmetry of (6.3) the letters $A$, $B$ and $C$ may be permuted in any way in the last sentence. The approximate posterior distribution of $\Psi$ is normal with mean equal to the same expression with $n_{ijk}$ for $\theta_{ijk}$, and variance $\sum_{i,j,k} n_{ijk}^{-1}$.

The parameter $\Psi$ can be used as a definition of the three-factor interaction in a $2 \times 2 \times 2$ table. This claim is easily substantiated since (6.3) is exactly the form of such an interaction in the analysis of variance of $\ln \theta_{ijk}$. It is therefore possible, by the methods described, to investigate a $2 \times 2 \times 2$ table using seven parameters: $\theta_{1\cdot\cdot}$, $\theta_{\cdot1\cdot}$, $\theta_{\cdot\cdot1}$; the three two-factor interactions corresponding to the measures of association defined in (5.4); and the three factor interaction $\Psi$. (The two-factor interactions can clearly be written in terms of $\phi_{ij\cdot}$, $\phi_{i\cdot k}$ and $\phi_{\cdot jk}$.) This is a complete breakdown analogous to that in the analysis of variance but it should be noted that the parameters are not in general independent and that the additivity present in the standard analysis, or in the analyses by $\chi^2$, Lancaster [13] or using information ideas, Kullback et al. [12] does not obtain here. If one or two of the classifications are non-random then only some of the parameters can be discussed.

The connection with the analysis of variance is an important one but there are often other parameters that are of interest besides the main effects and interactions already mentioned. For example, the above treatment is symmetric in the three classifications and yet it may be that the main purpose of the experiment that led to the contingency table was the investigation of the dependence of one classification on the others. If so, an unsymmetric analysis is clearly suggested. The Bayesian analysis provides, typically in an unmanageable form, the joint posterior distribution of all the parameters. What the statistician has to do is to extract from this material the main features that are likely to be of interest to the experimenter or of value in any decision problem for which the data are relevant. There must therefore be considerable flexibility in the type of analysis, or condensation of the posterior distribution that is used. Slavish attention to analysis of variance ideas is not likely to lead to many fruitful results. We now proceed to illustrate other approaches that are possible within the Bayesian log-odds framework, emphasizing that they are only illustrative and particular situations may demand other parameterizations. The discussion is within the framework of a $2 \times 2 \times 2$ table. Extensions to larger three-way tables are indicated at the end of the section.

Suppose that it is the dependence of the $C$-classification on the other two that is of interest. Then we may think of $A$ and $B$ as providing two factors whose influence on the dependent variable, represented by the $C$-classification, we wish to assess. If $A$ and $B$ are non-random then the only probabilities that are defined are

$$\Psi_{ij1} = p(C_1 \mid A_i, B_j), \qquad (i, j = 1, 2) \tag{6.4}$$

with $\Psi_{ij2} = 1 - \Psi_{ij1}$. These four probabilities may be compared in various ways using the log-odds

$$\ln p(C_1 \mid A_i, B_j) - \ln p(C_2 \mid A_i, B_j) \tag{6.5}$$

and one comparison is the expression (6.3). One possibility is to investigate whether

$$p(C_1 \mid B_j, A_1) = p(C_1 \mid B_j, A_2) \qquad (j = 1, 2). \tag{6.6}$$

If this obtains then, given the $B$-classification, the $C$- and $A$-classifications are independent. Or alternatively, given $B_j$, $A_i$ provides no further information about $A$. If this holds for both $B_1$ and $B_2$ then the $A$-classification provides no information not already provided by the $B$-classification. It is useful to refer to this as a *Markov* property, in analogy with the corresponding property studied in stochastic process theory where $A$, $B$ and $C$ refer to three points in time conveniently thought of as the past, present and future respectively.

The equalities (6.6) may be tested using the pair of log-odds

$$\{\ln p(C_1 \mid B_j, A_1) - \ln p(C_2 \mid B_j, A_1)\} \\ - \{\ln p(C_1 \mid B_j, A_2) - \ln p(C_2 \mid B_j, A_2)\} (j = 1, 2) \tag{6.7}$$

with means

$$\ln n_{1j1} - \ln n_{1j2} - \ln n_{2j1} + \ln n_{2j2}, \tag{6.8}$$

and variance

$$n_{1j1}^{-1} + n_{1j2}^{-1} + n_{2j1}^{-1} + n_{2j2}^{-1}. \tag{6.9}$$

The two expressions in (6.7) for $j = 1$ and 2 are independent and hence the Markov property may be tested by referring the sum for $j = 1$ and 2 of the squares of the means, (6.8), divided by the variances, (6.9) to $\chi^2$ on 2 degrees of freedom. (The null value of (6.7) is zero.) If the posterior distribution is sufficiently concentrated around zero then one might feel reasonably confident that one could proceed on the basis that the Markov property obtained. If so, then it is possible to discuss the simpler conditional probability $p(C_1 \mid B_j)$, which is otherwise undefined. It is not easy to see exactly what is meant by "sufficiently concentrated" in the sentence above. The degree of departure from the Markov property that is allowable will depend on the "robustness" of the analysis that results from using the property, to departures from it. This is not a subject that can be discussed here: some considerations of it in a different context are given in Lindley [14].

Consider next the case where it is still the dependence of $C$ on $A$ and $B$ that is of interest, but where one of the independent classifications, $A$ say, is random and the other, $B$, is not. Then, in addition to the conditional probabilities $p(C_k \mid A_i, B_j)$ already discussed, the probabilities $p(C_k \mid B_j)$ are also meaningful. The analysis of variance approach can be misleading in this context because the natural main effects and interactions are rather differently defined in contingency table analyses from the usual linear hypothesis situation. (The same point has already been discussed in the two-way table; compare equations (5.8) and (5.9).) The main effect of $B$ on $C$ is defined in terms of an average over the levels of $A$: in fact

$$p(C_k \mid B_j) = p(C_k \mid A_1, B_j)p(A_1 \mid B_j) + p(C_k \mid A_2, B_j)p(A_2 \mid B_j) \tag{6.10}$$

which is defined since the $A$-classification is random. Since $p(A_1 \mid B_j) \neq p(A_2 \mid B_j)$ in general this is a weighted average of the probabilities for each $A$-classification. The "factors" $A$ and $B$ are therefore "confounded" and the main effect is difficult to interpret. An example, due to Simpson [19], also quoted by Edwards [5] will clarify the issue. The data are as follows

(6.11)

| | Male | | Female | |
|---|---|---|---|---|
| | Untreated | Treated | Untreated | Treated |
| Alive | 4 | 8 | 2 | 12 |
| Dead | 3 | 5 | 3 | 15. |

The dependent $C$-classification is into Alive or Dead. The random independent $A$-classification is by sex, and the non-random independent $B$-classification is by treatment. Let us ignore sampling variations and suppose the proportions (out of a total of 52 persons) are the probabilities $\{\theta_{ijk}\}$. The $2 \times 2$ table for Males shows that the association between treatment and death is $-5/6$ when measured by the difference of log-odds (Equation (5.4)). The same value is obtained for the $2 \times 2$ table for Females. Consequently, as measured by the difference of log-odds the treatment has been equally beneficial for the two sexes. This is equivalent to saying that the three-factor interaction (6.3) is zero. But if the two tables for Males and Females are combined to produce results that do not refer to sex, we obtain

(6.12)

| | Untreated | Treated |
|---|---|---|
| Alive | 6 | 20 |
| Dead | 6 | 20 |

and the association between treatment and death is zero: or the two classifications are independent. The explanation is that the allocation of treatments and sex are confounded, as can be seen by considering the $2 \times 2$ table that results from ignoring the dependent classification into Alive or Dead. The result is

(6.13)

| | Male | Female |
|---|---|---|
| Untreated | 7 | 5 |
| Treated | 13 | 27 |

The females, with their higher death rate have been treated more often than the males. Consequently it is not easy to understand the separate effects of treatment and sex. In the usual analysis of variance situations it is possible to separate the effects to some extent but this method is not available with contingency tables since the association in (6.12) is not a linear function of those in the separate sex tables in (6.11). In (6.12) we use $\ln \theta_{\cdot jk}$ and not linear forms in $\ln \theta_{1jk}$ and $\ln \theta_{2jk}$ as in linear hypothesis situations (compare again equations (5.8) and (5.9)). Had the proportions of each sex undergoing treatment been the same at 40/52 the table corresponding to (6.12) would have had about the same measure of association as the separate $2 \times 2$ tables in (6.12). As it is, the only course is to consider the sexes separately.

The extension to general $r \times s \times t$ tables is straightforward. The interaction of all three factors may be defined in terms of the parameters

$$(6.14) \quad \{\ln \theta_{ijk} - \ln \theta_{rjk} - \ln \theta_{isk} + \ln \theta_{rsk}\} - \{\ln \theta_{ijt} - \ln \theta_{rjt} - \ln \theta_{ist} + \ln \theta_{rst}\}$$

for $i < r, j < s, k < t$. (Compare Equations (5.15).) If these are all zero then there is no interaction. The vanishing of (6.4) may be tested in the usual way, though it must be noticed that the log-odds in (6.14) for different values of $i, j$ and $k$ are not independent and the dispersion matrix needs to be found. The details of the calculations are substantially reduced by using the ideas of Goodman [9].

The Markov property that given $B$, the $A$- and $C$-classifications are independent may be based on consideration of the log-odds

$$\{\ln p(C_k \mid B_j, A_i) - \ln p(C_t \mid B_j, A_i)\} - \{\ln p(C_k \mid B_j, A_r) - \ln p(C_t \mid B_j, A_r)\} \tag{6.15}$$

for $i < r$, $k < t$ and all $j$. For different $j$ these are independent. For each $j$ there are $(r - 1)(t - 1)$ log-odds which are not independent, but the quadratic form yields a $\chi^2$-statistic having $(r - 1)(t - 1)$ degrees of freedom. The total number of degrees of freedom is therefore $(r - 1)s(t - 1)$.

**7. Relationship with previous work.** There is little earlier work known to the author on analysis of contingency tables that uses a prior distribution: exceptions are Jeffreys [10], in particular Section 5.11; and two papers by Good [6], [7]. Nevertheless much of the material that falls within the classical framework is relevant and illuminating for a Bayesian. It is well known and almost obvious that if one is dealing with a normal distribution of unknown mean and known variance, or generally with any location parameter, where the density is of the form $f(x - \theta)$ for the random variable $x$ and the parameter $\theta$, there is a close parallelism between the classical analyses based on $x$ for fixed $\theta$ and the Bayesian analysis using the likelihood function of $\theta$ for fixed $x$. As an example consider statements like (3.2) and (3.3). The same is broadly true in the multinomial context and corresponding to our Theorem 1 there is a result that the distribution of $\ln n_i$ for fixed $\theta_i$ is approximately normal with mean $\ln N\theta_i$ and variance $(N\theta_i)^{-1}$, and they are independent provided linear contrasts in the $\ln n_i$ are considered. This result has been used by Plackett [16] to provide classical analyses of contingency tables along lines closely similar to ours. In particular he has considered the problem of testing an interaction, a problem which had previously been studied by Roy and Kastenbaum [17]. Goodman [9] has shown how Plackett's computations can be simplified in the way outlined at the end of Section 5. The first mention of the method based on the property of $\ln n_i$ known to me is Woolf's use of it for the $2 \times 2 \times t$ table [20]. A paper by Darroch [4] is relevant in deciding which hypotheses might be considered. Good [8] has provided a general definition of interaction which is equivalent to ours in the case of a $2^n$ table but does not seem so easy to handle for tables involving classifications into more than two exclusive and exhaustive classes. Good has resolved a conjective made by Darroch. Birch [3] has discussed definitions of interactions similar to those proposed here.

The methods advocated by Lancaster [13] based on the breakdown of $\chi^2$ suffer from several defects. There is no justification for the use of the statistic, nor for the breakdown into additive components. The methods have been criticized by Plackett.

The methods of Kullback [11] and [12] do not appear to have a Bayesian interpretation despite the use of information-theoretic concepts. Again the breakdown into additive components is unnatural in many situations.

It is relevant to remark that the measure of association suggested for a $2 \times 2$ table, Equation (5.4), has been shown to have certain attractive properties by Edwards [5]. Specifically any measure of association which is a function of the conditional probabilities $p(A_i \mid B_j)$ and equally of the conditional probabilities $p(B_j \mid A_i)$ is necessarily a function of $\phi$.

**8. Future extensions.** The prior distribution used throughout this paper is the special one proportional to $\prod \theta_i^{-1}$. But as explained in the first section, the results are available whenever the prior distribution is equivalent, in the amount of information it contains, to data obtained from a contingency table of the same type as that to be analyzed, with the special prior. The two tables, hypothetical and real, may be combined and the analysis based on the combined table and the special prior.

It would be highly desirable to extend the analyses to more general priors. For example, one might have a $2 \times 2$ table with little prior knowledge of $\theta_{i \cdot}$ but substantial prior knowledge of $\phi_{ij}$ (for the notation see the beginning of Section 5). Such knowledge might be equivalent to few observations on the margin of the $2 \times 2$ table but many observations on the interior of the table for a selected set of marginal totals. Another example is a trinomial situation with classes $A_1$, $A_2$ and $A_3$. One may feel fairly sure that $\theta_1$ is around 0.20 whereas knowledge of $\theta_2$ and $\theta_3$ (apart from the fact that they add up to 0.80) is slight.

Another type of prior that needs consideration is one that allows correlations between the $\ln \theta_i$. All the analyses in the present paper hinge on independence of them. For example in a $2 \times 2$ table one may have prior knowledge that the two probabilities $p(A_1 \mid B_j)(j = 1, 2)$ are close in value, and it would be desirable to incorporate this into the analysis. Indeed, there are situations where this may be essential: as where $B_1$ corresponds to the "at-homes" and $B_2$ to the "not-at-homes" in a social survey. A second example of a correlated prior is where the multinomial distribution has arisen from a grouped frequency distribution: that is, a histogram. The smoothness of the underlying density produces a correlation between neighboring groups.

It is hoped to show in a future paper that such situations can be handled using log-odds and multivariate normal distributions.

REFERENCES

[1] Bartlett, M. S. (1935). Contingency table interactions. *J. Roy. Statist. Soc. Suppl.* **2** 248–252.

[2] Bartlett, M. S. and Kendall, D. G. (1946). The statistical analysis of variance-heterogeneity and the logarithmic transformation. *J. Roy. Statist. Soc. Suppl.* **8** 128–138.

[3] Birch, M. W. (1963). Maximum likelihood in three-way contingency tables. *J. Roy. Statist. Soc. Ser. B* **25** 220–233.

[4] Darroch, J. N. (1962). Interactions in multi-factor contingency tables. *J. Roy. Statist. Soc. Ser. B* **24** 251–263.

[5] EDWARDS, A. W. F. (1963). The measure of association for 2 × 2 tables. *J. Roy. Statist. Soc. Ser. A* **126** 109–113.

[6] GOOD, I. J. (1957). Saddle-point methods for the multinomial distributions. *Ann. Math. Statist.* **28** 861–881.

[7] GOOD, I. J. (1956). On the estimation of small frequencies in contingency tables. *J. Roy. Statist. Soc. Ser. B* **18** 113–124.

[8] GOOD, I. J. (1963). Maximum entropy for hypothesis formulation, especially for multidimensional contingency tables. *Ann. Math. Statist.* **34** 911–934.

[9] GOODMAN, L. A. (1963). On Plackett's test for contingency table interactions. *J. Roy. Statist. Soc. Ser. B* **25** 179–188.

[10] JEFFREYS, H. (1961). *Theory of Probability*. Oxford: Clarendon Press.

[11] KULLBACK, S. (1959). *Information Theory and Statistics*. Wiley, New York.

[12] KULLBACK, S., KUPPERMAN, M. and KU, H. H. (1962). Tests for contingency tables and Markov chains. *Technometrics* **4** 573–608.

[13] LANCASTER, H. O. (1951). Complex contingency tables treated by the partition of $\chi^2$. *J. Roy. Statist. Soc. Ser. B* **13** 242–249.

[14] LINDLEY, D. V. (1961). The robustness of interval estimates. *Bull. Inst. Internat. Stat.* **38** 209–220.

[15] LINDLEY, D. V. (1964). *Introduction to Probability and Statistics*. Cambridge Univ. Press.

[16] PLACKETT, R. L. (1962). A note on interactions in contingency tables. *J. Roy. Statist. Soc. Ser. B* **24** 162–166.

[17] ROY, S. N. and KASTENBAUM, M. A. (1956). On the hypothesis of no 'interaction' in a multi-way contingency table. *Ann. Math. Statist.* **27** 749–757.

[18] SCHEFFÉ, H. (1959). *Analysis of Variance*. Wiley, New York.

[19] SIMPSON, E. H. (1951). The interpretation of interaction in contingency tables. *J. Roy. Statist. Soc. Ser. B* **13** 238–241.

[20] WOOLF, B. (1955). On estimating the relation between blood group and disease. *Ann. Human Genetics* **19** 251–253.

# [3]

# INFERENCE ABOUT VARIANCE COMPONENTS IN THE ONE-WAY MODEL*

Bruce M. Hill
*The University of Michigan and Harvard University*

The estimation of variance components in the one-way model is considered from a subjective Bayesian point of view. The situation in which the classical unbiased estimate of the between variance component is negative is explored in some detail. Exact and approximate posterior distributions are obtained in both the balanced and unbalanced case. Common sense aspects of the problem are emphasized, and some contrasts with other approaches. For example, Bayesianly speaking, a large negative unbiased estimate of the between variance component indicates an uninformative experiment in which the effective likelihood for that variance component is extremely flat, instead of strong evidence that the variance component is nearly zero.

## 1. INTRODUCTION

We shall discuss some aspects of the estimation of variance components in a random model (Eisenhart Model II) analysis of variance from a subjective Bayesian point of view. Particular attention will be given to the situation in which the classical unbiased estimate is negative. In this paper we shall be concerned with the one-way model, balanced and unbalanced, extending the results to regression models and to higher dimensional arrays in a later report.

The analysis of variance opens a Pandora's box of problems which constitute a real challenge to any and all statisticians and theories of statistical inference. Although an enormous variety of models have been proposed for even the most simple arrays of observations, there are serious gaps in our understanding of the *analysis* of even the most simple of these models, not to mention design. Here the approach will be to focus attention on an extremely simple model, a model that can at best be regarded only as a rough approximation in situations of real interest. Nonetheless, if only because the real situations we are confronted with are so very much more complex, it seems essential to look closely at the simpler ones.

We work with the model

$$y_{ij} = \mu + \alpha_i + \epsilon_{ij}, \quad i = 1, \cdots, I, \quad j = 1, \cdots, J_i, \quad N = \sum_i J_i,$$

where $\alpha_i \sim N(0, \sigma_\alpha^2)$, $\epsilon_{ij} \sim N(0, \sigma^2)$, and the $\alpha_i$ and $\epsilon_{ij}$ are mutually independent. Although we shall return to this general unbalanced model it is convenient for the present to restrict attention to the balance case, $J_i = J$ for all $i$. In this case the statistics

$$\bar{y}_{..} = (IJ)^{-1} \sum_{i,j} y_{ij}, \qquad SSW = \sum_{i,j} (y_{ij} - \bar{y}_{i.})^2,$$

* This work was partially sponsored by the National Science Foundation, Grant #NSF-GP-12.

and

$$SSB = \sum_{i,j} (\bar{y}_{i\cdot} - \bar{y}_{\cdot\cdot})^2$$

are sufficient for $\theta = (\mu, \sigma^2, \sigma_\alpha^2)$, where

$$\bar{y}_{i\cdot} = (J)^{-1} \sum_{j=1}^{J} y_{ij}.$$

It is easily verified that $\bar{y}_{\cdot\cdot}$ is normally distributed about $\mu$ with variance $(IJ)^{-1}\sigma^2 + (I)^{-1}\sigma_\alpha^2$, and that $SSW$ and $SSB$ are distributed as multiples of chi-square random variables,

$$\sigma^2 \chi^2_{I(J-1)} \quad \text{and} \quad (\sigma^2 + J\sigma_\alpha^2)\chi^2_{(I-1)},$$

respectively, the three being mutually independent given $\mu$, $\sigma^2$ and $\sigma_\alpha^2$.

The customary approach to estimation is to view $\bar{y}_{\cdot\cdot}$ as an estimate of $\mu$, $MSW = [I(J-1)]^{-1}\, SSW$ as an estimate of $\sigma^2$, $MSB = [I-1]^{-1}\, SSB$ as an estimate of $\sigma^2 + J\sigma_\alpha^2$, and $(J)^{-1}\, [MSB - MSW]$ as an estimate of $\sigma_\alpha^2$, all being unbiased. However, the possibility of negative estimates of $\sigma_\alpha^2$ (which occur with substantial probability when $\sigma_\alpha^2$ is small) and the lack of exact and confusion about proposed approximate confidence intervals for $\sigma_\alpha^2$ (even in the balanced case) raise serious questions [3]–[7], [9]–[15], [18], [20], [24]–[26], [29]–[30].

The aim of this paper is to illustrate Bayesian inference in this problem, and to use the Bayesian formulation as a means to discuss certain common sense aspects of the situation. Since there is some disagreement as to just what is common sense here, such discussion does not seem out of place.

## 2. LIKELIHOOD FUNCTION AND POSTERIOR DISTRIBUTIONS

Based upon the above model, and writing Pr for what is in fact a density function, the likelihood function is

$$
\begin{aligned}
L(\mu, \sigma^2, \sigma_\alpha^2) &= L(\mu, \sigma^2, \sigma_\alpha^2 \mid \text{data } y_{ij}) = \Pr\{y_{ij} \mid \mu, \sigma^2, \sigma_\alpha^2\} \\
&= \int \cdots \int \Pr\{y_{ij} \mid \mu, \sigma^2, \sigma_\alpha^2, \alpha_i\} \Pr\{\alpha_i \mid \mu, \sigma^2, \sigma_\alpha^2\} \prod d\alpha_i \\
&\propto \int \cdots \int (\sigma^2)^{-N/2} (\sigma_\alpha^2)^{-I/2} \\
&\quad \cdot \exp\left[-\frac{1}{2}\left\{\sum_{i,j} (y_{ij} - \mu - \alpha_i)^2/\sigma^2 + \sum_i \alpha_i^2/\sigma_\alpha^2\right\}\right] \prod d\alpha_i \qquad (1) \\
&\propto (\sigma^2)^{-(N-I)/2} \exp\left[-\frac{1}{2}\frac{SSW}{\sigma^2}\right] \prod_i (\sigma^2 + J_i\sigma_\alpha^2)^{-1/2} \exp\left[-\frac{1}{2}\sum_i \frac{J_i(\bar{y}_i - \mu)^2}{\sigma^2 + J_i\sigma_\alpha^2}\right].
\end{aligned}
$$

Letting $\rho(\mu, \sigma^2, \sigma_\alpha^2)$ be a subjectively chosen prior density, the posterior density becomes $\rho''(\mu, \sigma^2, \sigma_\alpha^2) \propto L(\mu, \sigma^2, \sigma_\alpha^2)\, \rho(\mu, \sigma^2, \sigma_\alpha^2)$.

We shall be interested both in the situation where the likelihood function is sharp or highly concentrated relative to the prior distribution, in which case (some) Bayesian credible intervals typically differ but little from (some) classical confidence intervals, and also in those situations where although the likelihood function is expected to be relatively sharp (on the basis, say, of Fisherian information) before the experiment, it in actual fact is not, as for example when $MSB \ll MSW$. The following references discuss and illustrate Bayesian inference [1], [2], [8], [16], [17], [19], [21], [22], [23], [28].

All of our analysis will be based on the assumption of diffuse prior opinion for $\mu$ effectively independent of that for $(\sigma^2, \sigma_\alpha^2)$, or roughly $\rho(\mu, \sigma^2, \sigma_\alpha^2) = \rho(\sigma^2, \sigma_\alpha^2)$. This is not especially restrictive, and leads to the approximate marginal posterior density

$$
\begin{aligned}
\rho''(\sigma^2, \sigma_\alpha^2) &= \int_{-\infty}^{\infty} \rho''(\mu, \sigma^2, \sigma_\alpha^2)\, d\mu \\
&\propto \rho(\sigma^2, \sigma_\alpha^2)(\sigma^2)^{-(N-I)/2} \\
&\quad \cdot \exp\left[-\frac{SSW}{2\sigma^2}\right] \prod_{i=1}^{I} (\sigma^2 + J_i\sigma_\alpha^2)^{-1/2} \int_{-\infty}^{\infty} \exp\left[-\frac{1}{2}\sum_i \frac{J_i(\bar{y}_{i\cdot} - \mu)^2}{\sigma^2 + J_i\sigma_\alpha^2}\right] d\mu \\
&\propto \rho(\sigma^2, \sigma_\alpha^2)(\sigma^2)^{-(N-I)/2} \\
&\quad \cdot \exp\left[-\frac{SSW}{2\sigma^2}\right] \prod_{i=1}^{I} (\sigma^2 + J_i\sigma_\alpha^2)^{-1/2} \left\{\sum_{i=1}^{I} \frac{J_i}{\sigma^2 + J_i\sigma_\alpha}\right\}^{-1/2} \\
&\quad \cdot \exp\left[-\frac{1}{2}\sum_i \frac{J_i(\bar{y}_{i\cdot} - \hat{\mu})^2}{\sigma^2 + J_i\sigma_\alpha^2}\right]
\end{aligned} \tag{2}
$$

where

$$
\hat{\mu} = \frac{\sum_i J_i\bar{y}_{i\cdot}/(\sigma^2 + J_i\sigma_\alpha^2)}{\sum_i J_i/(\sigma^2 + J_i\sigma_\alpha^2)}
$$

is a function of $\sigma_\alpha^2/\sigma^2$. In the balanced case $J_i = J$ this simplifies to $\hat{\mu} = \bar{y}_{\cdot\cdot}$ and

$$
\begin{aligned}
\rho''(\sigma^2, \sigma_\alpha^2) \propto \rho(\sigma^2, \sigma_\alpha^2)(\sigma^2)^{-(N-I)/2} &\exp\left[-\frac{SSW}{2\sigma^2}\right](\sigma^2 + J\sigma_\alpha^2)^{-(I-1)/2} \\
&\cdot \exp\left[-\frac{SSB}{2(\sigma^2 + J\sigma_\alpha^2)}\right].
\end{aligned} \tag{3}
$$

It is to be noted that equation (3) expresses the fact that independent measurements have been made of $\sigma^2$ and $\sigma^2 + J\sigma_\alpha^2$, and that the problem of inference about $\sigma^2$ and $\sigma_\alpha^2$ is thus closely related to other problems, for example, inference about background and source radioactivity based upon independent readings with and without the source present. Thus if $Y_1$ = counts without source, $Y_2$ = counts with source, $E(Y_1) = \lambda_1$, $E(Y_2) = \lambda_1 + \lambda_2$, and $Y_1$ and $Y_2$ have independent Poisson distributions, then

$$\rho''(\lambda_1, \lambda_2)/\rho(\lambda_1, \lambda_2)$$
$$= \Pr\{Y_1, Y_2 \mid \lambda_1, \lambda_2\} \propto \lambda_1^{Y_1} \exp[-\lambda_1](\lambda_1 + \lambda_2)^{Y_2} \exp[-(\lambda_1 + \lambda_2)].$$

This is essentially equivalent to (3) as far as the likelihood function is concerned, although there is an interesting difference between inference about $\lambda_2$ and inference about $\sigma_\alpha^2$ which we will not go into here.

### 3. BALANCED CASE

A. *Inference on* $\sigma^2$

The marginal posterior density for $\sigma^2$ is

$$\rho''(\sigma^2) = \int_0^\infty \rho''(\sigma^2, \sigma_\alpha^2)d\sigma_\alpha^2$$
$$\propto (\sigma^2)^{-I(J-1)/2} \exp\left[-\frac{SSW}{2\sigma^2}\right] \int_0^\infty \rho(\sigma^2, \sigma_\alpha^2)(\sigma^2 + J\sigma_\alpha^2)^{-(I-1)/2}$$
$$\cdot \exp\left[-\frac{SSB}{2(\sigma^2 + J\sigma_\alpha^2)}\right] d\sigma_\alpha^2 .$$

Now it is instructive to view the factor

$$\rho'(\sigma^2) \equiv \int_0^\infty \rho(\sigma^2, \sigma_\alpha^2)(\sigma^2 + J\sigma_\alpha^2)^{-(I-1)/2} \exp\left[-\frac{SSB}{2(\sigma^2 + J\sigma_\alpha)}\right] d\sigma_\alpha^2 \quad (4)$$

as a partial posterior distribution based solely on the data $\bar{y}_{i.}$, and the factor

$$L(\sigma^2 \mid SSW) \equiv (\sigma^2)^{-I(J-1)/2} \exp\left[-\frac{SSW}{2\sigma^2}\right]$$

as a likelihood function based solely on the deviations $y_{ij} - \bar{y}_{i.}$. The over-all posterior density is thus $\rho''(\sigma^2) \propto \rho'(\sigma^2)L(\sigma^2 \mid SSW)$. The reason for this factorization is to focus attention on the information about $\sigma^2$ contained in the $\bar{y}_{i.}$, which seems often to be overlooked.

The subtleties involved in inference on $\sigma^2$ stem from the factor $\rho'(\sigma^2)$. When $\rho'(\sigma^2)$ can be regarded as gentle relative to $L(\sigma^2 \mid SSW)$ the principle of stable measurement applies and $\rho''(\sigma^2)$ will be nearly proportional to $L(\sigma^2 \mid SSW)$ [8, p. 15], [16], [19, Ch. 3], [22, p. 20], [23, Ch. 4]. For example, if $\rho'(\sigma^2)\alpha\sigma^{-2}$ (strictly speaking this is impossible, but when stable estimation occurs the exact choice of $\rho'(\sigma^2)$ is immaterial, as is discussed further at the end of this section) then

$$\rho''(\sigma^2) \propto (\sigma^2)^{-(I(J-1)/2)-1} \exp\left[-\frac{SSW}{2\sigma^2}\right] \quad (5)$$

so that $\sigma^2$ is distributed approximately like $SSW/\chi^2_{I(J-1)}$ where $\chi_f^2$ denotes a random variable having the chi-square distribution with $f$ degrees of freedom. This result is in harmony with the conventional approach under which $SSW/\sigma^2$ has the chi-square distribution with $I(J-1)$ degrees of freedom for given $\sigma^2$. From a Bayesian viewpoint, however, this "stable measurement posterior" is merely one (rather simple) posterior distribution that may arise, and we cannot regard it as having any innate significance, except at best as a reasonable approximation in some circumstances. Indeed, the factor $\rho'(\sigma^2)$ may not be gentle relative to $L(\sigma^2|SSW)$ even though sample sizes are large and $L(\sigma^2|SSW)$ is quite sharp, and we must be extremely cautious in invoking the stable measurement approximation.

Recall now that $\rho'(\sigma^2)$ represents the posterior density of $\sigma^2$ arising from measurement of $h^2=\sigma^2+J\sigma_\alpha^2$ (based only on the $\bar{y}_i$) in conjunction with the prior joint density $\rho(\sigma^2, \sigma_\alpha^2)$. It is clear that precise measurement of $h^2$ can lead to sharp upper bounds for tail probabilities of the form $\Pr'\{\sigma^2>C\}$ and $\Pr'\{\sigma_\alpha^2>C\}$ (where the single prime indicates posterior probability based only on the data $\bar{y}_{i.}$) since each is no larger than $\Pr'\{h^2>C\}$. On the other hand, within the set of $(\sigma^2, \sigma_\alpha^2)$ for which $\sigma^2+J\sigma_\alpha^2$ is in reasonable accord with our assumed precise measurement of $h^2$, the partial posterior density $\rho'(\sigma^2, \sigma_\alpha^2)$ based solely on the $\bar{y}_{i.}$ will be gentle *unless* our initial prior opinion as measured by $\rho(\sigma^2, \sigma_\alpha^2)$ is sharp. It is as though the precise measurement of $h^2$ tends to sharply restrict the space of $(\sigma^2, \sigma_\alpha^2)$ to a set in which $\sigma^2+J\sigma_\alpha^2$ is nearly constant and within that set preference is effectively governed by $\rho(\sigma^2, \sigma_\alpha^2)$.

To illustrate, let us suppose that measurement of $h^2$ is so sharp that we effectively determine a line $\sigma^2+J\sigma_\alpha^2=h^2=\hat{h}^2$ having probability unity in the $(\sigma^2, \sigma_\alpha^2)$ plane. Then the joint density of $(\sigma^2, \sigma_\alpha^2)$ given the data $\bar{y}_{i.}$ is

$$\rho'(\sigma^2, \sigma_\alpha^2) \propto \begin{cases} \rho(\sigma^2, \sigma_\alpha^2) & \text{if } \sigma^2 + J\sigma_\alpha^2 = \hat{h}^2, \\ 0 & \text{if not} \end{cases}$$

and

$$\rho'(\sigma^2) \propto \begin{cases} \rho(\sigma^2, J^{-1}(\hat{h}^2 - \sigma^2)) & \text{if } 0 \le \sigma^2 \le \hat{h}^2, \\ 0 & \text{if } \sigma^2 > \hat{h}^2 \end{cases}$$

Unless there is some rather special prior opinion, such as for example that it is highly probable that $\sigma_\alpha^2$ is close to some specified value or that $\sigma_\alpha^2<\sigma^2$, etc., one would anticipate that in many situations a rather typical prior opinion would have $\rho(\sigma^2, \sigma_\alpha^2)$ varying gently along the line $\sigma^2+J\sigma_\alpha^2=\hat{h}^2$. Thus it is natural to expect $\rho'(\sigma^2)$ to be nearly constant in the interval $0\le\sigma^2\le\hat{h}^2$, and similarly with respect to $\rho'(\sigma_\alpha^2)\propto\rho(\hat{h}^2-J\sigma_\alpha^2, \sigma_\alpha^2)$ for $0\le\sigma_\alpha^2\le\hat{h}^2/J$. Except in the face of rather special prior knowledge, the data $\bar{y}_{i.}$ for large $I$ will lead one to rather precise opinions about $h^2=\sigma^2+J\sigma_\alpha^2$, but one will be vague about $\sigma^2$ and $\sigma_\alpha^2$ outside of the fact that one will be nearly certain the first is not much larger than $\hat{h}^2$ and the second than $\hat{h}^2/J$.

Our discussion so far has been quite symmetric with respect to $\sigma^2$ and $J\sigma_\alpha^2$. An asymmetry is now introduced by bringing in the likelihood $L(\sigma^2|SSW)$ based on the deviations $y_{ij}-\bar{y}_{i.}$. This function tends to point sharply to a value $\hat{\sigma}^2$. If $\hat{\sigma}^2\ll\hat{h}^2$ and if $\rho(\sigma^2, \sigma_\alpha^2)$ is gentle along the line $\sigma^2+J\sigma_\alpha^2=\hat{h}^2$ then the over-all posterior $\rho''(\sigma^2)\propto\rho'(\sigma^2)L(\sigma^2|SSW)$ will be nearly proportional to $L(\sigma^2|SSW)$ within $[0, \hat{h}^2]$, and except for the truncation effect beyond $\hat{h}^2$ [which will be far out in the upper tail of $L(\sigma^2|SSW)$] we will have essentially the stable estimation posterior above (5). On the other hand, if $\hat{\sigma}^2\gg\hat{h}^2$, then

$\rho'(\sigma^2)$ and $L(\sigma^2 | SSW)$ tend to cancel one another in the sense that each is large where the other is small. Since $L(\sigma^2 | SSW)$ points roughly to $MSW$, while the measurement of $h^2$ points roughly to $MSB$, we see that the condition for such "cancellation" is roughly that $MSW \gg MSB$.

We shall now assume $\rho(\sigma^2, \sigma_\alpha^2) = \rho_1(\sigma^2)\rho_2(\sigma_\alpha^2)$. This certainly need not be the case, but, for example, if $\sigma^2$ reflects measurement error on an individual, while $\sigma_\alpha^2$ reflects variability of "true values" between individuals then it is not unnatural that opinions about $\sigma^2$ and $\sigma_\alpha^2$ should be roughly independent. We find

$$\begin{aligned}
\rho''(\sigma^2) &\equiv \int_0^\infty \rho''(\sigma^2, \sigma_\alpha^2)\, d\sigma_\alpha^2 \\
&\propto \int_0^\infty \rho(\sigma^2, \sigma_\alpha^2) L(\sigma^2 | SSW)(h^2)^{-(I-1)/2} \exp\left[-\frac{SSB}{2h^2}\right] d\sigma_\alpha^2 \\
&\propto \rho_1(\sigma^2) L(\sigma^2 | SSW) \int_0^\infty \rho_2(\sigma_\alpha^2)(\sigma^2 + J\sigma_\alpha^2)^{-(I-1)/2} \exp\left[-\frac{SSB}{2(\sigma^2 + J\sigma_\alpha^2)}\right] d\sigma_\alpha^2.
\end{aligned}$$

It is tempting at this point to take $\rho_2(\sigma_\alpha^2) \propto (\sigma_\alpha^2)^{-1}$ along the lines of Savage [23, p. 5.7] and others; however, the integral is then infinite because of the pole at zero. This is no real problem since the prior density $\rho_2(\sigma_\alpha^2) \propto (\sigma_\alpha^2)^{-1}$ would always be regarded as unrealistic near zero, and only be employed when this feature was not crucial in determining the posterior distribution. To avoid trouble near zero we shall give $\sigma_\alpha^2$ a proper prior distribution, taking

$$\rho_2(\sigma_\alpha^2) \propto (\sigma_\alpha^2)^{-((\lambda_\alpha/2)+1)} \exp\left[-\frac{C_\alpha}{2\sigma_\alpha}\right],$$

so that $(\sigma_\alpha^2)^{-1}$ has a gamma distribution with parameters $\lambda_\alpha/2$ and $C_\alpha/2$ chosen subjectively. Then

$$\begin{aligned}
\rho''(\sigma^2) &\propto \rho_1(\sigma^2) L(\sigma^2 | SSW) \int_0^\infty (\sigma_\alpha^2)^{-(\lambda_\alpha/2+1)} \exp\left[-\frac{C_\alpha}{2\sigma_\alpha^2}\right] (h^2)^{-((I-1)/2)} \\
&\quad \cdot \exp\left[-\frac{SSB}{2h^2}\right] d\sigma_\alpha^2 \\
&\propto \rho_1(\sigma^2) L(\sigma^2 | SSW) \int_0^\infty y^{-((\lambda_\alpha/2)+1)} \exp\left[-\frac{JC_\alpha}{2y}\right] (\sigma^2 + y)^{-((I-1)/2)} \\
&\quad \cdot \exp\left[-\frac{SSB}{2(\sigma^2 + y)}\right] dy, \qquad (6)
\end{aligned}$$

and the last integral may be regarded as proportional to the density of a random variable.

$$Z = X - Y = \{(\Gamma_{(I-3)/2, SSB/2})^{-1} - (\Gamma_{\lambda_\alpha/2, JC_\alpha/2})^{-1}\},$$

at least on the positive half-line, where $\Gamma_{\lambda,\theta}$ is a random variable having the density

$$f(t) \propto t^{\lambda-1}e^{-\theta t}, \qquad t \geq 0,$$

and $X$ and $Y$ are independent. If $C_\alpha = K_{\alpha\lambda\alpha}$, then

$$\lim_{\lambda_\alpha \to 0} (\sigma_\alpha^2)^{-(\lambda_\alpha/2+1)} \exp\left[-\frac{C_\alpha}{2\sigma_\alpha^2}\right] = (\sigma_\alpha^2)^{-1},$$

so we may represent diffuse prior opinion by a choice of $\lambda_\alpha$ near zero. However, the expression for $\rho(\sigma_\alpha^2)$, with suitable $\lambda_\alpha$ and $C_\alpha$, can be employed whether or not prior opinion about $\sigma_\alpha^2$ is diffuse.

The over-all posterior distribution of $\sigma^2$ thus may be regarded as a weighting of $\rho_1(\sigma^2)L(\sigma^2 | SSW)$ by the density of $Z$, the resulting function being normalized to have unit area on the positive half-line. Although $Z$ can clearly be negative, only the portion of its density for $\sigma^2 \geq 0$ is relevant to the over-all posterior distribution of $\sigma^2$. Since both the mean and mode of $Z$ are roughly $MSB - JC_\alpha/\lambda_\alpha$, this quantity locates the center of the distribution of $Z$. When it is negative, then the density of $Z$ will be monotonically decreasing for $\sigma^2 \geq 0$, and the effect of the information supplied by the $\bar{y}_{i.}$ is to modify the partial posterior $\rho_1(\sigma^2)$ $L(\sigma^2 | SSW)$ based upon the $y_{ij} - \bar{y}_{i.}$ alone by giving the greatest weight to small $\sigma^2$, and thus tending to move opinions downward. When $MSB - JC_\alpha/\lambda_\alpha > 0$ then the density of $Z$ will have only a portion of its left tail below zero, and there is a possibility of obtaining an extremely sharp posterior $\rho''(\sigma^2)$.

Now it is clear that a great variety of situations are possible, and that the posterior distributions are in general analytically cumbersome. Although there are some natural approximations, and one such will be described in Section 3B where the distributional form of $Z$ again arises in connection with $\sigma_\alpha^2$, the point of view taken here is that such approximations need only play a minor role in the *analysis* of data. For it is ordinarily a minor computing task to plot the posterior distributions corresponding to one or perhaps several choices of prior. After such plotting an approximation may be chosen in order to simplify the description of the posterior, but this is not essential. The real need for approximations in Bayesian theory is in the planning of an experiment, for example in the choice of sample sizes, which we shall not discuss in this paper.

We shall consider one other form of diffuse or gentle prior, namely $\rho(\sigma^2, \sigma_\alpha^2) \propto (\sigma^2)^{-1}(\sigma^2 + J\sigma_\alpha^2)^{-1}$ for $\sigma^2 \geq 0$, $\sigma_\alpha^2 \geq 0$, which as pointed out by Tiao and Tan [28] can be obtained by the method of Jeffreys. Although the $J$ in this prior depends upon the sample size and would thus seem inappropriate, our point of view is that this prior is simply a convenient analytical form only to be used when in fact it makes little difference what the exact choice of prior is because of the sharpness of the likelihood function. We find for this prior

$$\begin{aligned}
\rho''(\sigma^2) &\propto \int_0^\infty (\sigma^2)^{-1}(\sigma^2 + J\sigma_\alpha^2)^{-1} L(\sigma^2 | SSW)(h^2)^{-(I-1)/2} \exp\left[-\frac{SSB}{2h^2}\right] d\sigma_\alpha^2 \\
&\propto (\sigma^2)^{-1} L(\sigma^2 | SSW) \int_0^\infty (h^2)^{-((I-1)/2)-1} \exp\left[-\frac{SSB}{2h^2}\right] d\sigma_\alpha^2 \\
&\propto (\sigma^2)^{-1} L(\sigma^2 | SSW) F_{I-1}(SSB/\sigma^2), \qquad (7)
\end{aligned}$$

where $F_n(t)$ is the cumulative distribution function of chi-square with $n$ degrees of freedom. Note that this differs from the earlier stable estimation posterior only in the factor $F_{I-1}(SSB/\sigma^2)$, which is a monotonically decreasing function of $\sigma^2$, concave for $\sigma^2 < SSB/(I+1)$ and convex for $\sigma^2 > SSB/(I+1)$.

If in (7) we put $SSB=0$, then

$$\rho''(\sigma^2) \propto (\sigma^2)^{-(IJ-1)/2-1} \exp\left\{-\frac{SSW}{2\sigma^2}\right\},$$

so that posteriorly $\sigma^2 \sim SSW/\chi^2_{(IJ-1)}$, as was noted by Tiao and Tan [28]. This is a disturbing result. For if $I$ is large then an $SSB$ near zero would, based upon the $\bar{y}_{i.}$ alone, lead to a strong opinion that $\sigma^2$ is "small." On the other hand, a large $SSW$ would, based upon the $y_{ij}-\bar{y}_{i.}$ alone, lead to a strong opinion that $\sigma^2$ is "large." These two separate sources of information can thus lead to very different opinions about $\sigma^2$. Surely in the extreme case where $SSB$ is very small and $SSW$ is very large one is reluctant to accept $\sigma^2 \sim SSW/\chi^2_{(IJ-1)}$ as describing over-all posterior opinion about $\sigma^2$. It should be pointed out that it is not simply this result (which after all was based upon an improper prior) that is in question, but really any way of combining the two divergent sources of information, if we keep to the original model. We shall consider this question further in connection with the example of Section 3D.

A remark about the choice of prior may be appropriate here. In a rather abstract sense Bayesian theory supposes that an individual has a unique set of opinions at a given time, and thus a unique prior distribution of the parameters. Nonetheless the prior is never known with complete accuracy, and there is always a good deal of flexibility in the exact choice. Furthermore, when an informative experiment has been performed in the sense that the likelihood function is quite sharp relative to the prior, then the posterior distribution is essentially proportional to the likelihood function (i.e., we have stable estimation), and within wide limits the exact analytical specification of the prior is immaterial. In such circumstances certain conventional "diffuse" priors are convenient (like that of Jeffreys) although largely arbitrary. When the posterior distribution depends crucially upon the prior, as we shall see it does in inference upon $\sigma^2_\alpha$ when $\hat{\sigma}^2_\alpha < 0$, then no merely conventional prior can be used. In a certain sense this makes the prior seem to depend upon the data. In fact, however, it is only the degree of care we take in approximating our prior, not the prior itself, that depends upon the data.

## B. *Inference on $\sigma^2_\alpha$*

The posterior density for $\sigma^2_\alpha$ is

$$\rho''(\sigma^2_\alpha) \propto \int_0^\infty \rho(\sigma^2, \sigma^2_\alpha)(\sigma^2)^{-I(J-1)/2} \exp\left\{-\frac{SSW}{2\sigma^2}\right\} (h^2)^{-(I-1)/2} \exp\left\{-\frac{SSB}{2h^2}\right\} d\sigma^2.$$

Along the lines of the previous discussion we first take

$$\rho(\sigma^2, \sigma^2_\alpha) = \rho_1(\sigma^2)\rho_2(\sigma^2_\alpha)$$

so that

$$p''(\sigma_\alpha^2) \propto p_2(\sigma_\alpha^2) \int_0^\infty p_1(\sigma^2) L(\sigma^2 \mid SSW)(h^2)^{-(I-1)/2} \exp\left\{-\frac{SSB}{2h^2}\right\} d\sigma^2.$$

Further, taking $p_1(\sigma^2) \propto (\sigma^2)^{-1}$ (we could just as well have chosen a proper gamma density) we have

$$p''(\sigma_\alpha^2) \propto p_2(\sigma_\alpha^2) p'(\sigma_\alpha^2)$$

where

$$p'(\sigma_\alpha^2) \propto \int_0^\infty (\sigma^2)^{-(I(J-1)/2)-1} \exp\left\{-\frac{SSW}{2\sigma^2}\right\} (h^2)^{-(I-1)/2} \exp\left\{-\frac{SSB}{2h^2}\right\} d\sigma^2. \quad (8)$$

Let now $X = \Gamma_{I-3/2, SSB/2}$ and $Y = \Gamma_{I(J-1)/2, SSW/2}$ have independent gamma distributions. Then since $p'(\sigma_\alpha^2)$ is proportional to the density of $Z = J^{-1}(X^{-1} - Y^{-1})$ on the positive half-line, the posterior density $p''(\sigma_\alpha^2) \propto p_2(\sigma_\alpha^2) p'(\sigma_\alpha^2)$ may be regarded as a truncation from below at zero of the distribution of $Z$ together with a weighting by $p_2(\sigma_\alpha^2)$. We shall henceforth regard $p'(\sigma_\alpha^2)$ as proportional to the density of $Z$ on the whole real line, using the fact $p_2(\sigma_\alpha^2) = 0$ for $\sigma_\alpha^2 < 0$ to yield the required truncation at zero.

It is $p'(\sigma_\alpha^2)$ that plays the role of a likelihood function (actually a marginal likelihood) in the formation of the marginal posterior density $p''(\sigma_\alpha^2)$. Since the distribution of $Z$ depends upon $I$, $J$, $SSB$ and $SSW$, these four quantities determined by the data are parameters for $p'(\sigma_\alpha^2)$. The important thing to see is that the form of the posterior $p''(\sigma_\alpha^2)$ depends crucially upon the place where the truncation at zero occurs in the function $p'(\sigma_\alpha^2)$, and this in turn depends upon the above four parameters. In particular, since the random variable $Z$ has a unimodal density $p'(\sigma_\alpha^2)$ with mode roughly at $\hat{\sigma}_\alpha^2 = J^{-1}(MSB - MSW)$, it follows that when $\hat{\sigma}_\alpha^2 \ll 0$ the function $p'(\sigma_\alpha^2)$ will be monotonically decreasing in the interval $\sigma_\alpha^2 \geq 0$, with maximum at $\sigma_\alpha^2 = 0$. Since $p''(\sigma_\alpha^2) = 0$ for $\sigma_\alpha^2 < 0$ only this interval is of interest.

Thus we see that a negative $\hat{\sigma}_\alpha^2$ leads to a likelihood factor with maximum at $\sigma_\alpha^2 = 0$ and decreasing monotonically in $\sigma_\alpha^2$. The over-all import of such data is thus to give relatively more weight to small $\sigma_\alpha^2$ in the posterior than in the prior, and this is presumably in accord with some frequentistic interpretations of negative $\hat{\sigma}_\alpha^2$. However, when we ask further how the degree of sharpness with which $p'(\sigma_\alpha^2)$ is peaked at zero (which determines how strongly the data suggest that $\sigma_\alpha^2$ is small) depends upon the magnitude of the assumed negative $\hat{\sigma}_\alpha^2$, we come to a divergence of viewpoint. For when $\hat{\sigma}_\alpha^2 = J^{-1}(MSB - MSW) < 0$, then the larger is $SSW$ (and hence keeping $SSB$, $I$ and $J$ fixed, the more negative is $\hat{\sigma}_\alpha^2$) then the flatter is the function $p'(\sigma_\alpha^2)$. In fact, if all other quantities are held fixed,

$$\lim_{SSW \to \infty} p''(\sigma_\alpha^2) = p_2(\sigma_\alpha^2)$$

so that the posterior is the same as the prior, and in this sense the experiment has been completely uninformative about $\sigma_\alpha^2$!

To prove this we first note that

$$\left.\frac{\partial \ln \rho'(\sigma_\alpha^2)}{\partial \sigma_\alpha^2}\right|_{\sigma_\alpha^2=0} \approx \frac{(I-1)(J-1)}{2(MST)^2}[[1+2/I(J-1)]MSB - MSW] \qquad (9)$$

where

$$MST = \sum_{i,j}(y_{ij}-\bar{y}_{..})^2/(IJ-1). \text{ Thus } \left.\frac{\partial \rho'(\sigma_\alpha^2)}{\partial \sigma_\alpha^2}\right|_0 < 0$$

if and only if $MSW > [1+2/I(J-1)]MSB$ or $\hat{\sigma}_\alpha^2 < -2MSB/IJ(J-1)$, and when this occurs $\rho'(\sigma_\alpha^2)$ decreases monotonically from its value at zero. From (8) it follows easily that

$$\lim_{\sigma_\alpha^2 \uparrow \infty} \frac{\partial \ln \rho'(\sigma_\alpha^2)}{\partial \sigma_\alpha^2} = 0,$$

the limit being approached monotonically from below. Further, an asymptotic expansion as

$$SSW \uparrow \infty \text{ yields } \frac{\partial^2 \ln \rho'(\sigma_\alpha^2)}{\partial^2 \sigma_\alpha^2} > 0$$

simultaneously for all $\sigma_\alpha^2$ in any finite interval for sufficiently large $SSW$, and

$$\frac{\partial \ln \rho'(\sigma_\alpha^2)}{\partial \sigma_\alpha^2}$$

is then monotonically increasing within such intervals from its negative value (9) at $\sigma_\alpha^2 = 0$. Since also

$$\lim_{SSW \uparrow \infty} \left.\frac{\partial \ln \rho'(\sigma_\alpha^2)}{\partial \sigma_\alpha^2}\right|_0 = 0,$$

it follows that $\rho'(\sigma_\alpha^2)$ becomes "flat" (measured by logarithmic derivative) as $SSW \uparrow \infty$, and this in turn implies

$$\lim_{SSW \uparrow \infty} \rho''(\sigma_\alpha^2) = \rho_2(\sigma_\alpha^2), \text{ for all } \sigma_\alpha^2,$$

where both $\rho''(\sigma_\alpha^2)$ and $\rho_2(\sigma_\alpha^2)$ are here assumed normalized to have unit area. Convergence in distribution is then easily shown. It is worth observing that as $SSB \downarrow 0$,

$$\left.\frac{\partial \ln \rho'(\sigma_\alpha^2)}{\partial \sigma_\alpha^2}\right|_0$$

decreases to

$$-\frac{IJ}{2MSW}\,\frac{\left(1-\frac{1}{I}\right)\left(1-\frac{1}{IJ}\right)}{\left(1-\frac{1}{J}\right)} \quad \text{or} \quad \approx -\frac{IJ}{2MSW}$$

if both $I$ and $J$ are large, which indicates that the more negative $\hat{\sigma}_\alpha^2$ is for *fixed* $SSW$, then the stronger is the indication that $\sigma_\alpha^2$ is small.

One way of seeing the implications of these results is to compare $\rho'(\sigma_\alpha^2)$ with an exponential exp $[-f\sigma_\alpha^2]$. For fixed $SSW$ the decreasing function $\rho'(\sigma_\alpha^2)$ is most concentrated near zero when $SSB=0$, and in this case

$$\left.\frac{\partial \ln \rho'(\sigma_\alpha^2)}{\partial \sigma_\alpha^2}\right|_0 \approx -IJ/2MSW.$$

For large $SSW$,

$$\frac{\partial \ln \rho'(\sigma_\alpha^2)}{\partial \sigma_\alpha^2}$$

is an increasing function of $\sigma_\alpha^2$, so $\rho'(\sigma_\alpha^2)$ is "flatter" than

$$\exp\left[-\frac{IJ}{2MSW}\sigma_\alpha^2\right]$$

in the sense that

$$\rho'(0)/\rho'(\sigma_\alpha^2) \leq \exp\left[\frac{IJ}{2MSW}\sigma_\alpha^2\right].$$

When $IJ/2MSW$ is small we may thus be able to quickly appraise the over-all influence of the data upon our prior opinion about $\sigma_\alpha^2$, and see that the posterior is essentially the same as the prior.

We have so far discussed only the case $\hat{\sigma}_\alpha^2 \ll 0$, when the truncation leaves only the monotonically decreasing portion of $\rho'(\sigma_\alpha^2)$. Since $E(Z) \approx \hat{\sigma}_\alpha^2$, Var $(Z) \approx [2/IJ^2][MSB^2 + MSW^2/J]$ it follows that a positive $\hat{\sigma}_\alpha^2$ may lead to a situation in which only a small portion of the area under $\rho'(\sigma_\alpha^2)$ lies to the left of zero and is cut off by the truncation. In this case it will sometimes suffice to approximate $\rho'(\sigma_\alpha^2)$ by a normal density with the same mean and variance. Even when the truncation occurs precisely at the mode of $\rho'(\sigma_\alpha^2)$ a normal approximation may be suitable, although now only the half of the normal density above its mean will be used. As remarked earlier in connection with $\sigma^2$, such approximations are primarily useful in planning the experiment, since the likelihood factor $\rho'(\sigma_\alpha^2)$ and the posterior $\rho''(\sigma_\alpha^2)$ can always be obtained numerically.

We note that Var $(Z)$ increases with $MSW$, so that when $MSB > MSW$ (and thus $\hat{\sigma}_\alpha^2 > 0$), a large $MSW$ leads to a relatively spread out posterior distribution centered near $\hat{\sigma}_\alpha^2$. Both this and our earlier result that large $SSW$

leads to a flat $\rho'(\sigma_\alpha^2)$ when $\hat{\sigma}_\alpha^2<0$ are reflections of the fact that before the data are taken, the larger $\sigma^2$ is, the greater the variance of the random variable $\hat{\sigma}_\alpha^2$. Bayesianly, after the data are taken, $MSW$ plays much the same role as $\sigma^2$ did prior to the data.

Finally, if we use the Jeffreys prior $\rho(\sigma^2, \sigma_\alpha^2)=(\sigma^2)^{-1}(\sigma^2+J\sigma_\alpha^2)^{-1}$, then the posterior distribution of $\sigma_\alpha^2$ is that of the random variable

$$J^{-1}(SSB/\chi^2_{I-1} - SSW/\chi^2_{I(J-1)})$$

truncated from below at zero. This distribution, censored instead of truncated from below at zero, has also been proposed as a fiducial solution [11], [14]. Our viewpoint is that when $\hat{\sigma}_\alpha^2<0$, and particularly when $MSW$ is large, this distribution (even truncated, and certainly when censored) is inappropriate as a measure of posterior opinion. For we have seen that in this case the data are extremely uninformative, that $\rho''(\sigma_\alpha^2)\approx\rho_2(\sigma_\alpha^2)$, and since one is left with his prior density it is inappropriate to use a merely formal prior such as that of Jeffreys. From our point of view such a formal prior is only useful as a convenient but more or less arbitrary choice made from a collection of priors all of which essentially yield the likelihood function (normalized in some way) as posterior distribution. When, as in the present case with $\hat{\sigma}_\alpha^2\ll 0$, the likelihood factor is flat relative to the prior (the reverse of "stable estimation"), it becomes crucial to carefully assess the prior.

### C. *Inference on $\sigma_\alpha^2/\sigma^2$ and $\mu$*

We consider now inference on $\tau^2=\sigma_\alpha^2/\sigma^2$ and $\mu$ based upon the prior

$$\rho(\mu, \sigma^2, \sigma_\alpha^2) \propto (\sigma^2)^{-1}(\sigma_\alpha^2)^{(-\lambda\alpha/2)-1}\exp\left[-\frac{C_\alpha}{2\sigma_\alpha^2}\right] \tag{10}$$

We find (see Section 4 for details) the posterior density

$$g''(\tau^2) \propto \frac{(\tau^2)^{(-\lambda_\alpha/2)-1}\left(\tau^2+\dfrac{1}{J}\right)^{-(I-1)/2}}{\left[SSW+\dfrac{C_\alpha}{\tau^2}+\dfrac{SSB}{1+J\tau^2}\right]^{(N+\lambda_\alpha-1)/2}} \quad \text{for } \tau^2\geq 0.$$

Note that if we had chosen $\lambda_\alpha=0$, $C_\alpha=0$, corresponding to the prior $(\sigma_\alpha^2)^{-1}$, then

$$g''(\tau^2) \propto \frac{(\tau^2)^{-1}\left(\tau^2+\dfrac{1}{J}\right)^{-(I-1)/2}}{\left[SSW+\dfrac{SSB}{1+J\tau^2}\right]^{(IJ-1)/2}}$$

$$\propto (\tau^2)^{-1}\frac{\left(\tau^2+\dfrac{1}{J}\right)^{(I(J-1))/2}}{[SSB+SSW(1+J\tau^2)]^{(IJ-1)/2}}$$

would be improper. However, the difficulty arises only at $\tau^2$ near zero, where the prior $(\sigma_\alpha^2)^{-1}$ cannot be viewed as realistic, and sometimes can be avoided by the simple device of truncating $g''(\tau^2)$ at some $\epsilon>0$, or alternatively by approximating

$$(\tau^2)^{-1} \text{ by } \left(\tau^2 + \frac{1}{J}\right)^{-1},$$

which yields the same result as if we had chosen the Jeffreys prior to begin with. In the latter case we find

$$g''(\tau^2) \propto \frac{\left(\tau^2 + \frac{1}{J}\right)^{(I(J-1)/2)-1}}{[SSB + SSW(1 + J\tau^2)]^{(IJ-1)/2}}$$

which states that

$$\frac{SSW}{SSB}(1 + J\tau^2) = \frac{\chi^2_{I(J-1)}}{\chi^2_{I-1}}$$

truncated from below at SSW/SSB, or

$$(1 + J\tau^2) \sim \frac{MSB}{MSW} F_{I(J-1),I-1}, \quad \text{with } F \tag{11}$$

truncated from below at $MSW/MSB$, in some but not complete harmony with tradition. Here $F_{I(J-1),I-1}$ denotes a random variable having the truncated $F$ distribution with the indicated degrees of freedom. We also note that in the balanced case with $\lambda_\alpha = C_\alpha = 0$, that given $\tau^2$,

$$\left[SSW + \frac{SSB}{1 + J\tau^2}\right] \Big/ \sigma^2$$

has the chi-square distribution with $IJ-1$ degrees of freedom.

Now let us consider the situation $MSB \leq MSW$. To gain insight we shall make the approximation

$$1 + J\tau^2 \sim \frac{MSB}{MSW} F_{I(J-1),I-1}$$

with $F$ truncated from below at $MSW/MSB$ (the approximation of course may sometimes be poor) and find

$$\Pr''\{t_0 \leq \tau^2 \leq t_1\} = \frac{\Pr\left\{\frac{MSW}{MSB}(1 + Jt_0) \leq F^* \leq \frac{MSW}{MSB}(1 + Jt_1)\right\}}{\Pr\left\{\frac{MSW}{MSB} \leq F^* < \infty\right\}},$$

where here $F^*$ has an ordinary $F$ distribution with the above degrees of freedom. Now, depending on the degrees of freedom, unity may be either more or less than the 50th percentile of $F^*$. However, in the present case, with numerator degrees of freedom larger than denominator, unity is always less than the 50th percentile, the latter approaching unity as both degrees of freedom grow large. If $MSB \leq MSW$ then the probability $\Pr''\{t_0 \leq \tau^2 \leq t_1\}$ for intervals of given length will be maximal when $t_0 = 0$, and

$$\Pr''\{0 \leq \tau^2 \leq t_1\} = \frac{\Pr\left\{\frac{MSW}{MSB} \leq F^* \leq \frac{MSW}{MSB}(1 + Jt_1)\right\}}{\Pr\left\{\frac{MSW}{MSB} \leq F^* < \infty\right\}}.$$

Further, if we take the limit as $MSW/MSB$ goes to $\infty$, then

$$\lim_{MSW/MSB \uparrow \infty} \Pr''\{0 \leq \tau^2 \leq t_1\} = 1 - (1 + Jt_1)^{-(I-1)/2} \tag{12}$$

which is a proper probability distribution. When $I$ and $J$ are large it is thus clear that a large ratio $MSW/MSB$ leads to a strong opinion that $\tau^2$ is very small. Since in this case the truncation occurs far out in the tail of the $F$ distribution, in practice it would not be clear how seriously to take this limiting distribution in view of the strong dependence upon the tail of $F$.

When $MSW/MSB \approx 1$, $\Pr''\{0 \leq \tau^2 \leq t_1\}$ may be quite large for even small $t_1$. For example, if both degrees of freedom are large, then

$$\Pr''\{0 \leq \tau^2 \leq t_1\} \approx 2 \Pr\{1 \leq F^* \leq 1 + Jt_1\}$$

and if $t_1 = J^{-1}$ this is $2 \Pr\{1 \leq F^* \leq 2\}$ which will be very near unity.

Inference on $\mu$ is based on the result (see Section 4 for details) that, given $\tau^2$, $\sqrt{n}H(\mu - \bar{y}_{..})$ has the Student $t$ distribution with $n = N + \lambda_\alpha - 1$ degrees of freedom, and

$$\sqrt{n}H = [N + \lambda_\alpha - 1]^{1/2} \left[\frac{IJ/(1 + J\tau^2)}{SSW + \frac{C_\alpha}{\tau^2} + \frac{SSB}{1 + J\tau^2}}\right]^{1/2};$$

while, if further $\lambda_\alpha = C_\alpha = 0$, then

$$\mu - \bar{y}_{..} \sim \left[\frac{SST + J\tau^2 SSW}{N(N-1)}\right]^{1/2} t_{(N-1)} \tag{13}$$

given $\tau^2$, where here

$$SST = SSB + SSW = \sum_{i,j} (y_{ij} - \bar{y}_{..})^2,$$

and $t_{(N-1)}$ is a random variable having the $t$ distribution with $N - 1$ degrees of freedom. Thus, we obtain simple conditional credible intervals on $\mu$, given $\tau^2$, and obtain over-all posterior probabilities by integrating with respect to $g''(\tau^2)$. In any event, since for each given $\tau^2$ it is clear that the posterior expectation of $\mu$ is $\bar{y}_{..}$, this is also true unconditionally, and in fact the posterior distribution of $\mu$ is symmetric about $\bar{y}_{..}$.

### D. *An Example and Discussion*

We shall consider a rather extreme example in order to bring out most clearly the nature of our results. Suppose that $I=5$, $J=10$, $MSW=2500$, $MSB=.01$, and thus $\hat{\sigma}_\alpha^2=1/10\ (.01-2500)\approx -250$. Then

(a) Inference upon $\tau^2=\sigma_\alpha^2/\sigma^2$ can be based upon the limiting distribution (12) as $MSW/MSB$ goes to $\infty$, so that $\Pr''\{0\leq\tau^2\leq c|\text{data}\}\approx 1-(1+10c)^{-2}$, which is .99 if $c=.9$, and the data provide a strong indication that $\tau^2<1$.

(b) Since $-IJ/2MSW=-.01$ and $SSW$ is sufficiently large, the likelihood factor $\rho'(\sigma_\alpha^2)$ is "flatter" than the exponential $\exp[-.01\ \sigma_\alpha^2]$, and thus within the inverval $0\leq\sigma_\alpha^2\leq 10$ the likelihood ratio $\rho'(0)/\rho'(\sigma_\alpha^2)\leq\exp[.1]\approx 1.1$. Within this interval the likelihood factor and hence the data modify the prior $\rho_2(\sigma_\alpha^2)$ only slightly. Generally speaking in this kind of situation the prior density $\rho_2(\sigma_\alpha^2)$ must be carefully evaluated if the model is not rejected and decisions are to be based upon knowledge of $\sigma_\alpha^2$.

(c) Based upon (7) we find $\sigma^2\sim SSW/\chi_{49}^2=112500/\chi_{49}^2$ and the posterior expectation of $\sigma^2$ is about 2300. As discussed earlier the discrepancy between the inference based upon the $\bar{y}_{i.}$ alone and that based upon the $y_{ij}-\bar{y}_{i.}$ alone is disturbing. Thus $h^2\sim SSB/\chi_4^2=.04/\chi_4^2$, so that

$$\Pr'\{\sigma^2<c\,|\,\bar{y}_{i.}\}\geq\Pr'\{h^2<c\,|\,\bar{y}_{i.}\}=\Pr\{\chi_4^2>.04/c\}$$

which is quite near unity if $c>10$. But

$$\Pr'\{\sigma^2>c\,|\,y_{ij}-\bar{y}_{i.}\}=\Pr\{SSW/\chi_{45}^2>c\}=\Pr\{\chi_{45}^2<112500/c\}$$

which is also quite near unity if say $c<1000$. In a situation such as this one is forced to face up to the real uncertainties concerning the model. This is not at all inconsistent with Bayesian philosophy since real opinions about data never precisely correspond to any model, and even a very small prior doubt of the model can be enormously magnified by extreme data. When such data occur there is no alternative but to broaden the model, often along the lines suggested by the data. In this example it is natural to consider constraints, either deterministic or probabilistic, that tend to keep the $\bar{y}_{i.}$ nearly equal. It is important to see that extreme data cannot in themselves cause rejection of the model (except of course if the observations are literally impossible), but rather that they may give heavy weight to alternative models which were never before the experiment regarded as totally implausible. Although the Bayesian theory of models is in a primitive form, some useful ideas have been presented [1], [2], [17, Ch. V and VI].

(d) Using (13) we have, given $\tau^2$,

$$\mu\sim\bar{y}_{..}+\left[\frac{SST+J\tau^2SSW}{N(N-1)}\right]^{1/2}t_{(N-1)}$$

$\approx \bar{y}_{..}+\sqrt{45}(1+10\tau^2)^{1/2}t_{49}$, and the posterior distribution of $\mu.$ is symmetric about $\bar{y}_{..}$. Depending upon the amount of work we wish to do we can alternatively either integrate out $\tau^2$, replace $\tau^2$ by its posterior expectation making some allowance for uncertainty about $\tau^2$, or simply make use of rough bounds such as $45 \leq \text{Var}\ (\mu|\text{data}) \leq 11 \times 45$.

We emphasize that in this artificial example there can be no prior opinions about the parameters, since neither the data nor the parameters have any meaning. Our purpose was merely to indicate the form which inference will tend to take. In any real problem the prior can always be assessed, although perhaps with difficulty.

### 4. THE UNBALANCED CASE

We shall now deal with the estimation of $\sigma^2$, $\sigma_\alpha^2$, $\tau^2=\sigma_\alpha^2/\sigma^2$ and $\mu$ in the general unbalanced case. Our approach will be to work directly with the bivariate posterior distribution of $(\sigma^2, \tau^2)$ and $(\mu, \tau^2)$, since the conditional posterior distributions of $\sigma^2$, $\sigma_\alpha^2$ and $\mu$, given $\tau^2$, take on standard and relatively simple forms.

Based upon the approximate posterior density (2), and letting

$$p(\sigma^2, \sigma_\alpha^2) \propto (\sigma^2)^{-1}(\sigma_\alpha^2)^{-\lambda_\alpha/2-1} \exp[-C_\alpha/2\sigma_\alpha^2],$$

we obtain the posterior density

$$\begin{aligned} g''(\sigma^2, \tau^2) &\propto \sigma^2 p''(\sigma^2, \sigma^2\tau^2) \\ &\propto \sigma^2 p(\sigma^2, \sigma^2\tau^2)(\sigma^2)^{-(N-1)/2} \exp\left[-\frac{SSW}{2\sigma^2}\right] \prod_i (1+J_i\tau^2)^{-1/2} \\ &\quad \times \left[\sum_i \frac{J_i}{1+J_i\tau^2}\right]^{-1/2} \exp\left[-\frac{1}{2\sigma^2}\sum_i \frac{J_i(\bar{y}_{i.}-\hat{\mu})^2}{1+J_i\tau^2}\right] \\ &\propto (\sigma^2)^{-((N-1+\lambda_\alpha)/2)-1} \exp\left[-\frac{1}{2\sigma^2}\left(\frac{\tau^2 SSW + C_\alpha}{\tau^2} + \sum_i \frac{J_i(\bar{y}_{i.}-\hat{\mu})^2}{1+J_i\tau^2}\right)\right] \\ &\quad \times (\tau^2)^{-(\lambda_\alpha/2)-1} \prod_i (1+J_i\tau^2)^{-1/2}\left[\sum_i \frac{J_i}{1+J_i\tau^2}\right]^{-1/2}. \end{aligned}$$

The marginal posterior density of $\tau^2$ is then

$$g''(\tau^2) \propto \frac{(\tau^2)^{-\lambda_\alpha/2-1}\prod_i (1+J_i\tau^2)^{-1/2}\left[\sum_i \dfrac{J_i}{1+J_i\tau^2}\right]^{-1/2}}{\left\{SSW + \dfrac{C_\alpha}{\tau^2} + \sum_i \dfrac{J_i(\bar{y}_{i.}-\hat{\mu})^2}{1+J_i\tau^2}\right\}^{(N-1+\lambda_\alpha)/2}},$$

while the conditional posterior density of $\sigma^2$ given $\tau^2$ is

$$\frac{g''(\sigma^2, \tau^2)}{g''(\tau^2)} \propto (\sigma^2)^{-((N-1+\lambda_\alpha)/2)-1} \exp\left\{-\frac{1}{2\sigma^2}\left(SSW + \frac{C_\alpha}{\tau^2} + \sum_i \frac{J_i(\bar{y}_{i.}-\hat{\mu})^2}{1+J_i\tau^2}\right)\right\}.$$

Hence, given $\tau^2$, the variable

$$\left(SSW + \frac{C_\alpha}{\tau^2} + \sum_i \frac{J_i(\bar{y}_{i\cdot} - \hat{\mu})^2}{1 + J_i\tau^2}\right) \Big/ \sigma^2$$

is distributed like chi-square on $N+\lambda_\alpha-1$ degrees of freedom

$$\Pr''\{t_0 \leq \sigma^2 \leq t_1\} = \int_0^\infty \Pr''\{t_0 \leq \sigma^2 \leq t_1 \mid \tau^2\} g''(\tau^2) d\tau^2$$

$$= \int_0^\infty \Pr\left\{\frac{SSW + \frac{C_\alpha}{\tau^2} + \sum_i \frac{J_i(\bar{y}_{i\cdot} - \hat{\mu})^2}{1 + J_i\tau^2}}{t_1} \leq \chi^2_{(N+\lambda_\chi-1)} \leq \frac{SSW + \frac{C_\alpha}{\tau^2} + \sum_i \frac{J_i(\bar{y}_{i\cdot} - \hat{\mu})^2}{1 + J_i\tau^2}}{t_0}\right\} g''(\tau^2) d\tau^2$$

and

$$\Pr''\{t_0 \leq \sigma_\alpha^2 \leq t_1\} = \int_0^\infty \Pr''\{t_0 \leq \sigma_\alpha^2 \leq t_1 \mid \tau^2\} g''(\tau^2) d\tau^2$$

$$= \int_0^\infty \Pr''\left\{\frac{t_0}{\tau^2} \leq \sigma^2 \leq \frac{t_1}{\tau^2} \,\middle|\, \tau^2\right\} g''(\tau^2) d\tau^2$$

so that each of these posterior probabilities is the integral of the chi-square probability of an interval whose endpoints are functions of $\tau^2$, with respect to the posterior distribution of $\tau^2$. Various approximations are available, the simplest, which is appropriate when $g''(\tau^2)$ points sharply to a value $\hat{\tau}^2>0$, being

$$\Pr''\{t_0 \leq \sigma^2 \leq t_1\} \approx \Pr''\{t_0 \leq \sigma^2 \leq t_1 \mid \tau^2 = \hat{\tau}^2\}$$

and

$$\Pr''\{t_0 \leq \sigma_\alpha^2 \leq t_1\} \approx \Pr''\left\{\frac{t_0}{\hat{\tau}^2} \leq \sigma^2 \leq \frac{t_1}{\hat{\tau}^2} \mid \tau^2 = \hat{\tau}^2\right\},$$

the latter probabilities being evaluated from the chi-square distribution. Even when $g''(\tau^2)$ is not particularly sharp these approximations will be appropriate if the endpoints of the interval for $\chi^2{}_{(N-\lambda_\alpha-1)}$ are relatively gentle functions of $\tau^2$. In general, however, we must be prepared to perform the integration, perhaps numerically, and towards this end we shall now examine $g''(\tau^2)$ more closely.

Write

$$g''(\tau^2) \propto \frac{(\tau^2)^{-(\lambda_\alpha/2)-1} \prod (\tau^2 + L_i)^{-1/2} \left[\sum \frac{1}{\tau^2 + L_i}\right]^{-1/2}}{\left[SSW + \frac{C_\alpha}{\tau^2} + \sum \frac{(\bar{y}_{i\cdot} - \hat{\mu})^2}{\tau^2 + L_i}\right]^{(N+\lambda_\alpha-1)/2}} = g_1(\tau^2) g_2(\tau^2)$$

where $g_1(\tau^2)$ and $[g_2(\tau^2)]^{-1}$ are the numerator and denominator of $g''(\tau^2)$, and $L_i = J_i^{-1}$. Verify that

$$g_1(\tau^2) = (\tau^2)^{-((I+\lambda_\alpha-1)/2)-1} \prod (1 + L_i/\tau^2)^{-1/2} \left[ \sum_i \frac{1}{1 + L_i/\tau^2} \right]^{-1/2},$$

$$\frac{\partial \ln g_1(\tau^2)}{\partial \tau^2} < 0, \quad \text{and} \quad \lim_{\tau^2 \to \infty} (\tau^2)^{((I+\lambda_\alpha-1)/2)+1} g_1(\tau^2) = I^{-1/2},$$

so that $g_1(\tau^2)$ decreases montonically from $g_1(0) = \infty$ to $g_1(\infty) = 0$.

Next we find (see also Wald [27])

$$\frac{\partial}{\partial \tau^2} \left\{ \sum \frac{(\bar{y}_{i\cdot} - \hat{\mu})^2}{\tau^2 + L_i} \right\} = - \sum \frac{(\bar{y}_{i\cdot} - \hat{\mu})^2}{(\tau^2 + L_i)^2}$$

so that

$$g_2(\tau^2) = \left\{ SSW + \frac{C_\alpha}{\tau^2} + \sum_i \frac{(\bar{y}_{i\cdot} - \hat{\mu})^2}{\tau^2 + L_i} \right\}^{-(N+\lambda_\alpha-1)/2}$$

is monotonically increasing from

$$g_2(0) = \begin{cases} 0 & \text{if } C_\alpha > 0, \\ [SSW + \sum J_i(\bar{y}_{i\cdot} - \bar{y})^2]^{-(N-1)/2} & \text{if } C_\alpha = 0,\ \lambda_\alpha = 0, \end{cases}$$

to

$$g_2(\infty) = [SSW]^{-(N+\lambda_\alpha-1)/2},$$

where here

$$\bar{y} = (\sum J_i \bar{y}_{i\cdot}) / \sum J_i.$$

Thus $g''(\tau^2)$ is the product of monotonically increasing and decreasing factors, typically goes to zero as $\tau^2$ goes to zero or infinity, and has a unique mode say at $\hat{\tau}^2$. By various approximations we can judge the posterior probability of intervals about $\hat{\tau}^2$ (or better centered at the posterior mean or median), and thus determine approximate credible intervals. It is also natural to approximate $g''(\tau^2)$ by an $F$ density (with degrees of freedom to be fitted), since $g''(\tau^2)$ is of this form in the balanced case.

Now we turn to inference about $\mu$. If $g''(\mu, \sigma^2, \tau^2)$ is the joint posterior density of $(\mu, \sigma^2, \tau^2)$ then

$$\begin{aligned} g''(\mu, \tau^2) &= \int_0^\infty g''(\mu, \sigma^2, \tau^2) d\sigma^2 \\ &\propto (\tau^2)^{-\lambda_\alpha/2-1} \prod (1 + J_i \tau^2)^{-1/2} \int_0^\infty (\sigma^2)^{-((N+\lambda_\alpha)/2)-1} \\ &\quad \times \exp\left[ -\frac{1}{2\sigma^2} \left\{ SSW + \frac{C_\alpha}{\tau^2} + \sum \frac{J_i(\bar{y}_{i\cdot} - \mu)^2}{1 + J_i \tau^2} \right\} \right] d\sigma^2 \end{aligned}$$

$$\propto \frac{(\tau^2)^{-(\lambda_\alpha/2)-1} \prod (1 + J_i\tau^2)^{-1/2}}{\left\{SSW + \frac{C_\alpha}{\tau^2} + \sum \frac{J_i(\bar{y}_{i\cdot} - \mu)^2}{1 + J_i\tau^2}\right\}^{(N+\lambda_\alpha)/2}}.$$

Since

$$\sum \frac{J_i(\bar{y}_{i\cdot} - \mu)^2}{1 + J_i\tau^2} = \sum \frac{J_i(\bar{y}_{i\cdot} - \hat{\mu})^2}{1 + J_i\tau^2} + (\mu - \hat{\mu})^2 \sum \frac{J_i}{1 + J_i\tau^2},$$

then

$$g''(\mu \mid \tau^2) \propto \left\{1 + H^2(\mu - \hat{\mu})^2\right\}^{-(N+\lambda_\alpha)/2},$$

where

$$H = \left[\frac{\sum \frac{J_i}{1 + J_i\tau^2}}{SSW + \frac{C_\alpha}{\tau^2} + \sum \frac{J_i(\bar{y}_{i\cdot} - \hat{\mu})^2}{1 + J_i\tau^2}}\right]^{1/2}$$

Now let $\chi = \sqrt{n}H(\mu - \hat{\mu})$, $n = N + \lambda_\alpha - 1$. Then, given $\tau^2$, the conditional posterior density of $\chi$ is

$$\rho''(\chi \mid \tau^2) \propto \left\{1 + \frac{\chi^2}{n}\right\}^{-(n+1)/2},$$

so that $\sqrt{n}H(\mu - \hat{\mu})$ has (conditional on $\tau^2$) Student's distribution with $n = N + \lambda_\alpha - 1$ degrees of freedom.

By numerical integration or other approximation methods we can obtain the unconditional posterior distribution of $\mu$; for example, if $g''(\tau^2)$ is sharply concentrated at say $\hat{\tau}^2$, then we can act as though it were known that $\tau^2 = \hat{\tau}^2$ making some rough allowance for the residual uncertainty about $\tau^2$.

We note that the unbalanced case offers only more complexity in the form of the posterior distributions, and no fundamental difficulties.

REFERENCES

[1] Anscombe, F. J., "Tests of Goodness of Fit," *Journal of the Royal Statistical Society, Series B*, 25 (1963), 81–94.

[2] Box, G. E. P., Cox, D. R., "An Analysis of Transformations," *Journal of the Royal Statistical Society, Series B*, 26 (1964), 211–52.

[3] Bross, I., "Fiducial Intervals for Variance Component Analysis," *Biometrics* 6 (1950), 136–44.

[4] Bulmer, M. G., "Approximate Confidence Limits for Components of Variance," *Biometrika* 44 (1957), 159–67.

[5] Crump, S. L., "Estimation of Variance Components in Analysis of Variance", *Biometrics Bulletin*, Vol 2 (1946), p. 7.

[6] Crump, S. L., "The Present Status of Variance Component Analysis," *Biometrics* 7 (1951), 1–15.

[7] Daniels, H. E., "The Estimation of Components of Variance," *Journal of the Royal Statistical Society, Supplement* 6 (1939), 186–97.

[8] Edwards, W., Lindman, H., and Savage, L. J., "Bayesian Statistical Inference for Psychological Research," *Psychological Review* 70 (1963), 193–242.

[9] Eisenhart, C., "The Assumptions Underlying Analysis of Variance," *Biometrics* 3 (1947), 1–21.

[10] Fisher, R. A., *Statistical Method for Research Workers*. Thirteenth Edition. New York: Hafner Publishing Company, Inc., 1958.

[11] Fisher, R. A. "The Fiducial Argument in Statistical Inference," *Annals of Eugenics* 6 (1935), 391–8.

[12] Fisher, R. A., *The Design of Experiments*. Seventh Edition. New York: Hafner Publishing Company, Inc., 1960.

[13] Green, J. R., "A Confidence Interval for Variance Components," *Annals of Mathematical Statistics* 25 (1954), 671–85.

[14] Healy, M. J. R., "Fiducial Limits for a Variance Component," *Journal of the Royal Statistical Society, Series B*, 25 (1963), 128–30.

[15] Herbach, L. H., "Properties of Model II—Type Analysis of Variance Tests . . . ," *Annals of Mathematical Statistics*, 30 (1959), 939–59.

[16] Hill, B. M., "The Three Parameter Log-Normal Distribution and Bayesian Analysis of a Point-Source Epidemic," *Journal of the American Statistical Association*, 58 (1963), 72–84.

[17] Jeffreys, H., *Theory of Probability*. Third Edition, Oxford: Clarendon Press, 1961.

[18] Nelder, J. A., "The interpretation of negative components of variance," *Biometrika*, 41 (1954), 544–8.

[19] Raiffe H., and Schlaifer, R., *Applied Statistical Decision Theory*. Boston: Harvard Business School, 1961.

[20] Sattherthwaite, F. E., "An Approximate Distribution of Estimates of Variance Components," *Biometrics* 2 (1946), 110–14.

[21] Savage, L. J., *The Foundations of Statistics*. New York: John Wiley and Sons, Inc., 1954.

[22] Savage, L. J., *The Foundations of Statistical Inference, A Discussion*. London: Methuen and Co., Ltd., New York: John Wiley and Sons, Inc., 1962.

[23] Savage, L. J., "The Subjective Basis of Statistical Practice," University of Michigan Technical Report, 1961.

[24] Scheffé, H., *The Analysis of Variance*, New York: John Wiley and Sons, Inc., 1961.

[25] Searle, S. R., "Matrix Methods in Components of Variance and Covariance Analysis," *Annals of Mathematical Statistics*, 27 (1956), 737–48.

[26] Thompson, W. A., Jr., "The Problem of Negative Estimates of Variance Components," *Annals of Mathematical Statistics*, 33 (1962), 273–89.

[27] Wald, A., "A Note on the Analysis of Variance with Unequal Class Frequencies," *Annals of Mathematical Statistics*, 11 (1940), 96–100.

[28] Tiao, G. C., and Tan, W. Y., "Bayesian Analysis of Random-Effect Models . . . ," Technical Report No. 30, University of Wisconsin (1964).

[29] Yates, F., "The Analysis of Multiple Classifications with Unequal Numbers in the Different Classes," *Journal of the American Statistical Association*, 29 (1934), 51–66.

[30] Yates, F., Zacopanay, I., "The Estimation of the Efficiency of Sampling with Special Reference to Sampling for Yield in Cereal Experiments," *Journal of Agricultural Science*, 25 (1935), 545–77.

# [4]

Article from the *Journal of the Royal Statistical Society*, 1972, **B**, **34** (1), 1–41.

## Bayes Estimates for the Linear Model

By D. V. Lindley and A. F. M. Smith

*University College, London*

[Read before the Royal Statistical Society at a meeting organized by the Research Section on Wednesday, December 8th, 1971, Mr M. J. R. Healy in the Chair]

### Summary

The usual linear statistical model is reanalyzed using Bayesian methods and the concept of exchangeability. The general method is illustrated by applications to two-factor experimental designs and multiple regression.

*Keywords*: LINEAR MODEL; LEAST SQUARES; BAYES ESTIMATES; EXCHANGEABILITY; ADMISSIBILITY; TWO-FACTOR EXPERIMENTAL DESIGN; MULTIPLE REGRESSION; RIDGE REGRESSION; MATRIX INVERSION.

### Introduction

Attention is confined in this paper to the linear model, $E(\mathbf{y}) = \mathbf{A\theta}$, where $\mathbf{y}$ is a vector of observations, $\mathbf{A}$ a known design matrix and $\boldsymbol{\theta}$ a vector of unknown parameters. The usual estimate of $\boldsymbol{\theta}$ employed in this situation is that derived by the method of least squares. We argue that it is typically true that there is available prior information about the parameters and that this may be exploited to find improved, and sometimes substantially improved, estimates. In this paper we explore a particular form of prior information based on de Finetti's (1964) important idea of exchangeability.

The argument is entirely within the Bayesian framework. Recently there has been much discussion of the respective merits of Bayesian and non-Bayesian approaches to statistics: we cite, for example, the paper by Cornfield (1969) and its ensuing discussion. We do not feel that it is necessary or desirable to add to this type of literature, and since we know of no reasoned argument against the Bayesian position we have adopted it here. Nevertheless the reader not committed to this approach may like to be reminded that many techniques of the sampling-theory school are basically unsound: see the review by Lindley (1971b). In particular the least-squares estimates are typically unsatisfactory: or, in the language of that school, are inadmissible in dimensions greater than two. This follows since, by a well-known device in least-squares theory (see, for example, Plackett, 1960, p. 59), we may write the linear model after transformation in the form $E(z_i) = \xi_i$ for $i \leqslant p$ and $E(z_i) = 0$ for $i > p$. Here the $z$'s are transforms of the data, and the $\xi$'s of the parameters. Adding the assumption of normality, we can appeal to the results of Brown (1966), generalizing those of Stein (1956), which show that for a very wide class of loss functions the estimate of $\xi_i$ by $z_i$, for $i \leqslant p$ is inadmissible. In Section 1 of this paper we do comment on the admissibility of the Bayesian estimates and try to show, in a way that might appeal to an adherent of orthodox ideas, that they are likely to be superior, at least in some situations, to the least-squares estimates.

## 1. EXCHANGEABILITY

We begin with a simple example. Suppose, in the general linear model, that the design matrix is the unit matrix so that $E(y_i) = \theta_i$ for $i = 1, 2, \ldots, n$, and that $y_1, y_2, \ldots, y_n$ are independent, normally distributed with known variance $\sigma^2$. Such a simple model might arise if $y_i$ was the observation on the *i*th variety in a field trial, of average yield $\theta_i$. In considering the prior knowledge of the $\theta_i$ it may often be reasonable to assume their distribution *exchangeable*. That is, that it would be unaltered by any permutation of the suffixes: so that, in particular, the prior opinion of $\theta_7$ is the same as that of $\theta_4$, or any other $\theta_i$; and similarly for pairs, triplets and so on. Now one way of obtaining an exchangeable distribution $p(\boldsymbol{\theta})$ is to suppose

$$p(\boldsymbol{\theta}) = \int \prod_{i=1}^{n} p(\theta_i \mid \mu)\, dQ(\mu),$$

where $p(\theta_i \mid \mu)$, for each $\mu$, and $Q(\mu)$ describe arbitrary probability distributions. In other words, $p(\boldsymbol{\theta})$ is a *mixture*, by $Q(\mu)$, of independent and identical distributions, given $\mu$. Indeed, Hewitt and Savage (1955), in generalization of de Finetti's original result, have shown that if exchangeability is assumed for *every n*, then a mixture is the *only* way to generate an exchangeable distribution.

In the present paper we study situations where we have exchangeable prior knowledge and assume this exchangeability described by a mixture. In the example this implies $E(\theta_i) = \mu$, say, a common value for each *i*. In other words there is a linear structure to the *parameters* analogous to the linear structure supposed for the observations $\mathbf{y}$. If we add the premise that the distribution from which the $\theta_i$ appear as a random sample is normal, the parallelism between the two stages, for $\mathbf{y}$ and $\boldsymbol{\theta}$, becomes closer. In this paper we study the situation in which the parameters of the general linear model themselves have a general linear structure in terms of other quantities which we call *hyperparameters*.† In this simple example there is just one hyperparameter, $\mu$.

Indeed, we shall find it necessary to go further and let the hyperparameters also have a linear structure. This will be termed a *three-stage model* and is analysed in detail in the next section. There are straightforward extensions to any number of stages.

Returning to the simple example with $E(y_i) = \theta_i$, $E(\theta_i) = \mu$ and respective variances $\sigma^2$ and $\tau^2$, say, the situation will be completely specified once a prior distribution has been given for $\mu$. (Effectively this is the third stage just mentioned.) Supposing $\mu$ to have a uniform distribution over the real line—a situation usually described by saying there is vague prior knowledge of $\mu$—Lindley (1971a) has obtained the posterior distribution of $\theta_i$ and found its mean to be

$$E(\theta_i \mid \mathbf{y}) = \frac{y_i/\sigma^2 + y_./\tau^2}{1/\sigma^2 + 1/\tau^2}, \tag{1}$$

where $y_. = \sum y_i/n$. The detailed analysis has been given in the reference just cited, so we content ourselves with a brief discussion to serve as an introduction to the general theory in the next section.

The estimates, (1), will be referred to as *Bayes* estimates, and it is these that we propose as substitutes for the usual least-squares estimates. We denote them by $\theta_i^*$,

† We believe we have borrowed this terminology from I. J. Good but are unable to trace the reference.

and reserve the usual notation, $\hat{\theta}_i$, for the ordinary estimates. Notice that $\theta_i^*$ is a weighted average of $y_i = \hat{\theta}_i$ and the overall mean, $y_.$, with weights inversely proportional to the variances of $y_i$ and $\theta_i$. Hence the natural estimates are pulled towards a central value $y_.$, the extreme values experiencing most shift. We shall find the weighted average phenomenon will persist even within the general model. Of course the estimate (1) depends on $\sigma^2$ and $\tau^2$, which will typically be unknown, but their estimation presents no serious difficulties. If, for each $i$, there is replication of the $y_i$ then $\sigma^2$ may be estimated as the usual within variance. Since we have replication (from the distribution $N(\mu, \tau^2)$ underlying the exchangeability assumption) for the $\theta_i$, $\tau^2$ may be estimated. For example $\sum(\theta_i^* - \theta_.^*)^2/(n-1)$ might be a reasonable estimate of $\tau^2$, although in fact the reference just cited shows this can be improved upon. These estimates of $\sigma^2$ and $\tau^2$ can be used in place of the known values used in (1) and the cycle repeated.

Let us now digress from the Bayesian viewpoint and try to persuade an orthodox statistician that (1) is a sensible estimate for him to consider, and indeed is better than the least-squares estimate. Of course, $\theta_i^*$ is a biased estimate of $\theta_i$, so its merit cannot be judged by its variance. We use instead the mean-square error $E(\theta_i^* - \theta_i)^2$. This is just a criterion for judging the merit of one of the $n$ estimates, so let us look at the average mean-square error over the $n$ values. Simple, but tedious, calculations enable this to be found and compared with the corresponding quantity for $\hat{\theta}_i$, namely $\sigma^2$. The condition for the average m.s.e. for $\theta_i^*$ to be less than that for $\hat{\theta}_i$ is that

$$\sum(\theta_i - \theta_.)^2/(n-1) < 2\tau^2 + \sigma^2. \tag{2}$$

The m.s.e. for $\theta_i^*$ depends on $\theta_i$ and hence this condition does also. Consequently the Bayes estimates are not always superior to least-squares. But consider when (2) obtains. The $\theta_i$ are, by supposition, given $\mu, \tau^2$, a random sample from $N(\mu, \tau^2)$ so that the left-hand side of (2) is the usual estimate of $\tau^2$, had the $\theta_i$ been known. Hence the condition is that the estimate of $\tau^2$ be less than $2\tau^2 + \sigma^2$. The distribution of the estimate is a multiple of $\chi^2$ and simple calculations show that the chance—according to the $N(\mu, \tau^2)$ distribution—of (2) being satisfied is high for $n$ as low as 4 and rapidly tends to 1 as $n$ increases. But $\tau^2$, as we have seen, can itself be estimated, so with this in (1) we are almost certain to have a smaller m.s.e. for $\theta_i^*$ than for $\hat{\theta}_i$. In particular the expectation (over the $\theta$-distribution) is always in favour of the Bayes estimate.

That argument is heuristic. Our estimates are similar to those proposed by Stein (1956), which he rigorously showed to be superior (in the average m.s.e. sense) to the least-squares estimates. It has been pointed out to us by L. Brown (personal communication) that (1), with known $\sigma^2, \tau^2$, is an admissible estimate. Essentially this is because the impropriety in our prior distribution is confined to one dimension—in $\mu$. We digress to amplify this statement.

If a *proper* prior distribution (that is, one whose integral over the whole space is unity) and a *bounded* utility function are used, then the estimate obtained by using as an estimate that value which maximizes the expected (over the parameter distribution) utility is always admissible. This is easy to demonstrate since, under the two conditions stated, all the usual mathematical operations, such as reversals of order of integration, are valid. Difficulties arise if either of the italicized conditions above are violated. Quadratic loss, leading to m.s.e. is unbounded, but can conveniently be replaced by

$$1 - \exp\{-(\boldsymbol{\theta} - \mathbf{e})^{\mathrm{T}} \boldsymbol{\Lambda}(\boldsymbol{\theta} - \mathbf{e})\} \tag{3}$$

for estimate $\mathbf{e}$, where $\mathbf{\Lambda}$ is positive semi-definite and, in particular, a unit matrix. The use of vague prior knowledge, with a uniform, and therefore improper, prior distribution does cause difficulties and it is this feature, at least in dimensions higher than two, that gives rise to inadmissible estimates, as Stein was the first to show. In the general theory of the next section all our estimates will be admissible in terms of the bounded loss function (3) provided the prior distribution is proper; we conjecture admissibility if the impropriety is confined to at most two dimensions.

Returning, then, to the inequality (2), we see that there is good reason within the orthodox framework for preferring the new estimates to the old. Further justification may be found in papers by Hoerl and Kennard (1970a, b) who discuss a special case of the estimates that we shall develop in Section 5.3. We do not take these justifications very seriously, feeling that the Bayesian viewpoint is supported by so many general considerations in which criteria, like mean-square error, play little or no part, that the additional validation they provide is of small consequence.

Before proceeding to the general discussion one point must be emphasized. In the example we have assumed an exchangeable prior distribution. The estimates (1) are therefore only suggested when this assumption is practically realistic. It is the greatest strength of the Bayesian argument that it provides a formal system within which any inference or decision problem can be described. In passing from the real-world problem to its mathematical formulation it becomes necessary to make, and to expose, the assumptions. (This applies to any formalism, Euclidean geometry, for example, and not just to Bayesian statistics.) Here exchangeability is one such assumption, and its practical relevance must be assessed before the estimates based on it are used. For example, if, as suggested above, our model described the observed yields of $n$ varieties in an agricultural field trial, the exchangeability assumption would be inappropriate if one or more varieties were controls and the remainder were experimental. However, the assumption might be modified to one of exchangeability within controls and separately within experimental varieties. Similarly with a two-way classification into rows and columns, it might be reasonable to assume separately that the rows and the columns were exchangeable. In any application the particular form of the prior distribution has to be carefully considered.

It should be noted that in assigning a prior distribution to the $\theta_i$ of the above form, whilst we are effectively regarding them as a random sample from $N(\mu, \tau^2)$, we are not thereby passing to a Model II, random effects, situation such as has been discussed by Fisk (1967) and Nelder (1968). We are interested in the estimation of the *fixed* effects. One of us (A. F. M. S.) has studied the genuine Model II situation and obtained estimates for $\mu$ ($\boldsymbol{\theta}_2$ in the general model below) but this will be reported separately.

We now turn to the general theory. The mathematics is not difficult for someone familiar with matrix algebra, and the main result is stated as a theorem with corollaries. The results in Section 2 all assume *known* variances. The extensions to unknown variances will be described later.

## 2. GENERAL BAYESIAN LINEAR MODEL

The notation $\mathbf{y} \sim N(\boldsymbol{\mu}, \mathbf{D})$ means that the column vector $\mathbf{y}$ has a (multivariate) normal distribution with mean $\boldsymbol{\mu}$, a column vector, and dispersion $\mathbf{D}$, a positive semi-definite matrix.

*Lemma*. Suppose, given $\boldsymbol{\theta}_1$, a vector of $p_1$ parameters,

$$\mathbf{y} \sim N(\mathbf{A}_1 \boldsymbol{\theta}_1, \mathbf{C}_1) \tag{4}$$

and that, given $\boldsymbol{\theta}_2$, a vector of $p_2$ hyperparameters,

$$\boldsymbol{\theta}_1 \sim N(\mathbf{A}_2\boldsymbol{\theta}_2, \mathbf{C}_2). \tag{5}$$

Then (a) the marginal distribution of $\mathbf{y}$ is

$$N(\mathbf{A}_1\mathbf{A}_2\boldsymbol{\theta}_2, \mathbf{C}_1+\mathbf{A}_1\mathbf{C}_2\mathbf{A}_1^T), \tag{6}$$

and (b) the distribution of $\boldsymbol{\theta}_1$, given $\mathbf{y}$, is $N(\mathbf{Bb},\mathbf{B})$ with

$$\mathbf{B}^{-1} = \mathbf{A}_1^T\mathbf{C}_1^{-1}\mathbf{A}_1+\mathbf{C}_2^{-1} \tag{7}$$

and

$$\mathbf{b} = \mathbf{A}_1^T\mathbf{C}_1^{-1}\mathbf{y}+\mathbf{C}_2^{-1}\mathbf{A}_2\boldsymbol{\theta}_2. \tag{8}$$

(Here $\mathbf{y}$ is a vector of $n$ elements and $\mathbf{A}_1$, $\mathbf{A}_2$, $\mathbf{C}_1$ and $\mathbf{C}_2$ are known positive-definite matrices of obvious dimensions.)

The lemma is well known but we prove it here, both for completeness and because the proof has an unexpected byproduct.

To prove (a) we write (4) in the form $\mathbf{y} = \mathbf{A}_1\boldsymbol{\theta}_1+\mathbf{u}$, where $\mathbf{u}\sim N(\mathbf{0},\mathbf{C}_1)$ and (5) as $\boldsymbol{\theta}_1 = \mathbf{A}_2\boldsymbol{\theta}_2+\mathbf{v}$ where $\mathbf{v}\sim N(\mathbf{0},\mathbf{C}_2)$. Hence, putting these two equalities together, we have $\mathbf{y} = \mathbf{A}_1\mathbf{A}_2\boldsymbol{\theta}_2+\mathbf{A}_1\mathbf{v}+\mathbf{u}$. But, by the standard properties of normal distributions, $\mathbf{A}_1\mathbf{v}+\mathbf{u}$, a linear function of independent normal random variables, is $N(\mathbf{0},\mathbf{C}_1+\mathbf{A}_1\mathbf{C}_2\mathbf{A}_1^T)$ and the result follows.

To prove (b) we use Bayes's theorem,

$$p(\boldsymbol{\theta}_1|\mathbf{y}) \propto p(\mathbf{y}|\boldsymbol{\theta}_1)p(\boldsymbol{\theta}_1).$$

The product on the right-hand side is $e^{-\frac{1}{2}Q}$ where $Q$ is given by

$$\begin{aligned}(\mathbf{y}-\mathbf{A}_1\boldsymbol{\theta}_1)^T\mathbf{C}_1^{-1}(\mathbf{y}-\mathbf{A}_1\boldsymbol{\theta}_1)+(\boldsymbol{\theta}_1-\mathbf{A}_2\boldsymbol{\theta}_2)^T\mathbf{C}_2^{-1}(\boldsymbol{\theta}_1-\mathbf{A}_2\boldsymbol{\theta}_2)\\ = \boldsymbol{\theta}_1^T\mathbf{B}^{-1}\boldsymbol{\theta}_1-2\mathbf{b}^T\boldsymbol{\theta}_1+\{\mathbf{y}^T\mathbf{C}_1^{-1}\mathbf{y}+\boldsymbol{\theta}_2^T\mathbf{A}_2^T\mathbf{C}_2^{-1}\mathbf{A}_2\boldsymbol{\theta}_2\}\end{aligned}$$

on collecting the quadratic and linear terms in $\boldsymbol{\theta}_1$ together, and using the expressions (7) and (8) for $\mathbf{b}$ and $\mathbf{B}$. Completing the square in $\boldsymbol{\theta}_1$, $Q$ may finally be written

$$(\boldsymbol{\theta}_1-\mathbf{Bb})^T\mathbf{B}^{-1}(\boldsymbol{\theta}_1-\mathbf{Bb})+\{\mathbf{y}^T\mathbf{C}_1^{-1}\mathbf{y}+\boldsymbol{\theta}_2^T\mathbf{A}_2^T\mathbf{C}_2^{-1}\mathbf{A}_2\boldsymbol{\theta}_2-\mathbf{b}^T\mathbf{Bb}\}. \tag{9}$$

The term in braces is a constant as far as the distribution of $\boldsymbol{\theta}_1$ is concerned, and the remainder of the expression demonstrates the truth of (b).

The proof of the lemma is complete, but by combining the separate proofs of (a) and (b) an interesting result can be obtained. On integrating $e^{-\frac{1}{2}Q}$, with $Q$ given by (9), with respect to $\boldsymbol{\theta}_1$, the result is proportional to the density of $\mathbf{y}$, already obtained in (a). The integration does not affect the term in braces in (9) so that, in particular, the quadratic term in $\mathbf{y}$ in (9)—remembering that $\mathbf{b}$ contains $\mathbf{y}$—may be equated to the quadratic term obtained directly from (6), with the result that

$$\mathbf{C}_1^{-1}-\mathbf{C}_1^{-1}\mathbf{A}_1\mathbf{B}\mathbf{A}_1^T\mathbf{C}_1^{-1} = \{\mathbf{C}_1+\mathbf{A}_1\mathbf{C}_2\mathbf{A}_1^T\}^{-1}.$$

We therefore have the

*Matrix lemma.* For any matrices $\mathbf{A}_1$, $\mathbf{A}_2$, $\mathbf{C}_1$ and $\mathbf{C}_2$ of appropriate dimensions and for which the inverses stated in the result exist, we have

$$\mathbf{C}_1^{-1}-\mathbf{C}_1^{-1}\mathbf{A}_1(\mathbf{A}_1^T\mathbf{C}_1^{-1}\mathbf{A}_1+\mathbf{C}_2^{-1})^{-1}\mathbf{A}_1^T\mathbf{C}_1^{-1} = \{\mathbf{C}_1+\mathbf{A}_1\mathbf{C}_2\mathbf{A}_1^T\}^{-1}. \tag{10}$$

The result follows from the last equation on inserting the form for **B**, equation (7). It is, of course, easy to prove the result (10) directly once its truth has been conjectured: furthermore $\mathbf{C}_1$ and $\mathbf{C}_2$ do not have to be positive definite. It suffices to multiply the left-hand side of (10) by $\mathbf{C}_1+\mathbf{A}_1\mathbf{C}_2\mathbf{A}_1^T$ and verify that the result is a unit matrix. The above proof is interesting because it does not require an initial conjecture and because it uses a probabilistic argument to derive a purely algebraic result. The matrix lemma is important to us since it provides simpler forms than would otherwise be available for our estimates. This result has been given by Rao (1965, Exercise 2.9, p. 29).

We next proceed to the main result. As explained in Section 1, we are dealing with the linear model, which is now written in the form $E(\mathbf{y}) = \mathbf{A}_1\boldsymbol{\theta}_1$, the suffixes indicating that this is the first stage in the model. We generalize to an arbitrary dispersion matrix, $\mathbf{C}_1$, for $\mathbf{y}$. The prior distribution of $\boldsymbol{\theta}_1$ is expressed in terms of hyperparameters $\boldsymbol{\theta}_2$ as another linear model, $E(\boldsymbol{\theta}_1) = \mathbf{A}_2\boldsymbol{\theta}_2$ with dispersion matrix $\mathbf{C}_2$. This can proceed for as many stages as one finds convenient: it will be enough for us to go to three, supposing the mean, as well as the dispersion, known at the final stage. For our inferences, and in particular for estimation, we require the posterior distribution of $\boldsymbol{\theta}_1$. This is provided by the following result.

*Theorem.* Suppose that, given $\boldsymbol{\theta}_1$,

$$\mathbf{y}\sim N(\mathbf{A}_1\boldsymbol{\theta}_1, \mathbf{C}_1), \tag{11.1}$$

given $\boldsymbol{\theta}_2$,

$$\boldsymbol{\theta}_1\sim N(\mathbf{A}_2\boldsymbol{\theta}_2, \mathbf{C}_2) \tag{11.2}$$

and given $\boldsymbol{\theta}_3$,

$$\boldsymbol{\theta}_2\sim N(\mathbf{A}_3\boldsymbol{\theta}_3, \mathbf{C}_3). \tag{11.3}$$

Then the posterior distribution of $\boldsymbol{\theta}_1$, given $\{\mathbf{A}_i\}$, $\{\mathbf{C}_i\}$, $\boldsymbol{\theta}_3$ and $\mathbf{y}$ is $N(\mathbf{Dd}, \mathbf{D})$ with

$$\mathbf{D}^{-1} = \mathbf{A}_1^T\mathbf{C}_1^{-1}\mathbf{A}_1+\{\mathbf{C}_2+\mathbf{A}_2\mathbf{C}_3\mathbf{A}_2^T\}^{-1} \tag{12}$$

and

$$\mathbf{d} = \mathbf{A}_1^T\mathbf{C}_1^{-1}\mathbf{y}+\{\mathbf{C}_2+\mathbf{A}_2\mathbf{C}_3\mathbf{A}_2^T\}^{-1}\mathbf{A}_2\mathbf{A}_3\boldsymbol{\theta}_3. \tag{13}$$

(Here $\boldsymbol{\theta}_i$ is a vector of $p_i$ elements and the dispersion matrices, $\mathbf{C}_i$, are all supposed non-singular.)

The joint distribution of $\boldsymbol{\theta}_1$ and $\boldsymbol{\theta}_2$ is described in (11.2) and (11.3). The use of part (a) of the lemma enables the marginal distribution of $\boldsymbol{\theta}_1$ to be written down as

$$\boldsymbol{\theta}_1\sim N(\mathbf{A}_2\mathbf{A}_3\boldsymbol{\theta}_3, \mathbf{C}_2+\mathbf{A}_2\mathbf{C}_3\mathbf{A}_2^T). \tag{14}$$

(Notice that this is the prior distribution of $\boldsymbol{\theta}_1$ free of the hyperparameters $\boldsymbol{\theta}_2$. We could have expressed the prior in this way but in applications we find the hierarchical form more convenient.)

Then, with (14) as prior, (11.1) as likelihood, part (b) of the lemma shows that the posterior distribution of $\boldsymbol{\theta}_1$ is as stated.

In particular the mean of the posterior distribution may be regarded as a point estimate of $\boldsymbol{\theta}_1$ to replace the usual least-squares estimate. The form of this estimate is a generalization of the form noted in the example of Section 1; namely, it is a weighted average of the least-squares estimate $(\mathbf{A}_1^T\mathbf{C}_1^{-1}\mathbf{A}_1)^{-1}\mathbf{A}_1^T\mathbf{C}_1^{-1}\mathbf{y}$ and the prior mean $\mathbf{A}_2\mathbf{A}_3\boldsymbol{\theta}_3$ (equation (14)) with weights equal to the inverses of the corresponding dispersion matrices, $\mathbf{A}_1^T\mathbf{C}_1^{-1}\mathbf{A}_1$ for the least-squares values, $\mathbf{C}_2+\mathbf{A}_2\mathbf{C}_3\mathbf{A}_2^T$ for the prior distribution (14). For the simple example considered in Section 1 we produced an heuristic argument to show that, with respect to our prior distribution, we were confident of satisfying inequality (2) and thus achieving smaller mean square error than with the least-squares estimate. This result can be shown to hold generally for Bayes's estimates derived from hierarchical prior structures, as in (11.1)–(11.3), and will be presented in a future paper.

The matrix lemma enables us to obtain several alternative forms for the term in braces in (12), and hence for the posterior mean and variance, both of which involve this expression. These alternatives look more complicated than those already stated but are often useful in applications. Notice that a computational advantage of the matrix lemma is that its use reduces the order of the matrices to be inverted. The matrix on the right-hand side of (10) is of order $n$, whereas on the left-hand side, apart from $\mathbf{C}_1$ which is usually of a simple structure (often $\mathbf{C}_1 = \sigma^2 \mathbf{I}$), the matrix to be inverted is of order $p_1$, typically much less than $n$.

*Corollary* 1. An alternative expression for $\mathbf{D}^{-1}$ (equation (12)) is

$$\mathbf{A}_1^T \mathbf{C}_1^{-1} \mathbf{A}_1 + \mathbf{C}_2^{-1} - \mathbf{C}_2^{-1} \mathbf{A}_2 (\mathbf{A}_2^T \mathbf{C}_2^{-1} \mathbf{A}_2 + \mathbf{C}_3^{-1})^{-1} \mathbf{A}_2^T \mathbf{C}_2^{-1}. \tag{15}$$

This is immediate on applying (10), with the suffixes all increased by one, to the second term in (12).

In most applications of these results the design of the experiment rather naturally suggests the second stage, (11.2), in the hierarchy but at the third stage we find ourselves in a position where the prior knowledge is weak. (Least-squares results apply when the second-stage prior knowledge is weak.) It is natural to express this by supposing the third-stage dispersion matrix $\mathbf{C}_3$ to be large, or to let its inverse, the precision matrix, be zero. In the original form of (12) and (13) it is not easy to see what happens when $\mathbf{C}_3^{-1} = \mathbf{0}$, but (15) enables the form to be seen easily.

*Corollary* 2. If $\mathbf{C}_3^{-1} = \mathbf{0}$, the posterior distribution of $\boldsymbol{\theta}_1$ is $N(\mathbf{D}_0 \mathbf{d}_0, \mathbf{D}_0)$ with

$$\mathbf{D}_0^{-1} = \mathbf{A}_1^T \mathbf{C}_1^{-1} \mathbf{A}_1 + \mathbf{C}_2^{-1} - \mathbf{C}_2^{-1} \mathbf{A}_2 (\mathbf{A}_2^T \mathbf{C}_2^{-1} \mathbf{A}_2)^{-1} \mathbf{A}_2^T \mathbf{C}_2^{-1} \tag{16}$$

and

$$\mathbf{d}_0 = \mathbf{A}_1^T \mathbf{C}_1^{-1} \mathbf{y}. \tag{17}$$

The form for $\mathbf{D}_0^{-1}$ follows by direct substitution of $\mathbf{C}_3^{-1} = \mathbf{0}$ in (15). That for $\mathbf{d}_0$ follows by remarking that if the second and third terms in (15) are postmultiplied by $\mathbf{A}_2$ the result is zero, but such postmultiplication takes place in the original expression for $\mathbf{d}$, equation (13).

This corollary is the form we shall most often use in applications.

It is possible to extend the theorem to cases where some or all of the dispersion matrices $\mathbf{C}_i$ are singular. This can be accomplished using generalized inverses and will be the subject of a separate paper. Notice that we have not assumed, as in the usual least-squares theory, that $\mathbf{A}_1^T \mathbf{C}_1^{-1} \mathbf{A}_1$ is non-singular. (The case $\mathbf{C}_1 = \sigma^2 \mathbf{I}$ will be more familiar.) In the standard exposition it is usual to constrain the individual parameters in the vector $\boldsymbol{\theta}_1$ to preserve identifiability in the likelihood function. Identifiability problems do not arise in the Bayesian formulation since, provided the prior distribution is proper, so is the posterior, whether or not the parameters referred to in these two distributions are identifiable or not in the likelihood function. An example below will help to make this clear.

The situation described in Section 1 has already been discussed in detail by Lindley (1971a), though not within the general framework which was briefly described in Lindley (1969). The interested reader can easily fit the example into the argument of this section. Corollary 2 is relevant and it is an easy matter to perform the necessary matrix calculations. We proceed to the discussion of other examples.

## 3. EXAMPLES

### 3.1. *Two-factor Experimental Designs*

Consider $t$ "treatments" assigned to $n$ experimental units arranged in $b$ "blocks". If the $i$th treatment is applied within the $j$th block and yields an observation $y_{ij}$, the usual model is

$$E(y_{ij}) = \mu+\alpha_i+\beta_j \quad (1\leqslant i\leqslant t, 1\leqslant j\leqslant b)$$

with the errors independent $N(0,\sigma^2)$. In the general notation of (11.1)

$$\boldsymbol{\theta}_1^{\mathrm{T}} = (\mu, \alpha_1, \alpha_2, \ldots, \alpha_t, \beta_1, \beta_2, \ldots, \beta_b)$$

and $\mathbf{A}_1$ describes the design used.

For the second stage we argue as follows. It might be reasonable to assume that our prior knowledge of the treatment constants $\{\alpha_i\}$ was exchangeable, and similarly that of the block constants $\{\beta_j\}$, but that these were independent. We emphasize the word "might" in the last sentence. In repetition of the point made in Section 1, we remind the reader that this *assumption* is not always appropriate and our recipes below are not necessarily sensible when this form of exchangeability is unreasonable. For example, it may be known that the treatments are ordered, say $\alpha_1\leqslant\alpha_2\leqslant\ldots\leqslant\alpha_t$. In this case other forms of prior information are available and alternative estimates are sensible: these will be reported on in a separate paper.

Adding the assumptions of normality we therefore describe the second stage (11.2) by

$$\alpha_i\sim N(0,\sigma_\alpha^2), \quad \beta_j\sim N(0,\sigma_\beta^2), \quad \mu\sim N(\omega,\sigma_\mu^2),$$

these distributions being independent. The means of $\alpha_i$ and $\beta_j$ have been chosen to be zero. Any other value would do since the likelihood provides no information about them, but the choice of zero mean is convenient, since it leads to straightforward comparisons of the Bayes and (constrained) least-squares estimates as deviations from an average level. We shall consider the case where the prior knowledge of $\mu$ is vague, so that $\sigma_\mu^2\to\infty$; $\omega$ will then be irrelevant. A third stage is not necessary. We proceed to calculate expressions (12) and (13) for the posterior distribution of $\boldsymbol{\theta}_1$.

The matrix $\mathbf{C}_2$ is diagonal, so the same is true of $\mathbf{C}_2^{-1}$ and its leading diagonal is easily seen to be

$$(\sigma_\mu^{-2}, \sigma_\alpha^{-2}, \ldots, \sigma_\alpha^{-2}, \sigma_\beta^{-2}, \ldots, \sigma_\beta^{-2})$$

and as $\sigma_\mu^2\to\infty$, the first element tends to zero. $\mathbf{C}_1$ is the unit matrix times $\sigma^2$. We can therefore substitute these values into (12) and (13), remembering that $\mathbf{C}_3 = \mathbf{0}$ and $(\mathbf{A}_3\boldsymbol{\mu})^{\mathrm{T}} = (\omega, 0, \ldots, 0)$ and easily obtain

$$\mathbf{D}^{-1} = \sigma^{-2}\mathbf{A}_1^{\mathrm{T}}\mathbf{A}_1+\mathbf{C}_2^{-1}$$

and

$$\mathbf{d} = \sigma^{-2}\mathbf{A}_1^{\mathrm{T}}\mathbf{y}.$$

Hence $\boldsymbol{\theta}_1^*$, the Bayes estimate $\mathbf{Dd}$, satisfies the equations

$$(\mathbf{A}_1^{\mathrm{T}}\mathbf{A}_1+\sigma^2\mathbf{C}_2^{-1})\,\boldsymbol{\theta}_1^* = \mathbf{A}_1^{\mathrm{T}}\mathbf{y}. \tag{18}$$

These differ from the least-squares equations only in the inclusion of the extra term $\sigma^2 \mathbf{C}_2^{-1}$.

In the case of a complete randomized-block design where each treatment occurs exactly once in each block we have, on arranging the elements of $\mathbf{y}$ in lexicographical order,

$$(\mathbf{A}_1^T \mathbf{A}_1 + \sigma^2 \mathbf{C}_2^{-1}) = \begin{pmatrix} bt & b\mathbf{1}_t^T & t\mathbf{1}_b^T \\ b\mathbf{1}_t & (b+\sigma^2/\sigma_\alpha^2)\mathbf{I}_t & \mathbf{J}_{t,b} \\ t\mathbf{1}_b & \mathbf{J}_{b,t} & (t+\sigma^2/\sigma_\beta^2)\mathbf{I}_b \end{pmatrix}, \tag{19}$$

where $\mathbf{1}_m$ is a vector of $m$ 1's, $\mathbf{I}_m$ is the unit matrix of order $m$ and $\mathbf{J}_{m,n}$ is a matrix of order $m \times n$ all of whose elements are 1. As usual

$$(\mathbf{A}_1^T \mathbf{y})^T = (bty_{..}, by_{1.}, \ldots, by_{t.}, ty_{.1}, \ldots, ty_{.b}).$$

Notice that the matrix (19) is non-singular and the solution to (18) is easily seen to be

$$\mu^* = y_{..}, \quad \alpha_i^* = (b\sigma_\alpha^2 + \sigma^2)^{-1} b\sigma_\alpha^2 (y_{i.} - y_{..}), \quad \beta_j^* = (t\sigma_\beta^2 + \sigma^2)^{-1} t\sigma_\beta^2 (y_{.j} - y_{..}). \tag{20}$$

Consequently the estimators of the treatment and block effects (on being measured from the overall mean) are shrunk towards zero by a factor depending on the ratio of $\sigma^2$ to $\sigma_\alpha^2$ or $\sigma_\beta^2$ respectively. This is in agreement with the result, equation (1), quoted above. Because this is an orthogonal design the magnitude of the "shrinkage" of the treatment effect does not depend on the exchangeability for the blocks, and vice versa. With a non-orthogonal design, such as balanced incomplete blocks, the same remark is not true.

### 3.2. *Exchangeability Between Multiple Regression Equations*

The following practical example stimulated our extension from the example of Section 1 to the general model, and we shall report on its use in Section 5.2. The context was educational measurement where variables $x$ and $y$ were related with the usual linear regression structure. However the values of the regression parameters depended on the school the student had attended. Novick (personal communication) suggested to us that improved estimates might be obtained for any one school by combining the data for all schools. This is just what the Bayes estimates do, and would seem to be appropriate whenever exchangeability *between* regressions (schools) is a sensible assumption. The mathematics for $p$ regressor variables goes as follows.

Suppose

$$\mathbf{y}_j \sim N(\mathbf{X}_j \boldsymbol{\beta}_j, \mathbf{I}_{n_j} \sigma_j^2) \tag{21}$$

for $j = 1, 2, \ldots, m$ and $\boldsymbol{\beta}_j$ a vector of $p$ parameters: that is $m$ linear, multiple regressions on $p$ variables. In the notation of the Theorem, $\mathbf{A}_1$, expressed in terms of submatrices, is diagonal with $\mathbf{X}_j$ as the $j$th diagonal submatrix; $\boldsymbol{\theta}_1^T$ is $(\boldsymbol{\beta}_1^T, \boldsymbol{\beta}_2^T, \ldots, \boldsymbol{\beta}_m^T)$ of $mp$ elements. The exchangeability of the individual $\boldsymbol{\beta}_j$ added to normality gives us the second stage as

$$\boldsymbol{\beta}_j \sim N(\boldsymbol{\xi}, \boldsymbol{\Sigma}) \tag{22}$$

say. Here $\mathbf{A}_2$ is a matrix of order $mp \times p$, all of whose $p \times p$ submatrices are unit matrices, and $\boldsymbol{\theta}_2 = \boldsymbol{\xi}$. We shall suppose vague prior knowledge of $\boldsymbol{\xi}$ and use the special form of Corollary 2.

Simple calculations show that $(\mathbf{A}_2^T \mathbf{C}_2^{-1} \mathbf{A}_2)^{-1} = m^{-1}\boldsymbol{\Sigma}$ and then that

$$\mathbf{C}_2^{-1}\mathbf{A}_2(\mathbf{A}_2^T \mathbf{C}_2^{-1}\mathbf{A}_2)^{-1}\mathbf{A}_2^T \mathbf{C}_2^{-1}$$

is a matrix of order *mp all* of whose $p \times p$ submatrices are $m^{-1}\boldsymbol{\Sigma}^{-1}$. In the usual way $\mathbf{A}_1^T \mathbf{C}_1^{-1} \mathbf{A}_1$, expressed in terms of submatrices, is diagonal with $\sigma_j^{-2}\mathbf{X}_j^T \mathbf{X}_j$ as the $j$th diagonal submatrix. The equations for the Bayes estimates $\boldsymbol{\beta}_j^*$ are then found to be

$$\begin{pmatrix} \sigma_1^{-2}\mathbf{X}_1^T\mathbf{X}_1+\boldsymbol{\Sigma}^{-1} & \dots & \mathbf{0} \\ \dots & \sigma_2^{-2}\mathbf{X}_2^T\mathbf{X}_2+\boldsymbol{\Sigma}^{-1} & \dots \\ \vdots & \vdots & \vdots \\ \mathbf{0} & \dots & \sigma_m^{-2}\mathbf{X}_m^T\mathbf{X}_m+\boldsymbol{\Sigma}^{-1} \end{pmatrix}$$

$$\times \begin{pmatrix} \boldsymbol{\beta}_1^* \\ \boldsymbol{\beta}_2^* \\ \vdots \\ \boldsymbol{\beta}_m^* \end{pmatrix} - \boldsymbol{\Sigma}^{-1} \begin{pmatrix} \boldsymbol{\beta}_{.}^* \\ \boldsymbol{\beta}_{.}^* \\ \vdots \\ \boldsymbol{\beta}_{.}^* \end{pmatrix} = \begin{pmatrix} \sigma_1^{-2}\mathbf{X}_1^T\mathbf{y} \\ \sigma_2^{-2}\mathbf{X}_2^T\mathbf{y} \\ \vdots \\ \sigma_m^{-2}\mathbf{X}_m^T\mathbf{y} \end{pmatrix}, \qquad (23)$$

where $\boldsymbol{\beta}_{.}^* = \sum \boldsymbol{\beta}_i^*/m$. These equations are easily solved for $\boldsymbol{\beta}_{.}^*$ and then, in terms of $\boldsymbol{\beta}_{.}^*$, the solution is

$$\boldsymbol{\beta}_j^* = (\sigma_j^{-2}\mathbf{X}_j^T\mathbf{X}_j+\boldsymbol{\Sigma}^{-1})^{-1}(\sigma_j^{-2}\mathbf{X}_j^T\mathbf{y}+\boldsymbol{\Sigma}^{-1}\boldsymbol{\beta}_{.}^*), \qquad (24)$$

a compromise between the least-squares estimate and an average of the various estimates. The example of Section 1 is a special case with $p = 1$.

Noting that $\mathbf{D}_0^{-1}$, given in Corollary 2 (16), may, for this application, be written in the form,

$$\begin{pmatrix} \sigma_1^{-2}\mathbf{X}_1^T\mathbf{X}_1+\boldsymbol{\Sigma}^{-1} & \mathbf{0} \\ \vdots & \vdots \\ \mathbf{0} & \sigma_m^{-2}\mathbf{X}_m^T\mathbf{X}_m+\boldsymbol{\Sigma}^{-1} \end{pmatrix} - m^{-1}\begin{pmatrix} \boldsymbol{\Sigma}^{-1} \\ \vdots \\ \boldsymbol{\Sigma}^{-1} \end{pmatrix}$$

$$\times \begin{pmatrix} \boldsymbol{\Sigma} & \dots & \mathbf{0} \\ \vdots & \vdots & \vdots \\ \mathbf{0} & \dots & \boldsymbol{\Sigma} \end{pmatrix} (\boldsymbol{\Sigma}^{-1} \dots \boldsymbol{\Sigma}^{-1})$$

and thus may be inverted by the matrix Lemma (10), we can obtain an explicit form for $\boldsymbol{\beta}_j^*$. After some algebra we obtain the weighted form of (24) with $\boldsymbol{\beta}_{.}^*$ replaced by $\sum \mathbf{W}_i \hat{\boldsymbol{\beta}}_i$ where,

$$\mathbf{W}_i = \left[\sum_{j=1}^{m} (\mathbf{X}_j^T\mathbf{X}_j\sigma_j^{-2}+\boldsymbol{\Sigma}^{-1})^{-1}\mathbf{X}_j^T\mathbf{X}_j\sigma_j^{-2}\right]^{-1}(\mathbf{X}_i^T\mathbf{X}_i\sigma_i^{-2}+\boldsymbol{\Sigma}^{-1})^{-1}\mathbf{X}_i^T\mathbf{X}_i\sigma_i^{-2}.$$

This shows explicitly how the information from the $i$th regression equation is combined with the information from all equations.

### 3.3. *Exchangeability Within Multiple Regression Equations*

In contrast to the last section suppose that we have a single multiple regression situation

$$\mathbf{y} \sim N(\mathbf{X}\boldsymbol{\beta}, \mathbf{I}_n\sigma^2). \qquad (25)$$

In the educational context, the $p$ regressor variables might be the results of $p$ tests applied to students and the dependent variable, $y$, a measure of the students' performance after training. We are interested in the case where the individual regression coefficients in $\boldsymbol{\beta}^{\mathrm{T}} = (\beta_1, \beta_2, ..., \beta_p)$ are exchangeable. To achieve this it may be necessary to rescale the regressor variables: for example, to write (25) in correlation form in which the diagonal elements of $\mathbf{X}^{\mathrm{T}}\mathbf{X}$ are unity and the off-diagonals are the sample correlations. (Again we emphasize the point that this is an assumption and may not be appropriate). If the assumption is sensible then we may fit it into our general model by supposing

$$\beta_j \sim N(\xi, \sigma_\beta^2). \tag{26}$$

There are at least two useful possibilities: (i) to suppose vague prior knowledge for $\xi$ (Corollary 2), (ii) to put $\xi = 0$, reflecting a feeling that the $\beta_i$ are small.

In (i) simple but tedious calculations analogous to those of Section 3.2 show that

$$\boldsymbol{\beta}^* = \{\mathbf{I}_p + k(\mathbf{X}^{\mathrm{T}}\mathbf{X})^{-1}(\mathbf{I}_p - p^{-1}\mathbf{J}_p)\}^{-1}\hat{\boldsymbol{\beta}}, \tag{27}$$

where $k = \sigma^2/\sigma_\beta^2$. Similar calculations in (ii), using only a two-stage model, give

$$\boldsymbol{\beta}^* = \{\mathbf{I}_p + k(\mathbf{X}^{\mathrm{T}}\mathbf{X})^{-1}\}^{-1}\hat{\boldsymbol{\beta}}. \tag{28}$$

The estimates (27) and (28) are very similar to those proposed by Hoerl and Kennard (1970a). The main difference is that $k$ in their argument is a constant introduced for various intuitively sensible reasons, whereas here it is a variance ratio. Also the derivation is different: Hoerl and Kennard argue within the orthodox, sampling theory framework, whereas we use the formal theory. We do not attempt to reproduce their most convincing argument against the least-squares estimates and in favour of (27) and (28), merely referring the sampling-theorist to it and saying that we agree with its conclusions with the reservation that we feel that the estimates may not be so sensible if the exchangeability within the regression equation is inappropriate. We return to this example in Section 5.3 where the estimation of $k$ is discussed.

Examples 3.2 and 3.3 may be combined when there is exchangeability between *and* within regressions. We omit the details of this and many other extensions and instead consider how we might remove the major impediment to the application of the general theory, namely the assumption that all the variances are known. In the next section we show that the simple device of replacing the known variances by estimated values in the Bayes estimates is satisfactory.

## 4. ESTIMATION WITH UNKNOWN COVARIANCE STRUCTURE

For the purpose of the immediate exposition denote by $\theta$ the parameters of interest in the general model and by $\phi$ the nuisance parameters. The latter will include the dispersion matrices $\mathbf{C}_i$ when these are unknown. Consider how the Bayesian treatment proceeds. We first assign a joint prior distribution to $\theta$ and $\phi$—instead of just to $\theta$—and combine this with the likelihood function to provide the joint posterior distribution $p(\theta, \phi \mid \mathbf{y})$. This distribution then has to be integrated with respect to $\phi$, thus removing the nuisance parameters and leaving the posterior for $\theta$. Finally, if we are using quadratic loss or generally one of the forms given by (3), we shall require the mean of this distribution, necessitating another integration. The calculation of the mean will also require the constant of proportionality in Bayes's formula to be evaluated, involving yet another integration. Any reasonable prior distributions for $\phi$ that we have considered lead to integrals which cannot all be expressed in closed form and, as a result, the above argument is technically most complex to execute. We therefore consider an approximation to it which is technically much simpler and yet yields the bulk, though not unfortunately all, of the information required for the estimation.

The first approximation consists in using the *mode* of the posterior distribution in place of the *mean*. Secondly, we mostly use the mode of the *joint* distribution rather than that of the *θ-margin*. The modal values satisfy the equations

$$\frac{\partial}{\partial\theta}p(\theta,\phi\,|\,\mathbf{y}) = \frac{\partial}{\partial\phi}p(\theta,\phi\,|\,\mathbf{y}) = 0.$$

These equations may be re-written in terms of conditional and marginal distributions. In particular that for $\theta$ may be expressed as

$$\frac{\partial}{\partial\theta}p(\theta\,|\,\phi,\mathbf{y})\,p(\phi\,|\,\mathbf{y}) = 0$$

or, assuming $p(\phi\,|\,\mathbf{y}) \neq 0$, as

$$\frac{\partial}{\partial\theta}p(\theta\,|\,\phi,\mathbf{y}) = 0. \tag{29}$$

But the conditional density $p(\theta\,|\,\phi,\mathbf{y})$ in (29) is exactly what has been found in the general theory of Section 2, where it was shown to be normal, with mode consequently equal to the mean. Hence we have the result that the $\theta$-value of the posterior mode of the joint distribution of $\theta$ and $\phi$ is equal to the mode of the conditional distribution of $\theta$ evaluated at the modal value of $\phi$. Consequently all we have to do is to take the estimates derived in Section 2 and replace the unknown values of the nuisance parameters by their modal estimates. For example, the simple estimate (1) is replaced by

$$\frac{y_{i}/s^2 + y_{.}/t^2}{1/s^2 + 1/t^2},$$

where $s^2$ and $t^2$ are respectively modal estimates of $\sigma^2$ and $\tau^2$. This approach avoids the integrations referred to above. The modal estimates of $\phi$ may, analogous to (29), be found by supposing $\theta$ known, and then replacing $\theta$ in the result by their modes.

It is reasonably clear that the approximations are only likely to be good if the samples are fairly large and the resulting posterior distributions approximately normal. Also the approach does not provide information about the precision of the estimates, such as a standard error (of the posterior, not the sampling-theoretic distribution!) would provide. But as a first step on the way to a satisfactory description of the posterior distribution, it seems to go a long way and has the added merit of being intuitively sensible. In practice we shall find it convenient to proceed as follows. For an assumed $\phi$ calculate the mode $\theta^{(1)}$, say. Treating $\theta^{(1)}$ as known we can find the mode for $\phi$, $\phi^{(1)}$ say. This may be used to find $\theta^{(2)}$, and so on. This sequence of iterations typically converges and only involves equations for the modes of one parameter, knowing the value of the other.

We now proceed to apply these ideas to the situations discussed in Section 3. At the moment we have no general theory to parallel that of Section 2. The reason for this is essentially that we do not have an entirely satisfactory procedure for estimating the dispersion matrix of a multivariate normal distribution. This might appear an odd statement to make when there are numerous texts on multivariate analysis available that discuss this problem. But just as the usual estimates of the means are inadmissible, so are those of the variances and covariances (Brown, 1968), and are, in any case, obtained from unrealistic priors. We hope to report separately on this problem and defer discussion of a general theory.

## 5. EXAMPLES WITH UNKNOWN COVARIANCE STRUCTURE

### 5.1. *Two-factor Experimental Designs*

We saw in Section 3.1 that there were three variances in this situation: $\sigma^2$ the usual residual variance contributing to the likelihood function, and $\sigma^2_\alpha, \sigma^2_\beta$ being respectively the variances of the treatment and block effects. ($\sigma^2_\mu \to \infty$ so does not enter.) It is first necessary to specify prior distributions for these and this we do through the appropriate conjugate family, which is here inverse-$\chi^2$, assuming the three variances independent. This conjugate family involves two parameters and is sufficiently flexible for most applications. Specifically we suppose

$$\frac{\nu\lambda}{\sigma^2} \sim \chi^2_\nu, \quad \frac{\nu_\alpha \lambda_\alpha}{\sigma^2_\alpha} \sim \chi^2_{\nu_\alpha} \quad \text{and} \quad \frac{\nu_\beta \lambda_\beta}{\sigma^2_\beta} \sim \chi^2_{\nu_\beta}. \tag{30}$$

The joint distribution of all quantities involved can then be written down as proportional to

$$(\sigma^2)^{-\frac{1}{2}(n+\nu+2)} \exp\left[-\frac{1}{2\sigma^2}\{\nu\lambda + S^2(\mu,\alpha,\beta)\}\right]$$
$$\times (\sigma^2_\alpha)^{-\frac{1}{2}(t+\nu_\alpha+2)} \exp\left[-\frac{1}{2\sigma^2_\alpha}\{\nu_\alpha\lambda_\alpha + \sum \alpha_i^2\}\right]$$
$$\times (\sigma^2_\beta)^{-\frac{1}{2}(b+\nu_\beta+2)} \exp\left[-\frac{1}{2\sigma^2_\beta}\{\nu_\beta\lambda_\beta + \sum \beta_j^2\}\right], \tag{31}$$

where $S^2(\mu,\alpha,\beta)$ is the sum of squares $\sum(y_{ij}-\mu-\alpha_i-\beta_j)^2$.

If $\sigma^2$, $\sigma^2_\alpha$ and $\sigma^2_\beta$ are known, the mode of this distribution has been found—equation (18), or in the balanced case, equation (20). We have only to substitute the modal estimates of the three variances into these expressions. To find these modal estimates we can, reversing the roles of $\theta$ and $\phi$ in the general argument of the previous paragraph, suppose $\mu$, $\alpha$ and $\beta$ known. Using the corresponding Roman letters for these modes, we easily see them to be, from (31),

$$\left.\begin{aligned} s^2 &= \{\nu\lambda + S^2(\mu^*,\alpha^*,\beta^*)\}/(n+\nu+2), \\ s^2_\alpha &= \{\nu_\alpha\lambda_\alpha + \sum \alpha_i^{*2}\}/(t+\nu_\alpha+2), \\ s^2_\beta &= \{\nu_\beta\lambda_\beta + \sum \beta_j^{*2}\}/(b+\nu_\beta+2). \end{aligned}\right\} \tag{32}$$

These equations, together with (18) (or (20)), can now be solved iteratively. With trial values of $\sigma^2$, $\sigma^2_\alpha$ and $\sigma^2_\beta$, (18) can be solved for $\mu^*$, $\alpha^*$ and $\beta^*$. These values can be inserted into (32) to give revised values for $s^2$, $s^2_\alpha$ and $s^2_\beta$, which can again be used in (18). The cycle can be repeated until the values converge.

A few points about these solutions are worth noting. Firstly, the value of $S^2$ that occurs is not the usual residual sum of squares, which is evaluated about the least-squares value, but the sum about the Bayes estimates. Since the former minimizes the sum of squares, our $S^2$ is necessarily greater than the residual: $s^2$ could therefore be larger than the usual estimate. Secondly, whilst it would be perfectly possible to put $\nu = 0$ (referring to $\sigma^2$), so avoiding the specification of a value for $\lambda$ and thereby taking the usual vague prior for a variance, one cannot put $\nu_\alpha$ and $\nu_\beta$ zero. If this is done the modal estimates for the treatment and block effects are all zero. The point is discussed in detail in connection with the example of Section 1 in Lindley (1971a). Essentially the estimation of $\sigma_\alpha^2$ and $\sigma_\beta^2$ is difficult, in the sense that the data contain little information about them, when they are small in comparison with $\sigma^2$: the residual "noise" is too loud. In the contrary case where $\sigma_\alpha^2$ and $\sigma_\beta^2$ are large in comparison with $\sigma^2$, the actual values of $\nu_\alpha$, $\lambda_\alpha$, $\nu_\beta$ and $\lambda_\beta$ do not matter much provided the $\nu$'s are both small.

### 5.2. *Exchangeability Between Multiple Regression Equations*

We continue the discussion of Section 3.2 but mainly confine our attention to the homoscedastic case where $\sigma_j^2 = \sigma^2$, say, for all $j$. It is only necessary to specify prior distributions for $\sigma^2$ and $\mathbf{\Sigma}$, the dispersion matrix of the regression coefficients (equation (22)). As in the last example we suppose $\nu\lambda/\sigma^2 \sim \chi_\nu^2$. The conjugate distribution for $\mathbf{\Sigma}$ is to suppose $\mathbf{\Sigma}^{-1}$ has a Wishart distribution with $\rho$, say, degrees of freedom and matrix $\mathbf{R}$. We are not too happy with this assumption but at least it provides a large-sample solution (see the remarks at the end of Section 4). $\mathbf{\Sigma}$ and $\sigma^2$ are supposed independent.

The joint distribution of all the quantities is now

$$(\sigma^2)^{-\frac{1}{2}n} \exp\left\{-\frac{1}{2\sigma^2}\sum_{j=1}^{m}(\mathbf{y}_j-\mathbf{X}_j\boldsymbol{\beta}_j)^{\mathrm{T}}(\mathbf{y}_j-\mathbf{X}_j\boldsymbol{\beta}_j)\right\}$$
$$\times|\mathbf{\Sigma}|^{-\frac{1}{2}m}\exp\left\{-\frac{1}{2}\sum_{j=1}^{m}(\boldsymbol{\beta}_j-\boldsymbol{\xi})^{\mathrm{T}}\mathbf{\Sigma}^{-1}(\boldsymbol{\beta}_j-\boldsymbol{\xi})\right\}$$
$$\times|\mathbf{\Sigma}|^{-\frac{1}{2}(\rho-p-1)}\exp\{-\tfrac{1}{2}\operatorname{tr}\mathbf{\Sigma}^{-1}\mathbf{R}\}$$
$$\times(\sigma^2)^{-\frac{1}{2}(\nu+2)}\exp\{-\nu\lambda/2\sigma^2\}, \tag{33}$$

assuming $\boldsymbol{\xi}$ to have a uniform distribution over $p$-space. (The four lines of (33) come respectively from the likelihood, the distribution of $\boldsymbol{\beta}$, (22), the Wishart distribution for $\mathbf{\Sigma}^{-1}$ and the inverse-$\chi^2$ for $\sigma^2$.) The integration with respect to $\boldsymbol{\xi}$ is straightforward and effectively results in the usual loss of one degree of freedom. The joint posterior density for $\boldsymbol{\beta}$, $\sigma^2$ and $\mathbf{\Sigma}^{-1}$ is then proportional to

$$(\sigma^2)^{-\frac{1}{2}(n+\nu+2)}\exp\left[-\frac{1}{2\sigma^2}\sum_{j=1}^{m}\{m^{-1}\nu\lambda+(\mathbf{y}_j-\mathbf{X}_j\boldsymbol{\beta}_j)^{\mathrm{T}}(\mathbf{y}_j-\mathbf{X}_j\boldsymbol{\beta}_j)\}\right]$$
$$\times|\mathbf{\Sigma}|^{-\frac{1}{2}(m+\rho-p-2)}\exp\left[-\tfrac{1}{2}\operatorname{tr}\mathbf{\Sigma}^{-1}\left\{\mathbf{R}+\sum_{j=1}^{m}(\boldsymbol{\beta}_j-\boldsymbol{\beta}_.)(\boldsymbol{\beta}_j-\boldsymbol{\beta}_.)^{\mathrm{T}}\right\}\right], \tag{34}$$

where

$$\boldsymbol{\beta}_. = m^{-1}\sum_{j=1}^{m}\boldsymbol{\beta}_j.$$

The modal estimates are then easily obtained. Those for $\boldsymbol{\beta}_j$ are as before, equation (24), with $\boldsymbol{\Sigma}$ and $\sigma^2$ replaced by modal values. The latter are seen to satisfy

$$s^2 = \sum_{j=1}^{m} \{m^{-1}\nu\lambda + (\mathbf{y}_j - \mathbf{X}_j\boldsymbol{\beta}_1^*)^{\mathrm{T}}(\mathbf{y}_j - \mathbf{X}_j\boldsymbol{\beta}_j^*)\}/(n+\nu+2),$$

and

$$\boldsymbol{\Sigma}^* = \left\{\mathbf{R} + \sum_{j=1}^{m}(\boldsymbol{\beta}_j^* - \boldsymbol{\beta}_.^*)(\boldsymbol{\beta}_j^* - \boldsymbol{\beta}_.^*)^{\mathrm{T}}\right\} \Big/ (m+\rho-p-2). \tag{35}$$

It is possible in this case, as in Section 5.1 and 5.3, to proceed a little differently and obtain the posterior distribution of the $\boldsymbol{\beta}_j$'s, free of $\sigma^2$ and $\boldsymbol{\Sigma}$ and consider the modes of this. This is because the integration of (34) with respect to $\boldsymbol{\Sigma}^{-1}$ and $\sigma^2$ is possible in closed form. The result is

$$\left[\sum_{j=1}^{n}\{m^{-1}\nu\lambda + (\mathbf{y}_j - \mathbf{X}_j\boldsymbol{\beta}_j)^{\mathrm{T}}(\mathbf{y}_j - \mathbf{X}_j\boldsymbol{\beta}_j)\}\right]^{-\frac{1}{2}(n+\nu)} \times \left|\mathbf{R} + \sum_{j=1}^{m}(\boldsymbol{\beta}_j - \boldsymbol{\beta}_.)(\boldsymbol{\beta}_j - \boldsymbol{\beta}_.)^{\mathrm{T}}\right|^{-\frac{1}{2}(m+\rho-1)}. \tag{36}$$

The mode of this distribution can be used in place of the modal values for the wider distribution. The differentiation is facilitated by using the result that, with $\mathbf{V}$ equal to the matrix whose determinant appears in (36),

$$\frac{\partial}{\partial\boldsymbol{\beta}_i}\log|\mathbf{V}| = 2\mathbf{V}^{-1}(\boldsymbol{\beta}_i - \boldsymbol{\beta}_.).$$

It is then possible to verify that the modes for $\boldsymbol{\beta}_j$ satisfy the same equations as before, (24), with $\boldsymbol{\Sigma}$ and $\sigma^2$ replaced by values given by (35) except that the divisors on the right-hand sides are $(n+\nu)$ and $(m+\rho-1)$ rather than $(n+\nu+2)$ and $(m+\rho-p-2)$.

It is possible to extend this model significantly by reverting to the heteroscedastic case as originally considered, (21). Here we have to specify a joint distribution for the $\sigma_j^2$. A possible device is to suppose that, like the means, the $\sigma_j^2$ are exchangeable. A convenient distribution to generate the exchangeability is to suppose $\nu\lambda/\sigma_j^2 \sim \chi_\nu^2$. In the context of several means (Section 1) Lindley (1971a) has shown how the estimates of the variances get pulled towards a central value. The details are so similar here that we do not repeat them.

As explained in Section 3.2, it was Novick's suggestion to consider this problem in an educational context, and we conclude this section by briefly reporting on an application that he, in conjunction with Jackson, Thayer and Cole (1972), have made of these results. We are most grateful to them for permission to include the details here. Their analysis used data from the American College Testing Program on the prediction of grade-point average at 22 colleges from the results of 4 tests; namely, English, Mathematics, Social Studies and Natural Sciences. We therefore have the situation studied in this section with $p = 5$ (one variable corresponding to the mean), $m = 22$, and $n_j$ varying from 105 to 739. They used the heteroscedastic model but the basic equations (24) and (35) are essentially as here described. With the substantial amounts of data available the prior constants, $\nu$, $\lambda$, $\rho$ and $\mathbf{R}$ scarcely affect the analysis: the first three were taken to be small and changes of origin of the regressor variables effected to make the prior judgment that $\mathbf{R}$ was diagonal. With $\rho$ small the diagonal elements again play little role. We omit details of how the calculations were performed and refer the interested reader to their paper.

Data were available for 1968 and 1969. The approach was to use the 1968 data to estimate the regressions, to use these estimated equations on the 1969 $x$-data to estimate the $y$'s, the grade-point averages, and then to compare these predictions with the actual 1969 $y$-values, using as a criterion of prediction the mean of the squares of the differences. This operation was done twice; once with the full 1968 data, and once with a random 25 per cent sample from each College. The results are summarized in Table 1.

TABLE 1

*Comparison of predictive efficiency*

| | *Average mean-square error* | |
|---|---|---|
| | *Least-squares* | *Bayes* |
| All data | 0·5596 | 0·5502 |
| 25% sample | 0·6208 | 0·5603 |

The first row refers to the analysis of the whole data and shows that the Bayesian method only reduces the error by under 2 per cent. With such large samples there is little room for improvement. With the quarter sample, however, in the second row of the table, the reduction is up to 9 per cent and most strikingly the error is almost down to the value reached with the least-squares estimates for all the data. In other words, 25 per cent of the data and Bayes are as good as all the data and least squares: or the Bayesian method provides a possible 75 per cent saving in sample size. They also provide details of the comparisons between the two estimates of the regression coefficients. These tend to be "shrunk" towards a common value (for each regressor variable) and in some cases with the quarter sample the shrinkage is substantial.

It would be dangerous to draw strong conclusions from one numerical study but the analysis should do something to answer the criticism of those who have said that Bayesian methods are not "testable". We favour the method because of its coherence, but the pragmatists may like to extend the method of Novick *et al.* to other data sets, remembering, of course, that we have made an assumption of exchangeability, and the method cannot be expected to work when this is unreasonable.

### 5.3. *Exchangeability Within Multiple Regression Equations*

In this section we briefly indicate how the analysis of Section 3.3 proceeds when $\sigma^2$, the residual regression variance, and $\sigma^2_\beta$, the variance of the regression coefficients, are both unknown. As before, we assume that independently

$$\nu\lambda/\sigma^2 \sim \chi^2_\nu, \quad \nu_\beta \lambda_\beta/\sigma^2_\beta \sim \chi^2_{\nu_\beta}.$$

As in Section 5.2 the integration with respect to $\xi$, the mean of the $\beta_j$'s, may be performed and the result is that the posterior distribution of $\boldsymbol{\beta}$, $\sigma^2$ and $\sigma^2_\beta$ is proportional to

$$(\sigma^2)^{-\frac{1}{2}(n+\nu+2)} \exp\left[-\frac{1}{2\sigma^2}\{\nu\lambda + (\mathbf{y}-\mathbf{X}\boldsymbol{\beta})^{\mathrm{T}}(\mathbf{y}-\mathbf{X}\boldsymbol{\beta})\}\right]$$

$$\times (\sigma^2_\beta)^{-\frac{1}{2}(p+\nu_\beta+1)} \exp\left[-\frac{1}{2\sigma^2_\beta}\left\{\nu_\beta\lambda_\beta + \sum_{j=1}^{p}(\beta_j - \beta_{.})^2\right\}\right], \quad (37)$$

where

$$\beta_{.} = p^{-1}\sum_{j=1}^{p}\beta_j.$$

The modal equations are then easily seen to be (the first coming from (23))

$$\left.\begin{aligned}\boldsymbol{\beta}^* &= \{\mathbf{I}_p + k^*(\mathbf{X}^{\mathrm{T}}\mathbf{X})^{-1}(\mathbf{I}_p - p^{-1}\mathbf{J}_p)\}^{-1}\hat{\boldsymbol{\beta}},\\ s^2 &= \{\nu\lambda + (\mathbf{y}-\mathbf{X}\boldsymbol{\beta}^*)^{\mathrm{T}}(\mathbf{y}-\mathbf{X}\boldsymbol{\beta}^*)\}/(n+\nu+2),\\ s_\beta^2 &= \left\{\nu_\beta\lambda_\beta + \sum_{j=1}^{p}(\beta_j^* - \beta_{.}^*)^2\right\}\Big/(p+\nu_\beta+1).\end{aligned}\right\} \tag{38}$$

The value of $k^*$ is of course $s^2/s_\beta^2$. The marginal posterior distribution of $\boldsymbol{\beta}$ can be obtained in a manner similar to that described in the last section.

We are now in a position to compare our method with that of Hoerl and Kennard (1970b). We have taken the example of a 10-factor, non-orthogonal multiple regression summarized in Gorman and Toman (1966) and re-analysed by Hoerl and Kennard using their ridge regression method. The results are summarized in Table 2.

TABLE 2

*10-factor multiple regression example*

| *Estimate* | $\beta_1$ | $\beta_2$ | $\beta_3$ | $\beta_4$ | $\beta_5$ | $\beta_6$ | $\beta_7$ | $\beta_8$ | $\beta_9$ | $\beta_{10}$ | $k$ |
|---|---|---|---|---|---|---|---|---|---|---|---|
| Least-squares | −0·185 | −0·221 | −0·359 | −0·105 | −0·469 | 0·813 | 0·285 | 0·383 | 0·092 | 0·094 | 0·000 |
| Bayes | −0·256 | −0·178 | −0·326 | −0·086 | −0·289 | 0·592 | 0·195 | 0·349 | 0·117 | 0·116 | 0·039 |
| Ridge | −0·295 | −0·110 | −0·245 | −0·050 | −0·040 | 0·325 | 0·050 | 0·240 | 0·125 | 0·125 | 0·250 |

As already explained, the main difference between ridge regression and the Bayes approach lies in the choice of $k(=\sigma^2/\sigma_\beta^2)$ in equation (23). This has the value zero for least-squares, is chosen subjectively in the ridge method by selecting it so large that the regression estimates stabilize, and is estimated from the data in the Bayes method. In applying the Bayes method we started with $k^* = 0$ in (38), obtained estimates $\boldsymbol{\beta}^*$, which were then used in the other equations in (38) to obtain $s^2$ and $s_\beta^2$. It was found that 10 iterations were needed until the cycle converged. The solution is fairly insensitive to changes in the small, positive values of $\nu$ and $\nu_\beta$ and these were set to zero.

In the case of non-orthogonal data, the least-squares procedure has a tendency to produce regression estimates which are too large in absolute value, of incorrect sign and unstable with respect to small changes in the data. The ridge method attempts to avoid some of these undesirable features. The Bayesian method reaches the same conclusion but has the added advantage of dispensing with the rather arbitrary choice of $k$ and allows the data to estimate it. It will be seen from Table 2 that except for $\beta_1$, $\beta_9$ and $\beta_{10}$, all the estimates are pulled towards zero, the effect being greater with the ridge method than with Bayes, the latter choosing a considerably larger value of $k$ than the data suggest.

ACKNOWLEDGEMENTS

We are very grateful to Melvin R. Novick who first stimulated our interest in these problems, has continually made fruitful suggestions and, with his colleagues, has allowed us to include the example in Section 5.2. We are indebted to the referees for constructive comment on a first draft of this paper. The second author would like to thank the Science Research Council and the Central Electricity Generating Board for financial support during the course of this research.

REFERENCES

BROWN, L. D. (1966). On the admissibility of invariant estimators of one or more location parameters. *Ann. Math. Statist.*, **37**, 1087–1136.

—— (1968). Inadmissibility of the usual estimators of scale parameters in problems with unknown location and scale parameters. *Ann. Math. Statist.*, **39**, 29–48.

CORNFIELD, J. (1969). The Bayesian outlook and its application. *Biometrics*, **25**, 617–657.

DE FINETTI, B. (1964). Foresight: its logical laws, its subjective sources. In *Studies in Subjective Probability* (H. E. Kyburg, Jr and H. E. Smokler, eds), pp. 93–158. New York: Wiley.

FISK, P. R. (1967). Models of the second kind in regression analysis. *J. R. Statist. Soc.*, B, **29**, 266–281.

GORMAN, J. W. and TOMAN, R. J. (1966). Selection of variables for fitting equations to data. *Technometrics*, **8**, 27–51.

HEWITT, E. and SAVAGE, L. J. (1955). Symmetric measures on cartesian products. *Trans. Amer. Math. Soc.*, **80**, 470–501.

HOERL, A. E. and KENNARD, R. W. (1970a). Ridge regression: biased estimation for nonorthogonal problems. *Technometrics*, **12**, 55–67.

—— (1970b). Ridge regression: applications to nonorthogonal problems. *Technometrics*, **12**, 69–82.

LINDLEY, D. V. (1969). Bayesian least squares. *Bull. Inst. Internat. Statist.*, **43**(2), 152–153.

—— (1971a). The estimation of many parameters. In *Foundations of Statistical Inference* (V. P. Godambe and D. A. Sprott, eds), pp. 435–455. Toronto: Holt, Rinehart and Winston.

—— (1971b). *Bayesian Statistics, A Review*. Philadelphia: SIAM.

NELDER, J. A. (1968). Regression, model-building and invariance. *J. R. Statist. Soc.*, A, **131**, 303–315.

NOVICK, M. R., JACKSON, P. H., THAYER, D. T. and COLE, N. S. (1972). Estimating multiple regressions in $m$-groups; a cross-validation study. *Brit. J. Math. Statist. Psychology*, (to appear).

PLACKETT, R. L. (1960). *Principles of Regression Analysis*. Oxford: Clarendon Press.

RAO, C. R. (1965). *Linear Statistical Inference and its Applications*. New York: Wiley.

STEIN, C. (1956). Inadmissibility of the usual estimator for the mean of a multivariate normal distribution. *Proc. Third Berkeley Sympos.*, **1**, 197–206. Berkeley: University of California Press.

## DISCUSSION ON THE PAPER BY PROFESSOR LINDLEY AND DR SMITH

Dr J. A. NELDER (Rothamsted Experimental Station): I welcome this paper as shedding interesting new light on an old topic, the linear model with normal errors. The authors develop a thoroughly Bayesian argument, but this does not mean that we have to be Bayesians in order to make use of their ideas, as I shall try to show.

The authors have laid great stress on their estimates being *Bayesian* estimates, and have compared their estimates with least-squares estimates which they interpret only within a sampling-theoretic framework. Essentially what they are doing is incorporating in their model extra information about a set of (say) population means to the effect that they can be taken as a random sample from a "hyper-population". Such situations undoubtedly occur: e.g. the authors' own example about a random subset of the many lines produced in a breeding program. The same notion underlies designed experiments in incomplete blocks, where the blocks of the design are allocated at random to the finite population in the field. Here the Lindley-Smith estimates of the treatment effects when the block effects are assumed exchangeable but nothing is assumed about the treatments are the familiar ones obtained from the standard analysis using both inter-block and intra-block information. They have replaced the traditional one-stage model

$$y = \beta_i + t_j + e_{ij},$$

where

$$\beta_i \sim N(\mu, \sigma'^2) \quad \text{and} \quad e_{ij} \sim N(0, \sigma^2),$$

with a two-stage model

$$y_{ij} \sim N(\beta_i + t_j, \sigma^2),$$

where

$$\beta_i \sim N(\mu, \sigma'^2).$$

Here the Bayesian analysis of the two models gives identical estimates of $t_j$ and these are the same as the estimates first derived by Yates using maximum likelihood with the one-stage model. The interesting question now arises: what is the corresponding maximum-likelihood procedure for the two-stage model? I want to take a point of view neither Bayesian nor sample-theoretic and look at inferences from the likelihood function. How do we cope with two-faced parameters like the $\beta_i$ which are both random and fixed at the same time? The basic idea for incorporating prior information, due to Edwards (1969), is that we should express it by means of *prior likelihoods*, rather than by prior probabilities. The distinction is non-trivial; the use of prior likelihoods has the property that we do not close the universe of possible models, because we specify only the *relative* weights to be attached to the alternatives. Further alternatives can always be added. In the simple model with

$$y_i \sim N(\theta_i, \sigma^2),$$
$$\theta_i \sim N(\mu, \tau^2)$$

the log likelihood relevant to the location parameters is just

$$-\tfrac{1}{2}[\sum (y_i - \theta_i)^2/\sigma^2 + \sum (\theta_i - \mu)^2/\tau^2],$$

where the second term is the prior likelihood expressing the fact that the $\theta_i$'s are regarded as a random sample from a normal population of variance $\tau^2$. The ML estimators are just

$$\hat{\mu} = y_.,$$
$$\hat{\theta}_i = (y_i/\sigma^2 + y_./\tau^2)/(\sigma^{-2} + \tau^{-2}),$$

i.e. the same as the authors'. Applying this procedure to the incomplete-block example produces the same $t_j$ estimates as before and $\beta_i$ estimates equal to those of the authors.

This parallelism can be carried through a large part of the paper, and I conclude that though the sampling-theory school may be caught in the toils of inadmissibility, and perhaps deserve the authors' strictures, this does not mean that we have to be Bayesians to make use of their results.

I want to say something about ridge regression. The authors say of Hoerl and Kennard's papers that "they do not attempt to reproduce their most convincing argument against the least-squares estimates . . .". This is a pity because I cannot find any convincing arguments in those papers; they measure the deviation of $\hat{\boldsymbol{\beta}}$ from $\boldsymbol{\beta}$ in terms of the Euclidean distance, but it is hard to see why they should unless there is some underlying idea of exchangeability. When there is, the procedures of this paper can be used and provide a justification for their procedure; when there is not we have to accept the least-square estimates *with their associated information surface*. It is true that they may be very sensitive to small changes in the data, but this is not an indictment of least squares, rather it is an indictment of the *data* used for fitting that particular model. The data are generating information in what is almost a subspace of the parameter space, and it is important that this should be recognized by the experimenter and the necessary action taken. As for estimates with the wrong sign, this usually implies inadequate scaling of the parameters in the model. It is interesting in this context that Marquardt, who developed the equivalent of the ridge regression method in optimization algorithms to cope with near-singular surfaces, has recently concluded (Marquardt, 1970) that the use of generalized inverses, i.e. of acceptance of a subspace of estimability, is equally effective. As Marquardt says, "the generalized inverse method confines the solution to a linear subspace containing the origin, whereas the ridge method confines the regression vector to a sphere about the origin".

Finally a comment on the predictive efficiency of the new estimates as given in Table 1. It is hard to assess their success without some more knowledge of the internal consistency of the data in this respect, but taking them at their face value I would interpret them to mean that a better model gave better predictions. To paraphrase Dr Lehrer, mathematician and singer of songs at the piano, "a model is like a sewer, what you get out of it depends on what you put into it".

It will be clear, I hope, that I have very much enjoyed this paper, and so have real pleasure in proposing the vote of thanks.

Dr V. D. Barnett (University of Newcastle upon Tyne): It is a pleasure to second the vote of thanks to the authors, on an interesting and important paper. The linear model features as a central one for the application of statistical method, in unifying a vast range of practical problems including typically the various analysis of variance and regression situations. Much of what the authors might term our "orthodox" heritage of statistical methodology stems from applying the principle of least-squares to the linear model, or more particularly from linear hypothesis theory with an assumed normally distributed error structure. Anyone concerned with the comparative aspects of statistical inference, or simply wishing to explore the variety of methods which might be brought to bear to draw meaningful inferences in this area, must surely welcome this evening's detailed study of estimation for the linear model from the Bayesian viewpoint. It casts a new light on an old and important problem.

The authors themselves make it quite clear that they do not regard this work as the final word on the Bayesian analysis of estimation for the linear model. It provides a basic framework for extension in various respects, some of which are to be considered in promised further work. At the same time the basic assumptions underlying the present work are rather particular ones and demand careful scrutiny with regard both to practical application and formal justification. I should like to address a few remarks to these questions of basic premises and implementation.

The principle of least squares for parameter estimation has a rather unique universality, in making no distributional assumptions about the error structure. At this basic level its sample-specific nature may not be entirely satisfying, and we seek consolation in any asymptotic optimality properties that accrue when the error structure is normal. We have then a *general* principle of estimation; with inherited characteristics in a special case. We might ask to what extent it is possible to parallel this from the Bayesian viewpoint, in promoting a *general* principle of estimation reflected against the hierarchical normal structure? In initiating their study with a discussion of de Finetti's concept of *exchange ability*, it was tempting to believe that the authors would be placing crucial practical emphasis on what for long has seemed a vital idea. I must confess to being somewhat confused, however, in the later parts of the paper as to just how central this concept is in their work. To what extent, if any, can we attribute importance to exchangeability, *per se*, as an ingredient in Bayesian estimation? It figures in delimiting the normal hierarchical structure later used—and is offered as a *natural* expression of a certain form of prior information in certain practical situations—but it is in no way specific to the normal structure. Are any desirable features of the estimates to be credited to exchangeability, or to normality? In this respect some of the claims of Section 1 seem rather generous: in promising study of exchangeability as expressed by mixtures applied to linear models where the parameters themselves have linear structures in relation to hyperparameters. It does seem that only a very special form of this has been presented. Can we go further; for example without the normality assumption?

The bulk of the paper considers the known variance situation. Whilst much of the general theory of Section 2 has already appeared in Lindley (1971a) its fuller treatment is illuminating, and clearly illustrated in the market-place applications of Section 3. But in real-life applications we will not know the variance structure, and it seems that the major feature of this paper, in terms both of its novelty and applicability, must be its consideration of the case of unknown variance structure. The modal invariance property is a fascinating one, and a useful aid to Bayesian analysis of such multi-parameter problems. The suggestion, however, that estimation of unknown variances "presents no serious difficulties" seems optimistic and hardly fully substantiated; specifically the claim at the end of Section 3 that "the simple device of replacing the unknown variances by estimated values in the Bayes estimates is satisfactory". Even in the simple example of Section 1 the proposed iterative estimation of $\tau^2$ is not entirely convincing—even less so for the more complex situations described later. Two issues are involved which cause concern and surely need further comment or study. The first is the question of the *stability* of such iterative procedures. What do we really know about this in the context of the Bayesian estimation problems of this paper? Experience of iterative methods applied to "orthodox" multiparameter problems suggests that we should hardly be surprised to encounter anomalous behaviour. The second matter that seems to require attention is the question of how far we should proceed through the normal hierarchy before (that is, at what "stage") we call a halt to hyperparameterization in the unknown variances case.

A simple example (somewhat similar to that in Section 1) seems to highlight this. Suppose $\mathbf{y}$ is an independent random sample of size $n$ from $N(\theta_0, \sigma^2)$ where $\sigma^2$ is known. In the spirit of this paper we might take a two-stage model where $\theta_0$ has a prior $N(\mu, \tau^2)$ distribution. If $\mu$, $\tau^2$ are known, then $\theta_0$ has posterior distribution

$$N(\Delta\{(n\bar{y}/\sigma^2)+(\mu/\tau^2)\}, \Delta), \quad \text{with} \quad \Delta = (\tau^{-2}+n\sigma^{-2})^{-1}.$$

But suppose $\mu$, $\tau^2$ are not known and instead we have $k$ previous "true" values of $\theta$; $\theta_1, ..., \theta_k$ (a sample from $N(\mu, \tau^2)$). This could arise in industrial batch production problems, where batches of components are used in assembly and as a result precise values of the means for $k$ earlier batches are elicited. Here we do not even need to *estimate* $\theta_1, ..., \theta_k$. Following the authors we might approximate the posterior distribution of $\theta_0$ by substituting $t^2 = \sum(\theta_i-\theta_.)^2/(k-1)$ for $\tau^2$ and $\theta_.$ for $\mu$. But $t^2$ is only an *estimate* of $\tau^2$ and exhibits sampling fluctuations which are not taken into account in any way. When $k$ is small these can be important (if $k$ is large, less so); the estimation of $\theta_0$ surely needs to reflect this. Similar difficulties exist for the example of Section 1 but in more extreme form since here we do not know the $\theta_i$ but must estimate them iteratively. It might therefore seem more appropriate to introduce a third stage with a prior distribution for $(\mu, \tau^2)$ proportional to $\tau^{-2}$ (expressing prior ignorance) and to "up date" this by $\theta_1 ... \theta_k$ to yield a prior distribution for $\theta_0$ which in turn is augmented by $\mathbf{y}$ to produce the posterior distribution

$$\pi(\theta_0 \mid \mathbf{y}, \boldsymbol{\theta}) \quad \propto \quad \exp\left\{-\frac{n}{2\sigma^2}(\theta_0-\bar{y})^2\right\}\Big/\left\{1+\frac{k(\theta_0-\theta_.)^2}{(k^2-1)\,t^2}\right\}^{k/2}.$$

These two posterior distributions are quite different in form (though coincident when $k \to \infty$) and in certain cases, particularly when $k$ or $t^2$ are small, *resulting inferences differ greatly*; both modal estimates and Bayesian confidence intervals. (Modal estimates are surprisingly similar, however, unless $k$ or $t^2$ are very small.) How are we to choose between the two approaches? In a sense the three-stage model has more appeal. Why not always proceed to the stage where we can introduce prior ignorance assumptions, and avoid the iterative estimation?

There is one small point I should like to raise. On various occasions reference is made to the use of the *mean* of the posterior distribution as a natural estimator, and the *mode* is later proposed as an "approximation". Leaving aside their coincidence in normal distributions, is not the *mode* the more natural Bayesian estimator (as the "most likely" value) and the *mean* only supported if we superimpose a special form of loss structure?

Finally, I should like to comment on comparisons and pre-requisites. The authors firmly state at the outset their resolve to avoid further proliferation of comparative discussion of the relative merits of Bayesian and non-Bayesian methods. And yet they immediately appear to abandon this attitude, and the "orthodox" statistician finds himself frequently enlightened or instructed. I do not want to dwell on the "new lamps for old" arguments, but I cannot help feeling that the attributed or adopted criteria are not always entirely fair. I should not be surprised to find later discussants taking issue on some of these points. However, there is one constructive aspect of this matter on which I should appreciate the authors' comments. No statistician can afford to avoid recognizing the basic assumptions of his model or of the techniques he uses. Pragmatic assumptions may be necessary on occasions and concepts and criteria must have an element of arbitrariness about them. This is true of orthodox methods as well as Bayesian methods. Unfortunately, what is "unreasonable" to one person may be "obviously true" to another. But in so far as someone may reject outright a Bayesian approach there is undoubtedly an element of "preferring the devil they know"—in the sense of having built up a practical feeling for the dangers which may arise from the uncertain, or dubious, elements of the orthodox approach. Inevitably there is less experience of this from the Bayesian standpoint—and the authors would provide a valuable service if they could tell us something of the real nature of their "devils". They warn us to seriously beware of applying the methods of this paper if exchangeability, or normality, is unreasonable. But how do we assess this? In their examples even, I find it most difficult to assess the propriety of exchangeability or normal prior distributions on parameters or hyperparameters. What guidance have we on these matters?

In conclusion may I say how much I have enjoyed this paper. It provides much food for thought through cogent argument and compelling illustrations. Undoubtedly it is a vital contribution to the Bayesian study of an important topic and should stimulate useful further work in this area. I am most pleased to second the vote of thanks.

The vote of thanks was put to the meeting and carried unanimously.

Professor Cedric A. B. Smith (Galton Laboratory, University College London): Imagine that we are collecting information about the heights of men in various African tribes. In tribe $T_i$ the expected mean height is $\theta_i$, the expected variance $v_i$, the observed mean $\bar{y}_i$ in a sample of $m_i$ individuals. We are interested in the values of the $\theta_i$. We simplify the discussion by assuming all $m_i$ reasonably large and equal (with common value $m$), all distributions within tribes to be nearly Gaussian (as will be in any case true for the $\bar{y}_i$ by the central limit theorem), and (less realistically) that all variances $v_i$ are equal, with common value $v$ (which is known or estimated with sufficient accuracy). We can concentrate attention on the $\bar{y}_i$ which have variance $\sigma^2 = v/m$.

The $\theta_i$ themselves in the various tribes will have a distribution, strictly speaking discontinuous but representable nearly enough as $\phi(\theta)\,d\theta$. In the authors' treatment this is assumed Gaussian. But this seems less than realistic in many practical cases.

The question arises, can we regard the $\theta_i$ in the tribes actually examined as a random sample from the distribution $\phi$? Strictly speaking, this is almost never the case. We do not select tribes by formal randomization procedure. But we may feel justified in analysing the data *as if* it were a random sample. This may be called a "random effects model" or "exchangeability". Despite the authors' protestations, the distinction between these terms seems rather fine, but note that it involves an act of judgment, i.e. subjective and not objective probabilities, even in the random effects model.

At first sight the "pulling-in" of the estimates appears to contradict common sense; one would think it obvious that in each tribe the "best" estimate of $\theta_i$ is $\bar{y}_i$. If one could assume a very nearly flat prior distribution for $\phi$, this would indeed be so, and it is tempting to think that the "pulling-in" found by the authors is mainly due to a rather severe assumption of a Gaussian form for $\phi$. However, the following argument shows that this is not so. Note that if we have only a small number $n$ of tribes, we will not (in general) know the exact form of $\phi$, and we will have to proceed as the authors do by using a mixture of various plausible forms $\phi_j$ for $\phi$, each with its own prior probability (or density). However, for the sake of discussion, let us assume that we have examined a large number $n$ of tribes, so that we can identify $\phi$ with sufficient accuracy; also that we have a large number of observations in each tribe, so that $\sigma^2 = v/m$ is small. The posterior probability density for $\theta_i$ is then of the form

$$\text{const} \times \sigma^{-1} \exp\{-(\bar{y}_i - \theta_i)^2/2\sigma^2\}\, \phi(\theta_i).$$

Suppose that, near $\bar{y}_i$, we have $\ln \phi(\theta) = A_i + B_i(\theta - \bar{y}_i) + \text{negligible terms}$, where

$$B_i = \left\{\frac{d \ln \phi(\theta)}{d\theta}\right\}_{\theta=\bar{y}_i} \simeq \frac{\phi'\theta_i)}{\phi(\theta_i)}.$$

Then, on substituting this into the expression above, we see that to this order of approximation the posterior distribution is Gaussian with mean (and mode)

$$\theta_i^* = \bar{y}_i + \sigma^2\, \phi'(\theta_i)/\phi(\theta_i)$$

and with variance $\sigma^2$. The second term here represents the "pulling-in" effect.

What is the distribution of the $\theta_i^*$, taken over all $i$? Its expected mean is

$$E(\theta_i^*) = E(\bar{y}_i) + \sigma^2\, E\{\phi'(\theta_i)/\phi(\theta_i)\}.$$

But $E(\bar{y}_i) = E(\theta_i) = \mu$, the mean of the distribution $\phi$. And

$$E\left(\frac{\phi'}{\phi}\right) = \int_{-\infty}^{\infty} \frac{\phi'(\theta)}{\phi(\theta)}\, \phi(\theta)\, d\theta = [\phi(\theta)]_{-\infty}^{\infty} = 0.$$

So $E(\theta_i^*) = \mu$, i.e. the "pulling-in" does not affect the mean. It simplifies the calculation of the variance to take measurements from the mean $\mu$, so that we may assume $E(\theta_i^*) = 0$. Then, neglecting terms in $\sigma^4$, and setting $\text{var}\, \theta_i = \tau^2$,

$$\text{var}\, \theta_i^* = E(\theta_i^{*2}) \simeq E(\bar{y}_i^2) + 2\sigma^2\, E\{\bar{y}_i\, \phi'(\theta_i)/\phi(\theta_i)\}.$$

But $E(\bar{y}_i^2) = \text{var}\, \bar{y}_i^2 = \tau^2 + \sigma^2$. And, since $\bar{y}_i \simeq \theta_i$,

$$E\left\{\frac{\bar{y}_i\, \phi'(\theta_i)}{\phi(\theta_i)}\right\} \simeq \int_{-\infty}^{\infty} \theta \frac{\phi'(\theta)}{\phi(\theta)}\, \phi(\theta)\, d\theta = \int_{-\infty}^{\infty} \theta\phi'(\theta)\, d\theta.$$

Suppose that $\theta\phi(\theta) \to 0$ as $\theta \to \pm\infty$, as is true apart from very unusual distributions. Then on integrating by parts we find

$$\int_{-\infty}^{\infty} \theta\phi'(\theta)\, d\theta = 0 - \int_{-\infty}^{\infty} \phi(\theta)\, d\theta = -1.$$

Hence $\text{var}\, \theta_i^* = \tau^2 - \sigma^2$. Since $\sigma^2$ is not being neglected, this means that the variance of the $\theta_i^*$ is appreciably smaller than that of the $\bar{y}_i$, whatever the form of the distribution $\phi$. If we regard the $\theta_i^*$ as sensible estimates for the $\theta_i$, this means that for any form of the distribution $\phi$ there will be an appreciable "pulling-in" from the straightforward sample means $\bar{y}_i$; that is, this phenomenon cannot be attributed to the special Gaussian form used by Lindley and Smith.

I would like to add my agreement to Dr Nelder's suggestion that Lindley and Smith's estimates are, virtually speaking, maximum-likelihood estimates. This is true if one incorporates the density function for the $\theta_i$ given $\mu$ into the likelihood; and this seems reasonable from previous considerations (Smith, 1970) on a rather similar situation.

Mr T. LEONARD (University College London): I would like to make a few remarks about the possible extension of the excellent ideas expressed in this paper to situations where the exchangeable parameters cannot be considered to be normally distributed. In such circumstances, a good procedure is usually to transform the parameters in such a way that the normality assumption is more realistic for the new parameters.

A simple example occurs when the data $x_i$ are independent and binomially distributed, with parameters $\theta_i$ and indices $n_i$ $(i = 1, \ldots, m)$. It is convenient to transform to the log-odds $\alpha_i$, where

$$\alpha_i = \log\{\theta_i/(1-\theta_i)\}.$$

Under the exchangeability assumption, it is reasonable to suppose that, given $\mu$ and $\sigma^2$, the $\alpha_i$ are independent and normally distributed with common mean $\mu$ and variance $\sigma^2$, where $\mu$ is uniformly distributed, and $\sigma^2$ possesses an inverse $\chi^2$ distribution.

The prior to posterior analysis is straightforward, and when $\sigma^2$ is known, it is easily shown that the posterior modes of the $\alpha_i$ are given by the solutions to the equations

$$\frac{e^{\alpha_i}}{1+e^{\alpha_i}} = \frac{x_i}{n_i} - \frac{\alpha_i - \alpha_.}{n_i\,\sigma^2} \quad (i = 1, \ldots, m);$$

when $\sigma^2$ is unknown it is replaced by its modal estimate.

Unless $x_i$ is close to zero or $n_i$, approximations are given by

$$\alpha_i = \frac{\sigma^2\, l_i + v_i\, \alpha_.}{\sigma^2 + v_i},$$

where

$$l_i = \log\{x_i/(n_i - x_i)\},$$
$$v_i = x_i^{-1} + (n_i - x_i)^{-1}.$$

These provide similar weighted average forms to those described in the paper.

Logarithmic transformations and exchangeable normal prior distributions may also be employed in the analysis of Poisson or multinomial data. They may also be applied to the situation described on p. 3 in the main paper, where there are normal data with exchangeable variances. In all these cases, simple equations may be found for the exact modes, and weighted average forms may be approximated.

Professor M. R. NOVICK (University of Iowa): Within the time and space limitations imposed on them, Lindley and Smith have done a commendable job of laying out in proper mathematical form a model of intriguing complexity. It is a pity that we cannot carry on into the early hours of the morning, examining the remarkable theoretical implications of this model and, even more excitingly, the applications that can be made of these mathematical results. Such discussions are available, however, and I commend to the attention of this audience a series of reports on these topics. Some of these have or will shortly appear (Novick and Jackson, 1970; Jackson *et al.*, 1971; Novick *et al.*, 1971; Novick *et al.*, 1972). These papers contain elaborate praise for the remarkable work presented here this evening so that it seems appropriate that I concentrate now on whatever faults I can invent for tonight's paper.

First, Lindley and Smith have conveniently concealed the difficulties in actually getting out the required estimates. In the applications we have so far studied, modal estimates can be obtained by the iterative solution of what we have called *Lindley equations*. There are, however, definitely problems of bimodality. These can generally be handled, but only by giving careful attention to the prior distribution of the hyperparameters. It is not, however, a trivial exercise as Lindley and Smith have implied.

Secondly, I disagree with the choice of terminology as to whether all of this is Model I or Model II. I do not think that there is necessarily a right answer to this, but exchangeability does imply the equivalence of a random sampling assumption for parameters and this I (and Box and Tiao, 1968) would call the Random Effects Model.

Lindley and Smith have warned that we must have exchangeability in our prior beliefs to apply these methods. I would elaborate this by saying that we should have an informed exchangeability and not one implied by ignorance. In the cited empirical study, I personally went to great lengths to attain an informed exchangeability, with gratifying results. Later, for a very simple model, I was able to obtain an explicit formula showing that the increase in precision using this method was a function of the between group homogeneity of the parameters.

Let me further remark that this assumption of exchangeability is a very questionable one when imposed across variables rather than across groups. I think that application of the work discussed in Section 5.3, both Bayesian and classical, must be done very carefully. My own feeling is that further theoretical work needs to be done in this area, possibly with a principal components or orthogonal factor analytic model, in a *reduced* variable space. Here it may be more reasonable to make a valid assumption of exchangeability. If so, we may finally be able to do a really good job of estimating a single covariance matrix.

As a final remark, let me chide Lindley and Smith for not having referenced the work of Truman Kelley (1927) who, more than forty years ago used, in an educational measurement context, methods that are very similar to those later adopted by Robbins, by Stein (1966) and by those of us now working within a Bayesian framework with exchangeable prior distributions.

Professor D. R. Cox (Imperial College): Prior information enters regression analysis in the choice of explanatory variables and of the form of relation to be fitted. Once these are provisionally fixed, prior information may affect the estimation of parameter values, some of the main types of such information being (a) knowledge of the sign of one or more coefficients; (b) a relation with analogous parameters in other sets of data, e.g. that if $\beta_p$ and $\beta_p^*$ are the current and previous values respectively, then $\beta_p \geq \beta_p^*$ or $|\beta_p - \beta_p^*| \leq d_p$ or $\beta_p - \beta_p^*$ is a random variable of known distribution; (c) statements about the interrelations of the explanatory variables, such as used in path analysis; (d) the assumption that certain parameters are linked by having been generated by a common physical random mechanism; (e) the postulation of subjective prior distributions for parameters. Points (d) and (e) are linked mathematically, but conceptually are quite different, of course.

It seems to me highly desirable that if any such assumptions are introduced they should as far as possible be tested from the data, either graphically or by significance tests. Failure to do so may mean overlooking inconsistency arising from biased data, from the omission of important explanatory variables or from the prior assumption being misconceived.

In sampling theory, possibility (d) can often be represented as a fully parametrized empirical Bayes problem. That is, the vector observation Y has density $f_{\mathbf{Y}|\Theta}(\mathbf{y} \mid \theta; \phi)$ depending on unknown parameters $\phi$ and on parameters $\Theta$ with a density $f_\Theta(\theta; \psi)$ depending on further parameters $\psi$. We can thus find $f_{\mathbf{Y}}(\mathbf{y}; \phi, \psi)$ and obtain estimates $\hat{\phi}, \hat{\psi}$, for example by maximum likelihood. Now if we are interested in some components $\Theta_1$ of $\Theta$ we can, for given $\phi, \psi$, obtain the posterior density $f_{\Theta_1|\mathbf{Y}}(\theta_1 \mid \mathbf{y}; \phi, \psi)$. This suggests the estimate

$$\hat{f}(\theta_1) = f_{\Theta_1|\mathbf{Y}}(\theta_1 \mid \mathbf{y}; \hat{\phi}, \hat{\psi})$$

or of improved estimates based on refining this. The authors in a fully Bayesian approach have been able to integrate out $\phi, \psi$ instead of having to substitute point estimates.

Professor R. L. Plackett (University of Newcastle-upon-Tyne): I would like to emphasize the atmosphere of unity which has been such a welcome feature of this evening's discussion by deriving the authors' results from sampling theory without the use of prior distributions or prior likelihoods.

If we take the two-stage linear model with which they begin, then equations (7) and (8) can be obtained by combining the original set of observations $\mathbf{y} \sim N(\mathbf{A}_1 \boldsymbol{\theta}_1, \mathbf{C}_1)$ with an additional set $\mathbf{A}_2 \boldsymbol{\theta}_2 \sim N(\boldsymbol{\theta}_1, \mathbf{C}_2)$. It is no coincidence that equation (10), in the special case when $\mathbf{C}_2 = \mathbf{I}$, appears in the section of Plackett (1950) which is concerned with adjustments to the least-squares estimators, their dispersion matrix and the sum of squared residuals, due to the appearance of additional observations. If we extend this argument to the three-stage model, then we have to combine the last two stages. In that case, we are combining $\mathbf{y} \sim N(\mathbf{A}_1 \boldsymbol{\theta}_1, \mathbf{C}_1)$ with $\mathbf{A}_2 \mathbf{A}_3 \boldsymbol{\theta}_3 \sim N(\boldsymbol{\theta}_1, \mathbf{C}_2 + \mathbf{A}_2 \mathbf{C}_3 \mathbf{A}_2^T)$. When the authors come to apply the results of Section 2, their use of exchangeability implies that the additional data $\mathbf{A}_2 \boldsymbol{\theta}_2$ or $\mathbf{A}_3 \boldsymbol{\theta}_3$ have the form $\xi\mathbf{1}$. These relationships explain why the estimators take a weighted form throughout the paper, and why they are closer together than the least-squares estimators.

An attempt to estimate how much data is actually introduced in this way can be made from an analysis of Table 1. Consider the first column. Suppose that the variance of an individual observation is $\sigma^2$, and the variance of a predicted value from all the data is $a\sigma^2$. Then

$$\sigma^2(1+a) = 0{\cdot}5596 \quad \text{and} \quad \sigma^2(1+4a) = 0{\cdot}6208,$$

whence $a = 0{\cdot}04$, equivalent to 25 observations, and $\sigma^2 = 0{\cdot}54$. Let Bayes and all the data predict with variance $b\sigma^2$. Then $b = 0{\cdot}02$, equivalent to 50 observations. The conclusion from this posterior analysis is that the effect of using a Bayesian model is to double the amount of information in the data.

Professor P. Sprent (University of Dundee): The tendency to pull towards the central value gives the estimate in equation (1) a clear advantage over the least-squares estimate. As contributors to this discussion have already pointed out the least-squares estimates are too dispersed. This is evident if we write $y_i = \mu + \delta_i + \epsilon_i$, where $\theta_i = \mu + \delta_i$ and the $\epsilon_i$ are $N(0, \sigma^2)$ while the $\theta_i$ are $N(\mu, \tau^2)$; assuming independence, the $y_i$ are $N(\mu, \sigma^2 + \tau^2)$, and thus have greater dispersion than the $\theta_i$. The more extreme $y_i$ correspond to cases where $\delta_i$ and $\epsilon_i$ have the same sign and thus overestimate the magnitude of $\delta_i$.

If we take $\mu = 0$ for convenience and assume $n$ is large (but there is only one replicate for each $\delta_i$) then if $\tau^2 = \sigma^2 = 1$ and $\delta_i = 1$, the probability is approximately 0·95 that $y_i$ (i.e. $\hat{\theta}_i$) will lie between $-1$ and 3. Assuming $y_. = 0$; for large $n$ when $\mu = 0$, the Bayes estimate $\theta_i^*$ will lie between $-0{\cdot}5$ and $1{\cdot}5$ with the same probability. In this case, despite its bias, $\theta_i^*$ seems preferable to $\hat{\theta}_i$. If now $\tau^2 = 1$, $\sigma^2 = 9$ and $\delta_i = 1$, then with probability approximately 0·95, $\hat{\theta}_i$ will lie between $-5$ and 7 while $\theta_i^*$ will lie between $-0{\cdot}5$ and 0·7. It is clear that with a probability very close to unity $\theta_i^*$ will in this case be less than the true value. The practical man would, one hopes, faced with these values of $\tau^2$ and $\sigma^2$, realize the futility of trying to find out anything useful about $\delta_i$ from a single replicate corresponding to each $\delta_i$. The ratio $\lambda = \sigma^2/\tau^2$ is a useful guide both to this futility and the seriousness of bias in the Bayes estimate. Is it too frivolous to call it a *coefficient of stupidity*? If $\tau^2 = 9$ and $\sigma^2 = 1$, $\lambda$ is small; taking $\delta_i = 1$ we then have with probability approximately 0·95 that $\hat{\theta}_i$ lies between $-1$ and 3 while $\theta_i^*$ lies between $-0{\cdot}9$ and 2·7; the Bayes estimate again is reasonable.

Clearly when $\tau^2 = 0$ all $\delta_i = 0$ and $y_.$ is the appropriate estimate of $\mu$ in either case. Similarly when $\sigma^2 = 0$, both estimates reduce to $y_i$, the exact value of $\mu + \delta_i$.

I suggest that when bias in the Bayes estimate is large it cannot be dismissed lightly, but reflects a high value of the coefficient of stupidity, $\lambda$. If there are $r$ replicates for each $y_i$ in (1), having mean $y_{i.}$, we may conjecture that $\sigma^2$ in (1) should be replaced by $\sigma^2/r$, giving $\lambda = \sigma^2/(r\tau^2)$, and the Bayes estimates would approach more closely the least-squares estimates as $r$ increased. A study of degree of bias might prove illuminating in the more sophisticated examples in this paper.

Dr J. B. Copas (University of Essex): I am not sure what the "orthodox statistician" is supposed to be assuming when he is being convinced that the estimate labelled (1) is better than the least-squares estimate. Surely no one would disagree that if this special model of the random $\theta$'s is true, which implies amongst other things that the observations are positively correlated, then the estimates should be made to depend on each other in the way indicated. When the assumption of such a prior structure is absent, on the other hand, the situation is simply one in which an experiment is repeated $n$ times with different and unknown parameter values: in other words, a compound decision problem in the sense of Robbins. I attempted to show at one of our previous meetings (Copas, 1969) that it is very doubtful that compound decision rules, of which (1) is typical, are in fact better than the classical results, unless a relationship of some sort is assumed between the components, such a exchangeability. For one, only the *average* mean squared error over $i$ from 1 to $n$ is considered, whereas the risk of particular components may be unacceptably high, and in any case there would presumably be no particular reason for supposing that inequality (2) was satisfied unless the $\theta$'s were assumed to be some sort of random sample.

As Professor Cox has mentioned, the mathematical models described in this paper are almost identical to those dealt with in the completely non-Bayesian approach known as empirical Bayes. Thus this work could be described as giving Bayes solutions to the empirical Bayes problem. The main difference lies in the meaning of the mixing distribution $Q$ of Section 1. In empirical Bayes, $Q$ is simply a frequency model for a continuing series of repetitions of an experiment. In principle, any assumption in this model can be tested. If I understand correctly, we now see this distribution as a measure of prior beliefs operating through the assumption of exchangeability. Is it not remarkable that one's prior beliefs about treatment $i$ would be identical *in every way* to one's prior beliefs about treatment $j$? I hope that the elegant mathematics of this paper will not tempt us to rest assured in our ignorance and refrain from enquiring of our experimenter the differences which must inevitably exist between the various parts of his data.

Finally, two brief questions. What is meant by the statement that exchangeability is assumed for *every n*? Does this entail a hypothetical continuation of the number of components in the problem? Secondly, it is easy to show that the correlation between any pair of $\theta$'s in the mixture model must be non-negative. Do the authors have a simple representation of exchangeable priors with negative correlations? I have in mind a randomization approach to field trials in which the true plot yields are, in a sense, negatively associated.

Dr D. V. Hinkley (Imperial College): I have two points to make about this very interesting and useful paper. Both have to do with related unpublished work which I think can augment the solutions proposed by the authors.

The first point has to do with outliers—in the parameters, not the observations. Take the simple problem of estimating $\theta_1, \ldots, \theta_n$ when observations $y_i \sim N(\theta_i, 1)$ are available, and the $\theta_i$ are independently $N(0, \tau^2)$ with $\tau^2$ known; this is a simple version of the problem in Section 1. The Bayes estimate for quadratic loss is $\boldsymbol{\theta}^B = \hat{\boldsymbol{\theta}}\{\tau^2/(\tau^2+1)\}$, where $\hat{\boldsymbol{\theta}}$ is the vector of least-squares estimates, i.e. $\mathbf{y}$ in this case. The Bayes estimate minimizes the Bayes risk, but the risk for any given $\theta_i$ is unbounded. Thus the Bayes estimate is particularly vulnerable to atypical parameter values, that is outliers, which will be shrunk too much. Efron and Morris, in a series of unpublished papers at Stanford University and Rand, devise an estimate which limits risk with little relative increase in Bayes risk. For our simple problem of estimating $\theta_1, \ldots, \theta_n$ they propose estimates such as

$$\tilde{\theta}_i = \begin{cases} \hat{\theta}_i + M, & \hat{\theta}_i < -M(\tau^2+1) \\ \theta_i^B = \hat{\theta}_i\left(\dfrac{\tau^2}{\tau^2+1}\right), & -M(\tau^2+1) \leqslant \hat{\theta}_i \leqslant M(\tau^2+1), \\ \hat{\theta}_i - M, & M(\tau^2+1) < \hat{\theta}_i. \end{cases}$$

As an example, suppose that the Bayes risk of $\hat{\theta}$ is normalized to be 1·0, and that $r^2 = 1$. Then the Bayes risk of $\theta^B$ is 0·5. If $\tilde{\theta}_i$ are used with $M = 1$, the Bayes risk is only 0·525 and furthermore the maximum risk is 2 for any $\theta_i$. This approach can be generalized to the cases studied by the authors. Incidentally one suspects from Table 2 that $\beta_6$ is an outlier, particularly from a normal plot of the least-squares estimates; the Bayes estimate of $\beta_6$ is made normal.

My second point is of a somewhat mathematical nature, and has to do with use of variance estimates in Bayes estimates such as (1). Suppose now that we are estimating multivariate means $\boldsymbol{\theta}_i$, using independent observations $\mathbf{y}_i$ from $N(\boldsymbol{\theta}_i, \mathbf{I})$ where the $\boldsymbol{\theta}_i$ are independent samples from $N(\mathbf{0}, \boldsymbol{\Sigma})$. If we write $\mathbf{Y} = (\mathbf{y}_1, \ldots, \mathbf{y}_n)$, then the Stein-type estimate for $\boldsymbol{\theta}_i$ is of the form

$$\boldsymbol{\theta}_i^* = \{\mathbf{I} - k\mathbf{S}^{-1}\}\,\hat{\boldsymbol{\theta}}_i,$$

where $\hat{\boldsymbol{\theta}}_i$ is the maximum-likelihood estimate and $\mathbf{S} = \mathbf{Y}\mathbf{Y}^{\mathrm{T}}$. In a forthcoming paper by Efron and Morris (1972), reference is made to the fact that Stein has established a uniformly superior replacement for $k\mathbf{S}^{-1}$ in $\hat{\theta}_i^*$, namely $a\mathbf{S}^{-1} + b\mathbf{I}/\mathrm{tr}\,(\mathbf{S})$ for suitable $a$ and $b$. Does this help in the authors' search for a way of coping with their unknown variances?

Mr E. F. HARDING (Cambridge University): In estimating three means $\mu_1, \mu_2, \mu_3$ by normal samples with the loss function

$$(\hat{\mu}_1 - \mu_1)^2 + (\hat{\mu}_2 - \mu_2)^2 + (\hat{\mu}_3 - \mu_3)^2 \qquad (1)$$

the usual estimate $(\bar{x}_1, \bar{x}_2, \bar{x}_3)$ is, by Stein's so-called paradox, not admissible. To use an example I heard from Professor Barnard, if $\mu_1$ refers to butterflies in Brazil, $\mu_2$ to ball-bearings in Birmingham and $\mu_3$ to Brussels sprouts in Belgium, then an admissible estimator will cause the estimates of these three quite unrelated things to be related to each other, the largest $\hat{\mu}$ being on the whole pulled down and the smallest pulled up. Now, if you do not like that sort of thing, and in general I do not, then so much the worse for the loss function; one can, after all, always work a "paradox" backwards.

Coming now to exchangeability as such, take another instance in which the three $\mu$'s are the tax-allowable expenses of Professor Lindley, Dr Smith and myself, normally distributed data about these being available. The Tax Commissioners might, I suppose, regard the three of us as exchangeable *a priori*, and use the loss function (1); in which case Professor Lindley could find himself paying too much tax because one of the others of us was paying too little—fair to the Exchequer, perhaps, but not to him—and he could reasonably plead that they should use the estimator fairest *to him*: because he, after all, can hardly think himself exchangeable with, say, me. And this is the point: exchangeability is relative to the one who wants to do the exchanging.

Viewing statistical decision theory as "a game between the statistician and Nature" perhaps encourages the habit of imperiously acting according to principles of exchangeability, etc., since one hardly regards the "states of Nature" as having ethical rights; but when, in particular instances, it becomes a "game" between the statistician and other people, then one may expect one's hypotheses to demand respect.

I would not harp on this (the authors do, here and there, indicate that one ought, strictly, to *verify* exchangeability) had they not passed rather lightly over the matter of verification. Indeed, a first reading suggests that assuming exchangeability is a good thing because it leads to procedures with better statistical qualities. True: but they are better *relative to the loss function* (1), and if you dislike the ethical consequences then, again, so much the worse for the loss function. Of course, if $\mu_i$ were a sequence of batch means in a continuing industrial process and the profit was expressible in a form like (1), then exchangeability would lead to wholly desirable consequences. That would be a clear case, unlike, however, the authors' blithe and unqualified application of exchangeability to a case which is, I think, very like the imaginary tax example: their final example of estimating the merits of the different American colleges. If Federal funds were to be distributed according to merits, then estimates derived from exchangeability could well lead to the better colleges getting less, and the worse more, than their fair shares. In such cases I would prefer to solve a sequence of decision problems, in which the loss function for the $i$th mean would be $(\hat{\mu}_i - \mu_i)^2$, and I think that a special case has to be made out for doing otherwise, to show that the "unfairness' which must result is either justified or irrelevant. One might even be able to incorporate all this into a complicated loss function!

Mr A. P. DAWID (University College London): I should like to make two points and pose a problem.

Firstly, there seem to be close connections between the approach of tonight's paper and the work of Ericson (1969) on finite sampling theory, in which exchangeability also played a leading role. One might view Ericson's structure as a three-stage model with a first stage that is singular, because in finite sampling one usually observes the first-stage parameters, that is to say the values of the sampled units, without error. Ericson was interested in prediction for the unsampled units, and this raises the question of how we should deal with prediction in general for multi-stage models.

Secondly, let me take a fundamentalist approach, according to my understanding of de Finetti's ideas. Given the possibility of taking a set of observations, a subjectivist should be putting his prior distribution jointly on the set of possible (as yet unobserved) *outcomes*. If one chooses to work with "unknown parameters" it is not, generally, because of any physical structure of the problem, but because, given various assumptions of exchangeability and conditional exchangeability, such parameters are conjured into existence by invoking de Finetti's theorem. What, then, is the status of the first-stage parameter $\boldsymbol{\theta}_1$, and what assumptions about the distribution of $\mathbf{y}$ are we making when we express opinions of exchangeability among certain components of $\boldsymbol{\theta}_1$?

I am also interested to know the true mathematical expression of two-way exchangeability, as considered in the authors' first example, the two-factor design. We are interested in some first-stage parameters $\{\theta_{ij}\}$ measuring somehow a mean response for cell $(i, j)$. Lindley and Smith assume a singular second stage of the interaction-free form: $\theta_{ij} = \mu + \alpha_i + \beta_j$; and then take the $\alpha$'s and $\beta$'s as independently exchangeable. What I should like to know is this: suppose we have observations $\{y_{ij}\}$ which we are willing to regard as exchangeable both within rows and within columns (but not simultaneously). Is the above structure as general as possible? What about an interaction term: $\theta_{ij} = \mu + \alpha_i + \beta_j + \gamma_{ij}$? If $y_{ij} = \theta_{ij} + \epsilon_{ij}$, this is identical with the model

$$y_{ij} = \mu + \alpha_i + \beta_j + \eta_{ij},$$

where $\eta_{ij} = \gamma_{ij} + \epsilon_{ij}$, so without replicated $y$'s we cannot get much information about the $\gamma$'s. But even in this case there is a qualitative difference between the two models, and I am now asking for guidance on how to parametrize such a situation when such arbitrariness exists.

Finally, a problem: we have seen the application of multi-stage models to the standard normal theory linear set-up, and in the discussion Mr Leonard has mentioned approaches to similar problems with binomial and multinomial models. However, many situations have a more complicated structure, but the ideas presented tonight would be attractive to apply if they could be made tractable. Let me outline a problem I have in mind and ask whether Lindley and Smith can give me any help with it.

The situation is that a number of patients are asked questions a number of times by a number of doctors. When doctor $k$ asks patient $i$ the $j$th question for the $r$th time, he elicits a binary answer $S_{ijkr}$. We suppose that there is a true answer $T_{ij}$ which the doctors are trying to uncover, but some may be better at doing so than others. There are various exchangeabilities I should like to impose on this problem: firstly, $\{T_{ij}\}$ might be exchangeable over $i$, and perhaps in a restricted form over $j$. Then one could look at the conditional distribution of the $S$'s given the $T$'s, where we could introduce more symmetries. Perhaps we could make these conditional distributions depend on further exchangeable parameters. I cannot quite see what the multi-stage expression of such structure might be in general, but one can simplify at a number of points. However, the technical difficulties of inference in even the simplified models seem rather daunting. Does the normal theory analysis give any insights into how to tackle this problem?

The following contributions were received in writing, after the meeting.

Dr C. Chatfield (University of Bath): I had intended to attend this meeting, but finally decided that the discussion would probably involve yet another round in the endless controversy between Bayesians (subjectivists), Frequentists, Likelihoodlums, etc., which would probably not be profitable for an applied statistician like myself. Arguments about the foundations of statistics are, of course, extremely interesting from an academic point of view, but, as far as one can ascertain, the different approaches seem to give more or less the same answer in the vast majority of cases.

A major difficulty for applied statisticians is that the arguments involved tend to be extremely difficult to follow and centre around rather artificial examples. A further difficulty is that participants in these discussions tend to show scant respect for one another's position—for example, see the authors' statement that they "know of no reasoned argument against the Bayesian position". For my own part I agree with H. O. Hartley in the discussion of Cornfield's paper (1969) that no single theory of inference is entirely free from deficiencies and so, like most applied statisticians, I adopt the standard sampling theory approach for mainly practical reasons. (Of course, on occasions I also find it useful to look at likelihood surfaces or use Bayes theorem.)

Turning from these general matters to the paper, I wonder how many other people find the basic idea of exchangeability hard to swallow except possibly in a few special situations. I can imagine the reaction this assumption would receive from many scientists, and also the likely reaction to the Lindley-Smith estimates in their equation (1). Perhaps their estimates do have a smaller mean-square error than the ordinary least-squares estimates provided that the priors really are exchangeable and provided that one knows both $\sigma^2$ and $\tau^2$. However, in practice these variances will almost certainly be unknown and I am very doubtful about using the estimates (1) with estimates of $\sigma^2$ and $\tau^2$.

It is nice to see that the authors have actually performed a numerical study in which Bayes estimates do come out marginally superior to the ordinary least-squares estimates. However, as the authors say, one cannot draw strong conclusions from a single study. Moreover I am doubtful if the improvement they achieve justifies the extra effort involved. Scientists have trouble enough with straightforward regression without the complication of accepting the idea of subjective exchangeable priors.

Perhaps I am being slightly unfair in concentrating on the practical aspects of what is really a Research Section paper and which is clearly an important paper from a theoretical point of view. However, the meeting was billed as an ordinary meeting (organized by the Research Section), and I feel that ordinary meetings should be of broader interest and of more practical use than appears to be the case here.

Of course this last remark also applies to many other Research Section meetings, and so perhaps it would be a good idea to discontinue the recently introduced practice of calling all Research Section meetings "ordinary meetings".

Dr Stephen E. Fienberg (University of Chicago): It is a great pleasure for me to have the opportunity to discuss this paper by Professor Lindley and Dr Smith. I regret that I was unable to hear its presentation. The problems discussed here are ones on which I myself have expended considerable energies. It is more than faint praise for me to remark that I wish I had written this paper.

The presentation of the general theory in Sections 1 and 2 is lucid, and perhaps to the surprise of many of us, quite straightforward. Work on the Bayesian analysis of linear models abounds, but the use of a multi-stage Bayesian analysis with successive reductions in the number of parameters at each stage has considerable intuitive appeal.

I was especially glad to see Sections 3.1 and 5.1 dealing with the complete two-way layout, since I once worked my way through this example (Fienberg, 1967) to arrive at the simple generalization of expression (20) for the two-way layout with replications, i.e.

$$E(y_{ijk}) = \mu+\alpha_i+\beta_j+\gamma_{ij} \quad (1\leqslant i\leqslant t,\ 1\leqslant j\leqslant b,\ 1\leqslant k\leqslant n),$$

where the second stage is

$$\gamma_{ij}\sim N(0,\sigma_\gamma^2),\quad \alpha_i\sim N(0,\sigma_\alpha^2),$$

$$\beta_j\sim N(0,\sigma_\beta^2),\quad \mu\sim N(\omega,\sigma_\mu^2).$$

In this somewhat more general case, taking the posterior mean and letting $\sigma_\mu^2\to\infty$, one gets $\mu^* = y_{...}$,

$$\alpha_i^* = (y_{i..}-y_{...})\frac{bn\sigma_\alpha^2}{bn\sigma_\alpha^2+n\sigma_\gamma^2+\sigma^2},$$

$$\beta_j^* = (y_{.j.}-y_{...})\frac{tn\sigma_\beta^2}{tn\sigma_\beta^2+n\sigma_\gamma^2+\sigma^2}$$

and

$$\gamma_{ij}^* = (y_{ij.}-\mu^*-\alpha_i^*-\beta_j^*)\frac{n\sigma_\gamma^2}{n\sigma_\gamma^2+\sigma^2}.$$

Letting $n = 1$ and $\sigma_\gamma^2 = 0$ we get expression (20). The theorem and the simple form of expression (18) greatly simplify the amount of algebraic manipulation I was forced to carry out.

Going on to the case where the second-stage variances are unknown, it is interesting to note that, using the estimates

$$s^2 = \sum_{ijk}(y_{ijk}-y_{ij.})^2/\{2+bt(n-1)\},$$

$$s_\gamma^2 = n\sum_{ij}(y_{ij.}-y_{i..}-y_{.j.}+y_{...})^2/\{(b-1)(t-1)-2\},$$

$$s_\beta^2 = nt\sum_{j}(y_{.j.}-y_{...})^2/(b-3),$$

$$s_\alpha^2 = nb\sum_{i}(y_{i..}-y_{...})^2/(t-3)$$

for $\sigma^2$, $(\sigma^2+n\sigma_\gamma^2)$, $(\sigma^2+n\sigma_\gamma^2+nt\sigma_\beta^2)$ and $(\sigma^2+n\sigma_\gamma^2+nb\sigma_\alpha^2)$ respectively, in $\alpha_i^*$, $\beta_j^*$ and $\gamma_{ij}^*$ above, one arrives at the estimator of Stein (1966), provided negative estimates of variances are replaced by zeros, in keeping with the classical practice. (Here $t$ and $b$ are assumed to be $\geqslant 3$.) This estimator is of course similar to the one suggested by Lindley and Smith although no iteration is necessary. Stein points out that the anticipated contraction of the main effects is likely to be small relative to that for the interaction effects.

Rather than using point estimates for the prior parameters, Dempster (1971) suggests laying out intervals for them which are acceptably consistent with the data in the sense of passing simple significance tests, or equivalently of belonging in certain confidence regions. These intervals can then be translated into intervals for the posterior means, and in effect yield upper and lower bounds

I am somewhat troubled by the use of modal approximations proposed in Section 4 of this paper, and there are three items of concern. First, integrating out the nuisance parameters from the posterior distribution is not really as technically complex as implied by Lindley and Smith. One often gets posterior distributions of the $\theta$'s, which are products of multivariate $t$-distributions. Then, one can use an identity such as that given by Dickey (1968) to reduce the integral for the posterior mean to an integral of dimension one less than the number of $t$-like factors. For the one-way ANOVA model discussed in Section 1 this reduction yields a one-dimensional integral that can be handled easily by simple numerical techniques.

Second, the use of the mode of the posterior distribution in place of the mean is fine when the $\theta$ posterior margin is nearly normal. When the posterior distribution is unimodal but asymmetric, the use of the mode may lead to relatively large errors. Mosteller and Wallace (1964) suggest taking modes of transformed parameters relative to a weighted measure, so as to make the resulting mode and mean much closer than the mode and mean of the original $\theta$-margin.

Finally, the use of the mode of the joint posterior distribution in place of that of the $\theta$-margin is one which requires far more justification than that of computational expediency.

A problem not considered in the present paper is that of estimating parameters for linear models, when some or all of the observation vectors are incomplete. In a discussion of a paper by Hartley and Hocking (see Fienberg, 1971), I proposed a crude Stein-like improvement for a simple $k$-means problem where the maximum-likelihood estimator has a regression form. Extensions of the techniques presented by Lindley and Smith should yield substantial improvement over my estimators, which are, in turn, uniformly superior to the maximum-likelihood estimators according to the criterion of expected quadratic loss. Have the authors undertaken an investigation of such extensions?

Professor BRUCE M. HILL (University of Michigan): Professor Lindley and Dr Smith have presented an elegant and informative Bayesian treatment of an important class of problems. Although they apparently would not agree, I think they are dealing with a genuine Model II situation. The distinction they make in terms of where one is *interested* in fixed effects or variance components does not seem an important one to me. Nor do I think that exchangeability is really the appropriate basis for their results, but rather normality. Exchangeability of the $\theta_i$ is indeed a natural way to characterize subjective opinion in certain contexts, but there is no more reason (and in fact, I believe, much less) to expect robustness of normal theory prior distributions of the type they assume for the $\theta_i$ than to expect robustness of normal theory distributions for errors. Thus I expect their analysis is appropriate, even as an approximation, only for a small subclass of the exchangeable prior distributions for the $\theta_i$. This is not to argue against normal theory, which is often suggestive, as well as being important for more traditional reasons (i.e. central limit theorem mystique), but rather to emphasize that Bayesian inference need not be bound up with normality, either for likelihood or prior distributions, any more than classical theory as, for example, is shown in my paper (Hill, 1969).

Dr R. THOMPSON (Unit of Statistics, Edinburgh): The authors may care to comment on the relationship between the exchangeability model and the random effects Model II. Consider, for example, the two-factor experimental design discussed in Sections 3.1 and 5.1, when the variances $\sigma^2$, $\sigma_\alpha^2$ and $\sigma_\beta^2$ are known. Then equation (20) gives the same estimates of $\theta$ as the random effects model with the $\alpha$'s and $\beta$'s normally distributed with variances $\sigma_\alpha^2$ and $\sigma_\beta^2$. But estimation of the variances $\sigma^2$, $\sigma_\alpha^2$ and $\sigma_\beta^2$, by maximum likelihood, assuming a random model, does not in general give the same estimates as equations (32) for the exchangeable model. It seems that, when $\sigma_\alpha^2$ and $\sigma_\beta^2$ are large compared with $\sigma^2$, the estimates of the $\alpha$'s and $\beta$'s are pulled more to the general mean on a random effect model than on an exchangeable model.

Lindley and Smith assert in Section 5.1 that when there is vague prior knowledge of $\sigma^2$, $\sigma_\alpha^2$ and $\sigma_\beta^2$ the modal estimates of the treatment and block effects are zero (i.e. when $\nu = \nu_\alpha = \nu_\beta = 0$). Consideration of the following example suggests that this is not so.

Let

$$b = t = 4, \quad n = 16,$$

$$\sum_i (y_{i.}-y_{..})^2 = \sum_j (y_{.j}-y_{..})^2 = 40 \cdot 5,$$

$$\sum_i \sum_j (y_{ij}-y_{i.}-y_{.j}+y_{..})^2 = 60.$$

Then $s^2 = 4$, $s_\alpha^2 = s_\beta^2 = 8$ satisfy equations (32) (with the minor modification that if there is vague prior knowledge on $\sigma^2$, $\sigma_\alpha^2$ and $\sigma_\beta^2$ then presumably the 2 in the denominator of equations (32) should be deleted).

It would seem that a stronger form of the matrix lemma, (10), is needed for the within regression example (Section 3.3). For then the so-called inverse of $(\mathbf{C}_2+\mathbf{A}_2\,\mathbf{C}_3\,\mathbf{A}_2^T)$, found by application of (10), reduces to $(\mathbf{I}-p^{-1}\,\mathbf{J}_p)$ when $\mathbf{C}_3^{-1} = \mathbf{0}$, and this is a singular matrix.

Professor B. DE FINETTI (University of Rome): I think that the main point to stress about this interesting and important paper is its significance for the philosophical questions underlying the acceptance of the Bayesian standpoint as the true foundation for inductive reasoning, and in particular for statistical inference. So far as I can remember, the present paper is the first to emphasize the role of the Bayesian standpoint as a logical framework for the analysis of intricate statistical situations. So often in such situations the mathematical technicalities play an overwhelming role (usually suppressing any attempt at intellectual reflection), so that one is reduced to an empirical choice amongst various *ad hoc* procedures culled from standard texts.

The present paper not only rebuilds the usual procedures with the innovations of the Bayesian approach, but explains why such innovations are logically and practically necessary. These explanations are given both a Bayesian and "objectivistic" interpretation: for example, the digression that tries "to persuade an orthodox statistician that (1) is a sensible estimate *for him* to consider, and indeed is better than the least-squares estimate". Personally, I am particularly pleased to see the notion of exchangeability introduced into statistical techniques, for example, in between—and within—exchangeability. In this way the concept is not only applied but, through such different forms, its meaning and role should become clearer and more familiar to statisticians.

I would like to express my warmest congratulations to my friend Lindley and his valiant collaborator, Smith.

Professor OSCAR KEMPTHORNE (Statistical Laboratory, Iowa State University): The authors should be congratulated on their presentation. It will be informative to many. I have no detailed questions or remarks about the formal development, I wish mainly to comment on philosophical issues that underly the whole matter under discussion.

It seems *still* to need to be said that there is not, nor ever has been, a controversy about one type of use of Bayes's theorem. This is merely a statement of conditional probability and if the probabilities that enter into the computation are frequency probabilities, then the resultant conditional probability is a frequency probability. Any such conditional probability is verifiable by repetitions of the whole of the sampling process in just the same way that an ordinary probability is verifiable. While I have considerable difficulty understanding the idea of verifiability and literature search has not aided me, I assert that underlying all scientific knowledge is some concept of verifiability that is fairly generally accepted even if not well formalized, and that verification in experimental sciences is a matter of agreement of results among repetitions of an experiment, the repetitions being defined by the protocol of the experiment and done by different people. So, if, for instance, $\mathbf{y} = \mathbf{A}_1\,\boldsymbol{\theta}+\mathbf{e}$ conditionally on $\boldsymbol{\theta}$, $\boldsymbol{\theta} = \mathbf{A}_2\,\boldsymbol{\phi}+\mathbf{f}$ conditionally on $\boldsymbol{\phi}$ and so on, in which the parameters of the ultimate linear model are known, there is no problem in applying ideas of conditional probability to obtain the conditional distribution of $\boldsymbol{\theta}$ given y. It is useful that the authors have written out exactly how the mathematical computations proceed and the solution with particular priors.

It is worth noting also that the essential aspects of the Bayesian argument have been used widely in psychological testing theory for perhaps sixty years and in the adjustment of animal breeding records at a more complicated level for perhaps forty years (see, for example, Lush, 1937). I have drawn attention to this before (Kempthorne, discussion of Lindley, 1971a) but repetition seems necessary. I have some knowledge of the genetic case, and shall write what is done for a simple case. The situation is that we have records on cows, say, which we denote by $y_{ij}$, $i$ indexing the cow and $j$ the record within cow. The model used is

$$y_{ij} = \mu + c_i + e_{ij}, \quad j = 1, 2, \ldots, n_i,$$

with $\mu$ known—the breed average, say. The terms $e_{ij}$ arise from measurement error and are assumed to be uncorrelated with mean zero and variance $\sigma_e^2$. The terms $c_i$ arise from the Mendelian process and are assumed to be uncorrelated with mean zero and variance $\sigma_c^2$. Then it is completely standard knowledge (see, for example, Kempthorne, 1957) in these areas that the best mean-square error predictor of $c_i$ which is of the form $K_i(y_{i.} - \mu)$ is obtained with

$$K_i = \frac{\sigma_c^2}{\{\sigma_c^2 + (\sigma_e^2/n_i)\}} = \frac{n_i \rho}{1 + (n_i - 1)\rho},$$

where $\rho = \sigma_c^2/(\sigma_c^2 + \sigma_e^2)$. The "best" estimate then of $\mu + c_i$ is given by

$$\frac{\{(\sigma_e^2/n_i)\,\mu\} + \sigma_c^2\, y_{i.}}{\{\sigma_c^2 + (\sigma_e^2/n_i)\}} = \frac{(\mu/\sigma_c^2) + \{y_{i.}/(\sigma_e^2/n_i)\}}{1/\sigma_c^2 + 1/(\sigma_e^2/n_i)}.$$

That this is a weighted mean of two "estimates" $\mu$ and $y_{i.}$ weighting inversely as variance is clear and was mentioned by Alan Robertson in a short note (Robertson, 1955). The next stage of estimating $\mu$, $\sigma_c^2$, $\sigma_e^2$ by some way or other has been used *very widely*. There is a short description of this written by C. R. Henderson (an animal breeder) (Henderson *et al.*, 1959). It is worth noting that Henderson states and proves (using his notation)

$$(\mathbf{R} + \mathbf{ZDZ}')^{-1} = \mathbf{R}^{-1} - \mathbf{R}^{-1}\mathbf{Z}(\mathbf{Z}'\mathbf{R}^{-1}\mathbf{Z} + \mathbf{D}^{-1})^{-1}\mathbf{Z}'\mathbf{R}^{-1}.$$

This result grew out of a Bayesian process which I call legitimate because the model is based on prior knowledge and *not on lack of knowledge.* It is worth noting that Henderson advocates from the viewpoint of computation the maximization with regard to the $\theta_i$ of a sort of likelihood equal to $p(\theta_i)\, p(y \mid \theta_i)$, which gives the mode of the posterior distribution.

Let me turn to the philosophical issues.

First, the authors regard the result of Stein (1956) as compelling. I confess that I have failed to see the force of this result. The problem of treating a multivariate parameter estimation from the naive decision viewpoint is that even if one envisages a vector loss function one will eventually have to reduce the problem to one dimension by some norming in some general space. The loss function is usually a real-valued function and the result that one obtains depends critically on it. In the simple case of a random sample with a real unbounded parameter, the loss function $(\hat{\mu} - \mu)^2$ leads to the sample mean and the loss function $|\hat{\mu} - \mu|$ to the median. So the sample mean is inadmissible—and hence unsatisfactory—and so is the sample median. I may seem to be making a mere play on words here, because the authors know these facts as well as, and probably better than, I, but they say that because the multivariate mean $\boldsymbol{\mu}$ is inadmissible with regard to $(\hat{\boldsymbol{\mu}} - \boldsymbol{\mu})'(\hat{\boldsymbol{\mu}} - \boldsymbol{\mu})$ it is unsatisfactory. Is it not the fact that with a reasonable use of language, any estimate is "typically unsatisfactory" because it will at best be admissible only for a particular loss function or a small class of such functions? The Stein result for the case $\mathbf{y} \sim N(\boldsymbol{\mu}, \mathbf{I})$, where $\mathbf{y}$ and $\boldsymbol{\mu}$ are $p$-vectors, uses the loss function $(\hat{\boldsymbol{\mu}} - \boldsymbol{\mu})'(\hat{\boldsymbol{\mu}} - \boldsymbol{\mu})$, and implies that if this is your loss function you should not use $\hat{\mu} = y$. This, it appears, was a great surprise to the "sampling-theory school" to use the authors' phrase. It was so, I am told, because it had seemed previously that best invariant estimates were thought to be admissible. The informed reaction to the question of what is the "best" $\hat{\mu}$ in terms of a particular loss function surely has to be that one does not know because the mathematics is so difficult. But, in any case, even if $\hat{\mu} = y$ were the best, I would not regard the result as of great interest or binding practical relevance. I am of the opinion that the mathematical workers of the "sampling-theory school" are not in touch with real problems of learning from investigation. The investigator who runs an experiment on $k$ treatments, say $A, B, C, ..., K$ does not know what use he will make of the results. He may note, for instance, that another investigator has done an experiment with treatments $B$, $C$ and $E$ and others unlike those in $A, B, C, ..., K$. He will wish to have an estimate of $\mu_B, \mu_C, \mu_E$ to apply in his comparison. The loss function $(\hat{\mu}-\mu)'(\hat{\mu}-\mu)$ is simply irrelevant for that purpose. Furthermore, his use of $\hat{\mu}_B, \hat{\mu}_C, \hat{\mu}_E$ as given by a pulled-in estimator of $\mu' = (\mu_A, \mu_B, \mu_C, ..., \mu_K)$ is not necessarily admissible for $(\mu_B, \mu_C, \mu_E)$. The present authors indicate rather definitely that "inadmissible" implies "unsatisfactory". I seem to get the lesson from the past two decades that admissibility ideas have been unfruitful, except for the reduction of data to a sufficient statistic. There is a great communication gap, because a huge number of experiments are analysed by the techniques of the "methods" books, and neither the authors of these books nor the users seem to be bothered by the theoretical fact that they are using "inadmissible" procedures. Let me hasten to add that while I do not understand the deeply mathematical work that has been done, I know enough mathematics to form the opinion that it is very difficult and much of it is beautiful *qua* mathematics. I do not wish to denigrate the work or the workers. I merely question its relevance to problems of interpretation of a given set of data. I also would like to register the plaint that "inadmissible with respect to a particular loss function" becomes through journal space exigencies merely "inadmissible", and then it is quite an easy step to replace this word by "unsatisfactory". And this is just what the authors have done.

In connection with the Stein result, it is surely well realized by workers in the admissibility area that the result, if it is to be taken as having real logical force, involves the experimenter and the statistician in an embarrassing dilemma. Let us suppose that the experimenter has done an experiment comparing two treatments $A$ and $B$ and wishes to record numbers $\hat{\mu}_A, \hat{\mu}_B$. A little later he makes a test of treatment $C$ with independent errors. Then he is told by the admissibility workers that his answers for $\mu_A$ and $\mu_B$ are no longer good. Surely it is offensive to the experimenter to be told that his opinion (in whatever way he forms this) about $\mu_A$ and $\mu_B$ will be altered if it turns out later that he obtains data also on $\mu_C$. I believe this sort of idea will not "sell". I believe the idea will "butcher" the accumulation and condensation of investigational information. I doubt strongly that the authors have talked to workers in *experimental* science and tried to sell them the idea. I surmise, furthermore, that if our informed public were aware of the "pulling-in" of test records of their off-spring that they would not be happy. I am glad to see evidence that the public is concerned about the test procedures that are being used. It is high time. Let us try to persuade an "orthodox" Bayesian (e.g. Professor Lindley) that a scientist's opinions about measured physical properties of substance $X$ are *not* and should not be based on what he has observed for substance $Y$ unless errors of measurements are dependent. This is not to say that the scientist will not form opinions such as "substance $X$ is like substance $Y$ with regard to its phase diagram".

Let me also try to persuade the authors that the reporting only of Bayesian estimates, each based on the prior of the person who obtained them, will butcher the processes of science. Here I deliberately again use the word "butcher". This is not to say that workers should not report their own *private subjective* Bayesian estimates. But they should report sufficient statistics so that other workers can construct their own Bayesian estimates. Indeed, let me assert my opinion, with which I believe Professor Lindley must agree, that reported Bayesian estimates, which must be accompanied by reporting of the prior on which they are based, are useless to any other worker unless they have *not* been marginalized in any way, or unless the other worker has the same prior. Another worker will simply be unable to recover the sufficient statistic and construct his own Bayesian estimate. I believe this view is compelled by the Bayesians themselves, unless they argue (as perhaps some do and as Jeffreys does) that there is one and only one prior that is appropriate (and that one happens in the case of Jeffreys to be improper, so the logical status of any probability conclusions cannot but be obscure). Professor Lindley and the Bayesians should be happy that the conventional "inadmissible" and "unsatisfactory" statistics are reported, because they can then apply whatever prior they wish.

I can well surmise the attitude of scientists whose only data input from other workers consists of other workers' Bayesian estimates. Surely the answer will be, "I do not care what Joe thinks about the parameter: I want to know the observational facts or a good condensation of them". *I believe our present authors are not in touch with the processes of science.* They are not aware of the need for the development of interpersonal validity to opinions. For them life is very lonely. They are concerned about how they alone should make a terminal decision. They are not concerned with the acceptability of their experimental conclusions to other workers. In contrast, I would report the standard statistics with estimates of error and so on. I might also compute my own posterior distribution, but I would be hesitant about burdening the literature with it. If a compelling case could be made for a Jeffreys prior I would use it and report corresponding Bayesian estimates, but the case presented has not been found compelling. In just the same way, if a compelling case could be made for a particular loss function, I would base my thinking on it. But I would not, I hope, merely try to rationalize a loss function that fits with the probability structure for the data that I am using, which the authors do seem to do in connection with equation (3). What loss functions are candidates in the case of the multivariate normal mean? Very little thought suggests $\sum(\hat{\mu}_i-\mu_i)^2$, or $\sum(\hat{\mu}_i-\mu_i)^4$, or $\max_i|\hat{\mu}_i-\mu_i|$ to go to the extreme power. But also, I can readily imagine there being interest in selecting the component of $\mu$ that is highest. There are huge possibilities for very interesting and difficult mathematical work.

On quite a different point, why should one, *as a Bayesian*, be interested in average mean-square error? The concept of repetition in the sample space is dismissed, it appears, by the Bayesians, but like the present authors, they like to discuss averages in a population of repetitions in the sample space. But, at the same time, they castigate others who use the concept of repetitions, namely those of the "sampling theory" school, for so doing. If the purpose of a Bayesian analysis is solely to produce estimators that have good mean-square error properties, there is no basis for argument. If, loosely speaking, standing on one's head aids one in getting an estimator with good repetitive properties, one should stand on one's head. No one has ever questioned this. It is very curious to me that the appeal to a Bayesian process is made on the basis of hard logic from axioms of reasonable behaviour, but often recourse is made for justification to repeated sampling. Or am I alone in thinking there is a grave discrepancy here? On the same point, what is the status of $P(\text{Data}\,|\,\theta)$? Is this a frequency probability? Where did it come from?

I have to comment on the sentence: "It is the greatest strength of the Bayesian argument that it provides a formal system within which any inference or problem can be described". I would like to turn it around and say: "It is the greatest defect of the Bayesian argument that it provides a formal system according to which you can believe what you wish and, furthermore, without any data". I believe the search for the sort of panacea envisaged is a false one, which is based on a *total* misunderstanding of the nature of language and the nature of knowledge. Here again I believe some homework is desirable.

Why, incidentally, should one be interested in the standard error of the posterior distribution? Does Bayesian inference have any predictive value? The posterior distribution, its mean, or median or mode, or whatever, is merely a statistic. Most of the world likes to have some idea of the reliability of a statistic and I believe correctly so. The Bayesians believe quite the opposite and, I believe, incorrectly so.

The use of the exchangeability assumption is warned against. "The estimates (1) are therefore suggested only when this assumption is practically realistic", it is said. But how does one check the assumption? What does it mean to say that an assumption is "practically realistic"? The authors use the phrase. What do they mean? Furthermore, let us turn to the American College Testing Program. Is it "practically realistic" to use an exchangeable prior? Information is available in the records to show that schools differ widely, students of different social and ethnic backgrounds perform differentially on tests, and so on. Information on students is available to show, I think, that exchangeable priors are "practically unrealistic" whatever that means. Are they being used in high school and college testing? It seems that many of our (U.S.A.) societal processes merit outside examination.

I close with brief remarks. First, the paper presented has very interesting material and material that may well be useful to a "legitimate" Bayesian, e.g. an animal breeder. Second, even though I disagree strongly with the underlying background I value the presentation and the discussion. Third, I believe the authors and most Bayesians (but not all) may rather easily be "hoisted with their own petard", to use a delightful Anglicism. Finally, I regard Professor Lindley as a personal friend and have a Bayesian feeling that this is reciprocated. The type of criticism I have voiced is possible, intrinsically, only on some basis of this sort. The same applies to my discussion of the paper by Cornfield that is cited, though I regret to add that the discussion by others and by me in that paper and in the Waterloo paper did not apparently merit discussion. It is only by hard arguing that obscurities in the minds of all of us can be removed and that is why I must push hard and be treated in the same way.

Professor LINDLEY and Dr SMITH replied briefly at the meeting and subsequently in writing as follows:

So many interesting points have been raised in the discussion that we cannot, within reasonable space limitations, give them all the attention they deserve. We therefore hope that contributors will not take a one-sentence response to an idea to imply lack of interest on our part: quite the contrary.

Dr Nelder is not correct when he says that our estimates are the same as those derived by Yates. Using an exchangeable prior for the blocks and vague prior for the treatments, the estimates are the same when all variances are *known*. Even with this last condition, an exchangeable prior for the treatments produces different estimates. In both cases when we turn to the *unknown* variance situation, the techniques described in Section 5 lead to estimates different from those usually employed for balanced incomplete blocks. We do not understand methods based on likelihoods (whether prior or otherwise) since they seem to fail in one of the most basic of all statistical problems, namely sampling from a finite population (Godambe, 1966). The remark of Marquardt that was quoted is not strictly correct since the ridge estimates are not *confined* to a sphere. They are a *compromise* between least-squares estimates and others so confined. The Bayesian method does not invoke such restrictions whether to a sphere or a linear subspace: in this case it allows a compromise, the magnitude of which can be estimated from the data.

Dr Barnett and Professor Hill both have doubts concerning our assumption of normality at the second and subsequent stages. We share them. As a partial justification we would say that a prior distribution can sometimes be reasonably taken to be normal, and it may be better to have an approximate answer than none at all. In answer to Dr Barnett's query regarding the relative importance of exchangeability, *per se*, or normality, *per se*, as the basis for Bayesian estimation, we note the following. The representation of exchangeability by the mixture form is by no means confined to the normal hierarchy: we cite as an example the discussion of heteroscedasticity in Lindley (1971a), where it is applied to gamma distributions. On the other hand, the normal hierarchy can be used to incorporate meaningful, non-exchangeable priors; for example, if the means in a one-way model correspond to different levels of a single factor, we might expect them to lie on a linear or quadratic surface, and this can be represented within the structure (11.1)–(11.3). This will be reported on elsewhere by Dr Smith.

Professor C. A. B. Smith puts forward an interesting argument that suggests the assumption of normality is robust in respect of a shift effect. We note that in the case of a normal likelihood the form of the prior can have a substantial influence on the magnitude of the shift. Let data $x \sim N(\theta, 1)$ and let $\theta$ have a Cauchy prior with density proportional to $(1+\theta^2)^{-1}$. The posterior distribution has turning points at the roots of the cubic $(x-\theta) = 2\theta(1+\theta^2)^{-1}$. If $x$ is large an approximate root is $x$, a better value is $x-(2/x)$. In particular, the shift of the natural estimate, $x$, towards the prior value, zero, tends to zero as $x \to \infty$. This is quite different from the behaviour of our estimates. We have tried using $t$-like priors and hope to report on this elsewhere. Our guess is that the methods of the paper are not robust to outliers. Dr Hinkley's suggestions are valuable here but we do not like the abrupt change of form of the estimates that he describes.

Dr Barnett asks about Bayesian devils. Is not the whole point of the Ramsey–Savage–de Finetti approach, leading to the Bayesian position, that it bars them from the beginning? Axiom one is essentially "there shall be no devils". His query about the two posterior distributions does not revolve around which is right and which is wrong, but rather on their appropriateness to the practical situation. A subjectivist Bayesian assesses each problem on its merits. The formal framework makes it clear what has to be specified and what calculations will lead to the required answer. The choice of whether $\mu$ is known or not has to be dictated by the practical situation.

Mr Leonard's remarks are valuable. The method shows promise of being capable of extension to any member of the exponential family with its associated conjugate family.

Professors Novick and Kempthorne take us to task for omitting important references. Their strictures are justified and we apologize to all the authors they cite for failing adequately to acknowledge their contributions. (It is particularly remiss of us to have omitted the work of Box and his colleagues since it is so close to ours and appears in the "usual statistical journals"). The fact is that practitioners in psychology and genetics have been ahead of statisticians in appreciating the significance of related information. Perhaps the only novelty in the paper is the suggestion of using these estimates in wider fields. Both Novick and Fienberg alert us to the difficulties concerned with the estimation of the variances (and covariances). Our guess is that faulty estimation of these does not matter too much when the primary object is to determine the means (just as the correct weights are not too important in standard least squares), but they are surely correct when the variances themselves are of interest. Dickey's work is important here and again we humbly apologize for omitting to refer to it.

Our reason for not using Model II terminology, a point raised by Novick and Hill, is that Model II typically has a decision space involving the variance components whereas ours incorporates the means. Indeed, by looking at the latter we obtain estimates for the former. We agree with the remark made somewhere by Yates that the distinctions between the models are largely irrelevant.

Professor Cox and Dr Copas discuss an empirical Bayes approach. Cox's estimates are clearly unsatisfactory in some situations; for example, Model II analysis of variance where a maximum-likelihood estimate is rather absurdly zero. However, it is interesting to note that in the general linear model with $\mathbf{C}_3^{-1} = 0$ the posterior mean for $\boldsymbol{\theta}_1$, given effectively by equations (16) and (17), can be written explicitly in the form

$$(\mathbf{A}_1^T \mathbf{C}_1^{-1} \mathbf{A}_1 + \mathbf{C}_2^{-1})^{-1} (\mathbf{A}_1^T \mathbf{C}_1^{-1} \mathbf{A}_1 \hat{\boldsymbol{\theta}}_1 + \mathbf{C}_2^{-1} \mathbf{A}_2 \hat{\boldsymbol{\theta}}_2)$$

where $\hat{\boldsymbol{\theta}}_1$ is the standard least-squares estimator derived from equation (11.1), and $\hat{\boldsymbol{\theta}}_2$ is the least-squares estimator of $\boldsymbol{\theta}_2$ which obtains if (11.1) and (11.2) are combined to give the distribution of $\mathbf{y}$ conditional on $\boldsymbol{\theta}_2$. This has an obvious interpretation in Professor Cox's empirical Bayes formulation.

Our limited experience suggests that scientists do often perform experiments with reasonably homogeneous treatments and would be prepared to take similar bets on each of the treatments being used. Dr Copas says the assumption can, *in principle*, be tested. Our assumptions can *actually* be tested—in terms of gambles for example. The point concerning negative correlations is interesting. We do not know the complete answer to his question but point out that the correlation must exceed $-1/n$ in the symmetric case of $n$ means, and so must be small.

Professor Plackett's contribution is totally delightful. It may be dangerous to express prior knowledge in terms of equivalent numbers of observations, but his is a valuable exercise in doing this and throws considerable light on the approach.

We cannot agree to Professor Sprent's term, coefficient of stupidity. The use of prior knowledge becomes even more important when the data are providing relatively little information. In his example ($\tau = 1$, $\sigma = 3$) it is certainly true that the estimate will be less than the true value with high probability, but on the other hand it is more likely to be within $\pm 1$ of the true value (63 per cent for our estimate, 26 per cent for the least squares).

We are most interested in Dr Hinkley's final remarks about Stein's results. One of us (D. V. L.) has obtained similar estimates for a multivariate normal dispersion matrix using a Wishart prior, but there are several snags in its use. Our only reply to Dr Harding can be that we would not be willing to exchange butterflies for ball-bearings, nor is the loss function he quotes the only one for which the usual estimate is inadmissible—the paradox holds for almost all losses.

Why did not Professor Fienberg's memorandum get published? Since writing this paper D. V. L. has obtained estimates closely similar to those he mentions, by extending the arguments of Sections 3.1 and 5.1 to include an interaction component. The method is particularly attractive because it provides estimates of the cell means, $\mu+\alpha_i+\beta_j+\gamma_{ij}$, which depend on estimates of the relative sizes of the main effects and interactions, thereby quite avoiding the usual significance tests. The estimates of variances that he proposes do not seem so sensible because they use the *data* $y_{ij.}$ etc. rather than the *estimates* of the cell means. Stein's famous estimate $x_i[1+(n-2)/\sum x_i^2)$ can probably be improved by using $Z_i = x_i[1+(n-2)/\sum Z_i^2]$ in accordance with the iteration procedure suggested in the paper. In answer to his query concerning incomplete observation vectors, some work has been done on this and although, as always, the Bayesian procedure is clear, its implementation presents technical problems that we have not been able to surmount. As a partial justification of the modal approximation we cite the work of Tiao and Zellner (1964) which shows that posterior distributions which are proportional to products of $t$-kernels are well approximated by normal distributions, when reasonable amounts of data are available. We note that Dickey's reduction in dimensionality of the integral becomes rather intractable when more than two such products are involved.

We do not understand Dr Thompson's numerical solution to equation (32) since he does not appear to have calculated the values of $\alpha_i^*$ and $\beta_j^*$. Nor do we understand his point concerning singular matrices for the within-regression example. The estimates are found directly from equations (16) and (17).

Mr Dawid makes an important point in relating our work to that of Ericson. In a personal communication he has shown how, using a result of Ericson, it is possible in some cases to write down the posterior mean quite simply, thereby avoiding the tedious matrix manipulations of this paper. To answer his second point about the status of $\boldsymbol{\theta}_1$, consider the case where $n$ varieties of wheat are being tested in a trial. The components of $\boldsymbol{\theta}_1$ are essentially measures of the qualities of the varieties. In a sense they are never needed since all we want to know is how a variety will perform in a specified situation. Nevertheless it is practically convenient to have such a concept as an average yield. We do not know the answer to his problem of two-way exchangeability, nor to that of error-rates.

It was good to have Professor de Finetti's kind remarks since all Bayesian work owes so much to his important and original views on probability.

The Royal Statistical Society is well known for the discussions that take place at its meetings and for the invective that often accompanies them. We are particularly grateful for the kindness that contributors have shown towards our ideas. Most of them have been prepared to look at the estimates and judge them as estimates, whether from a likelihood, least squares or other approach, and not get involved in continual arguments on philosophy. This is a healthy approach. Let us see what the results amount to and temper our theory with a little pragmatism.

The two exceptions to this are Dr Chatfield and Professor Kempthorne. If a reduction by 75 per cent is not of practical importance then we would like to know what does measure up to Dr Chatfield's peculiar criteria. Professor Kempthorne twice complains that his previous discussions of our work and Cornfield's have been ignored. The reason is very simple. His fellow Iowan, Aydelotte (1966), has said, "an imprecise or slipshod formulation is impregnable: a statement that has no exact meaning cannot be disproved". Professor Kempthorne is impregnable because he will not say precisely what he would do. He knocks everyone about, Bayesians, sampling theorists, the lot, because we make our arguments precise. He does none of this and then has the effrontery to tell us we are not scientific. Precision is an essential ingredient of science.

On the one hand, he tells us that we are doing what some scientists, namely geneticists, have been doing for a long time; and later he says we are out of touch with scientific thinking. He cannot have it both ways. He casts doubts on Stein's results on the grounds that the range of means that might be involved when considering one of them is uncertain. We quote a little example that came our way recently from a physicist. He had four sample means, 79, 82, 89 and 126. He was puzzled by the last and thought it overestimated. He then obtained other means: 83, 91, 104, 111 and 112. He then revised his opinion about the 126: a high value certainly, but not excessive. Is not this just what a Stein estimate does in precise terms.

It would impose unreasonably upon our Society's space to answer every one of his points in detail. We conclude by remarking that his penultimate paragraph reveals that he has not, despite de Finetti's expectation, understood the idea of exchangeability (it does not mean the units are the same), and by expressing the hope that the American College Testing Program can deal with his accusations.

## References in the Discussion


Aydelotte, W. O. (1966). Quantification in history. *Amer. Hist. Rev.*, **71**, 814–833.

Box, G. E. P. and Tiao, G. C. (1968). Bayesian estimation of means for the random effects model. *J. Amer. Statist. Ass.*, **63**, 174–181.

Copas, J. B. (1969). Compound decisions and empirical Bayes. (With Discussion). *J. R. Statist Soc.* B, **31**, 397–425.

Dempster, A. P. (1971). Model searching and estimations in the logic of inference. In *Foundations of Statistical Inference* (V. P. Godambe and D. A. Sprott, eds) pp. 56–77. Toronto: Holt, Rinehart and Winston.

Dickey, J. M. (1968). Three multidimensional-integral identities with Bayesian applications. *Ann. Math. Statist.*, **39**, 1615–1627.

Edwards, A. W. F. (1969). Statistical methods in scientific inference. *Nature, Lond.* **222**, 1233–1237.

Efron, B. and Morris, C. (1972). Empirical Bayes on vector observations. *Biometrika*, **59** (in press).

Ericson, W. A. (1969). Subjective Bayesian models in sampling finite populations (with Discussion). *J. R. Statist. Soc.*, B, **31**, 195–233.

Fienberg, S. F. (1967). Cell estimates for one-way and two-way analysis of variance tables. Memorandum NS-69, Department of Statistics, Harvard University.

—— (1971). Discussion of a paper by H. O. Hartley and R. R. Hocking on incomplete data analysis. *Biometrics*, **27**, 813–817.

Godambe, V. P. (1966). A new approach to sampling from finite populations. I. Sufficiency and linear estimation. *J. R. Statist. Soc.*, B, **28**, 310–319.

HENDERSON, C. R., KEMPTHORNE, O., SEARLE, S. R. and VON KROSIGK, C. M. (1959). The estimation of environmental and genetic trends from records subject to culling. *Biometrics*, **15**, 192–218.

HILL, B. M. (1969). Foundations for the theory of least squares. *J. R. Statist. Soc.* B, **31**, 89–97.

JACKSON, P. H., NOVICK, M. R. and THAYER, DOROTHY T. (1971). Estimating regressions in *m*-groups. *Brit. J. Math. Statist. Psychol.*, **24**, 129–153.

KELLEY, T. L. (1927). *The interpretation of Educational Measurements.* New York: World Books.

KEMPTHORNE, O. (1957). *An Introduction to Genetic Statistics.* New York: Wiley. Reprinted in 1968 by the Iowa State University Press.

LUSH, J. L. (1937). *Animal Breeding Plans.* Ames, Iowa: Iowa State University Press.

MARQUARDT, D. W. (1970). Generalized inverses, ridge regression, biased linear estimation and nonlinear estimation. *Technometrics*, **12**, 591–612.

MOSTELLER, F. and WALLACE, D. L. (1964). *Inference and Disputed Authorship: the Federalist Papers*, Chapter 4. Reading, Mass.: Addison-Wesley.

NOVICK, M. R. and JACKSON, P. H. (1970). Bayesian guidance technology. *Rev. Educ. Res.*, **40**, 459–494.

NOVICK, M. R., JACKSON, P. H. and THAYER, DOROTHY T. (1971). Bayesian inference and the classical test theory model: reliability and true scores. *Psychometrika*, **36**, 207–328.

NOVICK, M. R., JACKSON, P. H., THAYER, DOROTHY T. and COLE, NANCY S. (1972). Estimating multiple regressions in *m*-groups; a cross-validation study. *Brit. J. Math. Statist. Psychol.*, **25**, (in press).

PLACKETT, R. L. (1950). Some theorems in least squares. *Biometrika*, **37**, 149–157.

ROBERTSON, A. (1955). Prediction equations in quantitative genetics. *Biometrics*, **11**, 95–98.

SMITH, C. A. B. (1970). Discussion of a paper by A. W. F. Edwards. *J. R. Statist. Soc.*, B, **32**, 165–166.

STEIN, C. (1966). An approach to the recovery of inter-block information in balanced incomplete block designs. In *Festschrift for J. Neyman*: *Research Papers in Statistics* (F. N. David, ed.), pp. 351–366. New York: Wiley.

TIAO, G. C. and ZELLNER, A. (1964). Bayes's theorem and the use of prior knowledge in regression analysis. *Biometrika*, **51**, 219–230.

# [5]

# Bayes Methods for Combining the Results of Cancer Studies in Humans and Other Species

WILLIAM H. DuMOUCHEL and JEFFREY E. HARRIS*

We propose a class of Bayesian statistical methods for interspecies extrapolation of dose-response functions. The methods distinguish formally between the conventional sampling error within each dose-response experiment and a novel error of uncertain relevance between experiments. Through a system of hierarchical prior distributions similar to that of Lindley and Smith (1972), the dose-response data from many substances and species are used to estimate the interexperimental error. The data, the estimated error of interspecies extrapolation, and prior biological information on the relations between species or between substances each contribute to the posterior densities of human dose-response. We apply our methods to an illustrative problem in the estimation of human lung cancer risk from various environmental emissions.

KEY WORDS: Interspecies extrapolation; Carcinogenesis; Mutagenesis; Lung cancer; Empirical Bayes; Exchangeability.

## 1. INTRODUCTION

The assessment of cancer risks from environmental agents has been plagued by an abundance of precise data in nonhuman species, but little accurate information on direct effects in humans. While it is desirable to use all the available information to reduce the uncertainty about human health risks, the relevance of the nonhuman data to man is itself uncertain.

In this article we propose a class of Bayesian statistical methods for interspecies extrapolation of dose-response functions. The approach provides a general theory within which previous studies of interspecies extrapolation can be incorporated.

Our statistical methods are founded on the standard Bayesian paradigm for the analysis of covariance components models. We distinguish formally between the conventional measurement error within each dose-response experiment and a novel error of uncertain relevance between experiments. Through a system of hierarchical prior distributions similar to that of Lindley and Smith (1972), we use dose-response data from many substances and species to estimate the interexperimental error. From the data, the estimated error of interspecies extrapolation, and prior biological information on the relations between species or between substances, we calculate posterior densities of human dose-response.

We assume the availability of a set of independently performed studies on various substances in humans and other species. Each "experiment," we further assume, has already been analyzed according to a specific dose-response model. We do not address the problem of selecting dose-response models to analyze individual experiments. Nor do we belabor the well-known fact that the choice of dose-response model critically influences the predicted risk of cancer at very low exposures. However, as we discuss in Section 7, there is a formal connection between errors of model misspecification within individual experiments and errors of extrapolation between experiments.

The main message of this article is that convincing estimates of human cancer risk from nonhuman studies require evidence from both multiple substances and multiple species, including at least some informative data in humans. Human risk extrapolations based solely on the results of one compound in one nonhuman species will be highly sensitive to prior beliefs about the relevance of the nonhuman species to man.

Our strategy of presentation is to analyze a real problem in the assessment of human lung cancer risks from various environmental emissions. While we regard our specific quantitative results as a meaningful contribution, our main purpose is to use the data to point out generally the statistical and biological questions surrounding interspecies extrapolation.

The next section presents the data. In Section 3 we formally develop our approach; in Section 4 we apply our model to the data, drawing distinctions between a fully Bayesian analysis and various computational shortcuts. In Section 5 we show how prior biological information can be incorporated in our Bayesian framework, and in

* William H. DuMouchel is Associate Professor, Department of Mathematics, Massachusetts Institute of Technology, Cambridge, MA 02139. Jeffrey E. Harris is Associate Professor, Department of Economics, Massachusetts Institute of Technology, and Clinical Associate, Massachusetts General Hospital, Boston. DuMouchel's research was supported by National Science Foundation Grant MCS-80-05483. Harris's research was supported by Public Health Service Research Grant DA-02620 and by Research Career Development Award DA-00072. A preliminary version of this paper was presented at the 21st meeting of the NBER-NSF Seminar on Bayesian Inference in Econometrics, Chicago, October 31, 1980, and appeared as Technical Report No. 24, Department of Mathematics, Massachusetts Institute of Technology, February 1981. The authors gratefully acknowledge the comments and suggestions of H. Chernoff, D. Freedman, R. Olshen, J. Koziol, D. Krantz, R. Solow, J. Yellin, and six referees.

© Journal of the American Statistical Association
June 1983, Volume 78, Number 382
Invited Papers Section

Section 6 we explore the problem of deciding which experiments are most relevant to human cancer risks.

## 2. AN ILLUSTRATIVE PROBLEM

Table 1 displays the results of five types of studies on nine related environmental agents, arranged in a two-way table. Experimental data are available in 36 of the 45 cells in the table. The first row of experiments represents the results of epidemiological studies of human lung cancer incidence in relation to occupational exposure to roofing tar emissions (Hammond et al. 1976); occupational exposure to coke oven emissions (Lloyd 1971; Mazumdar et al. 1975; U.S. Environmental Protection Agency 1979); and cigarette smoking (Kahn 1966). The remaining rows represent the results of various laboratory experiments on the dichloromethane extracts of roofing tar emissions, coke oven emissions, particulate emissions from four different diesel engines and one gasoline-powered engine; on the polyaromatic hydrocarbon benzo(a)pyrene; and on whole cigarette smoke condensate. The laboratory experiments include skin tumor initiation in Sencar mice (Nesnow et al. 1981); enhancement of viral oncogenic transformation in Syrian hamster embryo (SHE) cells (Casto et al. 1980); and mutagenesis experiments in L5178Y mouse lymphoma cells performed with and without metabolic activator (Mitchell et al. 1980). The laboratory experiments in each row were conducted under identical conditions, as part of the U.S. Environmental Protection Agency's diesel emission research program (Huisingh et al. 1980).

The choice of experiments in Table 1 reflected the limited availability of experimental data comparing diesel emissions to other related environmental emissions in the same bioassay. An assay was included in the table only if (a) it was considered to have been performed reliably and reproducibly; (b) sufficient data were provided to estimate a dose-response relation; and (c) the experiment was regarded as a valuable, quantitative measure of carcinogenicity or mutagenicity in mammalian systems (Harris 1981,1983).

The entries in each nonempty cell represent the results of fitting a linear dose-response model to the observed experimental data. For each experiment, three numbers are given: the estimated slope of the dose-response relation; its coefficient of variation (the ratio of the standard error of the estimated slope to the estimate itself); and the natural logarithm of the estimated slope. (The role of these log slopes will become apparent shortly.) In our analysis of these data, we shall interpret the dose-response slope to be a measure of the potency of a given agent in a given species. The slopes and their standard errors were estimated by maximum likelihood methods, as described in Harris (1981,1983). Appendix A briefly summarizes the dose-response models and their estimation.

Table 1 provides relatively precise data on the human effects of cigarette smoke and coke oven emissions. But the effect of roofing tar is not precisely estimated, and for the remaining agents we have no human data. The main question is how to use all of the evidence in Table 1 to obtain more precise estimates of the human lung cancer risks.

To answer this question we need to posit an underlying common mechanism or hypothesis that generates all the data in the table. Specifically, we shall examine the hy-

*Table 1. Two-Way Table of Epidemiological Studies and Laboratory Experiments on Nine Environmental Mixtures[a]*

| | Roofing Tar Emissions | Coke Oven Emissions | Diesel Engine Emissions | | | | Gasoline Engine Emissions | Benzo(a) pyrene | Cigarette Smoke[f] |
|---|---|---|---|---|---|---|---|---|---|
| | | | A | B | C | D | | | |
| Lung Cancer | 1.64 | 4.40 | | | | | | | .03 |
| (Humans)[b] | 1.41 | .34 | | | | | | | .15 |
| | .50 | 1.48 | | | | | | | −3.46 |
| Skin Tumor | .54 | 2.10 | .53 | .16 | | .01 | .03 | 85.28 | .00 |
| Initiation (Sencar | .04 | .04 | .04 | .22 | | .82 | .26 | .03 | 1.30 |
| Mice)[c] | −.63 | .74 | −.64 | −1.86 | | −4.51 | −3.61 | 4.45 | −5.88 |
| Enhancement of | 2.07 | .86 | .65 | .07 | .13 | .04 | .20 | 540.00 | .58 |
| Viral Transform. | .18 | .10 | .15 | .33 | .18 | .59 | .12 | .04 | .08 |
| (SHE cells)[d] | .73 | −.15 | −.44 | −2.70 | −2.06 | −3.24 | −1.59 | 6.29 | −.54 |
| Mutagenesis −MA | .31 | .73 | 1.66 | .27 | 2.55 | .16 | .35 | | .59 |
| (Mouse 5178Y | .39 | .21 | .31 | .43 | .16 | .24 | .11 | | .23 |
| Lymphoma Cells)[e] | −1.17 | −.32 | .51 | −1.31 | .93 | −1.86 | −1.06 | | −.53 |
| Mutagenesis +MA | 9.56 | 9.96 | 1.87 | .76 | 1.01 | .05 | .99 | | .45 |
| (Mouse 5178Y | .16 | .07 | .26 | .14 | .20 | .43 | .10 | | .13 |
| Lymphoma Cells)[e] | 2.26 | 2.30 | .63 | −.27 | .01 | −3.02 | −.01 | | −.79 |

[a] Each nonempty cell contains the observed dose-response slope, its coefficient of variation, and the natural logarithm of the dose-response slope.
[b] Units of measurement (for all agents except cigarette smoke) are the increment in the relative risk per $10^4$ $\mu g/m^3$ of organic extractables per year.
[c] Units of measurement are papillomas per mouse per mg of organic extractables at 27 weeks.
[d] Units of measurement are transformations per $2 \times 10^6$ surviving cells per $\mu g/ml$ of organic extractables; "SHE" = Syrian hamster embryo.
[e] Units of measurement are average mutant colonies per $10^6$ survivors per $\mu g/ml$ of organic extractables; "−MA" = without metabolic activator; "+MA" = with metabolic activator.
[f] Units of measurement for cigarette smoke refer to whole smoke condensate rather than organic extractables.

pothesis that the ratio of dose-response slopes (that is, the relative potency) of any two environmental emissions is preserved across species. This hypothesis arises naturally from the data of Table 1 because the dose-response slopes in each species are measured in the same units, and therefore their ratios are comparable unitless quantities. (Although the results in the last column are measured per unit of cigarette smoke condensate rather than per unit of extractable organics, the ratios of the slopes for cigarette smoke to the other slopes are similarly in identical units.) We therefore avoid the need to specify potentially complex or implausible conversions of dosage units between species.

Even a cursory examination of Table 1 suggests that the constant relative potency hypothesis does not fit the data all that well. Thus, coke oven emissions are apparently carcinogenic in humans, while the coke oven extract initiates skin tumors in mice. Since the diesel A extract also initiates skin tumors, with about one-quarter the potency of the coke oven extract, then diesel A emissions should likewise be human carcinogens, with about one-quarter the potency of coke oven emissions. But cigarette smoke is carcinogenic in man, while its condensate is at best a weak skin tumor initiator in mice. Moreover, in the mutagenesis experiments on mouse lymphoma cells, the addition of metabolic activator markedly increases the effect of the coke oven extract, whereas the response of the diesel extract is less consistent.

These findings are not surprising. Carcinogenic processes vary considerably across environmental agents and species. There are differences in chemical composition, bioavailability, target site of action, metabolism, and genetic repair mechanisms. Any distinctive action of a particular agent in a particular species would constitute a deviation from the hypothesis of constant relative potency. Such an hypothesis at best approximates the relation between the experiments. The question is: can we use the data of Table 1 to gauge the accuracy of the approximation?

A partial answer is given in Figure 1, which depicts on a logarithmic scale the means and standard errors of the dose-response slopes from the upper left 3 × 3 submatrix of Table 1. The error bars correspond to the standard errors of the log slopes, which have been approximated by the coefficients of variation in the table. Since the three vertical axes are drawn to the same scale, the hypothesis of constant relative potency requires that the lines connecting successive pairs of estimates be parallel across species. We have therefore inserted the weighted least squares parallel lines, the weights having been chosen according to a procedure to be described in Section 3. The deviations of the fitted lines from the observed data points in Figure 1 suggest that the hypothesis of constant relative potency, as a model of the relation between experiments, could be no more accurate than a factor of 2.

Underlying this conclusion is a critical assumption. Since the nonhuman data in Figure 1 are more precise

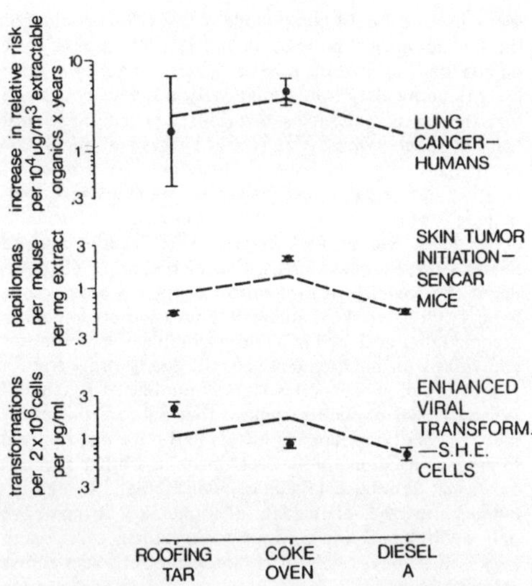


*Figure 1. ACCURACY OF THE CONSTANT RELATIVE POTENCY HYPOTHESIS FOR THE UPPER LEFT 3 × 3 SUBMATRIX OF TABLE 1. The means and standard error bars for the dose-response slopes are plotted on a logarithmic scale. Superimposed are the weighted least squares parallel lines, as described in the text.*

than the epidemiological results, our test of the relevance of the nonhuman experiments to the humans relies heavily on the relevance of one nonhuman experiment to another. Put differently, any observed species-emission interaction gives us some information on the probability of a distinctive effect in humans.

In extrapolating nonhuman results to man, some scientists might place greater reliance on in vivo studies in mammals than on in vitro studies in mammalian cells or unicellular organisms. But the argument that whole animal experiments, such as skin tumor initiation in mice, are better models of carcinogenicity in man is not based on systematic empirical study (Purchase 1980). In fact, it is hardly obvious that the skin painting studies in Figure 1 are any more or less relevant to humans than the cell transformation experiments, or that the human epidemiologic studies are any more or less relevant to mouse skin than the hamster embryo cell experiments. Nor is it apparent a priori that the coke oven results are any more or less relevant to diesel emissions than the roofing tar results. To go beyond these expressions of ignorance requires sharp prior information on chemical composition, metabolism, and other biological characteristics.

Assumptions of minimal prior information have been implicit in previous studies of interspecies comparison of carcinogenicity. And most have implicitly used the notion of constant relative potencies. Thus, Meselson and Russell (1977) compared the log mutagenic potency in the Salmonella Ames test against the log carcinogenic potency in mice for 14 compounds. Clive (1977) compared the log mutagenic potency in the L5178Y mouse lymphoma assay against the log carcinogenic potency in mice for 23 chemicals. Crouch and Wilson (1979) compared log carcinogenic potencies of various chemicals in several pairs of species, most extensively in rats and mice. Numerous investigators have studied the concordance of positive and negative test results among many test compounds in various species (McCann et al. 1975; Purchase et al. 1978; Rinkus and Legator 1979; Poirier and de Serres 1979; Purchase 1982). Freireich et al. (1966) compared the toxicity of anti-cancer agents in mouse, rat, hamster, dog, monkey, and man. These authors implicitly assumed that each cell in a two-way table of experiments constitutes an independent, equally likely draw from a population of species-chemical interactions. Crouch and Wilson (1981) explicitly suggest that data on the variability of relative potencies in rats and mice can be used to assess variations between humans and other species. Likewise, Purchase (1980) concludes that beyond detailed knowledge of the fate of a chemical in man, the only additional evidence supporting human carcinogenicity is the consistency of carcinogenic response across other species.

In what follows, therefore, we put forth the notion that the experiments in Table 1 are all related through some unifying biological hypothesis. More formally, we assume that the results of each experiment are summarized by a single number, such as the slope of the dose-response relation. Each slope is subject to conventional measurement error. The actual slopes, we posit, lie near the response surface of an underlying regression model. The unknown parameters of this model correspond to the uncertain relative potencies among emissions and the uncertain conversion factors between species. Moreover, since some agents have distinctive effects in some species, such a model necessarily entails some error. These errors gauge the uncertain relevance of one experiment to another. The critical factor linking the experiments is the scientist's prior information on the exchangeability (or otherwise) of these errors of interspecies extrapolation.

## 3. THE STATISTICAL MODEL

Let $y_{kl}$ denote the logarithm of the estimated dose-response slope for the experiment in species $k$ on environmental agent $l$, where $k = 1, \ldots, K$ and $l = 1, \ldots, L$. Although some $y_{kl}$ may be missing from the two-way table of experiments, we assume that the set of observed $y_{kl}$ is connected. (That is, any observed $(k, l)$ can be reached from any other observed $(k', l')$ by a series of moves from one observed experiment to another in which each move is along a single row or column. Although connectedness is not strictly required by the theory, in practice unless strong prior information is used, two unconnected experiments cannot each contribute information about any single parameter and thus may as well be analyzed separately.)

Each $y_{kl}$, we further assume, is normally distributed with unknown mean $\theta_{kl}$ and known standard error $c_{kl}$. This assumption is motivated by the fact that the observed $y_{kl}$ in Table 1 are maximum likelihood estimates derived from relatively large experiments.

The quantities $\theta_{kl}$ are the true log slopes, the primary parameters of interest. Each $\theta_{kl}$, we assume, has a prior distribution with mean value

$$E[\theta_{kl} \mid \mu, \alpha_k, \gamma_l] = \mu + \alpha_k + \gamma_l, \qquad (3.1)$$

where the parameters $\{\mu, \alpha_k, \gamma_l\}$ represent the overall mean effect, species-specific effects, and agent-specific effects, respectively. Equation (3.1) embodies the hypothesis that the relative potency of the two emissions is on average preserved across species. Although the various $\theta_{kl}$ are measured in different units, the additive model (3.1) is meaningful so long as $(\theta_{kl} - \theta_{kl'}) - (\theta_{k'l} - \theta_{k'l'})$ is a dimensionless quantity, a condition satisfied by the data of Table 1. The units of measurement for $\mu$, $\alpha_k$, and $\gamma_l$ can then be chosen so that the species-agent interaction effects $\delta_{kl} = \theta_{kl} - \mu - \alpha_k - \gamma_l$ are similarly dimensionless.

Conditional on the value of the parameter $\sigma$, we further assume that the $\delta_{kl}$ are a priori independently and identically normally distributed with zero mean and variance $\sigma^2$. Accordingly, the interaction effects $\delta_{kl}$ are exchangeable (de Finetti 1937). Although certain agents may have distinctive actions in particular species, we cannot identify a priori which entry in our two-way table of experiments is more likely to exhibit a distinctive action. Nor do we have any prior information that any pair of interaction effects are in the same direction.

The parameter $\sigma$ gauges the accuracy of the equal relative potency model. Since the quantities $\exp(\delta_{kl})$ represent dimensionless, multiplicative deviations from the model, 95 percent of such deviations would be within $\exp(\pm 2\sigma)$. For example, a value of $\sigma = 2.3$ would imply that with probability .95, the model would be accurate to a factor of $\exp(2\sigma) \cong 100$ in either direction.

Conditional on $\sigma$, the observed $y_{kl}$ are thus generated by

$$y_{kl} = \mu + \alpha_k + \gamma_l + \delta_{kl} + \epsilon_{kl}, \qquad (3.2)$$

where the three sets of variables $\{\mu, \alpha_k, \gamma_l\}$, $\{\delta_{kl}\}$ and $\{\epsilon_{kl}\}$ are independent a priori, and where the $\epsilon_{kl}$ are independent $N(0, c_{kl}^2)$. Following the notation for the general linear model, we replace the expressions $\mu + \alpha_k + \gamma_l$ in (3.2) by $X\beta$, where $\beta$ is a column vector of parameters and $X$ is an appropriately chosen design matrix. Of the $K + L + 1$ parameters in $\{\mu, \alpha_k, \gamma_l\}$, at most $K + L - 1$ are independently estimable in the classical sense. So long as we use an informative full rank prior distribution on all $K + L + 1$ parameters, no restrictions on the parameters are necessary. However, in certain cases to be considered below (Proposition 2 and the empirical Bayes estimates), we shall assume that $\beta$ corresponds only to the independently estimable components of

$\{\mu, \alpha_k, \gamma_l\}$ and that $X$ is the corresponding full rank design matrix.

Finally, we assume that $\beta$ is a priori multivariate normal with mean vector $b$ and nonsingular covariance matrix $V$, which are independent of $\sigma$, and that $\sigma$ has a prior distribution $\Pi$ with density $\pi(\sigma)$. Now let $i = 1, \ldots, n$ index the experiments, replacing the paired indices $(k, l)$. Let $m$ be the rank of $X$. Let $Y$, $\theta$, $\delta$, and $\epsilon$ be $n \times 1$ column vectors replacing $\{y_{kl}\}$, $\{\theta_{kl}\}$, $\{\delta_{kl}\}$, and $\{\epsilon_{kl}\}$, respectively. Let $I$ be the $n \times n$ identity matrix, and let $C = \text{diag}(c_1^2, \ldots, c_n^2)$. Our model can be formulated generally as $Y = X\beta + \delta + \epsilon$, where $\theta = X\beta + \delta$, and

$$\sigma \sim \Pi, \tag{3.3a}$$

$$(\beta \mid \sigma) \sim N(b, V), \tag{3.3b}$$

$$(\theta \mid \beta, \sigma) \sim N(X\beta, \sigma^2 I), \tag{3.3c}$$

$$(Y \mid \theta, \beta, \sigma) \sim N(\theta, C). \tag{3.3d}$$

The experimental data $Y$ and $C$, as well as $b$, $V$, and the distribution $\Pi$, are assumed to be known. The choices of a prior distribution $\Pi$ and of $b$ and $V$ are left unspecified for now.

This model possesses a hierarchical structure similar to that formulated by Lindley and Smith (1972). Since estimation of the parameters of (3.3) proceeds along familiar Bayesian lines, we state the following proposition without proof.

*Proposition 1*. For the model (3.3), the posterior density of $\sigma$ is

$$\pi(\sigma \mid Y) \propto \pi(\sigma) \mid C + \sigma^2 I + XVX' \mid^{-1/2} \times \exp\{-\tfrac{1}{2}(Y - Xb)'[C + \sigma^2 I + XVX']^{-1}(Y - Xb)\} \tag{3.4}$$

where $\mid A \mid$ is the determinant of $A$. Moreover, the posterior density of $\beta$ is $f(\beta \mid Y) = \int_0^\infty f(\beta \mid Y, \sigma)\pi(\sigma \mid Y)d\sigma$, where $f(\beta \mid Y, \sigma)$ is the multivariate normal density $N(\tilde{\beta}, \tilde{V})$, and

$$\tilde{\beta} = \tilde{V}[X'WY + V^{-1}b],$$
$$\tilde{V} = [X'WX + V^{-1}]^{-1}, \tag{3.5}$$

where $W = (C + \sigma^2 I)^{-1}$. Finally, the posterior density of $\theta$ is $g(\theta \mid Y) = \int_0^\infty g(\theta \mid Y, \sigma)\pi(\sigma \mid Y)d\sigma$, where $g(\theta \mid Y, \sigma)$ is the multivariate normal density $N(\tilde{\theta}, \tilde{C})$, and

$$\tilde{\theta} = \tilde{C}[C^{-1}Y + (XVX' + \sigma^2 I)^{-1}Xb],$$
$$\tilde{C} = [C^{-1} + (XVX' + \sigma^2 I)^{-1}]^{-1}. \tag{3.6}$$

The posterior distributions of both $\beta$ and $\theta$ are mixtures of multivariate normal distributions, with mixing probabilities given by (3.4), the posterior density of $\sigma$. From (3.5), $\tilde{\beta}$ is the precision-weighted average of its least squares estimate and its prior mean $b$. From (3.6), $\tilde{\theta}$ is the precision-weighted average of the original data $Y$ and the corresponding prior mean $Xb$.

The results in Proposition 1 depend upon our specifying a proper prior distribution for $\beta$. Specifying a prior distribution on $\beta$ requires knowledge not only about the relative potencies of agents $\{\gamma_l\}$, but also about the conversion factors of dosage units among species $\{\alpha_k\}$. In many situations, such information could be extremely vague, and a more uninformative prior that avoids complex or implausible conversions of interspecies dosage units would be desirable. To investigate such cases in detail, we offer the following proposition, which is proved in Appendix B.

*Proposition 2*. If $V$ is replaced by $tV$, where $t$ is a scalar, then as $t \to \infty$ the posterior density of $\sigma$ approaches

$$\pi(\sigma \mid Y) \propto \pi(\sigma) \mid W \mid^{1/2} \mid X'WX \mid^{-1/2} \exp\{-\tfrac{1}{2}Y'SY\} \tag{3.7}$$

where $S = W - WX(X'WX)^{-1}X'W$, and where $X$ is the full rank design matrix. Moreover, the posterior density of $\beta$ approaches $f(\beta \mid Y) = \int_0^\infty f(\beta \mid Y, \sigma)\pi(\sigma \mid Y)d\sigma$, where $f(\beta \mid Y, \sigma)$ is multivariate normal $N(\tilde{\beta}, \tilde{V})$ and

$$\tilde{\beta} = \tilde{V}X'WY,$$
$$\tilde{V} = (X'WX)^{-1}. \tag{3.8}$$

Finally, the posterior density of $\theta$ approaches $g(\theta \mid Y) = \int_0^\infty g(\theta \mid Y, \sigma)\pi(\sigma \mid Y)d\sigma$, where $g(\theta \mid Y, \sigma)$ is multivariate normal $N(\tilde{\theta}, \tilde{C})$ and

$$\tilde{\theta} = \tilde{C}C^{-1}Y = (I + CR)^{-1}Y,$$
$$\tilde{C} = (C^{-1} + R)^{-1}, \tag{3.9}$$

where $R = \sigma^{-2}[I - X(X'X)^{-1}X']$.

We also consider empirical Bayes approaches to estimating $\theta$. For these estimates, we retain (3.3c) and (3.3d), but use the data $Y$ to construct the prior distributions on $\sigma$ and $\beta$. Several options are available. First, we could estimate both $\sigma$ and $\beta$ from the data $Y$, by maximum likelihood or other methods, and then assume that the prior density for $\sigma$ is concentrated at the estimate $\hat{\sigma}$ and the prior density of $\beta$ is concentrated at the estimate $\hat{\beta}$. For a given $\sigma$, the maximum likelihood estimate of $\beta$ is the least squares estimate

$$\hat{\beta}_{\text{MLE}}(\sigma) = (X'WX)^{-1}X'WY,$$

where $W$ depends on $\sigma$. The concentrated likelihood of $\sigma$, evaluated at $\beta = \hat{\beta}_{\text{MLE}}(\sigma)$, is proportional to

$$\hat{L}(\sigma) = \mid W \mid^{1/2}\exp\{-\tfrac{1}{2}Y'SY\}$$

where $W$ and $S$ depend on $\sigma$. Let $\hat{\sigma}_{\text{MLE}}$ be the value of $\sigma$ that maximizes $\hat{L}(\sigma)$. Then the resulting empirical Bayes posterior distribution of $\theta$, which treats both $\sigma = \hat{\sigma}_{\text{MLE}}$ and $\beta = \hat{\beta}_{\text{MLE}}(\hat{\sigma}_{\text{MLE}})$ as if they were known with certainty, is $N(\hat{\theta}_{\text{MLE}}, \hat{C}_{\text{MLE}})$, where

$$\hat{\theta}_{\text{MLE}} = \hat{C}_{\text{MLE}}[C^{-1}Y + \hat{\sigma}_{\text{MLE}}^{-2}X\hat{\beta}_{\text{MLE}}(\hat{\sigma}_{\text{MLE}})],$$
$$\hat{C}_{\text{MLE}} = [C^{-1} + \hat{\sigma}_{\text{MLE}}^{-2}I]^{-1}. \tag{3.10}$$

Alternatively, we could assume a diffuse prior on $\beta$ and estimate only $\sigma$ from the data $Y$. In that case, the appropriate likelihood function for $\sigma$ is equation (3.7) with the prior density $\pi(\sigma)$ omitted, that is,

$$L^*(\sigma) = | W |^{1/2} | X'WX |^{-1/2} \exp\{-\tfrac{1}{2}Y'SY\}.$$

Let $\hat{\sigma}_{EB}$ be the value of $\sigma$ that maximizes $L^*(\sigma)$. The corresponding empirical Bayes posterior distribution of $\theta$, which treats $\sigma = \hat{\sigma}_{EB}$ as if it were known with certainty, is $N(\hat{\theta}_{EB}, \hat{C}_{EB})$, where

$$\hat{\theta}_{EB} = [I + \hat{\sigma}_{EB}^{-2}C(I - X(X'X)^{-1}X')]^{-1}Y$$
$$\hat{C}_{EB} = [C^{-1} + \hat{\sigma}_{EB}^{-2}(I - X(X'X)^{-1}X')]^{-1}. \tag{3.11}$$

(In case $\hat{\sigma}_{EB}^2$ is near or equal to zero, the lemma of Appendix B can be used to show that the limiting values in (3.11) are $\hat{\theta}_{EB} = X\hat{\beta}_{MLE}(0)$ and $\hat{C}_{EB} = X(X'C^{-1}X)^{-1}X'$.)

Comparing (3.10) and (3.11), we note that $\hat{C}_{MLE} < \hat{C}_{EB}$, in the sense that $\hat{C}_{MLE} - \hat{C}_{EB}$ is negative definite. (To establish this result, it is enough to show that $\hat{\sigma}_{MLE} < \hat{\sigma}_{EB}$. The latter follows from the fact that $\hat{\sigma}_{MLE}$ and $\hat{\sigma}_{EB}$ maximize $\hat{L}(\sigma)$ and $L^*(\sigma)$, respectively, and that the ratio $\hat{L}(\sigma)/L^*(\sigma) = | X'(C + \sigma^2 I)^{-1}X |^{1/2}$ is a decreasing function of $\sigma$.) Thus, the assumption that $\beta = \hat{\beta}_{MLE}$ with certainty leads to a smaller estimated posterior variance for $\theta$ than does the assumption of a diffuse prior for $\beta$.

If we wish to avoid the computational burden of determining $\hat{\sigma}_{MLE}$ or $\hat{\sigma}_{EB}$, we could begin with $\hat{\beta}_{MLE}(0) = (X'C^{-1}X)^{-1}X'C^{-1}Y$. The residual sum of squares for this estimate, RSS $= \sum_{i=1}^{n} (y_i - x_i\hat{\beta})^2/c_i^2$, has expectation

$$E[\text{RSS}] = n - m + \sigma^2\left[\sum_{i=1}^{n} c_i^{-2} - \text{tr}\, C^{-1}X(X'C^{-1}X)^{-1}X'C^{-1}\right]$$

which suggests the estimate

$$\hat{\sigma}_{RSS}^2 = [\text{RSS} - (n - m)] \Big/ \left[\sum_{i=1}^{n} c_i^{-2} - \text{tr}\, C^{-1}X(X'C^{-1}X)^{-1}X'C^{-1}\right] \tag{3.12}$$

where we take $\hat{\sigma}_{RSS}^2 = 0$ if RSS $< n - m$. (The value of $E$[RSS] was derived for us by H. Chernoff.) In the analysis that follows, we report the empirical Bayes estimate $\hat{\theta}_{RSS}$, derived by substituting $\hat{\sigma}_{RSS}$ for $\hat{\sigma}_{EB}$ in (3.11).

For future reference, we also define the posterior expectation of $\sigma^2$ as

$$\sigma^{*2} = \int_0^\infty \sigma^2 \pi(\sigma \mid Y) d\sigma, \tag{3.13}$$

where $\pi(\sigma \mid Y)$ is defined either by (3.4) or by (3.7). The quantity $\sigma^{*2}$ is the mean squared error of prediction (on the log scale) for an experiment yet to be performed.

Results similar to Propositions 1 and 2 have been derived by other authors. Lindley and Smith (1972) and Smith (1973) describe a "modal Bayesian" estimation procedure, which corresponds in our model to the use of $\hat{\sigma}_{MLE}$ for $\hat{\sigma}_{EB}$ in (3.11). Our (3.9) is a special case of equation (A.1) in Smith (1973). Our likelihood function $L^*(\sigma)$ is equivalent to that derived by Harville (1974, eq. 3) for the mixed variance components model. Our empirical Bayes estimate $\hat{\sigma}_{EB}$ (the modal value of $L^*$) corresponds to the "restricted" maximum likelihood estimate of variance components proposed by Patterson and Thompson (1971). (See also Harville 1977.) Dempster, Rubin, and Tsutakawa (1981) analyze a covariance components model similar to ours, in which the fixed effects can assume a diffuse prior. With suitable modification, their EM-algorithm estimate of the variance of the random effects is equivalent to our $\hat{\sigma}_{EB}^2$.

Our analysis has relied upon the assumption that, conditional upon $\sigma$, the vector $\delta$ is distributed a priori as $N(0, \sigma^2 I)$. In anticipation of Section 5, however, we note that the covariance matrix $\sigma^2 I$ could be replaced by a more general matrix $\Sigma$. In place of (3.3a), the parameters in $\Sigma$ would then have joint prior distribution $\Pi$. In Proposition 1, expression (3.4) would give the joint posterior density of these parameters. This density would serve as the mixing probabilities for the posterior distributions of $\beta$ and $\theta$. Proposition 2 would follow analogously, with $W = (C + \Sigma)^{-1}$ and $R = \Sigma^{-1} - \Sigma^{-1}X(X'\Sigma^{-1}X)^{-1}X'\Sigma^{-1}$. Although the derivations of $\hat{\theta}_{EB}$ and $\hat{\theta}_{MLE}$ would follow a similar generalization, there is no analog for $\hat{\theta}_{RSS}$.

We have also assumed that $\beta$ is a priori independent of $\sigma$. But our results are easily generalized to the case where $b$ and $V$ are replaced by $b(\sigma)$ and $V(\sigma)$. Finally, we assumed that each $\epsilon_i$ has known variance $c_i^2$. Yet our results are similarly generalizable to the case where the $c_i$ are also uncertain and have prior distributions.

## 4. APPLYING THE MODEL

In the following illustrative calculations, we assume a diffuse prior distribution on the parameters $\beta$. The use of prior biological information to formulate proper priors on $\beta$ is deferred to Section 5.

For the critical parameter $\sigma$, we consider two alternative prior distributions.

*Prior 1*: log $\sigma$ is uniformly distributed on the interval $.05 \le \sigma \le 5.0$.

This prior distribution implies that within a range of one normal standard deviation, that is, with probability .68, the underlying constant relative potency model could be accurate within a multiplicative factor ranging from $\exp(.05) \cong 1.05$ up to $\exp(5) \cong 150$. Given our uncertainty about even the order of magnitude of error, we have assumed that log $\sigma$, rather than $\sigma$, is uniformly distributed. Although Prior 1 assigns zero probability outside the interval [.05, 5], this restriction, as we shall see shortly, does not significantly affect our main conclusions. Prior 1 has the property that for $.05 \le \sigma \le 5$, the posterior density of log $\sigma$ is proportional to the likelihood function $L^*$.

*Prior 2*: log $\sigma$ is normally distributed with mean and standard deviation both equal to log 2.

This prior distribution has a modal value of $\sigma = 2$, with 95 percent probability in the interval $.5 \leq \sigma \leq 8$. Accordingly, with probability .68, the underlying constant relative potency model could be accurate within a multiplicative factor ranging roughly from $\exp(.5) \cong 1.7$ to $\exp(8) \cong 3000$, with an error factor of $\exp(2) \cong 7.4$ being most likely. Prior 2 is therefore more conservative than Prior 1 about the accuracy of interspecies extrapolation.

To simplify the computations, we evaluate Priors 1 and 2 and their corresponding posterior distributions only at discrete points equally spaced on the log scale. The posterior distributions of $\beta$ and $\theta$ are then finite mixtures of normal distributions.

Figure 2 displays the posterior densities $\pi(\sigma \mid Y)$ based upon Priors 1 and 2. For both priors, virtually all of the posterior probability mass of $\sigma$ is concentrated in the interval [.8, 1.6]. From the Bayesian estimate $\sigma^* \cong 1.1$ we calculate that, with 95 percent credibility, extrapolation will likely to be accurate only within a multiplicative factor of $\exp(2\sigma^*) \cong 9$.

Table 2 shows the posterior means and standard deviations of the log carcinogenic potencies of roofing tar emissions, coke oven emissions, and diesel A emissions in humans. Although the marginal posterior distributions of the $\theta_i$ are mixtures of normals, the distributions in this example are adequately characterized by their means $\theta_i^*$

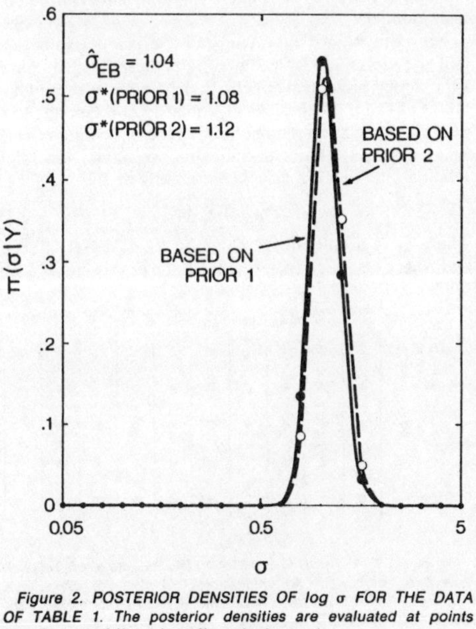


*Figure 2. POSTERIOR DENSITIES OF log σ FOR THE DATA OF TABLE 1. The posterior densities are evaluated at points equally spaced (by .1 log 10) on the log scale.*

*Table 2. Bayes and Empirical Bayes Estimates of Log Slopes for Lung Cancer Risk in Humans: 5 × 9 Data Matrix*

| *Environmental Emission* | *Posterior Mean* | *Posterior Standard Deviation* | *Lower Tail* | *Upper Tail* |
|---|---|---|---|---|
| Roofing Tar | | | | |
| Original Data | .50 | 1.41 | | |
| $\theta_{Bayes\ 1}$ | .12 | 1.02 | .010 | .011 |
| $\theta_{Bayes\ 2}$ | .14 | 1.03 | .010 | .011 |
| $\theta_{EB}$ | .12 | 1.01 | | |
| $\theta_{RSS}$ | .06 | .95 | | |
| $\theta_{MLE}$ | −.01 | .70 | | |
| Coke Oven | | | | |
| Original Data | 1.48 | .34 | | |
| $\theta_{Bayes\ 1}$ | 1.38 | .33 | .010 | .010 |
| $\theta_{Bayes\ 2}$ | 1.38 | .33 | .010 | .010 |
| $\theta_{EB}$ | 1.38 | .33 | | |
| $\theta_{RSS}$ | 1.35 | .33 | | |
| $\theta_{MLE}$ | 1.30 | .31 | | |
| Diesel Engine A | | | | |
| $\theta_{Bayes\ 1}$ | −.46 | 1.45 | .011 | .012 |
| $\theta_{Bayes\ 2}$ | −.45 | 1.50 | .011 | .012 |
| $\theta_{EB}$ | −.46 | 1.40 | | |
| $\theta_{RSS}$ | −.51 | 1.25 | | |
| $\theta_{MLE}$ | −.57 | .80 | | |

NOTE: $\theta_{Bayes\ 1}$ assumes Prior 1 for $\pi(\sigma)$, and diffuse prior on $\beta$. $\theta_{Bayes\ 2}$ assumes Prior 2 for $\pi(\sigma)$, and diffuse prior on $\beta$. $\theta_{EB}$ assumes prior $\pi(\sigma)$ concentrated at $\hat{\sigma}_{EB} = 1.04$, and diffuse prior on $\beta$. $\theta_{RSS}$ assumes prior $\pi(\sigma)$ concentrated at $\hat{\sigma}_{RSS} = 0.93$, and diffuse prior on $\beta$. $\theta_{MLE}$ assumes prior $\pi(\sigma)$ concentrated at $\hat{\sigma}_{MLE} = 0.80$, and prior on $\beta$ concentrated at $\hat{\beta}_{MLE}$. Lower Tail = $\Pr\{\theta \leq \theta^* - 2.326c^* \mid Y\}$, Upper Tail = $\Pr\{\theta \geq \theta^* + 2.326c^* \mid Y\}$.

and standard deviations $c_i^*$. That is, the tail probabilities $\Pr\{\theta_i \geq \theta_i^* + 2.326c_i^* \mid Y\}$ and $\Pr\{\theta_i \leq \theta_i^* - 2.326c_i^* \mid Y\}$ do not deviate substantially from the value of .01 predicted for the normal density.

Because the original coke oven data were relatively precise, the means and standard deviations of the coke oven log slope do not differ much from their original values. For roofing tar, however, there is a substantial improvement in precision. For the Bayes estimate based on Prior 1, the 95 percent credible interval for the carcinogenic potency of roofing tar (that is, $e^{\theta}$) is narrowed from [.097, 27.8] to [.148, 8.66]. From the dosage measure given in Appendix A, the 95 percent credible interval for the relative risk of lung cancer from 20 years of occupational exposure to roofing tar is narrowed from [1.11, 31.6] with mode 2.80, to [1.16, 10.5] with mode 2.24.

The estimates for diesel engine A illustrate the dependence of the posterior precision of $\theta$ on our prior assumptions about $\sigma$ and $\beta$. Thus, the Bayes estimate based on the conservative Prior 2 has a slightly larger posterior standard deviation than that based on Prior 1. Similarly, $\hat{\theta}_{EB}$ and $\hat{\theta}_{RSS}$, which assume fixed, known values of $\sigma$, have smaller posterior variances than the Bayes estimates. The estimate $\hat{\theta}_{MLE}$, based on the dubious assumption that both $\beta$ and $\sigma$ are fixed and known, yields on overly optimistic estimate of precision.

Despite the divergence of Priors 1 and 2, the posterior densities in Figure 2 are almost identical. This finding is to be contrasted with Figure 3, which displays the pos-

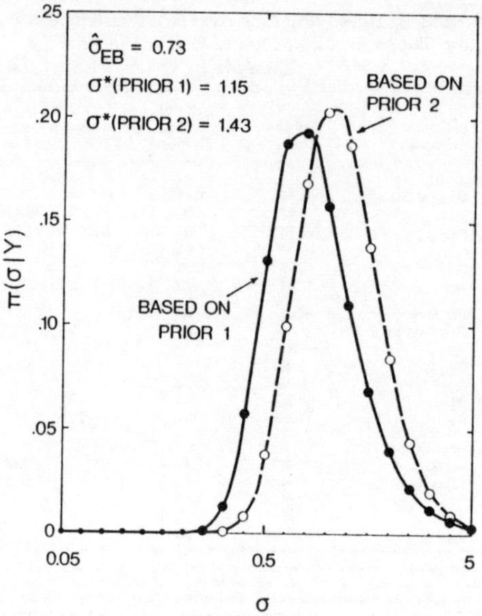


*Figure 3. POSTERIOR DENSITIES OF log σ FOR THE 3 × 3 SUBMATRIX OF EXPERIMENTS IN FIGURE 1. The posterior densities are evaluated at points equally spaced (by .1 log 10) on the log scale.*

terior densities of $\sigma$ based solely on the upper left 3 × 3 submatrix of experiments, as displayed in Figure 1. With considerably fewer experiments, the posterior densities of $\sigma$ now diverge substantially. The posterior standard deviations of the diesel A log potency are: 1.22 for the empirical Bayes estimate based on $\hat{\sigma}_{EB}$; 1.82 for the Bayes estimate based on Prior 1; and 2.24 for the Bayes estimate based on Prior 2. The corresponding 95 percent credible intervals for the diesel A dose-response slope are, respectively, [.139, 18.3], [.041, 59.3], and [.017, 132.].

With even fewer experiments, the precision of the human risk estimates depends almost entirely on prior beliefs. Suppose that we predict the lung cancer risk of diesel A emissions solely from the human data on coke oven emissions in combination with data on their relative potencies in mice. The maximum likelihood estimate $\hat{\sigma}_{MLE}$ and the empirical Bayes estimate $\hat{\sigma}_{EB}$ would be identically zero. For the empirical Bayes estimate, the 95 percent credible interval for the diesel A dose-response slope ($e^{\theta}$) would be [.549, 2.23]. For Prior 1, however, it would be [.0008, 1451] and, for Prior 2, it would be [3.3 × $10^{-6}$, 3.7 × $10^{5}$].

In our statistical model, we assumed that the standard errors $c_i$ of the observed log slopes were known. To check the sensitivity of our results to this assumption, we repeated our analysis on the data of Table 1, arbitrarily doubling the $c_i$ for all the nonhuman experiments. For Prior 1, we obtained $\sigma^* = .99$; the posterior mean and standard deviation of the human carcinogenic potency of diesel A were −.61 and 1.35, respectively. The other estimated human carcinogenic potencies were likewise only slightly different from those in Table 2.

## 5. INCORPORATING PRIOR INFORMATION

Figure 4 diagrams a 5 × 10 array of experiments. In the figure, the columns of Table 1 have been rearranged so that the diesel experiments *A* through *D* are on the right. In addition, we have appended the results of an epidemiological study of men exposed to a fifth type of diesel emission (Diesel E). The estimation for this additional dose-response slope, taken from Harris's (1981,1983) analysis of lung cancer incidence among diesel bus workers, is summarized in Appendix A.

Figure 4 exemplifies the case where there are only nonhuman data on some agents and only epidemiological data on another related agent. Unless we have some prior information on the relation among the agents, the inclusion of the epidemiological data cannot improve our estimates of human cancer risk. Thus, in the present example, if we continued to assume a diffuse prior distribution on the parameters $\beta$, our inclusion of the epidemiological study on diesel E would not affect the results in Table 2. The extra column effect corresponding to diesel E would merely be set so that the observed data on diesel E in humans exactly fit the underlying constant relative potency model.

Accordingly, we now consider the prior assumption that the emissions from the various diesel engines have very similar biological effects. In the double-subscript notation of (3.1), this biological similarity corresponds to a positive correlation among the five diesel column effects $\{\gamma_6, \ldots, \gamma_{10}\}$. More specifically, we model the prior relation between the diesel column effects by

$$\gamma_l = \gamma_0 + \eta_l, \quad l = 6, \ldots, 10 \qquad (5.1)$$

where $\gamma_0$ is a component common to all diesel effects, and where $\{\eta_6, \ldots, \eta_{10}\}$ represent the deviations of each

| | ROOF TAR | COKE OVEN | GAS ENGINE | BaP | CIG. | DIESEL A | DIESEL B | DIESEL C | DIESEL D | DIESEL E |
|---|---|---|---|---|---|---|---|---|---|---|
| LUNG CANCER | ● | ● | | | ● | | | | | ● |
| SKIN TUMOR INIT. | ● | ● | ● | ● | ● | ● | ● | | ● | |
| VIRAL TRANSFORM. | ● | ● | ● | ● | ● | ● | ● | ● | ● | |
| MUTAGENESIS −MA | ● | ● | ● | | ● | ● | ● | ● | ● | |
| MUTAGENESIS +MA | ● | ● | ● | | ● | ● | ● | ● | ● | |
| | 1 | 2 | 3 | 4 | 5 | 6 | 7 | 8 | 9 | 10 |

*Figure 4. 5 × 10 DATA ARRAY. The columns of Table 1 have been permuted. An epidemiological study of diesel bus workers (Diesel E) has been appended. A filled circle means data available.*

*Table 3. Covariance Matrix V Under Informative Prior for Diesel Effects*

$$\begin{bmatrix} \begin{matrix} 100 & 0 & \bullet & \bullet & 0 \\ 0 & 100 & & & \bullet \\ \bullet & & \bullet & & \bullet \\ \bullet & & & \bullet & 0 \\ 0 & \bullet & \bullet & 0 & 100 \end{matrix} & 0 \\ 0 & \begin{matrix} 100 & v_0 & \bullet & \bullet & v_0 \\ v_0 & 100 & & & \bullet \\ \bullet & & \bullet & & \bullet \\ \bullet & & & \bullet & v_0 \\ v_0 & \bullet & \bullet & v_0 & 100 \end{matrix} \end{bmatrix}$$

diesel effect from the common mean. Each $\eta_l$, we assume, has prior mean zero, is independent of the other $\eta_l$ and of $\gamma_0$, and has prior variance $v_\eta$. We denote the prior variance of $\gamma_0$ by $v_0$.

Since the covariance among the diesel column effects is our main interest, we specify an uninformative prior for the remaining parameters. Specifically, we set the prior covariances of all the parameters $\{\mu, \alpha_k, \gamma_l\}$ equal to 100. (Hence, $v_0 + v_\eta = 100$.) Moreover, except for the covariances $v_0$ among the diesel effects $\{\gamma_6, \ldots, \gamma_{10}\}$, we set the covariances among all the parameters $\{\mu, \alpha_k, \gamma_l\}$ equal to zero. The prior covariance matrix $V$ thus takes the form given in Table 3, where the upper left $11 \times 11$ submatrix is the prior covariance of $\{\mu, \alpha_1, \ldots, \alpha_5, \gamma_1, \ldots, \gamma_5\}$ and the lower right $5 \times 5$ submatrix is the prior covariance of $\{\gamma_6, \ldots, \gamma_{10}\}$.

Thus, given $\sigma$, each $\theta_i$ has a prior variance $300 + \sigma^2$ or, equivalently, with probability 0.68, each dose-response slope could vary by a factor of $\exp \sqrt{300} \cong 3 \times 10^7$. (It is possible, as Smith 1973 shows in a simpler case, to let the variances $(v_0 + v_\eta)$ of the diesel effects go to infinity while $v_\eta$ remains fixed.) Since the prior variances of the $\theta_i$ are so large, and since all the observed $y_i$ are within $\pm 5$, we can arbitrarily set $b = 0$ without affecting the conclusions below.

The prior relation between the diesel column effects depends critically upon $v_\eta$, for which we consider several possible values.

First, we consider $v_\eta = 100$, so that $v_0 = 0$ and $V = 100I$, where $I$ is the $16 \times 16$ identity matrix. A priori, the diesel effects would be uncorrelated and diesel emissions would resemble each other no more than the other emissions.

Second, we consider $v_\eta = 0$, so that $v_0 = 100$ and $V$ has rank 12. The diesel column effects would be a priori identical, and the number of distinct column effects would be reduced to 6. Any two diesel log potencies $\theta_i$ and $\theta_{i'}$ in the same species would have the same model predictions (that is, $x_i\beta = x_{i'}\beta$) and differ only in their model deviations (that is, $\delta_i$ and $\delta_{i'}$). This case resembles that analyzed by Yates and Cochran (1938) and Cochran (1980), in which experiments of very similar design differed primarily in their date and location. We might adopt this form of prior information when the various diesel emissions are thought to contain essentially the same biologically active components, but variations in engine type or operating conditions could produce idiosyncratic effects in some species.

Finally, we consider $v_\eta = 1$ (so that $v_0 = 99$) and $v_\eta = 2$ (so that $v_0 = 98$). In these cases, the diesel effects would be a priori correlated. Although the prior correlations are .99 and .98, respectively, our setting $v_\eta = 1$ or 2 in fact permits a substantial range of relative potencies. To see this, note that for a given $\sigma$, the difference between any two diesel log slopes in the same species has variance $2(v_\eta + \sigma^2)$. Provided that the prior means of the diesel column effects are equal, the relative potency of any two diesel emissions is a priori $\exp[z(2(v_\eta + \sigma^2))^{1/2}]$, where $z$ is distributed $N(0, 1)$. Since $\exp(z\sqrt{2}) \cong 10$ for $z = 1.6$, our setting $v_\eta = 1$ already allows for more than a 10 percent probability that one of the diesels is 10 times more potent than the other. For $v_\eta = 2$, this probability increases to more than 25 percent. We might adopt these less extreme forms of prior information when the diesel emissions are thought to share many biologically active components, but the overall composition of the mixture varies with engine type and operating conditions.

Table 4 shows the resulting estimates of $\sigma$ and of the $\theta$'s for roofing tar, diesel A, diesel E, gasoline engine emissions, and benzo(a)pyrene. Since the assumption that $v_\eta = 100$ (diesel effects uncorrelated) approximates complete ignorance about $\beta$, the results in that column are quite close to those obtained in the diffuse prior case in Table 2. As expected, the diesel E epidemiological study contributes only minimally to the estimation of $\sigma$ and to the remaining $\theta$'s.

The assumption that $v_\eta = 0$ (identical diesel column effects) results in a marked increase in $\sigma^*$. Any variation in the diesel $\theta$'s that is in fact due to the $\gamma$'s is forced to be fit to the corresponding $\delta$'s. Hence, the posterior variance of $\delta$ increases. Although the posterior standard de-

*Table 4. Bayes Estimates of Log Slopes for Human Lung Cancer Risk under Alternative Assumptions on the Relation Between Diesel Column Effects*

| $v_\eta$ | *100* | *0* | *1* | *2* | *Original Data* |
|---|---|---|---|---|---|
| $\sigma^*$ | 1.08 | 1.39 | 1.10 | 1.09 | |
| Roofing Tar | | | | | |
| $\theta^*$ | .13 | .53 | .38 | .32 | .50 |
| $c^*$ | 1.02 | 1.11 | 1.02 | 1.01 | 1.41 |
| Diesel Engine A | | | | | |
| $\theta^*$ | −.44 | −.98 | −.28 | −.25 | |
| $c^*$ | 1.45 | 1.68 | 1.43 | 1.42 | |
| Diesel Engine E | | | | | |
| $\theta^*$ | 1.87 | .31 | .35 | .64 | 1.92 |
| $c^*$ | 1.50 | 1.16 | 1.17 | 1.23 | 1.51 |
| Gasoline Engine | | | | | |
| $\theta^*$ | −2.01 | −1.29 | −1.59 | −1.68 | |
| $c^*$ | 1.44 | 1.79 | 1.46 | 1.44 | |
| Benzo(a)pyrene | | | | | |
| $\theta^*$ | 5.85 | 6.30 | 6.01 | 5.92 | |
| $c^*$ | 1.55 | 1.94 | 1.57 | 1.55 | |

viation for diesel E is reduced, the precisions of the other estimates deteriorate.

The assumptions that $v_\eta = 1$ and $v_\eta = 2$ (correlated diesel effects) do not have these limitations. The estimate $\sigma^*$ is not substantially affected. The posterior standard deviations of the log potencies of roofing tar, diesel A emissions, gasoline engine emissions, and benzo(a)pyrene are not much different from those in the first column of the table. But the precision of the estimate for diesel E is improved, and the potencies of diesel A and diesel E are drawn toward each other. As expected, the effect of going from $v_\eta = 0$ to $v_\eta = 1$ is much more dramatic than the effect of going from $v_\eta = 1$ to $v_\eta = 2$. Comparing the original data for diesel E with the estimate under $v_\eta = 2$, and using the dosage measure given in Appendix A, we find that the 95 percent credible interval for the relative risk of lung cancer for a diesel bus garage worker with 20 years' occupational exposure is narrowed from [1.007, 3.95] to [1.003, 1.31].

These illustrative results by no means exhaust the potential uses of prior biological information in our statistical framework. The inclusion of several groups of chemically similar environmental agents, for example, entails a straightforward generalization of (5.1). Such an analysis would permit us to examine whether the observed strong correlations between short-term bioassays and animal carcinogenesis studies result merely from the comparison of many chemically similar compounds within only a few chemical classes (Rinkus and Legator 1979).

We could also incorporate prior information on the relation between species or bioassays. Suppose, for example, that the experimental data matrix included several biologically related mutagenesis tests (Skopek et al. 1978). Prior correlations among the row effects $\{\alpha_k\}$ might be appropriate when, say, some strains test frameshift mutations and others test base-substitution mutations.

Although the exchangeable prior on $\delta$ could be criticized as an overly naive expression of prior ignorance, we stress that this assumption is the central starting point in any approach to interspecies comparisons. Nevertheless, as we noted in Section 3, nearly all of our results can be generalized to the case where $\delta$ is distributed as $N(0, \Sigma)$. Thus, suppose that one species is considered to be more heterogeneous than the others. For example, the subjects observed in the epidemiological studies were surely more genetically diverse than Sencar mice, Syrian hamster cells, or mouse lymphoma cells. Then we might specify $\delta_i \sim N(0, \tau^2)$ for all the human studies and $\delta_{i'} \sim N(0, \sigma^2)$ for the remaining experiments, where the $\delta_i$ and $\delta_{i'}$ are independent and $\tau$ is a priori likely to exceed $\sigma$.

Moreover, a simple correlation of row or column effects (that is, a prior on $\beta$) will not capture instances where two related species are thought to be especially sensitive to one class of compounds and especially insensitive to another. In those cases, the off-diagonal elements of $\Sigma$ will be nonzero. As we explained in Section 3, the assumption that $\beta$ is a priori uncorrelated with $\sigma$ can also be relaxed. Such a generalization might be appropriate when the overall accuracy of interspecies extrapolation is thought to be related to the relative potency of two agents.

## 6. SELECTING THE MOST RELEVANT EXPERIMENTS

So far, we have assumed that the data of Table 1 are given in advance. Although the experiments were selected according to reasonable scientific criteria, it is not obvious that these data are most relevant for predicting human lung cancer risks. Even if we regard Table 1 as an exhaustive listing of available data, it is unclear whether a subset of experiments might not perform better. The question is: which of all possible subsets should we choose and upon what basis?

One possible starting point is to examine the residuals from the full model. Accordingly, Figure 5 depicts the posterior mean values of the residuals $\delta = \theta - X\beta$ derived from the Bayes estimates in Table 2. (We return to the case of a diffuse prior on $\beta$ in order to keep the interpretation simple.) As Figure 5 shows, the posterior mean residuals for cigarette smoke are in four of five cases relatively large in absolute value. Further, the range of the mean residuals is largest for the direct mutagenesis experiments in the absence of metabolic activator. By

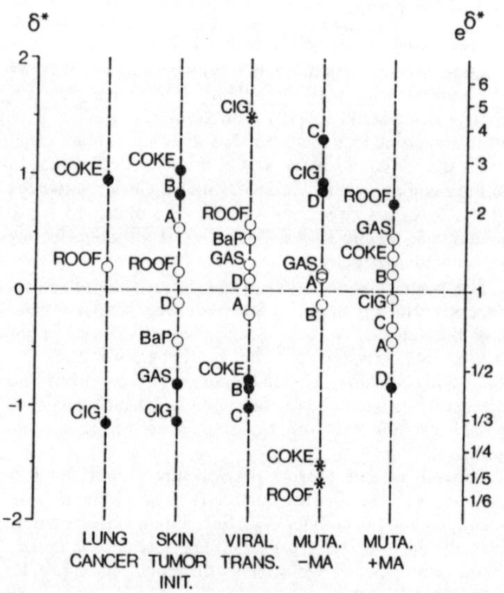


*Figure 5. Posterior Mean Residuals $\delta_i^* = \theta_i^* - x_i\beta^*$ Derived From Bayes Estimates of Table 1. Species-emission pairs with no original data, and therefore zero residuals, are omitted. An open circle means $|\delta_i^*| < \omega_i^*$; a closed circle means $\omega_i^* \le |\delta_i^*| < 2\omega_i^*$; and a star means $|\delta_i^*| \ge 2\omega_i^*$, where $\omega_i^*$ is the posterior standard deviation of $\delta_i$.*

contrast, the residuals for indirect mutagenesis with activator are more concentrated around the origin. A similar finding applies to those viral transformation residuals other than cigarette smoke.

Such an informal examination of residuals suggests that we should merely exclude those species-emission pairs with large values of $|\delta_i^*|$. But these individual interactions are what determine the relatedness of the various species and emissions. It would be more appropriate to assess whether a specific species or a specific agent is more or less relevant to the others.

Moreover, confining our attention solely to the model deviations $\delta$ ignores uncertainty in the model predictions $X\beta$. Consider a particular human log slope $\theta_i = x_i\beta + \delta_i$. The presence of less relevant experiments may accentuate the error $\delta_i$, but it may also improve the posterior precision of $x_i\beta$. As we successively remove less relevant experiments, the variance of $\delta_i$ declines, but the variance of $x_i\beta$ increases and, at some point, the variance of $\theta_i$ also increases. In fact, when we have no initial data on $\theta_i$, as in the case of Diesel A, the elimination of every conceivably irrelevant experiment leaves us with exactly what we had at the start—no information on $\theta_i$ at all.

We therefore consider the following procedure. Let $Y_{k-}$ be the vector of log slopes formed by exclusion of all experiments involving species $k$, and define $\sigma^*_{k-} = E[\sigma^2 \mid Y_{k-}]^{1/2}$. Analogous definitions apply to $Y_{-l}$ and $\sigma^*_{-l}$ for each environmental agent $l$. Given an initial set of experiments $Y$, a prior density $\pi(\sigma)$, and a particular $\theta_i$ of interest, we first calculate $\sigma^*_{k-}$ and $\sigma^*_{-l}$ for each $k$ and $l$, and determine the species $k'$ or emission $l'$ for which $\sigma^*_{k-}$ or $\sigma^*_{-l}$ is minimized. We then eliminate species $k'$ or emission $l'$ and iterate our procedure on the reduced set of experiments so long as (a) there exists a species $k$ or emission $l$ for which $\sigma^*_{k-} < \sigma^*$ or $\sigma^*_{-l} < \sigma^*$; (b) the species $k'$ or emission $l'$ does not correspond to $\theta_i$; and (c) the elimination of species $k'$ or emission $l'$ reduces the posterior standard deviation of $\theta_i$. If conditions (a), (b), and (c) are not satisfied, the procedure terminates.

We applied this procedure to the data of Table 1. We assumed Prior 1 for $\pi(\sigma)$ and focused on the human risks for roofing tar and diesel emission A. Table 5 shows the minimized values of $\sigma^*$ and the corresponding values of $\theta_i^*$ and $c_i^*$ for each iteration. Successive elimination of the row for mutagenesis without metabolic activator ($M-$), the column for cigarette smoke (CIG), and the row for skin tumor initiation in mice (SKIN) resulted in a reduction of $\sigma^*$ from 1.08 to .48. Each of these steps also reduced $c_i^*$ for both roofing tar and diesel A. In the final step, however, removal of the column for coke oven emissions (COKE) reduced $\sigma^*$ to .395, but the posterior standard deviations of both roofing tar and diesel A increased. Hence, the procedure was terminated. As Table 5 indicates, the posterior mean values $\theta_i^*$ varied somewhat with successive iterations, but these variations were within $\pm c_i^*$.

*Table 5. Results of Diagnostic Procedure*[a]

| | 5 × 9 | 4 × 9 | 4 × 8 | 3 × 8 | 3 × 7 | Original Data |
|---|---|---|---|---|---|---|
| Row/Column Eliminated | | M− | CIG | SKIN | COKE | |
| $\sigma^*$ | 1.08 | .93 | .73 | .48 | .40 | |
| $\chi^{*2}$ | | 13.5 | 11.5 | 15.8 | 2.5 | |
| d.f. | | 7 | 3 | 6 | 2 | |
| $P$ | | .06 | .009 | .01 | .28 | |
| $P$ | | .001 | .003 | .0004 | .28 | |
| Roofing Tar | | | | | | |
| $\theta^*$ | .12 | .31 | .96 | 1.53 | .50 | .50 |
| $c^*$ | 1.02 | .96 | .92 | .74 | 1.41 | 1.41 |
| Coke Oven | | | | | | |
| $\theta^*$ | 1.38 | 1.37 | 1.46 | 1.42 | | 1.48 |
| $c^*$ | .33 | .33 | .33 | .33 | | .34 |
| Diesel Engine A | | | | | | |
| $\theta^*$ | −.46 | −.71 | .21 | .33 | −.84 | |
| $c^*$ | 1.45 | 1.30 | 1.16 | .87 | 1.58 | |

[a] The columns labeled 5 × 9, 4 × 9, 4 × 8, 3 × 8, and 3 × 7 refer to successive iterations of the diagnostic procedure.
NOTE: M− denotes mutagenesis experiments in L5178Y mouse lymphoma cells without metabolic activator. CIG denotes experiments on cigarette smoke. SKIN denotes skin tumor initiation experiments in Sencar mice. COKE denotes experiments on coke oven emissions. See text for definitions of $\chi^{*2}$, $P$, and $p$.

The elimination of a particular subset of experiments is equivalent to estimating an augmented model of the form

$$Y = (X\ X_0)\begin{pmatrix}\beta \\ \beta_0\end{pmatrix} + \delta + \epsilon \qquad (6.1)$$

where the columns in the appended matrix $X_0$ are indicator vectors corresponding to all but one of the experiments in the subset to be eliminated, and where a diffuse prior is assumed on the corresponding parameters $\beta_0$. The hypothesis that a particular subset of experiments is less relevant to the others amounts to the assertion that $\beta_0$ is far from the origin. Let $\beta_0^*$ and $V_0^*$ denote, respectively, the posterior mean and covariance matrix of $\beta_0$, and let $\chi^{*2} = \beta_0^{*\prime} V_0^{*-1} \beta_0^*$. Since the posterior distribution of $\beta_0$ is approximately normal, the quantity $p = \exp -\frac{1}{2}\chi^{*2}$ is the approximate posterior density ratio comparing $\beta_0 = 0$ to $\beta_0 = \beta_0^*$. Moreover, $P = \text{Prob}\{\chi^2 \geq \chi^{*2}\}$ is the approximate posterior probability that $\beta_0$ is at least as far from $\beta_0^*$ as $\beta_0^*$ is from 0 (where distance is defined by the metric $V_0^{*-1}$, and where $\chi^2$ is a chi-squared variable with degrees of freedom (df) equal to the number of omitted experiments minus 1). As a check on our results, we have therefore displayed these statistics for each iteration in Table 5. Note that the $\chi^2$ tail probability $P$ markedly increased when we eliminated the column for coke oven emissions.

The resulting tradeoff between predictive bias ($\sigma^*$) and predictive efficiency ($c_i^*$) is depicted graphically in Figure 6. Beyond the reduced 3 × 8 array, any further reduction in $\sigma^*$ is at the cost of a marked loss of precision. To be sure, we cannot unequivocally conclude that the predictions resulting from the 3 × 8 matrix are most pre-

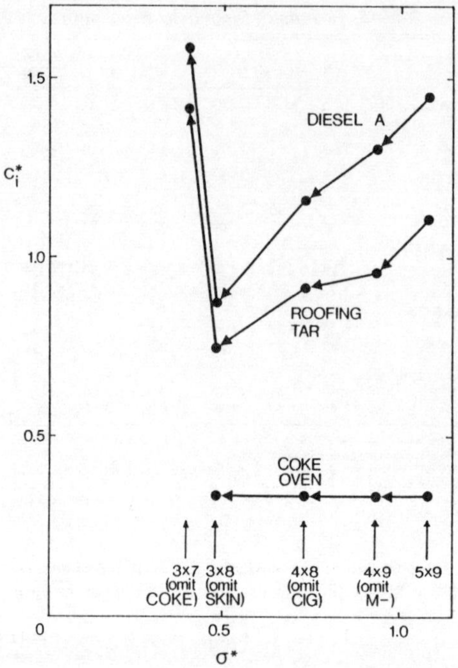


*Figure 6. TRADEOFF BETWEEN $\sigma^*$ AND $c_i^*$ FOR EACH ITERATION OF THE DIAGNOSTIC PROCEDURE. The excluded experiments and the resulting two-way tables at each iteration are given at the bottom of the figure.*

ferred. For many public health and environmental policy applications, however, a reduction in the extent of uncertainty about human risks is desired. To seek to eliminate less relevant experiments, so long as predictive efficiency is improved, appears to be appropriate for such applications.

Cigarette smoke, Figure 5 suggests, is a weaker human lung carcinogen, a weaker mouse skin tumor initiator, a more potent cell transforming agent, and a more potent direct mutagen than would be predicted from the underlying constant relative potency model. Peto (1977) has also noted that the mutagenic potency of cigarette smoke appears to be greater than would be expected from its systemic carcinogenicity. Such a finding may reflect chemical differences between cigarette smoke and the other combustion products in Table 1. (For example, cigarette smoke contains tobacco-specific nitrosamines and various heterocyclics.) Whatever the interpretation, our procedure leads us to eliminate a column containing the most precise human data. As a consequence, our estimate of $\sigma$ is derived primarily from the nonhuman data. We see no completely satisfactory response to this limitation other than to suggest, where possible, the inclusion of other precise human data.

Our results necessarily depend upon the choice of prior density $\pi(\sigma)$. However, this dependence should be important only when there are few experiments. If we assume Prior 2 for $\sigma$, then the row for mutagenesis without activator, the column for cigarette smoke, and the row for skin tumor initiation are likewise successively eliminated, and $\sigma^*$ is reduced from 1.12 to .73. Thereafter, however, no further reduction in $\sigma^*$ is possible. Prior 2 is so conservative that any fewer experiments diminish our confidence in interspecies extrapolation.

Our procedure represents only one possible strategy of stepwise search through the subsets of a given data set. In the current example, there is in fact no row or column in the $3 \times 8$ array whose elimination reduces both $\sigma^*$ and the $c_i^*$. But there could be a more efficient way of arriving at the frontier of the ($\sigma^*$, $c_i^*$) tradeoff. Adapting the methods in Efron and Morris (1973), we could replace our assumption that $V(\delta_i) = \sigma^2$ (for all $i$) with the assumption that $V(\delta_i) = \tau^2$ for some subset of experiments and then employ a joint prior distribution for ($\sigma$, $\tau$) to derive the posterior distribution of $\theta$. Unless the suspicious subset can be identified a priori, the present method appears to be much simpler in practice.

Despite such qualifications, we find the results of our diagnostic procedure intriguing. Assays for direct mutagenicity and tumor initiation have been excluded as less relevant. The remaining laboratory bioassays are designed to gauge an agent's interference with gene replication and cell differentiation. For the polyaromatic-containing emissions remaining in the $3 \times 8$ table, these biological processes could be the most relevant to human carcinogenesis.

## 7. CONCLUDING COMMENTS

The covariance components model of this article distinguishes between the measurement error of each experiment and an error of imperfect relevance among experiments. For the illustrative data set of Table 1, $\sigma^*$, the standard deviation of the error of relevance, was about 1.1 on a log scale. That is, with 95 percent credibility, extrapolations from nonhuman data to human cancer risks can be accurate at best to a factor of $\exp(2\sigma^*) \cong 9$ in either direction. From this result, we computed credible intervals for the human carcinogenic potencies of various environmental agents for which there were little or no direct human observations. We next applied the Bayesian statistical framework to incorporate prior information on the structural similarity of various compounds. Finally, we explored the problem of deciding which experiments are most relevant to human cancer risks. Mammalian cell transformation studies and indirect mutagenesis studies appeared to be most relevant for the class of polyaromatic hydrocarbon-containing emissions considered here.

A Bayesian approach makes explicit the genuine uncertainties in the problem at hand. When the scientist has a large body of data, as illustrated by Table 1, these uncertainties are likely to be reduced substantially. But when a compound's human cancer risk is predicted solely from experiments with that compound in a single nonhuman species such as rats (in effect, a 2 × 1 table with an empty cell), then the resulting human risk estimates will depend entirely on one's prior beliefs about the conversion factor between rats and humans or about the relevance of rat to man.

Moreover, performing many different nonhuman laboratory studies on only one environmental agent does not necessarily reduce the uncertainty about that agent's human health risks. Such an exercise corresponds to a $K \times 1$ case, where our statistical specification (3.2) reduces to $y_k = \mu + \alpha_k + \delta_k + \epsilon_k$. In this situation, profitably combining the experiments requires strong prior assumptions about the interspecies conversion factors $\{\alpha_k\}$. In the absence of such sharp prior information, one needs experiments in many species and many agents.

The results of combining experiments unavoidably depend upon the dose-response models used for each experiment. For example, we were constrained by the available data to apply a relative risk model to the human studies, where dosage was measured as cumulative lifetime exposure. We would have liked to estimate an additive risk model that distinguishes between the duration and intensity of exposure (Doll and Peto 1978). But our illustrative relative risk models did fit the original data reasonably (Harris 1981,1983).

Nor do we attach any special limitation to our reliance on the slope of a linear dose-response relationship. There are alternative methods of summarizing the results of an experiment, such as $TD_{50}$, the dose at which 50 percent of subjects develop tumor (Meselson and Russell 1977; Ames and Hooper 1978). A multivariate generalization could summarize each experiment by a vector of numbers. In this way, one could incorporate nonlinearities of dose-response or possible synergistic actions with other agents.

The validity of individual dose-response models is closely tied to the idea of exchangeability. The misspecification of the dose-response model for a particular experiment may affect the accuracy of the model relating all the experiments. Accordingly, exchangeability means that it is unknown a priori which dose-response model is more likely to be subject to misspecification. Perhaps misspecification is more probable in epidemiological studies based upon occupational exposures to relatively high doses. But empirical studies to support this presumption are not available.

Since there is uncertainty whether any single experiment on agent $l$ in species $k$ is representative of all such experiments on the same agent in the same species, it is arguable that the variances $c_{kl}^2$ are understated. The question is whether to allocate this source of uncertainty to the within-experiment errors $\epsilon_{kl}$ or the between-experiment errors $\delta_{kl}$. We prefer the latter. When multiple experiments on a given agent in a given species have been performed in different laboratories, informative priors like those of Section 5 would apply. In any case, when we arbitrarily doubled the $c_{kl}$ for all nonhuman experiments in Table 1, the resulting estimates of human carcinogenic potency were not substantially changed.

Is the normal prior structure of our model particularly objectionable? Deviations from the underlying constant relative potency model may certainly arise from biological processes that are non-Gaussian. But since the normal distribution has smaller tails than other likely candidates, any outliers from the underlying normal model will have a stronger contribution to the overall estimate of $\sigma$. In this respect, the use of the normal model is more conservative. On the other hand, the model may be misleading about the posterior probabilities of extreme values of the log carcinogenic potencies $\theta$, such as $\Pr\{\theta > X\beta + 3\sigma\}$.

The statistic $\sigma^{*2}$ corresponds to the mean squared error of prediction (on the log scale) for an experiment yet to be performed. It could thus be interpreted as the Bayesian risk of interspecies extrapolation under the loss function $(\theta_i - x_i\beta^*)^2$, where $\theta_i$ is estimated by $x_i\beta^*$. Although this loss function is symmetric with respect to proportional errors of prediction, it attaches greater weight to absolute underprediction of cancer risks. We regard this characteristic of $\sigma^*$ as appropriately conservative.

The present statistical model resolves the difficulty, encountered by Crouch and Wilson (1979,1981), of having to perform separate comparisons of carcinogenic potency in different pairs of species. It also satisfies these authors' desire for a systematic method of identifying potential exceptions to the underlying extrapolative model. Since the use of informative priors permits us to include multiple experiments on the same agent in the same species, we avoid the problem, encountered by these authors, of deciding which of several experiments to incorporate in the analysis. The use of informative priors could also incorporate information about the faulty design or execution of an experiment. Furthermore, a multivariate generalization of our model could incorporate the incidences of tumors of different sites. This would avoid the additional difficulty, encountered by Crouch and Wilson, of deciding which of several endpoints to choose.

The Bayesian paradigm provides a unified statistical theory for previous studies of interspecies extrapolation, in particular, comparisons of carcinogenic potency in mammals with genotoxic potency in short-term tests (Ames and Hooper 1978; Clive 1977; McCann et al. 1975,1981; Meselson and Russell 1977; Poirier and de Serres 1979; Purchase et al., 1978; Rinkus and Legator 1979). Instead of individual two-way comparisons of, say, a particular tumorigenesis model with a particular strain of Salmonella, it is possible to analyze multiple tests of carcinogenicity and genotoxicity simultaneously. The use of informative priors can take account of known differences in route of exposure or organ sensitivity in carcinogenesis studies and source of metabolic activator or target genetic locus in mutagenesis studies. A variant of the diagnostic procedure in Section 6 might be valuable in determining which genotoxic test is most relevant to animal carcinogenicity.

Finally, it is very often infeasible to conduct a precise epidemiological study of the human cancer risks of a particular agent. In such cases, the Bayesian model can help to decide which of many other unperformed experiments might be most informative about the agent's carcinogenicity.

## APPENDIX A: DOSE-RESPONSE MODELS

The slopes in Table 1 were estimated from linear dose-response models. In the nonhuman experiments, we observed $n_j$, the number of positive responses (skin papillomas, transformed cells, mutant colonies), and $N_j$, the number of surviving experimental sites (surviving mice, surviving SHE cells in culture, surviving L5178Y cells in culture), at each of several doses $d_j$. For each experiment, we assumed that the $n_j$ were independently Poisson distributed with mean values $N_j(\zeta + \xi d_j)$, where $\zeta$ and $\xi$ are unknown parameters.

In the epidemiological studies, we assumed that the relative risk of lung cancer was a linear function of cumulative lifetime exposure. For both the roofers study and the diesel bus workers study introduced in Section 5, we observed $n_{ij}$, the number of lung cancer cases, $E_{ij}$, the number of expected cases based on direct age standardization, and $d_{ij}$, the corresponding cumulative lifetime dose for subjects observed during calendar period $i$ in exposure group $j$. (The exposure groups reflected varying durations of union membership in the roofers study and diverse job categories in the bus workers study.) For these two studies, we assumed that, conditional upon $n_{i\cdot} = \sum_j n_{ij}$, the $n_{ij}$ were multinomially distributed with mean values $\zeta_i E_{ij}(1 + \xi d_{ij})$, where the $\zeta_i$ are constants of proportionality and $\xi$ is an unknown parameter. For both the coke oven worker and cigarette smoking studies, we observed $n_{ij}$, the number of lung cancer cases, $N_{ij}$, the number of person-years at risk, and $d_{ij}$, the corresponding cumulative lifetime dose for subjects in age group $i$ and exposure category $j$. (The exposure categories reflected various job groups in the coke oven worker study and differences in smoking habits in the cigarette smoking study.) For these two studies, we assumed that, conditional upon $n_{i\cdot} = \sum_j n_{ij}$, the $n_{ij}$ were multinomially distributed with mean values $\zeta_i N_{ij}(1 + \xi d_{ij})$ where, again, the $\zeta_i$ are constants of proportionality and $\xi$ is an unknown parameter.

The data for our analysis are based on the maximum likelihood estimates and standard errors of the dose-response slopes $\xi$. The estimation algorithms are given in DuMouchel (1981). Except for the column corresponding to cigarette smoke, the results are reported in Harris (1981,1983).

In our analysis of these experiments, the information about the log slope $i(\theta) = \partial^2 L(\theta)/\partial\theta^2$ (where $L(\theta)$ is the log likelihood for the experiment) always varied less, as a function of $\theta$, than did the information about the slope, namely $i(\theta)e^{-2\theta}$. We therefore found it more plausible that the estimates of $\theta$, rather than of $e^\theta$, have known variances. The parameterization in terms of log slopes excludes negative carcinogenic potencies. Unless an agent is suspected to be anti-carcinogenic, we do not regard this restriction as especially serious.

If $\theta$ denotes the log carcinogenic potency of a particular emission in humans, and if $d$ is the cumulative lifetime exposure to extractable organics (or whole condensate in the case of cigarettes), then the relative risk of lung cancer is $1 + de^\theta$. In units of $10^4$ $\mu g/m^3 \times$ years, typical values of the cumulative dose $d$ are: .30 for a coke oven worker aged 45–54 years; 1.1 for a roofer with 20 years' occupational exposure; .021 for a diesel bus garage worker with 20 years' exposure (diesel E introduced in Section 6); and about 210 for a smoker of one pack per day for 30 years. For an increase in the market share of light-duty diesel vehicles (such as diesel A) to 25 percent over a 20-year period, an urban resident would typically receive an ambient cumulative dose $d$ equal to about $5 \times 10^{-4}$. These dosage values are included solely to facilitate interpretation of our results. They are not intended as precise estimates of exposure to be used in computing excess cancer rates in selected populations.

## APPENDIX B: PROOF OF PROPOSITION 2

*Lemma.* Let $U$ be an $n \times m$ matrix of rank $m < n$, $I$ be the $n \times n$ identity martix, and $t$ be a scalar. Then as $t \to \infty$,

$$(I + tUU')^{-1} = I - U(U'U)^{-1}U' + t^{-1}(UU')^{+} + O(t^{-2}), \quad \text{(B.1)}$$

$$|I + tUU'| = t^m |U'U| [1 + t^{-1}\text{tr}(UU')^{+} + O(t^{-2})], \quad \text{(B.2)}$$

where $A^+$ is the Moore-Penrose pseudoinverse of $A$.

*Proof.* The $n \times n$ matrix $UU'$, which has rank $m$, can be represented as

$$UU' = \sum_{j=1}^{m} \lambda_j u_j u_j',$$

where $\{\lambda_j\}$ are the strictly positive characteristic roots of $UU'$ and $\{u_j\}$ are the corresponding characteristic vectors. The $n \times n$ identity matrix can be represented as

$$I = \sum_{j=1}^{m} u_j u_j' + \sum_{j=m+1}^{n} v_j v_j',$$

where the unit vectors $\{v_j\}$ are all orthogonal to the characteristic vectors $\{u_j\}$. Combining these two expressions, we have

$$I + tUU' = \sum_{j=1}^{m} (1 + t\lambda_j)u_j u_j' + \sum_{j=m+1}^{n} v_j v_j'. \quad \text{(B.3)}$$

Hence

$$(I + tUU')^{-1} = \sum_{j=1}^{m} (1 + t\lambda_j)^{-1} u_j u_j' + \sum_{j=m+1}^{n} v_j v_j'.$$

As $t \to \infty$,

$$\sum_{j=1}^{m} (1 + t\lambda_j)^{-1} u_j u_j' = t^{-1} \sum_{j=1}^{m} \lambda_j^{-1} u_j u_j' + O(t^{-2}).$$

Equation (B.1) now follows from our recognition that $\sum_{j=m+1}^{n} v_j v_j'$ is the orthogonal projection operator which maps $R^n$ onto the subspace of $R^n$ orthogonal to the columns of $U$ (namely $I - U(U'U)^{-1}U'$), while $\sum_{j=1}^{m} \lambda_j^{-1} u_j u_j'$ defines the pseudoinverse. Similarly, from (B.3),

$$|I + tUU'| = \prod_{j=1}^{m} (1 + t\lambda_j)$$

$$= t^m \prod_{j=1}^{m} \lambda_j \left\{ 1 + t^{-1} \sum_{j=1}^{m} \lambda_j^{-1} + O(t^{-2}) \right\}.$$

Equation (B.2) follows from our recognition that $|U'U| = \prod_{j=1}^{m} \lambda_j$ and $\mathrm{tr}(UU')^{+} = \sum_{j=1}^{m} \lambda_j^{-1}$.

The proof of Proposition 2 requires successive applications of this lemma. To show that $(Y - Xb)'[C + \sigma^2 I + XVX']^{-1}(Y - Xb)$ in (3.4) reduces to $Y'SY$ in (3.7), we use expansion formula (B.1), setting $U = W^{1/2}XV^{1/2}$ and $S = W^{1/2}[I - U(U'U)^{-1}U']W^{1/2}$. (The quadratic form $Y'SY$ in (3.7) is the sum of squared residuals of the weighted least squares regression of $Y$ on the columns of $X$, where the weights are the diagonal elements of $W$.) To show that the determinant $|C + \sigma^2 I + XVX'|^{-1/2}$ in (3.4) becomes proportional to $|W|^{1/2} |X'WX|^{-1/2}$ in (3.7), we use expansion formula (B.2) under the same definition of $U$. Expression (3.8) follows from our setting $V^{-1} = 0$ in (3.5). That expression (3.6) reduces to (3.9) is a result of formula (B.1), where we set $U = XV^{1/2}$ in order to evaluate the terms $(XVX' + \sigma^2 I)^{-1}$ in (3.6).

*[Received February 1981. Revised June 1982.]*

## REFERENCES

AMES, B.N., and HOOPER, K. (1978), "Does Carcinogenic Potency Correlate with Mutagenic Potency in the Ames Assay?" *Nature*, 274, 19–20.

CASTO, B.C., HATCH, G.C., HSUNG, S.L., HUISINGH, J.L., NESNOW, S., and WATERS, M.S. (1980), "Mutagenic and Carcinogenic Potency of Extracts of Diesel and Related Environmental Emissions: In Vitro Mutagenesis and Oncogenic Transformation," *Health Effects of Diesel Engine Emissions: Proceedings of an International Symposium* (2 vols.) eds. W.E. Pepelco, R.M. Danner, and N.A. Clarke, Cincinnati: U.S. Environmental Protection Agency.

CLIVE, D. (1977), "A Linear Relationship Between Tumorigenic Potency *In Vivo* and Mutagenic Potency at the Heterozygous Thymidine Kinase ($TK^{+/-}$) Locus of L5178Y Mouse Lymphoma Cells Coupled with Mammalian Metabolism," *Progress in Genetic Toxicology*, eds. D. Scott, B.A. Bridges, and F.H. Sobels, Amsterdam: Elsevier/North-Holland.

COCHRAN, W.G. (1980), "Summarizing the Results of a Series of Experiments," *Proceedings of the 25th Conference on Design of Experiments in Army Research, Development and Testing*, ARD Report 80-2, Washington, D.C.: U.S. Department of Defense, pp. 21–34.

CROUCH, E., and WILSON, R. (1979), "Interspecies Comparison of Carcinogenic Potency," *Journal of Toxicology and Environmental Health*, 5, 1095–1118.

——— (1981), "The Regulation of Carcinogens," *Risk Analysis*, 1: 47–57.

DE FINETTI, B. (1937), "La Prévision: Ses Lois Logiques, Ses Sources Subjectives," *Annales de l'Institut Henri Poincaré*, 7, 1–68. Reprinted in English in *Studies in Subjective Probability* (1964), eds. H.E. Kyburg, Jr. and H.E. Smokler, New York: John Wiley.

DEMPSTER, A.P., RUBIN, D.B., and TSUTAKAWA, R.K. (1981), "Estimation in Covariance Components Models," *Journal of the American Statistical Association*, 76: 341–353.

DOLL, R., and PETO, R. (1978), "Cigarette Smoking and Bronchial Carcinoma: Dose and Time Relationships Among Regular Smokers and Lifelong Nonsmokers," *Journal of Epidemiology and Community Health*, 32, 302–313.

DuMOUCHEL, W.H. (1981), "Documentation for CATDATA," Cambridge, Massachusetts: Massachusetts Institute of Technology, Statistics Center Technical Report.

EFRON, B. and MORRIS, C. (1973), "Combining Possibly Related Estimation Problems" (with Discussion), *Journal of the Royal Statistical Society*, Ser. B, 35, 379–421.

FREIREICH, E.J., GEHAN, E.A., RALL, D.P., SCHMIDT, L.H., and SKIPPER, H.E. (1966), "Quantitative Comparison of Toxicity of Anti-Cancer Agents in Mouse, Rat, Hamster, Dog, Monkey, and Man," *Cancer Chemotherapy Reports*, 50, 219–245.

HAMMOND, E.C., SELIKOFF, I.J., LAWTHER, P.L., and SEIDMAN, H. (1976), "Inhalation of Benzpyrene and Cancer in Man," *Annals of the New York Academy of Sciences*, 271, 116–124.

HARRIS, J.E. (1981), *Potential Risk of Lung Cancer from Diesel Engine Emissions*, Washington, D.C.: National Academy Press.

——— (1983), "Diesel Emissions and Lung Cancer" (with Discussion)," *Risk Analysis*, 3, 1, in press.

HARVILLE, D.A. (1974), "Bayesian Inference for Variance Components Using Only Error Contrasts," *Biometrika*, 61, 383–385.

——— (1977), "Maximum Likelihood Approaches to Variance Components Estimation and to Related Problems" (with Discussion), *Journal of the American Statistical Association*, 72, 320–340.

HUISINGH, J.L., BRADOW, R.L., JUNGERS, R.H, HARRIS, B.D., ZWEIDINGER, R.B., BUSHING, K.M., GILL, B.E., and ALBERT, R.E. (1980), "Mutagenic and Carcinogenic Potency of Extracts of Diesel and Related Environmental Emissions: Study Design, Sample Generation, Collection, and Preparation," *Health Effects of Diesel Engine Emissions: Proceedings of an International Symposium* (2 vols.), eds. W.E. Pepelco, R.M. Danner, and N.A. Clarke, Cincinnati: U.S. Environmental Protection Agency.

KAHN, H.A. (1966), "The Dorn Study of Smoking and Mortality Among U.S. Veterans: Report on Eight and One-Half Years of Observations," *Epidemiological Approaches to the Study of Cancer and other Chronic Diseases*, ed. W. Haenszel, National Cancer Institute Monograph No. 19, pp. 1–125.

LINDLEY, D.W., and SMITH, A.F.M. (1972), "Bayes Estimates for the Linear Model" (with Discussion), *Journal of the Royal Statistical Society*, Ser. B, 34, 1–41.

LLOYD, W.J. (1971), "Long-Term Mortality Study of Steelworkers v. Respiratory Cancer in Coke Plant Workers," *Journal of Occupational Medicine*, 13, 53–68.

MAZUMDAR, S., REDMOND, C., SOLLECITO, W., and SUSSMAN, N. (1975), "An Epidemiological Study of Exposure to Coal Tar Pitch Volatiles Among Coke Oven Workers," *Journal of the Air Pollution Control Association*, 25, 382–389.

McCANN, J., CHOI, E., YAMASAKI, E., and AMES, B.N. (1975), "Detection of Carcinogens as Mutagens in the Salmonella/Microsome Test: Assay of 300 Chemicals," *Proceedings of the National Academy of Sciences, U.S.A.*, 72, 5135–5139.

McCANN, J., HORN, L., LITTON, G., KALDOR, J., MAGAW, R., BERNSTEIN, L., and PIKE, M. (1981), "Short-Term Tests for Carcinogens and Mutagens: A Data-Base Designed for Comparative, Quantitative Analysis," *Structure Activity Correlation as a Predictive Tool in Toxicology: Fundamentals, Methods, and Applications*, ed. L. Goldberg, Washington, D.C.: Hemisphere Press.

MESELSON, M., and RUSSELL, K. (1977), "Comparisons of Carcinogenic and Mutagenic Potency," *Origins of Human Cancer*, eds. H.H. Hiatt, J.D. Watson, and J.A. Winsten, Cold Spring Harbor: Cold Spring Harbor Laboratory.

MITCHELL, A.D., EVANS, E.L., JOTZ, M.M., RICCIO, E.S., MORTELMANS, K.E., and SIMMON, V.F. (1980), "Mutagenic and Carcinogenic Potency of Extracts of Diesel and Related Environmental Emissions: In Vitro Mutagenesis and DNA Damage," *Health Effects of Diesel Engine Emissions: Proceedings of an International Symposium* (2 vols.) eds. W.E. Pepelco, R.M. Danner, and N.A. Clarke, Cincinnati: U.S. Environmental Protection Agency.

NESNOW, S., TRIPLETT, L.L., and SLAGA, T.J. (1981), "Tumorigenesis of Diesel Exhaust, Gasoline Exhaust, and Related Emission Extracts on Sencar Mouse Skin," *Short-Term Bioassays in the Analysis of Complex Environmental Mixtures II*, eds. M.D. Waters, S.S. Sandhu, J.L. Huisingh, L. Claxton, and S. Nesnow, New York: Plenum Press.

PATTERSON, H.D., and THOMPSON, R. (1971), "Recovery of Interblock Information When Block Sizes Are Unequal," *Biometrika*, 58, 545–554.

PETO, R. (1977), "Epidemiology, Multistage Models, and Short-Term Mutagenicity Tests," *Origins of Human Cancer*, eds. H.H. Hiatt, J.D. Watson, and J.A. Winsten, Cold Spring Harbor: Cold Spring Harbor Laboratory.

POIRIER, L.A., and DE SERRES, F.J. (1979), "Initial National Cancer Institute Studies on Mutagenesis as a Prescreen for Chemical Carcinogenesis: An Appraisal," *Journal of the National Cancer Institute*, 62, 919–926.

PURCHASE, I.F.H. (1980), "Inter-Species Comparisons of Carcinogenicity," *British Journal of Cancer*, 41, 454–468.

——— (1982), "An Appraisal of Predictive Tests for Mutagenicity," *Mutation Research*, in press.

PURCHASE, I.F.H., LONGSTAFF, E., ASHBY, J., STYLES, J.A., ANDERSON, D., LEFEVRE, P.A., and WESTWOOD, F.R. (1978), "An Evaluation of Six Short-Term Tests for Detecting Organic Chemical Carcinogens," *British Journal of Cancer*, 37, 873–903.

RINKUS, S.J., and LEGATOR, M.S. (1979), "Chemical Characterization of 465 Known or Suspected Carcinogens and Their Correlation with Mutagenic Activity in the Salmonella Typhimurium System," *Cancer Research*, 39, 3289–3318.

SKOPEK, T.R., LIBER, H.L., KADEN, D.A., and THILLY, W.G. (1978), "Relative Sensitivities of Forward and Reverse Mutation Assays in *Salmonella typhimurium*," *Proceedings of the National Academy of Sciences, U.S.A.*, 75, 4465–4469.

SMITH, A.F.M. (1973), "Bayes Estimates in One-Way and Two-Way Models," *Biometrika*, 60, 319–329.

U.S. ENVIRONMENTAL PROTECTION AGENCY (1979), "The Carcinogen Assessment Group's Final Risk Assessment on Coke Oven Exposure," Technical Report, Washington, D.C.

YATES, F., and COCHRAN, W.G. (1938), "The Analysis of Groups of Experiments," *Journal of Agricultural Science*, 28, 556–580, reprinted in F. Yates (1970), *Experimental Design: Selected Papers of Frank Yates*, London: Griffin.

# Comment

## Carcinogenic Risk Assessment

DANIEL KREWSKI*

Increasing public concern over the presence of potentially carcinogenic agents present in the environment has led to a tremendous increase in research on carcinogenesis and in efforts towards the development of new strategies for the regulation of carcinogens (Krewski and Brown 1981). In Canada, cancer currently accounts for about 19 and 22 percent of all deaths among males and females, respectively (HWC 1979). Working with United States statistics, Doll and Peto (1981) recently speculated that environmental contamination may account for only about two percent of the total cancer burden, while occupational exposure was thought to be responsible for a further two to eight percent. Other opinions, however, have been somewhat higher (Epstein 1979).

In pragmatic terms, the process of carcinogenic risk assessment may be described in terms of four distinct phases (Whyte and Burton 1980). Initially, *risk identification* involves demonstrating that a well-defined hazard exists. Once a carcinogenic agent has been clearly identified, *risk estimation* techniques may in theory be used to determine the actual magnitude of the risk posed to man. Next, *risk evaluation* procedures may be employed to assess the societal significance of this risk in light of any perceived benefits associated with the agent in question. Finally, suitable strategies for *risk management* may be invoked in order to control exposure to environmental carcinogens to the extent possible.

The four major methods for identifying chemical carcinogens are (a) examination of the physical and chemical properties of the test compound, (b) short-term in vitro and in vivo tests designed to detect certain critical events of the carcinogenic process, (c) long-term bioassay experiments using laboratory animals, and (d) epidemiological studies of human populations (OTA 1981). Risk identification based solely on an analysis of chemical structure and activity is unfortunately of only limited value since slight alterations in molecular configuration can have marked effects on the carcinogenic potential of a compound (Lawley 1976). Although over 100 rapid and relatively inexpensive short-term tests for specific mutagenic effects have been described in the literature (Hollstein and McCann 1979), the predictive value of such tests with respect to carcinogenicity in mammalian species remains to be more clearly established (Krewski et al. 1982). Currently, the most widely employed method of identifying chemical carcinogens is the long-term carcinogen bioassay in small rodents (Gart et al. 1979; IARC 1980a). Although epidemiological investigations can provide the most direct evidence of carcinogenic effects in man, studies of human populations are often of limited sensitivity due to low levels of human exposure and may be subject to unknown biases due to the presence of con-

* Daniel Krewski is Chief, Biostatistics and Computer Applications Division, Environmental Health Directorate, Health and Welfare Canada, Ottawa, Ontario, Canada K1A 0L2.

founding risk factors (IRLG 1980). Since human exposure necessarily precedes the initiation of epidemiological studies, moreover, reliance solely on epidemiological data is inconsistent with preventitive regulatory philosophy (Munro et al. 1983). To date, epidemiological studies have identified some 36 chemical agents or processes as being human carcinogens (IARC 1980b), whereas a much larger number of compounds has been shown to induce cancer in animals (Sax 1981).

Once the existence of a carcinogenic hazard has been demonstrated in qualitative terms, the next step in the risk assessment process requires a quantitative estimate of the magnitude of risk posed to man. For carcinogens identified on the basis of bioassay results obtained in animal models for man, this involves extrapolation from the accelerated doses necessarily used in animal tests to the lower doses typical of human exposure to environmental contaminents (Armitage 1982; Krewski et al. 1983). In addition, the results obtained in animals must somehow be converted to the human situation (Tomatis 1979; Munro and Krewski, 1983).

Because of the uncertainties involved in both low-dose and interspecies extrapolation, several rough yet robust measures of potency which may be used to at least rank carcinogenic agents according to their strength have recently been proposed (Purchase 1982; Squire 1981). This pioneering article by DuMouchel and Harris, which introduces a Bayes procedure for combining potency measures obtained in diverse species under the assumption of constant relative potency of two carcinogenic agents across species, represents a significant advance in this regard. By clearly distinguishing between sampling and nonsampling errors and examining the exchangeability of the species-agent interaction effects, moreover, DuMouchel and Harris are able to identify those experiments that best satisfy the constant relative potency hypothesis. Other notable features of the proposed approach include the apparent robustness of the Bayes estimators with respect to the prior information on the adequacy of the constant relative potency model embodied in $\Pi(\sigma)$ for moderately large data bases and the potential for incorporating available knowledge concerning differences in metabolic and pharmacokinetic processes between species (Dedrick and Bischoff 1980) in the prior distribution on species-agent interaction effects.

Although the application of a hierarchical Bayesian model to the problem of combining relevant information on carcinogenic potency is conducted with great care, the reliance on linear dose-response relationships as in Crouch and Wilson (1981) seems an unnecessary oversimplification (Brown 1981). If the estimates of potency are to be interpreted in a quantitative biological sense, moreover, this may lead to a model misspecification error not accounted for in the posterior standard deviations of the estimates (Hoel 1981). Likewise, the biological rationale offered in Appendix 1 in support of the linearity assumption does not fully exploit our current knowledge of the possible underlying mechanisms of mutagenesis and carcinogenesis (Margolin et al. 1981; Kalbfleisch et al. 1983). These problems could be avoided by the use of more flexible models providing for nonlinear dose-response curves and a corresponding potency measure such as the $TD_{50}$ suggested by the authors. The use of a lower quantile such as the $TD_{10}$, which might still be estimated with a reasonable degree of accuracy and yet come closer to the low dose region of most interest, might also be considered. In either case, however, a suitable means of correcting the quantiles for the presence of background response is required (Hoel 1980).

Regardless of the summary statistic used to define potency, the estimates of potency based on similar studies with different agents should be comparable with respect to their context of observation. The slopes of the near linear dose response curves for liver neoplasia induced in mice as a result of chronic exposure to the experimental carcinogen 2-acetylaminofluorene, for example, appear to vary between 18, 24, and 33 months on test (Littlefield et al. 1980). By ensuring that data obtained from similar experiments on different agents are standardized with respect to any differences in design parameters such as time on test, one may avoid artifactual species-agent interactions resulting from differences in experimental protocol.

The reliance on linear dose-response relationships notwithstanding, the procedure of DuMouchel and Harris represents a potentially useful tool for the estimation of carcinogenic potency. Future refinements of this technique to exploit sharp prior information on biological differences between species responsible for certain species-agent interactions may further enhance its applicability. The derived potency estimates will no doubt be useful at the risk evaluation stage, in which formal decision procedures ranging from simple applications of the zero risk principle to more elaborate risk/benefit analyses may be employed (Lave 1981). In the final stage of the risk assessment process, risk management strategies ranging from public education to the promulgation of specific regulations may be adopted as circumstances dictate (Munro and Krewski 1981).

## REFERENCES

ARMITAGE, P. (1982), "The Assessment of Low-Dose Carcinogenicity," *Biometrics*, 38, (Supplement on Current Topics in Biostatistics and Epidemiology), 119–129.

BROWN, C. (1981), "Comment on *The Regulation of Carcinogens* by Crouch and Wilson," *Risk Analysis*, 1, 105–106.

CROUCH, E., and WILSON, R. (1981), "The Regulation of Carcinogens," *Risk Analysis*, 1, 47–57.

DEDRICK, R.L., and BISCHOFF, K.B. (1980), "Species Similarities in Pharmacokinetics," *Federation Proceedings*, 39, 54–59.

DOLL, R., and PETO, R. (1981), "The Causes of Cancer: Quantitative Estimates of Avoidable Risks of Cancer in the United States Today," *Journal of the National Cancer Institute*, 66, 1193–1308.

EPSTEIN, S. (1979), *The Politics of Cancer*, Garden City, New York: Anchor Press.

GART, J.J., CHU, K.C., and TARONE, R. (1979), "Statistical Issues in the Interpretation of Chronic Bioassay Tests for Carcinogenicity," *Journal of the National Cancer Institute*, 62, 957–974.

HOEL, D.G. (1980), "Incorporation of Background Response in Dose-Response Models," *Federation Proceedings*, 39, 73–75.

——— (1981), "Carcinogenic Risk," *Risk Analysis*, 1, 63–64.

HOLLSTEIN, M., and McCANN, J. (1979), "Short-Term Tests for Carcinogens and Mutagens," *Mutation Research*, 65, 133–226.
HWC: HEALTH and WELFARE CANADA (1979), *Cancer Patterns in Canada 1931–1974*. Bureau of Epidemiology, Health & Welfare Canada, Ottawa.
IARC: INTERNATIONAL AGENCY FOR RESEARCH ON CANCER (1980a), *Long-Term and Short-Term Screening Assays for Carcinogens: A Critical Appraisal*. IARC Monographs, Supplement 2. IARC, Lyon.
——— (1980b), "An Evaluation of Chemicals and Industrial Processes Associated with Cancer in Humans Based on Human and Animal Data: IARC Monographs Volumes 1 to 20," *Cancer Research*, 40, 1–12.
IRLG: INTERAGENCY REGULATORY LIAISON GROUP (1980), "Scientific Bases for the Identification of Potential Carcinogens and Estimation of Risks," *Annual Review of Public Health*, 1, 345–393.
KALBFLEISCH, J.D., KREWSKI, D., and VAN RYZIN, J. (1983), "Dose Response Models for Time to Response Toxicity Data," *Canadian Journal of Statistics*, To appear.
KREWSKI, D., and BROWN, C. (1981), "Carcinogenic Risk Assessment: A Guide to the Literature," *Biometrics*, 37, 353–366.
KREWSKI, D., CLAYSON, D., COLLINS, B., and MUNRO, I. (1982), "Toxicological Procedures for Assessing the Carcinogenic Potential of Agricultural Chemicals," in *Genetic Toxicology: An Agricultural Perspective*, (eds. R.A. Fleck and A. Hollaender), New York: Plenum, 461–497.
KREWSKI, D., CRUMP, K., FARMER, J., GAYLOR, D., HOWE, R., PORTIER, C., SALSBURG, D., SIELKEN, R.L., and VAN RYZIN, J. (1983), "A Comparison of Statistical Methods for Low Dose Extrapolation Using Time to Tumour Data," *Fundamental and Applied Toxicology*, To appear.
LAVE, L. (1981), *The Strategy of Social Regulation: Decision Frameworks for Policy*, Washington, D.C.: Brookings Institution.
LAWLEY, P.D. (1976), "Carcinogenesis by Alkylating Agents," *Chemical Carcinogens* (C.E. Searle, ed.), ACS Monograph 173, Washington, D.C.: American Chemical Society, 83–244.
LITTLEFIELD, N.A., FARMER, J.H., GAYLOR, D.W., and SHELDON, W.G. (1980), "Effects of Dose and Time in a Long-Term, Low-Dose Carcinogenic Study," *Journal of Environmental Pathology and Toxicology*, 3, 17–34.
MARGOLIN, B., KAPLAN, N., and ZEIGER, E. (1981), "Statistical Analysis of the Ames Salmonella/Microsome Test," *Proceedings of the National Academy of Science*, 78, 3779–3783.
MUNRO, I., and KREWSKI, D. (1981), "Risk Assessment and Regulatory Decision Making," *Food and Cosmetics Toxicology*, 19, 549–560.
——— (1983), "Regulatory Considerations in Risk Management," in *Regulatory Considerations in Risk Management*, ed. H. Stich, Boca Raton: CRC Press, in press.
MUNRO, I.C., MILLER, C.T., and KREWSKI, D. (1983), "Regulatory Control of Environmental Chemicals: A Canadian Viewpoint," *Journal of the American College of Toxicology*, 2, in press.
OTA: OFFICE OF TECHNOLOGY ASSESSMENT (1981), *Assessment of Technologies for Determining Cancer Risks from the Environment*, U.S. Government Printing Office, Washington, D.C.
PURCHASE, I.F.H. (1982), "Quantitative Estimation of Risk to Human Health: A Comparison of the Potency of the Mutagenic Effect of Chemicals in Short-term Tests With Their Carcinogenic Effect in Rodent Carcinogenicity Experiments," Proceedings of the Workshop on Quantitative Estimation of Risk to Human Health from Chemicals, Rome, July 12, 1982, to appear.
SAX, N.I. (1981), *Cancer Causing Chemicals*, New York: Van Nostrand Reinhold.
SQUIRE, R. (1981), "Ranking Animal Carcinogens: A Proposed Regulatory Approach," *Science*, 214, 877–880.
TOMATIS, L. (1979), "The Predictive Value of Rodent Carcinogenicity Tests in the Evaluation of Human Risks," *Annual Review of Pharmacology and Toxicology*, 19, 511–530.
WHYTE, A.V., and BURTON, I. (eds.) (1980), *Environmental Risk Assessment*, New York: John Wiley.

# Comment

ADRIAN F.M. SMITH*

## INTRODUCTION

Anyone who engages in the process of extracting and summarizing information from complex data sets by means of statistical models is painfully aware of the dilemma involved in choosing the level of detail to be incorporated in the model. As Bellman (1957, Preface) has remarked: "The Scientist, like the Pilgrim, must wind a straight and narrow path between the Pitfalls of Oversimplification and the Morass of Overcomplication." The authors of this paper have succeeded brilliantly in doing just that and are to be congratulated on an elegant and stimulating piece of work.

I have no specific comments to make on the biological background to the problem or on the choice of the data sets. I would, however, like to discuss a couple of issues related to the hierarchical Bayesian methodology (presented by Lindley and myself in 1972 in the general context of linear models, but extensively used much earlier than that in contingency table analysis; see Good 1980 for a review of such applications).

* Adrian F.M. Smith is Professor, Department of Mathematics, University of Nottingham, University Park, Nottingham NG72RD, England.

## HIERARCHICAL MODELS AND "EMPIRICAL BAYES"

A general hierarchical model can be written in the form

$$\begin{cases} p(\mathbf{x} \mid \boldsymbol{\theta}) \\ p(\boldsymbol{\theta} \mid \boldsymbol{\phi}) \\ p(\boldsymbol{\phi}), \end{cases}$$

where the first stage relates observables, $\mathbf{x}$, to parameters $\boldsymbol{\theta}$; the second stage specifies the nature of the relationship existing among the parameters $\boldsymbol{\theta}$ (with the dimension of $\boldsymbol{\phi}$ typically much less than that of $\boldsymbol{\theta}$); and the third stage incorporates further information, if any, about the general form of relationship identified in the second stage.

If interest centers on inference for $\theta$, we have

$$p(\theta \mid \mathbf{x}) = \int p(\theta \mid \mathbf{x}, \phi) p(\phi \mid \mathbf{x}) d\phi,$$

where $p(\phi \mid \mathbf{x})$ is obtained, via Bayes theorem, from the likelihood given by collapsing the first two stages of the hierarchy into $p(\mathbf{x} \mid \phi)$. In many situations, $\phi$ is rather low-dimensional compared with $\theta$ and, given substantial data $\mathbf{x}$, $p(\phi \mid \mathbf{x})$ will often be relatively sharply peaked. This implies that

$$p(\theta \mid \mathbf{x}) \approx p(\theta \mid \mathbf{x}, \hat{\phi}),$$

where $\hat{\phi}$ is the marginal posterior mode for $\phi$ (which is, more or less, the maximum likelihood estimate of $\phi$ from $p(\mathbf{x} \mid \phi)$ provided $p(\phi)$ is fairly uninformative).

The approximation $p(\theta \mid \mathbf{x}, \hat{\phi})$ thus has the form that *would have arisen* if (a) we had estimated $\phi$ from the first two stages of the hierarchy; (b) we had then formed the posterior for $\theta$ in the usual way on the basis of the prior specification $p(\theta \mid \hat{\phi})$. It is thus *as though* "the prior is estimated from the data"; and this, no doubt, is what leads people to use the term "empirical Bayes" (as the authors do in Section 3). I rather wish that such usage could be avoided. The analysis outlined above makes clear that we are actually dealing with a particular kind of *approximation in the context of a wholeheartedly Bayesian approach*. No distinct "empirical Bayes" philosophy or methodology is involved.

In fact, the so-called "empirical Bayes" approximation might be further refined by noting that, considering $\hat{\phi} \approx E(\phi \mid \mathbf{x})$, we are essentially using the approximation

$$p(\theta \mid \mathbf{x}) = E[p(\theta \mid \mathbf{x}, \phi)] \approx p(\theta \mid \mathbf{x}, \hat{\phi}),$$

where $E$ denotes expectation with respect to $p(\phi \mid \mathbf{x})$. This suggests—taking the case of a single parameter $\phi$, for convenience of exposition—that the form

$$p(\theta \mid \mathbf{x}) \approx p(\theta \mid \mathbf{x}, \hat{\phi}) + \tfrac{1}{2} V_\phi p''(\theta \mid \mathbf{x}, \hat{\phi})$$

might be adopted, where $V_\phi = \text{var}(\phi \mid \mathbf{x})$ and $p''$ denotes the second derivative of $p(\theta \mid \mathbf{x}, \phi)$ with respect to $\phi$.

## MORE FLEXIBLE FORMS OF HIERARCHICAL MODEL

There seem to be some contexts in which the assumption of a particular form for the second stage of a hierarchy is unobjectionable and commands widespread assent (as, for example, in certain applications of the "random effects" model in one-way ANOVA—equivalent, for a subjectivist, to the assumption of a straightforward form of exchangeability). There are other contexts, however—and that discussed in this paper is surely one!—where a critical evaluation of the particular form of assumption adopted is of central importance in the analysis. In such situations, I think it is important to search, in the appropriate stage of the hierarchy, for "elaborated" forms of representation (as suggested, for example, in Box 1980), which enable a certain amount of probing of "the assumption we first thought of" to take place.

Let me illustrate this idea by considering the authors' form of prior specification for the $\delta_{kl}$ parameters (in Section 3). Changing to a single subscript notation, $\Delta_1, \ldots, \Delta_N$, for convenience, the particular kind of *exchangeability* assumption adopted by the authors leads to a joint prior density of the form

$$p^{(0)}(\Delta_1, \ldots, \Delta_N) = \int \prod_{i=1}^{N} p(\Delta_i \mid \sigma)\pi(\sigma) d\sigma,$$

where $p(\Delta_i \mid \sigma)$ denotes an $N(0, \sigma^2)$ density for $\Delta_i$.

However, it is important to realize that there are many other representations leading to an exchangeable form. For example, we could take, for any $k > 1$,

$$p^{(1)}(\Delta_1, \ldots, \Delta_N) = \sum_{j=1}^{N} (N)^{-1} p_j(\Delta_1, \ldots, \Delta_N),$$

where

$$p_j(\Delta_1, \ldots, \Delta_N) = \int p(\Delta_j \mid k\sigma) \prod_{i \neq j} p(\Delta_i \mid \sigma)\pi(\sigma) d\sigma,$$

and thus explicitly consider the possibility of (an unknown) one of the $\Delta_j$ being rather different from the others. Clearly, such an idea could be extended in an obvious way to more than one "outlier" among the $\Delta_j$. Moreover, if $p^{(r)}(\Delta_1, \ldots, \Delta_N)$ were the density corresponding to an assumption of $r$ "outliers" the form

$$p^*(\Delta_1, \ldots, \Delta_N) = \sum_{r=0}^{M} \pi_r p^{(r)}(\Delta_1, \ldots, \Delta_N),$$

with $\pi_0 + \cdots + \pi_M = 1$, provides an exchangeable prior representation allowing for *up to M* "outliers." Prior to posterior analysis yields the form

$$p^*(\Delta_1, \ldots, \Delta_N) \mid \text{data}) = \sum_{r=0}^{M} \pi_r(\text{data}) p^{(r)}(\Delta_1, \ldots, \Delta_N \mid \text{data}),$$

where the posterior weights, $\pi_r$(data), provide the basis for a rather detailed assessment to be made of the heterogeneity of the $\Delta_j$. In a similar manner, we could calculate posterior probabilities of "outlyingness" for specified subsets of the $\Delta_j$'s.

Ideas of this kind are currently being explored by L.I. Pettit in a Ph.D. thesis being written for submission at Nottingham University. Despite the added complexity of the prior formulation, such analyses are relatively tractable (since linear combinations of conjugate forms are also conjugate), and it would be interesting to see how such an extended hierarchical methodology performed on the data given in this paper.

## REFERENCES


BELLMAN, R. (1957), *Dynamic Programming*, Princeton University Press.

BOX, G.E.P. (1980), "Sampling and Bayes' Inference in Scientific Modelling and Robustness (with discussion)," *Journal of the Royal Statistical Society*, Ser. A, 143, 383–430.

GOOD, I.J. (1980), "Some History of the Hierarchical Bayesian Methodology," in *Bayesian Statistics*, ed. J.M. Bernardo et al., Valencia University Press.

# Comment

ROBERT E. KASS*

Laboratory experiments often provide strong evidence about carcinogenesis or mutagenesis at high doses of a toxic agent in certain species. With specific assumptions about the relevance of this evidence, could some of its strength be borrowed, so to Tukey-speak, in order to better estimate the effect of exposure to the agent on the probability of cancer in humans? DuMouchel and Harris suggest that it can and they propose Bayesian analysis of covariance components models as an appropriate technology for accomplishing the task.

The great benefit of the Bayesian approach to this problem is that it makes precise the assessments of relevance and uncertainty in related results. Just as the purpose of a Bayesian theory of scientific inference is, as Jeffreys (1967, p. 8) put it, to "tidy up" the process, so too might Bayesian methods improve what advisors to policy makers do informally. In principle, the paradigm is simple and, in its ability to accommodate a variety of beliefs, it is comprehensive. Yet the Bayesian approach has its difficulties, for while it is surely desirable to express beliefs explicitly, in particular through models, it is often difficult to do so accurately. Lurking beside each analysis are the interrelated dangers of oversimplification, overstated precision, and neglect of beliefs other than the analyst's.

In the authors' model (3.2), the normal exchangeable prior on $\delta$ is probably the most important cause for concern. Interspecies comparisons begin with the assumption that results among different species and toxic agents are somehow related, and that the data carry information about the relationships. The normal exchangeable prior, however, represents a firm belief in a very simple situation; its use entails a strong statement about the behavior of the interactions. It is plausible that some groups of species are sufficiently similar with respect to the toxic action of some groups of agents, that, within those groups, model (3.2) would not be grossly inaccurate. In principle, the model could be modified to allow for grouping and, in Section 5, the authors provide an illustration of this sort of modification. However, it is not clear how Table 1 should be refined so as to improve the accuracy of the model. Thus, there is a serious complication associated with missing information as to group membership, and I would hesitate to apply the model without additional theoretical or empirical knowledge.

Overstated precision is a potential problem in the authors' model. Not only does each estimate "shrink" toward others in its row and column, but its variance is reduced, as well. If there is doubt about the accuracy of the model, then the resulting variance is inappropriately small. Consider, for example, the value $c^* = 1.02$ for roofing tar in Table 2. My own reaction to that number is that it is too small and, to interpret it, I crudely but quickly modify model (3.2) by mixing it with the model that assumes no relationship among the entries in Table 1. In addition, since the original data come from an observational study, the standard deviation of 1.41 might be considered too small, and, before mixing, it could be replaced with a value considered more realistic. Thus, my modified posterior, based on viewing Table 2, would be roughly equal to a mixture of the reported posterior and the variance-inflated original (normally distributed) data.

Even greater overstated precision results from the model selection described in Section 6. Evaluation of precision is worrisome whenever the same data are used both to select the model and to make inferences. Here, the problem is compounded. Not only is each estimate affected by the selection of entries in its row and column, but both the estimate and its variance are affected by increased homogeneity of the interactions. In particular, I would not agree with the conclusion that knowledge of the human roofing tar log slope may be summarized by a Normal$(1.53, (.74)^2)$ distribution.

The concern expressed here about difficulties in applying the authors' methodology raises a general question of policy analysis: how much inferential and decision-theoretic formalism should be used? A second issue that arises in this and many other policy problems is, at what stage in the process should enter the desire for adequate caution with regard to human risk? In arriving at a policy decision, scientific evidence must be assessed and social values must be evaluated and considered. Some argue that it is important to distinguish, as much as possible, the assessment phase from the evaluation and decision-making phase of policy formulation. A thorough assessment would then include descriptions of the consequences of proposed actions, the likelihoods of their occurrence, and the actions that should be taken under each possible set of assignments of value to the consequences. The use of Bayesian decision theory in this context is described in detail by Keeney and Raiffa (1976). A complication is that probabilities under various assumptions may differ widely (as they usually do for dose-response models). While, in principle, the probability of an outcome may be expressed as a mixture over the assumptions, in practice the probabilities of the validity of the assumptions may be extremely difficult to obtain. One solution to this problem would further complicate an already elaborate scheme by providing assessments according to each of the many possible combinations of

* Robert E. Kass is Assistant Professor, Department of Statistics, Carnegie-Mellon University, Pittsburgh, PA 15213.

assumptions that could reasonably be entertained. This seems to me to be the best way of facing the dangers mentioned above when sufficient time and resources are available. (Other, non-Bayesian, methodologies for expressing belief attempt to deal with some of these problems in the inferential context; see, e.g., Shafer 1982.)

On the other hand, varying degrees of formal structure may be thought suitable: where precise expressions of knowledge are dubious, informal assessment may be preferred. Furthermore, by anticipating value judgments, assessment may be simplified. A conservative strategy is often adopted: at each stage in modeling and analysis the route that leads to the greater estimate of human risk is taken. This approach is more restrictive than the Bayesian decision-theoretic procedure referenced above, in which appropriate prudence would be applied during the evaluation phase of the process.

The authors seem to have mixed sympathies with regard to conservativeness in analysis. Their methodology is developed as an alternative to the conservative procedure that uses only the worst case among species for a particular agent (see Crouch and Wilson 1981). Meanwhile, their analysis begins with the conservative assumption of linear dose-response curves and they consider some conservative features of their model appropriate (see Section 7).

The paper by DuMouchel and Harris gives much insight into the problem of combining information about human cancer from different but related sources, and it offers an advance from the method recommended by Crouch and Wilson. Here, as elsewhere, the Bayesian approach is especially helpful in forming a conceptual foundation for inference and decision making. As data on interspecies comparisons accumulate, formal methods that utilize the available information will become increasingly useful. The scheme laid out and discussed by the authors will then provide a Bayesian path from cognitive framework to policy analysis.

### REFERENCES

CROUCH, EDMUND, and WILSON, RICHARD (1981), "Regulation of Carcinogens (with discussion)," *Risk Analysis*, 1, 47–66.

JEFFREYS, HAROLD (1967), *Theory of Probability* (3rd ed.), London: Oxford University Press.

KEENEY, RALPH, and RAIFFA, HOWARD (1976), *Decision Making With Multiple Objectives: Preferences and Value Tradeoffs*, New York: John Wiley.

SHAFER, GLENN (1982), "Lindley's Paradox (with discussion)," *Journal of the American Statistical Association*, 77, 325–351.

# Rejoinder

WILLIAM H. DuMOUCHEL and JEFFREY E. HARRIS

We are grateful to Messrs. Krewski, Smith, and Kass for an insightful and stimulating discussion. Krewski's synopsis of the literature on carcinogenic risk assessment is especially welcome.

Smith and Kass inquire whether our exchangeable prior for the species-agent interactions might be too strong an assumption. Smith suggests more general forms of exchangeability. Kass finds exchangeability itself to be an oversimplification and urges prior grouping of agents or species along the lines of our Section 5. Kass cautions, however, that prior information about group membership may be sparse.

The exchangeability assumption treats distinctive effects symmetrically. Thus, in an assessment of the relevance of laboratory experiments to humans, inconsistent responses among a battery of animal tests are just as informative as inconsistencies between human and animal responses. It may be desirable to refine the exchangeability assumption by reference to current knowledge of species differences in bioavailability, metabolism, genetic repair mechanisms, and the like. In terms of strategy of presentation, however, we regarded exchangeability as an appropriate starting point.

In our illustrative example of Table 1, we could have assumed that whole animal experiments in mice were more relevant to humans than were mammalian cell experiments. But we recognized that the tumor initiation endpoint gauged in the Sencar mice studies might be no more relevant to the origin of human lung cancer from combustion mixtures than cell transformation or mutagenesis. Likewise, we had a vague inkling that cigarette smoke was less relevant to the other combustion mixtures because of its distinct chemical profile. But we were not sure if the presence of tobacco-specific nitrosamines or the relative paucity of polyaromatics in cigarette smoke were really critical factors. Our adoption of the naive exchangeability assumption, as well as our choice of a diffuse prior distribution for the relative potencies, reflected our desire not to impose vague and controversial beliefs on the reader at the outset.

Once the consequences of exchangeability were detailed, we were in a better position to study the effects of prior information in Section 5. In that section, we specifically assumed that the diesel emissions had biologically similar effects—a belief that we regarded as straightforward and uncontroversial. We showed how grouping of the diesel emissions was translatable into informative priors on the parameters $\beta$ without reference to the species-agent interactions $\delta$. Still, Kass expresses hesitation about applying the model until additional prior information is marshalled. We would certainly agree that better information is always better, but feel compelled to note that a paucity of such prior information hardly precludes the use of our statistical methods.

There are ways to test exchangeability. As suggested in Section 6, we could examine other data sets, particularly those with epidemiological evidence. Among the agents that have been labeled as carcinogenic on the basis of epidemiologic evidence, many have not been subject to extensive whole animal and short-term testing. This is an important point. If the current abundance of toxicological testing is to be informative about human cancer risks, then such testing needs to be performed on agents for which human evidence already exists.

An alternative method, suggested by Smith, is to examine the consequences of relaxing our exchangeability assumption. In doing so, however, we need to be careful that the substitute assumption is truly less restrictive than that under scrutiny. A sharp prior belief that nonhuman experiments are completely irrelevant to humans (which Kass wishes to mix with our exchangeable prior) would indeed yield different results. But sensitivity to such an extreme prior restriction hardly constitutes a test of our model. Surely the mammalian experiments have some relevance to human cancer risks, and the biological effects of coke oven and other emissions have some relevance to the effects of diesel engine emissions. The question is exactly how much relevance.

We performed additional calculations within the following framework. Suppose that a subset of agents and species adheres less closely to the underlying regression model, but we remain uncertain which, if any, species or agents belong to the suspicious subset. In contrast to Smith's specific suggestion, and in accordance with Section 6, we allow only species or agents, rather than individual experiments, to be the exceptions. Suppose further that each species and agent has an independent, identical probability $p$ of belonging to the subset of exceptions. (The species-agent interactions, in that case, still have identical univariate marginal prior distributions, but are no longer fully exchangeable.) Suppose finally that $\delta_{kl} \sim N(0, \tau^2)$ when species $k$ or agent $l$ is an exception, and $\delta_{kl} \sim N(0, \sigma^2)$ otherwise. We studied the case where $p = .1$ and $\tau - \sigma = \frac{1}{2} \log 10$, so that $\exp(2\tau) = 10 \exp(2\sigma)$. (In effect, each species or agent has a one-tenth chance of being ten times farther from the underlying regression model.) When we applied these specific assumptions to the $5 \times 9$ array of Table 1, retaining a diffuse prior on $\beta$, we obtained posterior densities of human carcinogenic potency quite close to those reported in Table 2, with increases in posterior standard deviation ranging from 0 to 10 percent. Some subsets of species and agents indeed exhibited high likelihoods of being exceptional. For example, the Bayes factor for the model in which cigarette smoke (CIG), mutagenesis without activator (M-), and skin tumor initiation (SKIN) constituted the suspicious subset was 33 times as large as that for the model with no exceptions. But because each species and agent had an independent, equal prior probability of being exceptional, the posterior estimates from the high likelihood models contributed little to the overall posterior estimates.

From the strict Bayesian viewpoint, therefore, the posterior standard deviations for the $3 \times 8$ array in Table 5 (where we dropped CIG, M-, and SKIN through a stepwise search) are overly optimistic. Such optimism is a difficulty with any stepwise search procedure, as Kass properly notes. Still, our diagnostic procedure can help decide which animal tests are most relevant to human cancer risks. It can also help identify which of many short-term tests are most relevant to carcinogenicity in laboratory animals. Such information would be valuable not only for "risk estimation," but also for "risk identification" (as Krewski uses the terms).

Krewski and Kass raise concerns about our statistical analysis of the individual cancer studies. As noted from the outset, we did not purport to solve the problem of selecting dose-response models for individual experiments. But we did note how errors of model misspecification enter into our analysis, and we specifically examined the effects of increasing the standard deviations of the nonhuman data in Table 1. Moreover, we explained how a multivariate generalization of our model would accommodate nonlinear dose-response curves, in which the results of an experiment would be summarized by a vector of numbers. Krewski warns that tests in the same species need to be performed under similar conditions. In fact, the data in Table 1 satisfy this requirement. (For example, the skin tumor counts in Sencar mice were all performed at 27 weeks.)

Nor did we purport to offer an extensive analysis of the uncertainties in the epidemiologic method. Although not noted in our article, Harris (1981,1983) had already increased the standard deviations for the epidemiologic studies in Table 1 in an attempt to account for uncertainties in occupational exposure and other possible confounding factors.

We apologize to Smith for labeling certain computational shortcuts as "empirical Bayes." The terminology was convenient and consistent with the literature. Our point was that such computational shortcuts may conceal the genuine uncertainty that arises when there are relatively few experiments. Perhaps Smith's suggested refinement, involving a second-order approximation for the density of $(\theta \mid Y, \sigma)$ about $\hat{\sigma}$, would avoid this difficulty. But our message should still be clear. Human cancer risk assessment requires data on many agents in many species. In the absence of strong prior information on cancer mechanisms, one good rat study is just not enough.

Our goal here was to impose a formal theoretical structure on the previously ill-defined problem of interspecies extrapolation. In this effort, three ideas were central.

First, there is a crucial distinction between the error of measurement within each dose-response study and the error of imperfect relevance among the studies. Second, the uncertain relevance between experiments can be formalized by reference to a hypothetical common mechanism that generates all the experimental data. Relevance is then quantifiable, roughly speaking, by the fit of the data to the underlying model. Third, the available biological information on species differences, chemical characteristics, disease mechanisms, and the like enters the analysis in the form of prior assumptions about the parameters of the underlying model relating the experiments. If we have made any contribution, we hope that these ideas persist in future research on interspecies comparisons. To be somewhat more presumptuous, we hope that our viewpoint finds its way to other problems where combining diverse forms of scientific evidence is at issue.

# [6]

*The Annals of Statistics*
1974, Vol. 2, No. 4, 615-629

## SPECIAL INVITED PAPER

## PRIOR DISTRIBUTIONS ON SPACES OF PROBABILITY MEASURES[1]

BY THOMAS S. FERGUSON

*University of California, Los Angeles*

Methods of generating prior distributions on spaces of probability measures for use in Bayesian nonparametric inference are reviewed with special emphasis on the Dirichlet processes, the tailfree processes, and processes neutral to the right. Some applications are given.

**0. Introduction.** Recently there has been active research in a narrow but important area, that of construction of prior distributions on spaces of probability measures for use in deriving Bayesian decision rules in nonparametric statistical problems. There are two desirable properties of such prior distributions: (1) the support of the prior with respect to some suitable topology on the space of probability measures should be large, and (2) the posterior distribution given a sample from the true probability measure should be manageable analytically.

One of the drawbacks of decision theory in general and of the Bayesian approach to it in particular is the difficulty of putting the cost of the computation into the model. This drawback is particularly severe in Bayesian nonparametric problems. There are no doubt examples in which "quick and easy" rules are preferable to "optimal" rules for a Bayesian simply because it costs less to perform the computations. On the other hand, Bayes rules are certainly desirable since generally they are admissible and have nice large sample properties [12]. Therefore, it behooves the statistician to suggest large classes of easily computable Bayes rules in the hope that users may find some rules to their liking.

It is the purpose of this paper to review the literature in this area and the somewhat limited success to date. Because the earlier papers of Freedman [19] and Fabius [14] were concerned mainly with other problems, their value in Bayesian nonparametric statistics has not been generally appreciated. It is hoped that this paper serves to recognize this work and to clarify the relationship with later work.

For ease of exposition, we restrict attention unless otherwise specified to prior distributions on the space of all probability measures on $(\mathbb{R}, \mathscr{B})$ where $\mathbb{R}$ is the

Received January 1973; revised October 1973.

[1] Special invited address at the annual meeting of the I.M.S., Hanover, August 28–September 1, 1972. This work was supported in part by NSF Grant No. GP-33431X.

*AMS* 1970 *subject classifications.* Primary 62C10; Secondary 60B05.

*Key words and phrases.* Bayesian nonparametric inference, prior distributions, Dirichlet process, tailfree processes, characterization of distributions, neutral to the right, adaptive sampling with recall, adaptive investment.

real line and $\mathscr{B}$ is the $\sigma$-algebra of Borel subsets of $\mathbb{R}$. Let

$$\mathscr{F} = \{P: P \text{ is a probability measure on } (\mathbb{R}, \mathscr{B})\}.$$

(In this paper, the word "measure" refers to a nonnegative $\sigma$-additive set function.) Let $\mathscr{A}$ denote some suitable $\sigma$-algebra of subsets of $\mathscr{F}$, for example the Borel sets with respect to the topology of weak convergence. We use $\mathscr{P}$ to denote a probability measure on $(\mathscr{F}, \mathscr{A})$, and $\mathscr{E}$ to denote the expectation with respect to $\mathscr{P}$. We let $P$ denote the random probability measure chosen according to $\mathscr{P}$, so that $\int x\,dP(x)$ and $P(B)$ for $B \in \mathscr{B}$ are random variables. We let $X_1, X_2, \cdots, X_n$ denote a random sample chosen according to $P$. The information $X_1, X_2, \cdots, X_n$ is to be used to make inferences about the true value of $P$.

We may restate the problem. Find $\mathscr{P}$ so that (1) the support of $\mathscr{P}$ with respect to weak convergence, say, is $\mathscr{F}$, and (2) the posterior distribution of $P$ given $X_1, \cdots, X_n$ is manageable analytically.

**1. The Dirichlet process.** The simplest of the prior distributions $\mathscr{P}$ described in this paper are the Dirichlet processes. There are a large number of Dirichlet processes on the real line, one for each finite non-null measure $\alpha$ on $(\mathbb{R}, \mathscr{B})$. We refer to $\alpha$ as the parameter of the process. There are various ways to describe this process, denoted by $\mathscr{D}(\alpha)$, of which two are presented below. The main source of the results is [17].

We use $\mathscr{G}(\alpha, \beta)$ for $\alpha > 0$ and $\beta > 0$ to represent the gamma distribution with density $\Gamma(\alpha)\beta^{-\alpha}e^{-x/\beta}x^{\alpha-1}I_{(0,\infty)}(x)$, where $I$ denotes the indicator function. If $\alpha = 0$, $\mathscr{G}(\alpha, \beta)$ is defined to be degenerate at zero. We use $\mathscr{B}e(\alpha, \beta)$ for $\alpha > 0$ and $\beta > 0$ to denote the beta distribution with density $\Gamma(\alpha + \beta)(\Gamma(\alpha)\Gamma(\beta))^{-1}x^{\alpha-1}(1 - x)^{\beta-1}I_{(0,1)}(x)$. If $\alpha = 0$ and $\beta > 0$, $\mathscr{B}e(\alpha, \beta)$ is defined to be degenerate at zero, while if $\alpha > 0$ and $\beta = 0$, $\mathscr{B}e(\alpha, \beta)$ is defined to be degenerate at one.

For the purposes of this paper, it is convenient to define the $m$-dimensional Dirichlet distribution with parameter $(\alpha_1, \alpha_2, \cdots, \alpha_m)$, where $\alpha_i \geqq 0$ and $\sum_1^m \alpha_i > 0$, as the distribution of $(Z_1/S, Z_2/S, \cdots, Z_m/S)$, where $Z_1, Z_2, \cdots, Z_m$ are independent random variables with $Z_i \in \mathscr{G}(\alpha_i, 1)$ $i = 1, \cdots, m$, and $S = \sum_1^m Z_i$. The one-dimensional marginal distributions of the Dirichlet are beta, for example, $Z_1/S \in \mathscr{B}e(\alpha_1, \sum_2^m \alpha_i)$. A convenient source of information on the Dirichlet distribution is Wilks [27].

DEFINITION 1. Let $\alpha(\cdot)$ be a finite non-null measure on $(\mathbb{R}, \mathscr{B})$, and let $P(\cdot)$ be a stochastic process indexed by elements of $\mathscr{B}$. We say $P$ is a Dirichlet process with parameter $\alpha$ and write $P \in \mathscr{D}(\alpha)$, if for every finite measurable partition $\{B_1, \cdots, B_m\}$ of $\mathbb{R}$ (i.e. the $B_i$ are measurable, disjoint, and $\bigcup_1^m B_i = \mathbb{R}$), the random vector $(P(B_1), \cdots, P(B_m))$ has a Dirichlet distribution with parameter $(\alpha(B_1), \cdots, \alpha(B_m))$.

In particular, for every $B \in \mathscr{B}$, $P(B) \in \mathscr{B}e(\alpha(B), \alpha(\mathbb{R}) - \alpha(B))$ and therefore $\mathscr{E}P(B) = \alpha(B)/\alpha(\mathbb{R})$.

For the second definition of the Dirichlet process on $(\mathbb{R}, \mathscr{B})$, it is convenient to use the distribution function form of the measures. Let

$$\alpha(t) = \alpha((-\infty, t]) \quad \text{and} \quad F(t) = P((-\infty, t])\,.$$

DEFINITION 2. We say $P \in \mathscr{D}(\alpha)$ (or $F \in \mathscr{D}(\alpha)$), if the process $F(t)$ may be written as $Z_t/Z_\infty$ where $Z_t$ is a process with independent increments, $Z_t \in \mathscr{G}(\alpha(t), 1)$ and $Z_\infty = \lim_{t\to\infty} Z_t \in \mathscr{G}(\alpha(\mathbb{R}), 1)$.

To contrast these definitions, note that although the existence of the process of Definition 1 requires a demonstration, the existence of the process of Definition 2 is immediate since the existence of the independent increment process with gamma distributions, sometimes called the gamma process, is well known. On the other hand, Definition 1 may be used to define the Dirichlet process on an arbitrary measurable space, whereas the extension of the gamma process to arbitrary spaces is not so well known. Nevertheless,

*Fact* 1. These two definition are equivalent.

An alternative definition, that views the Dirichlet process as a limit of Polya urn schemes, may be found in Blackwell and MacQueen [4].

*Fact* 2. If $F \in \mathscr{D}(\alpha)$, then with probability one $F$ is discrete.

It is well known that the separable version of the gamma process on the real line increases only in jumps with probability one. Therefore $F$ with probability one is a discrete distribution function. The gamma process, and hence the Dirichlet process on an arbitrary measurable space, also gives probability one to sums of point masses. This may be seen easily using a result of Ferguson and Klass [18]. Using the first definition, results of Blackwell [3] show that Dirichlet processes on arbitrary spaces concentrate on discrete distributions.

*Fact* 3. The support of $\mathscr{D}(\alpha)$ with respect to the topology of weak convergence is the set of all distributions whose support is contained in the support of $\alpha$.

This is a version of desirable property 1 mentioned in the introduction. If the support of $\alpha$ is $\mathbb{R}$, then the support of $\mathscr{D}(\alpha)$ with respect to convergence in law is $\mathscr{F}$.

*Fact* 4. For any nonnegative measurable function $g$, $\int g(t)\, d\alpha(t) < \infty$ if, and only if, $\int g(t)\, dF(t) < \infty$ with probability one.

This exhibits a strong connection between the parameter $\alpha$ of the process and the random distribution function $F$. In particular, a $k$th moment of $\alpha$ exists if and only if a $k$th moment of $F$ exists with probability one. The main result, that shows the Dirichlet process satisfies the second desirable property, is the following.

THEOREM 1. *If $F \in \mathscr{D}(\alpha)$ and if $X_1, \cdots, X_n$ is a sample from $F$, then the posterior distribution of $F$ given $X_1, \cdots, X_n$ is $\mathscr{D}(\alpha + \sum_1^n \delta_{X_i})$, where $\delta_x$ is the measure giving mass one to $x$.*

In distribution function form, the posterior parameter of the process is

$$\alpha(t) + \sum_1^n I_{[X_i,\infty)}(t) .$$

EXAMPLES. Consider the problem of estimating an unknown distribution function $F$ by a distribution function $\hat{F}$ with loss function $L(F, \hat{F}) = \int (F(t) - \hat{F}(t))^2 \, dW(t)$, where $W(t)$ is a given non-random weight function (finite measure). Suppose as a prior distribution, we take $F \in \mathscr{D}(\alpha)$. If we have no observations from $F$, then since $F(t) \in \mathscr{B}e(\alpha(t), \alpha(\mathbb{R}) - \alpha(t))$, the Bayes estimate is

$$\hat{F}(t) = \mathscr{E} F(t) = \alpha(t)/\alpha(\mathbb{R}) =_{\mathrm{def}} F_0(t) . \tag{1.1}$$

We may consider $F_0(t)$ as our prior guess at $F(t)$. The Bayes estimate based on a sample $X_1, \cdots, X_n$ from $F$ is therefore

$$\begin{aligned} \hat{F}_n(t) &= \mathscr{E}(F(t) \mid X_1, \cdots, X_n) = (\alpha(t) + \textstyle\sum_1^n I_{[X_i,\infty)}(t))/(\alpha(\mathbb{R}) + n) \\ &= p_n F_0(t) + (1 - p_n) F_n(t) \end{aligned} \tag{1.2}$$

where $F_n$ is the sample distribution function,

$$F_n(t) = \frac{1}{n} \sum_1^n I_{[X_i,\infty)}(t) \tag{1.3}$$

and

$$p_n = \alpha(\mathbb{R})/(\alpha(\mathbb{R}) + n) . \tag{1.4}$$

The Bayes estimate $\hat{F}_n$ is thus a mixture of the prior guess $F_0$ and the sample distribution function $F_n$: If $\alpha(\mathbb{R})$ is large compared to $n$, $\hat{F}_n$ gives most of its weight to $F_0$, while if $\alpha(\mathbb{R})$ is small compared to $n$, $\hat{F}_n$ gives most of its weight to $F_n$. Thus, one may consider $\alpha(\mathbb{R})$ as a measure of the strength of belief in the prior guess, measured in units of sample size. The parameter of a Dirichlet prior is specified by the function $F_0$, and the real parameter $\alpha(\mathbb{R})$.

Similarly, one may estimate the mean of an unknown distribution $F$ with loss function $L(F, \hat{\mu}) = (\int t \, dF(t) - \hat{\mu})^2$. For the prior distribution $F \in \mathscr{D}(\alpha)$ with $\int t^2 \, d\alpha(t) < \infty$, the Bayes estimate based on a sample $X_1, \cdots, X_n$ from $F$ is

$$\begin{aligned} \hat{\mu}_n &= \mathscr{E}(\textstyle\int t \, dF(t) \mid X_1, \cdots, X_n) = \int t \, d\mathscr{E}(F(t) \mid X_1, \cdots, X_n) \\ &= \textstyle\int t \, d\hat{F}_n(t) = p_n \mu_0 + (1 - p_n)\bar{X}_n \end{aligned}$$

where $\bar{X}_n$ is the sample mean and $\mu_0$ is the prior guess at the mean

$$\mu_0 = \int t \, dF_0(t) .$$

It is interesting that the Bayes estimate of the mean depends on the parameter of the Dirichlet prior only through the values of $\mu_0$ and $\alpha(\mathbb{R})$.

Similar results are obtained by these methods for the following problems: (1) Estimating moments, or a variance or covariance; (2) Estimating a median or other quantiles; (3) Estimating $P(X > Y)$ in a two-sample problem; (4) Estimating a quantile by a "tolerance" region; (5) Testing one-sided hypotheses concerning quantiles. Brunk and Pierce [5] have applied these methods to the estimation of a cumulative regression. G. J. Hall, Jr. [20] has discussed the

Dirichlet prior for use in an adaptive sequential search problem. Two other applications are presented in the last section of this paper.

These methods may be extended, though with less success, usually because of the relative difficulty of computing the Bayes decision rules, to such problems as bio-assay, regression or observation with error, and empirical Bayes. In these more complex problems, even if the prior is a Dirichlet process, the posterior often turns out to be a mixture of Dirichlet processes, wherein the parameter of the process $\alpha_u$ is indexed by a variable $u$ taken to be random. This subject is treated in detail by Antoniak [1].

The bio-assay problem has also been treated by Ramsey [26] and by Kraft and van Eeden [23], the latter using a tailfree process (discussed in the next section) for a prior.

For certain other problems, particularly hypothesis testing problems, these methods turn out to be unsuitable. For example, consider the goodness-of-fit problem of testing the hypothesis that a distribution on [0, 1] is uniform. If for the alternative hypothesis we take $\mathscr{D}(\alpha)$ for a prior where $\alpha$ is uniform, $\alpha(t) = ct$ on [0, 1] for some $c > 0$, then the only non-trivial non-randomized Bayes rule is to reject the null hypothesis if and only if some two observations are exactly equal. This is really a test of continuity against discreteness.

**2. Tailfree processes.** The limitations of the Dirichlet process stem mainly from the fact that it chooses discrete distributions with probability one, so that we expect to see some observations repeated exactly. To avoid these limitations, we should try to find workable priors that choose continuous distributions with probability one. There are some among the tailfree processes of Freedman [19] and Fabius [14], which we now describe.

Let $\{\pi_m; m = 1, 2, \cdots\}$ be a tree of measurable partitions of $(\mathbb{R}, \mathscr{B})$; that is, let $\pi_1, \pi_2, \cdots$ be a sequence of measurable partitions such that $\pi_{m+1}$ is a refinement of $\pi_m$ for each $m = 1, 2, \cdots$, and $\bigcup_1^\infty \pi_m$ generates $\mathscr{B}$.

DEFINITION 3. The distribution of a random probability $P$ on $(\mathbb{R}, \mathscr{B})$ is said to be tailfree with respect to $\{\pi_m\}$ if there exists a family of nonnegative random variables $\{V_{m,B}; m = 1, 2, \cdots, B \in \pi_m\}$ such that

(1) the families $\{V_{1,B}; B \in \pi_1\}, \{V_{2,B}; B \in \pi_2\}, \cdots$ are independent, and

(2) for every $m = 1, 2, \cdots$, if $B_j \in \pi_j$, $j = 1, \cdots, m$ is such that $B_1 \supset B_2 \supset \cdots \supset B_m$, then $P(B_m) = \prod_{j=1}^m V_{j,B_j}$.

Corresponding to Theorem 1 for Dirichlet processes we have the following for tailfree processes.

THEOREM 2. *If the distribution of $P$ is tailfree with respect to $\{\pi_m\}$ and if $X_1, \cdots, X_n$ is a sample from $P$, then the posterior distribution of $P$ given $X_1, \cdots, X_n$ is tailfree with respect to $\{\pi_m\}$.*

In addition, the posterior distributions of the $V$'s of Definition 3 given $X_1, \cdots, X_n$

may easily be computed. A simple example illustrates the main ideas. We construct tailfree processes on the interval (0, 1] with respect to $\{\pi_m\}$ where $\pi_m$ is the set of all dyadic intervals of length $1/2^m$, $\pi_m = \{((i-1)/2^m, i/2^m]; i = 1, \cdots, 2^m\}$.

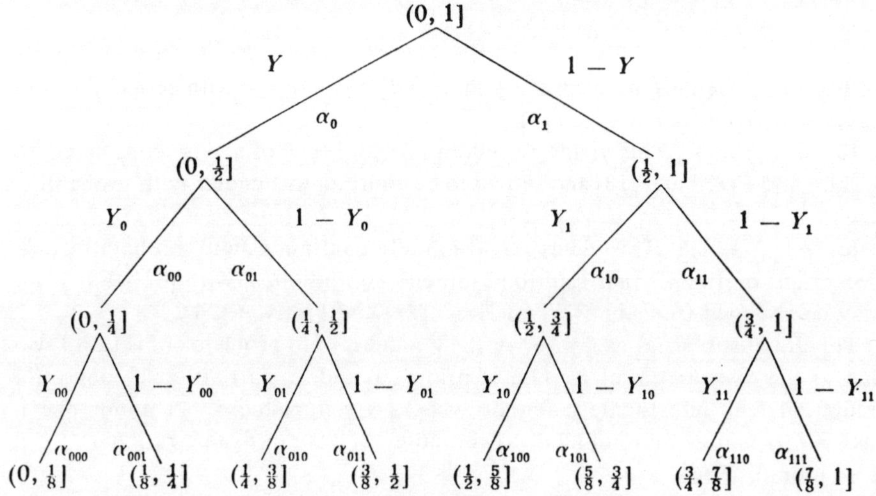


FIG. 1.

A simpler notation for the variables $V_{m,B}$ of Definition 3 is more appropriate for the tree of dyadic intervals. Let $.\varepsilon_1\varepsilon_2 \cdots \varepsilon_m$ denote the binary expansion of the dyadic rational $\sum_1^m \varepsilon_j 2^{-j}$, where each $\varepsilon_j$ is zero or one. If $B \in \pi_m$ is of the form $(.\varepsilon_1 \cdots \varepsilon_m, .\varepsilon_1 \cdots \varepsilon_m + 2^{-m}]$, then for $\varepsilon_m = 0$ we use $Y_{\varepsilon_1\cdots\varepsilon_m}$ to denote $V_{m,B}$, while for $\varepsilon_m = 1$ we use $1 - Y_{\varepsilon_1\cdots\varepsilon_m}$ to denote $V_{m,B}$. This is possible since $P$ is assumed to be a random probability. Then $P(B)$ is the product of all the variables associated with the path in the tree from (0, 1] to $B$, so that

$$(2.1) \qquad P(B) = (\textstyle\prod_{j=1;\varepsilon_j=0}^m Y_{\varepsilon_1\cdots\varepsilon_{j-1}})(\prod_{j=1;\varepsilon_j=1}^m (1 - Y_{\varepsilon_1\cdots\varepsilon_{j-1}}))\,.$$

For example $P((\frac{3}{8}, \frac{1}{2}]) = Y(1 - Y_0)(1 - Y_{01})$. The independence hypothesis of Definition 3 requires that the $Y$ variables be independent between rows in Figure 1.

If we choose *all* the $Y$ variables independently, with $Y_{\varepsilon_1\cdots\varepsilon_{m-1}} \in \mathscr{B}e(\alpha_{\varepsilon_1\cdots\varepsilon_{m-1}0}, \alpha_{\varepsilon_1\cdots\varepsilon_{m-1}1})$, then the posterior distributions of the $Y$ variables given a sample $X_1, \cdots, X_n$ from $P$ again has the same structure—independent with beta distributions. The posterior distribution of $Y_{\varepsilon_1\cdots\varepsilon_{m-1}}$ given $X_1, \cdots, X_n$ is $\mathscr{B}e(\alpha_{\varepsilon_1\cdots\varepsilon_{m-1}0} + M, \alpha_{\varepsilon_1\cdots\varepsilon_{m-1}1} + N)$, where $M$ is the number of $X_i$'s that fall in $(.\varepsilon_1 \cdots \varepsilon_{m-1}0, .\varepsilon_1 \cdots \varepsilon_{m-1}1]$ and $N$ is the number of $X_i$'s that fall in $(.\varepsilon_1 \cdots \varepsilon_{m-1}1, .\varepsilon_1 \cdots \varepsilon_{m-1}1 + 2^{-m}]$.

The Dirichlet process is tailfree with respect to every tree of partitions. The above scheme is a Dirichlet process if the parameters add in the following way. For every $\varepsilon_1 \cdots \varepsilon_m$,

$$(2.2) \qquad \alpha_{\varepsilon_1\cdots\varepsilon_m} = \alpha_{\varepsilon_1\cdots\varepsilon_m 0} + \alpha_{\varepsilon_1\cdots\varepsilon_m 1}\,.$$

Thus, the tailfree processes are much more flexible than the Dirichlet. There are twice as many parameters at your disposal for each row in Figure 1. With this extra freedom, we can choose the parameters ($\alpha$'s) so that the random probability $P$ is continuous singular or absolutely continuous with probability one. It is worthwhile to investigate three cases.

(a) $\alpha_{\epsilon_1 \cdots \epsilon_m} = 2^{-m}$. This yields a Dirichlet process and $P$ is discrete with probability one. Conditions on the $\alpha$'s that lead to discrete $P$ with probability one have been given by Blackwell [3].

(b) $\alpha_{\epsilon_1 \cdots \epsilon_m} = 1$. This yields a random probability $P$ of a type considered by Dubins and Freedman [13] and shown to be continuous singular with probability one.

(c) $\alpha_{\epsilon_1 \cdots \epsilon_m} = m^2$. This yields a $P$ absolutely continuous with probability one. Conditions on the $\alpha$'s that lead to absolutely continuous $P$ with probability one may be obtained from the work of Kraft [22] and Metivier [24].

The drawback of (a) as a prior in the goodness of fit problem of Section 1 was due to the discreteness of $P$. Use of priors (b) and (c) partially overcomes this limitation, but unfortunately new drawbacks are introduced. A minor one is that $\mathscr{E}F(t)$ is now more difficult to compute. If $t$ is not dyadic rational, $\mathscr{E}F(t)$ is an infinite series.

The main drawback is that the dyadic points of subdivision play a strong role in the posterior distributions. For each of (a), (b), and (c), the prior guess at $F$ is, from symmetry, the uniform distribution, $\mathscr{E}F(t) = t$, $t \in (0, 1]$. The posterior expectation given a sample of size 1, $\mathscr{E}(F(t) \mid X = x)$ is still uniform in the dyadic intervals in which $x$ does not lie, but it has corners at the dyadic rationals near $x$ in cases (b) and (c). At $x$, $\mathscr{E}(F(t) \mid X = x)$ has a discontinuity in case (a), infinite slope in case (b), and bounded slope in case (c). We further note in case (c) that even though the density with respect to Lebesgue measure exists with probability one, the density has discontinuities at all the dyadic rationals with probability one.

It should be considered a liability that the points used to describe the process appear strongly in the almost sure properties or in the posterior expectations.

**3. Characterization of the Dirichlet process.** Are there tailfree processes other than the Dirichlet for which the points of subdivision chosen for the tree of partitions do not play an essential role in the behavior of the process? Except for three trivial types of processes, the answer is no. This and two other characterizations of the Dirichlet process given here are due to Doksum [11] and Fabius [15]. The three trivial types of processes seem to appear in all these characterizations. They are

$T_1$. $P$ non-random ($F \equiv F_0$).

$T_2$. $P$ degenerate at a random point ($F = I_{[X,\infty)}$ where $X$ has distribution $F_0$).

$T_3$. $P$ concentrated on two non-random points ($F = UI_{[a,\infty)} + (1 - U)I_{[b,\infty)}$ where $U$ has an arbitrary distribution on $[0, 1]$, and $a < b$).

Types $T_1$ and $T_2$ are limits of the Dirichlet process as $\alpha(\mathbb{R}) \to \infty$ and $\alpha(\mathbb{R}) \to 0$, respectively, with $F_0$ fixed.

*Characterization* 1. If $P$ is tailfree with respect to every tree of partitions, then $P$ is either a Dirichlet process or of types $T_1$, $T_2$ or $T_3$.

From Definition 1 it follows that the Dirichlet process is the only random probability measure for which the distribution of $(P(B_1), \cdots, P(B_m))$ is Dirichlet for every measurable partition $\{B_1, \cdots, B_m\}$. Therefore, characterizations of the (finite-dimensional) Dirichlet distributions lead to characterizations of the Dirichlet process. The characterization of the Dirichlet distribution given by Darroch and Ratcliff [8] may be so used. If the Darroch–Ratcliff characterization of the Dirichlet distribution is strengthened as in Fabius [16], the resulting characterization of the Dirichlet process contains Characterization 1.

A related characterization of the Dirichlet distribution based on a different independence condition, called neutrality, leads to a second characterization of the Dirichlet process. The concept of neutrality is due to Connor and Mosimann [7]. The independence condition used below is essentially in the form given by Fabius [15], and weakened slightly in Fabius [16].

*Characterization* 2. If $P$ is neutral with respect to every finite measurable partition (that is, if for every measurable partition $\{B_1, \cdots, B_m\}$, $P(B_1)$ and the vectors $(P(B_2)/(1 - P(B_1)), \cdots, P(B_m)/(1 - P(B_1)))$ are conditionally independent given $P(B_1) \neq 1$), then $P$ is either a Dirichlet process or of types $T_1$, $T_2$ or $T_3$.

Although the above characterizations may be considered as attractive properties of the Dirichlet process, the next characterization may be considered as a drawback.

*Characterization* 3. If for every measurable set $B$, the posterior distribution of $P(B)$ given a sample $X_1, \cdots, X_n$ from $P$, depends on $X_1, \cdots, X_n$ only through the number of observations that fall in $B$, then $P$ is either a Dirichlet process or of types $T_1$, $T_2$ or $T_3$.

One would like to have a prior distribution for $P$ with the property that if $X$ is a sample from $P$ and $X = x$, then the posterior guess at $P$ gives more weight to values close to $x$ than the prior guess at $P$ does. For the Dirichlet process prior, the posterior guess at $P$ gives more weight to the point $x$ itself, but it treats all other points equally. In particular, the posterior guess at $P$ actually gives less weight to points near $x$ but not equal to $x$. From this point of view, the tailfree process prior that chooses absolutely continuous distributions with probability one would seem to be more appropriate.

**4. Processes neutral to the right.** A large class of prior distributions on $\mathscr{F}$ that are not dependent on the partition points used to describe the process has recently been discovered by Doksum [11].

DEFINITION 4. A random distribution function $F(t)$ on the real line is said to be neutral to the right if for every $m$ and $t_1 < t_2 < \cdots < t_m$, there exist independent random variables $V_1, V_2, \cdots, V_m$, such that $(1 - F(t_1), 1 - F(t_2), \cdots, 1 - F(t_m))$ has the same distribution as $(V_1, V_1V_2, \cdots, \prod_1^m V_i)$.

Essentially, this says that $F$ is neutral to the right if $1 - F(t_1), (1 - F(t_2))/(1 - F(t_1)), \cdots, (1 - F(t_m))/(1 - F(t_{m-1}))$ are independent when $t_1 < t_2 < \cdots < t_m$. Because of the possibility of the denominators being zero with positive probability, we prefer to use Definition 4 instead. Note that if $F$ is neutral to the right then $Y_t = -\log(1 - F(t))$ has independent increments. Doksum gives a simple characterization of processes neutral to the right.

Let $Y_t$ be a process with independent increments non-decreasing a.s., right continuous a.s., $\lim_{t\to-\infty} Y_t = 0$ a.s. and $\lim_{t\to+\infty} Y_t = +\infty$ a.s. We allow the possibility that $Y_t = \infty$ with positive probability for finite $t$. Then $F(t) = 1 - e^{-Y_t}$ is a random distribution function neutral to the right.

For such a process $Y_t$ there exist at most countably many fixed points of discontinuity $t_1, t_2, \cdots$. The corresponding jumps $S_1, S_2, \cdots$, are independent non-negative (possible infinite-valued) random variables with corresponding densities $f_{t_1}, f_{t_2}, \cdots$ (with respect to some convenient measure). The difference

$$(4.1) \qquad Z_t = Y_t - \textstyle\sum_j S_j I_{[t_j,\infty)}(t)$$

is a non-decreasing process with independent increments without fixed points of discontinuity and therefore has Lévy formula

$$(4.2) \qquad \log Ee^{uZ_t} = ub(t) + \textstyle\int_0^\infty (e^{uz} - 1)\, dN_t(z)$$

where $b$ is a non-decreasing continuous function with $\lim_{t\to-\infty} b(t) = 0$, and where $N_t$ is a continuous Lévy measure, that is, for every $B \in \mathscr{B}$, $N_t(B)$ is non-decreasing and continuous in $t$, and for each $t$, $N_t$ is a measure on the Borel sets of $(0, \infty)$. We define $Z_t$ to be $+\infty$ unless

$$(4.3) \qquad \textstyle\int_0^\infty z/(1 + z)\, dN_t(z) < \infty\,.$$

The process neutral to the right is specified by the four quantities $\{t_1, t_2, \cdots\}$, $\{f_{t_1}, f_{t_2}, \cdots\}$, $b$, and $N_t$.

The function $b$ corresponds to the non-random part of the process $Y_t$. If $b \equiv 0$, then $Y_t$, and hence $F(t)$, increase only in jumps a.s., so that $F$ is discrete with probability one.

Corresponding to Theorems 1 and 2, the main result for processes neutral to the right is

THEOREM 3. *If $F$ is neutral to the right, and if $X_1, \cdots, X_n$ is a sample from $F$, then the posterior distribution of $F$ given $X_1, \cdots, X_n$ is neutral to the right.*

The paper of Doksum also contains a description of the posterior distributions of $F$ in terms of the prior distribution of $F$. We give below an alternative description of Doksum's result in terms of the distribution of the process $Y_t$. It is sufficient to give this description for $n = 1$ since the description for arbitrary sample size would follow by repeated application.

Let the prior distribution of $Y_t$ be specified by $\{t_1, t_2, \cdots\}$, $\{f_{t_1}, f_{t_2}, \cdots\}$, $b$, $N_t$, and let $X$ be a sample from $F(t) = 1 - e^{-Y_t}$. The posterior distribution of $Y_t$ given $X = x$ is best treated in two cases.

*Case* 1. $x$ is one of the prior fixed points of discontinuity, say $x = t_k$. The posterior distribution of $Y_t$ given $X = x = t_k$ is specified by

(1) the same set of fixed points of discontinuity and the same deterministic component function, $b$,

(2) the posterior Lévy measure

$$\begin{aligned} dN_t(z \mid x) &= e^{-z}\, dN_t(z) && \text{for } t < x \\ &= e^{-z}\, dN_x(z) + d[N_t(z) - N_x(z)] && \text{for } t > x\,, \end{aligned}$$

(3) for $i \neq k$, the posterior distribution of the jump at $t_i$,

$$\begin{aligned} f_{t_i}(s \mid x) &= ce^{-s} f_{t_i}(s) && \text{for } t_i < x \\ &= f_{t_i}(s) && \text{for } t_i > x\,, \end{aligned}$$

(4) the posterior distribution of the jump at $x = t_k$

$$f_{t_k}(s \mid t_k) = c(1 - e^{-s}) f_{t_k}(s)$$

(where $c$ represents a normalizing constant).

*Case* 2. $x$ is not one of the prior points of discontinuity. The posterior distribution of $Y_t$ given $X = x$ is the same as in Case 1 (1), (2), (3) except that $x$ may now be a fixed point of discontinuity, and (4) is replaced by

(4′) define the measure $\mu_B$ on $(\mathbb{R}, \mathscr{B})$ for each fixed Borel subset $B$ of $[0, \infty)$ to satisfy

$$\mu_B((-\infty, t]) = \int_B (1 - e^{-z})\, dN_t(z) + b(t) I_B(0)$$

where as usual $I_B$ denotes the indicator function. Note that $\mu_B \ll \mu_{[0,\infty)}$. The posterior distribution of the jump $S$ at $x$ is given by $P(S \in B \mid x) = \varphi_B(x)$, where $\varphi_B$ is the Radon–Nikodym derivative of $\mu_B$ with respect to $\mu_{[0,\infty)}$.

One may state conditions (1), (2) and (3) more simply as the condition that the distribution of the increments $Y_t - Y_{t-\varepsilon}$ for $t < x$ and $\varepsilon > 0$ are changed by multiplying the density by $e^{-z}$ and renormalizing, while the distribution of the increments $Y_{t+\varepsilon} - Y_t$ for $t > x$ and $\varepsilon > 0$ remain unchanged.

It is best to take an example of the use of (4′). Suppose there are no fixed points of discontinuity, and suppose $dN_t(z) = \gamma(t)N(z)$ where $N(z)$ is a fixed measure such that $\int_0^\infty z/(1 + z)\, dN(z) < \infty$, and where $\gamma(t)$ is non-decreasing and continuous with $\gamma(t) \to 0$ as $t \to -\infty$ and $\gamma(t) \to \infty$ as $t \to \infty$. Suppose further that $b(t)$ and $\gamma(t)$ are absolutely continuous. Then

$$\mu_B((-\infty, t]) = \gamma(t) \int_B (1 - e^{-z})\, dN(z) + b(t) I_B(0)\,, \tag{4.4}$$

so that

$$\varphi_B(t) = (1 - q(t)) \frac{\int_B (1 - e^{-z})\, dN(z)}{\int_0^\infty (1 - e^{-z})\, dN(z)} + q(t) I_B(0) \tag{4.5}$$

where

(4.6) $$q(t) = b'(t)/(\gamma'(t) \int_0^\infty (1 - e^{-z})\, dN(z) + b'(t))$$

represents the probability that the jump at the new fixed point of discontinuity is zero. If $b \equiv 0$, then $q \equiv 0$ and the distribution of the jump is independent of where it occurs. Note that the distribution of the new jump is not necessarily infinitely divisible.

It is interesting to view the Dirichlet process as a process neutral to the right. If $F \in \mathscr{D}(\alpha)$, then $F$ is neutral to the right, and if $\alpha$ is continuous then $Y_t = -\log(1 - F(t))$ has no fixed points of discontinuity. This implies that if $X \in \mathscr{B}e(\alpha, \beta)$ then $Y = -\log(1 - X)$ is infinitely divisible. The density of $Y$ is

(4.7) $$f_Y(y) = \frac{\Gamma(\alpha + \beta)}{\Gamma(\alpha)\Gamma(\beta)} e^{-\beta y}(1 - e^{-y})^{\alpha-1} I_{(0,\infty)}(y)$$

and the moment generating function of $Y$ is

(4.8) $$M_Y(u) = Ee^{uY} = \frac{\Gamma(\alpha + \beta)\Gamma(\beta - u)}{\Gamma(\beta)\Gamma(\alpha + \beta - u)} \qquad \text{for } u < \beta ,$$

an unlikely looking function to be the moment generating function of an infinitely divisible distribution. The follow lemma gives the Lévy representation of this moment generating function.

LEMMA 1. *If $Y = -\log(1 - X)$ where $X \in \mathscr{B}e(\alpha, \beta)$, then*

$$\log M_Y(u) = \int_0^\infty (e^{uz} - 1) \frac{e^{-\beta z}(1 - e^{-\alpha z})}{(1 - e^{-z})z}\, dz .$$

PROOF. Using the formula $\Gamma(x) = x^{-1}\Gamma(x + 1)$, one may write (4.8) as

(4.9) $$M_Y(u) = \left(\prod_{k=0}^{n-1} \frac{(\beta + k)(\alpha + \beta - u + k)}{(\alpha + \beta + k)(\beta - u + k)}\right) \cdot \frac{\Gamma(\alpha + \beta + n)\Gamma(\beta - u + n)}{\Gamma(\beta + n)\Gamma(\alpha + \beta - u + n)} .$$

Stirling's formula, $\Gamma(x) \sim (2\pi x)^{\frac{1}{2}}(x/e)^x$ for large $x$, implies that the term involving the gamma function on the right side of (4.9) tends to 1 as $n \to \infty$. Hence,

$$\log M_Y(u) = \sum_{k=0}^\infty \log \frac{(\beta + k)(\alpha + \beta - u + k)}{(\alpha + \beta + k)(\beta - u + k)} \qquad \text{for } u < \beta .$$

Consequently, the Lévy representation

$$\log \frac{\lambda}{\lambda - u} = \int_0^\infty (e^{uz} - 1) \frac{e^{-\lambda z}}{z}\, dz \qquad \text{for } u < \lambda$$

for the moment generating function of the negative exponential distribution with parameter $\lambda$ implies

$$\log M_Y(u) = \sum_{k=0}^\infty \int_0^\infty (e^{uz} - 1) \frac{e^{-\beta z}(1 - e^{-\alpha z})}{z} e^{-kz}\, dz \qquad \text{for } u < \beta$$

from which the lemma follows immediately.

Returning to the Dirichlet process as a process neutral to the right, let $F \in \mathscr{D}(\alpha)$ and $Y_t = -\log(1 - F(t))$. Then $F(t) \in \mathscr{B}e(\alpha(t), \alpha(\mathbb{R}) - \alpha(t))$ so that the Lévy measure for $Y_t$ is

$$dN_t(z) = \frac{e^{-(\alpha(\mathbb{R})-\alpha(t))z}(1 - e^{-\alpha(t)z})}{(1 - e^{-z})z}\,dz$$
$$= e^{-\alpha(\mathbb{R})z}(e^{\alpha(t)z} - 1)(1 - e^{-z})^{-1}z^{-1}\,dz$$

provided $0 < \alpha(t) < \alpha(\mathbb{R})$. Therefore,

$$\mu_B((-\infty, t]) = \textstyle\int_B e^{-\alpha(\mathbb{R})z}(e^{\alpha(t)z} - 1)z^{-1}\,dz$$

so that

$$\varphi_B(t) = \alpha(\mathbb{R}) \textstyle\int_B e^{-\alpha(\mathbb{R})z}\,dz\,.$$

Thus, the distribution of the jump in $Y_t$ at $t = x$, the new fixed point of discontinuity, is independent of where it occurs and is negative exponential with density $\alpha(\mathbb{R})e^{-\alpha(\mathbb{R})s}I_{(0,\infty)}(s)$.

If in further samples from $F$, $k$ more observations fall at $x$, $r$ more observations fall above $x$, and the rest of the observations fall below $x$, the posterior distribution of the jump in $Y_t$ at $x$ has density $ce^{-(\alpha(\mathbb{R})+r)s}(1 - e^{-s})^k I_{(0,\infty)}(s)$, a density of the form (4.7).

It is useful to note that when $F$ is neutral to the right, $\mathscr{E}F(t)$ can be computed directly from the moment generating functions of the variables involved in $Y_t$.

$$\mathscr{E}F(t) = 1 - \mathscr{E}\exp\{-Y_t\} = 1 - \mathscr{E}\exp\{-Z_t - \textstyle\sum_{t_j \le t} S_j\}$$
$$= 1 - M_{Z_t}(-1)\textstyle\prod_{t_j \le t} M_{S_j}(-1)\,.$$

Furthermore, the moment generating functions for the posterior distributions of the increments $Y_t - Y_{t-\varepsilon}$ for $t < x$ and $\varepsilon > 0$ and $Y_{t+\varepsilon} - Y_t$ for $t > x$ and $\varepsilon > 0$ may easily be found from the corresponding moment generating functions of the prior. However, to compute the distribution of the jumps at new fixed points of discontinuity it is useful to have knowledge of the Lévy form of the moment generating function of $Z_t$.

**5. Applications.** 1. *Adaptive sampling with recall.* Let $F \in \mathscr{D}(\alpha)$ where $\int t^2\,d\alpha(t) < \infty$, and let $X_1, X_2, \cdots$ be independent identically distributed observations from $F$. Consider the problem of finding a stopping rule $N$ to maximize

$$\mathscr{E}(\max_{0 \le j \le N} X_j - Nc)$$

where $c > 0$ and where $X_0 = 0$.

The interpretation of this problem is as follows. You have an object to sell. Bids for the object come in one at a time chosen independently from some distribution $F$ you do not know exactly. A Dirichlet process with parameter $\alpha$ expresses your prior knowledge of $F$. It costs you an amount $c$ to wait from one bid to the next. You may stop viewing bids at any time and either sell the object for the maximum of the bids you have seen so far, or throw the object away and receive zero. Looking at a bid is costly, but it always gives information about the true value of $F$ and it may be a large bid. When do you stop looking?

The condition $\int t^2 \, d\alpha(t) < \infty$ implies that $\mathscr{E} X_i^2 < \infty$, and Theorem 1, page 352 of De Groot [9] implies that there is an optimal rule. Since the problem is monotone (see Chow, Robbins and Siegmund [6]), and one can show that the expected return can be approximated by the expected returns of truncated problems, this rule is the one-stage look-ahead rule. It is optimal to stop at the first $n$ for which your present return is greater than or equal to your conditional expected return if you continue one more stage and stop. That is, stop at the first $n$ for which

$$\max_{0 \leq j \leq n} X_j - nc \geqq \mathscr{E}(\max_{0 \leq j \leq n+1} X_j - (n+1)c \mid X_1, \cdots, X_n) .$$

Let $M_n$ denote $\max_{0 \leq j \leq n} X_j$, and rewrite this condition as: Stop at the first $n$ for which

$$\begin{aligned} c &\geqq \mathscr{E}(\max(0, X_{n+1} - M_n) \mid X_1, \cdots, X_n) \\ &= \int \max(0, x - M_n) \, d\hat{F}_n(x) \\ &= p_n \int \max(0, x - M_n) \, dF_0(x) + (1 - p_n) \int \max(0, x - M_n) \, dF_n(x) \\ &= p_n \int \max(0, x - M_n) \, dF_0(x) \end{aligned}$$

where $\hat{F}_n$, $F_0$, $F_n$ and $p_n$ are as in Section 1. This application was pointed out to me by J. B. MacQueen. A similar application may be made to the exponential version of this model in which $Y_n = \beta^n \max_{i \leq n} X_i$. The optimal rule is the one-stage look-ahead rule. It stops at the first $n$ for which $(1 - \beta)M_n \geqq \beta p_n \int \max(0, x - M_n) \, dF_0(x)$.

2. *An adaptive investment model.* Consider an investor with initial resources $X_0$ who makes investments during discrete time periods in $m$ different investment opportunities. Denote by $\mathbf{b}_i$ the $m$-vector whose $j$th component $b_{ij}$ is the amount invested during time period $i$ in investment opportunity $j$. If $X_{i-1}$ denotes the investor's fortune at the beginning of the $i$th period, we require of $\mathbf{b}_i$ that

$$\sum_{j=1}^{m} b_{ij} \leqq X_{i-1} \quad \text{and} \quad b_{ij} \geqq 0 \tag{5.1}$$

for all $i$. Denote by $\mathbf{Y}_i$ the random $m$-vector of returns, whose $j$th component $Y_{ij} \geqq 0$ is the return per unit invested during time period $i$ in investment opportunity $j$. Given $\mathbf{Y}_i$ we may compute $X_i$ from $X_{i-1}$ and $\mathbf{b}_i$ by the formula

$$X_i = (X_{i-1} - \sum_{j=1}^{m} b_{ij}) + \sum_{j=1}^{m} b_{ij} Y_{ij} . \tag{5.2}$$

It is the objective of the investor to maximize the expected value of the utility of his fortune at the end of $n$ time periods. For logarithmic utility, adaptive problems can be handled. Therefore we desire to maximize

$$\mathscr{E} \log X_n . \tag{5.3}$$

We assume that $\mathbf{Y}_1, \mathbf{Y}_2, \cdots, \mathbf{Y}_n$ are independent identically distributed from some unknown distribution function $F$. As a prior distribution for $F$, we take the Dirichlet process whose parameter $\alpha$ is a finite non-null measure on $m$-dimensional Euclidean space, $F \in \mathscr{D}(\alpha)$.

In addition to (5.1) there are further constraints on the choice of the $\mathbf{b}_i$ that reflect the requirement that the rule be non-anticipatory; that is, $\mathbf{b}_i$ may be a function only of $X_0, \mathbf{b}_1, \mathbf{Y}_1, \mathbf{b}_2, \mathbf{Y}_2, \cdots, \mathbf{b}_{i-1}, \mathbf{Y}_{i-1}$, and not of $\mathbf{Y}_i, \cdots, \mathbf{Y}_n$. The problem then is to find a sequence $\mathbf{b}_1, \cdots, \mathbf{b}_n$ subject to these constraints to maximize $\mathscr{E} \log X_n$.

This problem when $F$ is known and multinomial ($\mathbf{Y}_j$ takes only the values $(1, 0, \cdots, 0), (0, 1, \cdots, 0), \cdots, (0, 0, \cdots, 1)$) was introduced by Kelly [21]. That Kelly's model could be extended to the adaptive case was noticed by Bellman and Kalaba [2]. Further results along these lines were obtained by Murphy [25].

The optimal rule for these problems is the rule that, at each stage, maximizes the expected log of the resources one stage ahead. The proof as in Kelly is by backward induction. At the beginning of the $n$th stage one is to choose $\mathbf{b}_n$ to maximize $\mathscr{E}(\log X_n \mid \mathbf{Y}_1, \cdots, \mathbf{Y}_{n-1})$. Defining $\mathbf{c}_n$ by $\mathbf{b}_n = \mathbf{c}_n X_{n-1}$, we may compute

$$\begin{aligned} &\mathscr{E}(\log X_n \mid \mathbf{Y}_1, \cdots, \mathbf{Y}_{n-1}) \\ (5.4)\qquad &= \log X_{n-1} + \mathscr{E}(\log(1 - \textstyle\sum_{j=1}^m c_{nj} + \sum_{j=1}^m c_{nj} Y_{nj}) \mid \mathbf{Y}_1, \cdots, \mathbf{Y}_{n-1}) \\ &= \log X_{n-1} + \textstyle\int \log(1 - \sum_{j=1}^m c_{nj}(y_j - 1))\, d\hat{F}_{n-1}(\mathbf{y}) \end{aligned}$$

where $\hat{F}_{n-1}$ is $\mathscr{E}(F \mid \mathbf{Y}_1, \cdots, \mathbf{Y}_{n-1})$ as in Section 1. The optimal rule chooses $\mathbf{c}_n$ to minimize this integral subject to the constraints $\sum_1^m c_{nj} \leqq 1$ and $c_{nj} \geqq 0$. Therefore, at the beginning of stage $n - 1$, the investor wants to choose $\mathbf{b}_{n-1}$ to minimize the expectation of (5.4) given $\mathbf{Y}_1, \cdots, \mathbf{Y}_{n-2}$. Since the last term of (5.4) does not depend on $\mathbf{b}_{n-1}$, this is equivalent to minimizing $\mathscr{E}(\log X_{n-1} \mid \mathbf{Y}_1, \cdots, \mathbf{Y}_{n-2})$. This procedure obviously continues down to the first stage. The optimal rule is therefore: Choose $\mathbf{c}_i$, $i = 1, \cdots, n$ to maximize

$$\int \log(1 + \textstyle\sum_{j=1}^m c_{ij}(y_j - 1))\, d\hat{F}_{i-1}(\mathbf{y})$$

subject to the constraints $\sum_{j=1}^m c_{ij} \leqq 1$ and $c_{ij} \geqq 0$, and invest $\mathbf{b}_i = \mathbf{c}_i X_{i-1}$. This is a convex programming problem for each $i = 1, \cdots, n$.

## REFERENCES

[1] Antoniak, C. (1969). Mixtures of Dirichlet processes with application to Bayesian nonparametric problems. Ph. D. dissertation, Univ. of California, Los Angeles.

[2] Bellman, R. and Kalaba, R. (1958). On communication processes involving learning and random duration. *IRE Nat. Convention Record* Part IV 16-21.

[3] Blackwell, David. (1973). Discreteness of Ferguson selections. *Ann. Statist.* **1** 356-358.

[4] Blackwell, D. and MacQueen, J. B. (1973). Ferguson distributions via Polya urn schemes. *Ann. Statist.* **1** 353-355.

[5] Brunk, H. D. and Pierce, D. A. (1972). Note on Bayesian approaches to estimation of a cumulative regression. Technical Report, Oklahoma State Univ.

[6] Chow, Y. S., Robbins, H. and Siegmund, D. (1971). *Great Expectations: The Theory of Optimal Stopping*. Houghton-Mifflin, Boston.

[7] Connor, R. J. and Mosimann, J. E. (1969). Concepts of independence for proportions with a generalization of the Dirichlet distribution. *J. Amer. Statist. Assoc.* **64** 194-206.

[8] Darroch, J. N. and Ratcliff, D. (1971). A characterization of the Dirichlet distribution. *J. Amer. Statist. Assoc.* **66** 641-643.

[9] DeGroot M. H. (1970). *Optimal Statistical Decisions*. McGraw-Hill, New York.

[10] Doksum, K. A. (1972). Decision theory for some nonparametric models. *Proc. Sixth Berkeley Symp. Math. Statist. Prob.* **1** 331–343.

[11] Doksum, K. A. (1974). Tailfree and neutral random probabilities and their posterior distributions. *Ann. Probability* **2** 183–201.

[12] Doob, J. L. (1949). Application of the theory of martingales. *Colloq. Internat. du CNRS* 23–27.

[13] Dubins, L. E. and Freedman, D. A. (1966). Random distribution functions. *Proc. Fifth Berkeley Symp. Math. Statist. Prob.* **2** 183–214.

[14] Fabius, J. (1964). Asymptotic behavior of Bayes estimates. *Ann. Math. Statist.* **35** 846–856.

[15] Fabius, J. (1972). Neutrality and Dirichlet distributions. *Trans. Sixth Prague Conference on Information Theory, Statistical Decision Functions, and Random Processes.* To appear.

[16] Fabius, J. (1973). Two characterizations of the Dirichlet distribution. *Ann. Statist.* **1** 583–587.

[17] Ferguson, T. S. (1973). A Bayesian analysis of some nonparametric problems. *Ann. Statist.* **1** 209–230.

[18] Ferguson, T. S. and Klass, M. J. (1972). A representation of independent processes without Gaussian components. *Ann. Math. Statist.* **43** 1634–1643.

[19] Freedman, D. A. (1963). On the asymptotic behavior of Bayes estimates in the discrete case. *Ann. Math. Statist.* **34** 1386–1403.

[20] Hall, G. J., Jr. (1973). Sequential search with random overlook probabilities. Ph. D. thesis in mathematics, Univ. of California, Los Angeles.

[21] Kelly, J. L. Jr. (1956). A new interpretation of information rate. *Bell System Tech. J.* **35** 917–926.

[22] Kraft, C. H. (1964). A class of distribution function processes which have derivatives. *J. Appl. Probability* **1** 385–388.

[23] Kraft, C. H. and van Eeden, C. (1964). Bayesian bio-assay. *Ann. Math. Statist.* **35** 886–890.

[24] Metivier, M. (1971). Sur la construction de mesures aléatoires presque sûrement absolument continues par rapport à une mesure donnée. *Z. Wahrscheinlichkeitstheorie und Verw. Gebiete* **20** 332–344.

[25] Murphy, R. E. Jr. (1965). *Adaptive Processes in Economic Systems.* Academic Press, New York.

[26] Ramsey, F. L. (1972). A Bayesian approach to bio-assay. *Biometrics* **28** 841–858.

[27] Wilks, S. S. (1962). *Mathematical Statistics.* Wiley, New York.

Department of Mathematics
University of California
Los Angeles, California 90024

# Part II
# Asymptotics

# [7]

# Accurate Approximations for Posterior Moments and Marginal Densities

LUKE TIERNEY and JOSEPH B. KADANE*

This article describes approximations to the posterior means and variances of positive functions of a real or vector-valued parameter, and to the marginal posterior densities of arbitrary (i.e., not necessarily positive) parameters. These approximations can also be used to compute approximate predictive densities. To apply the proposed method, one only needs to be able to maximize slightly modified likelihood functions and to evaluate the observed information at the maxima. Nevertheless, the resulting approximations are generally as accurate and in some cases more accurate than approximations based on third-order expansions of the likelihood and requiring the evaluation of third derivatives. The approximate marginal posterior densities behave very much like saddle-point approximations for sampling distributions. The principal regularity condition required is that the likelihood times prior be unimodal.

KEY WORDS: Bayesian inference; Laplace method; Asymptotic expansions; Computation of integrals.

## 1. INTRODUCTION

A user of Bayesian methods in practice needs to be able to evaluate various characteristics of posterior and predictive distributions, especially their densities, means, and variances. If the problem under consideration does not involve a conjugate prior-likelihood pair, these tasks cannot be performed in closed form; analytical or numerical approximation methods are needed. In these cases it would often be useful to have approximations that are more accurate than the usual normal approximation, yet not as computationally intensive as numerical integration methods (Naylor and Smith 1983) or Monte Carlo methods (Kloek and Van Dijk 1978; Zellner and Rossi 1982). Lindley (1980) has proposed approximations for moments that capture the first-order error terms of the normal approximation. This is generally accurate enough, but, as Lindley points out, the required evaluation of third derivatives of the posterior can be rather tedious—in particular, in problems with several parameters. Mosteller and Wallace (1964, sec. 4.6C) suggested a similar approach but introduced a transformation of the parameters to avoid the need for the direct use of third derivatives. The proposed transformation, however, depends on the second derivatives of the log-likelihood. A numerical maximization routine for locating the posterior mode of the transformed parameters will therefore require third derivatives of the log-likelihood unless a more complicated derivative-free algorithm is used.

In this article we introduce an easily computable approximation for the posterior mean and variance of a nonnegative parameter or, more generally, of a smooth function of the parameter that is nonzero on the interior of the parameter space. For definiteness we will take this function to be positive. In this introduction we give a brief, informal description of the basic approximation and its properties. Section 2 contains a more careful statement.

Let $g$ be a smooth, positive function on the parameter space. The posterior mean of $g(\theta)$ can be written as

$$E_n[g] = E[g(\theta) \mid X^{(n)}] = \frac{\int g(\theta)e^{\mathcal{L}(\theta)}\pi(\theta)d\theta}{\int e^{\mathcal{L}(\theta)}\pi(\theta)d\theta}, \quad (1.1)$$

where $\mathcal{L}$ is the log-likelihood function and $\pi$ is the prior density. It is common practice to approximate the denominator integral in (1.1) by an integral of an approximating normal curve centered at the posterior mode and having variance equal to minus the inverse of the second derivative of the log posterior density at its mode. This approximation, used by Lindley and by Mosteller and Wallace, can be viewed as an application of the Laplace method for integrals, as described in De Bruijn (1961). It will produce reasonable results as long as the posterior is unimodal or at least dominated by a single mode; we will assume that the sample size is large enough for this to be the case. The new feature in the approximation proposed in the present article is in its approach to the numerator integral in (1.1). Rather than expand the integrand of this integral about the posterior mode, we propose to locate the mode of the numerator integrand itself, find the second derivative at this new mode, and then approximate this integral by a second application of Laplace's method.

The computational requirements of this approach are rather minimal: We only need to be able to evaluate first and second derivatives and maximize the two integrands, which can be viewed as slightly modified likelihood functions. These computations will be feasible for any problems in which maximum likelihood estimates and the observed information can be computed. Nevertheless, the resulting approximations are quite accurate. An intuitive explanation is this: If the function $g$ is bounded away from zero near the posterior mode, then the two integrands will be similar in shape. Thus by applying the same approximation technique, Laplace's method, to the numerator and the denominator we will be making similar errors, and in taking the ratio some portion of these errors will cancel. In asymptotic terms, Laplace's method typically has an error of order $O(n^{-1})$. Used in this ratio form, the error is of order $O(n^{-2})$—the order $O(n^{-1})$ error terms in the numerator and the denominator cancel.

* Luke Tierney is Associate Professor, School of Statistics, University of Minnesota, Minneapolis, MN 55455. Joseph B. Kadane is Professor, Departments of Statistics and Social Sciences, Carnegie-Mellon University, Pittsburgh, PA 15213. Kadane's research was sponsored in part by Office of Naval Research Contract N00014-82-K-0622. The research of both authors was sponsored in part by National Science Foundation Grant DMS-8503019. The authors thank Rob Kass for many helpful conversations, for suggesting the argument leading to (4.3), and for bringing the results of Leonard and of Mosteller and Wallace to their attention.


Journal of the American Statistical Association
March 1986, Vol. 81, No. 393, Theory and Methods

In the next section we give a more formal statement of the proposed approximation for the posterior mean of a positive function $g$ and an approximation for its variance. Section 3 briefly discusses the approximation of predictive densities. The use of Laplace's method for approximating marginal posterior densities is discussed in Section 4, and an application to data from the Stanford heart-transplant experiment is given in Section 5. Section 6 presents some closing remarks, and the Appendix contains some technical details on the asymptotic error rates.

In the derivations given in this article we have chosen to sacrifice formal rigor for clarity of exposition. Explicit statements of regularity conditions have been omitted. A technical report including some of the omitted details is available from us.

## 2. POSTERIOR MEANS AND VARIANCES

Setting $L = \log \pi + \mathcal{L}/n$ and $L^* = \log g + \log \pi + \mathcal{L}/n$, expression (1.1) for the posterior mean of $g$ can be written as

$$E_n[g] = \int e^{nL^*}d\theta \Big/ \int e^{nL}d\theta. \tag{2.1}$$

Let $\hat{\theta}$ be the posterior mode, the mode of $L$, and let $\sigma^2 = -1/L''(\hat{\theta})$. Then Laplace's method produces the approximation

$$\int e^{nL(\theta)}d\theta \approx \int \exp[nL(\hat{\theta}) - n(\theta - \hat{\theta})^2/(2\sigma^2)]d\theta = \sqrt{2\pi}\sigma n^{-1/2}e^{nL(\hat{\theta})} \tag{2.2}$$

for the denominator integral in (2.1). Similarly, if $\hat{\theta}^*$ is the mode of $L^*$ and $\sigma^{*2} = -1/L^{*''}(\hat{\theta}^*)$, then the Laplace approximation to the numerator integral in (2.1) is $\sqrt{2\pi}\sigma^* n^{-1/2}\exp\{nL^*(\hat{\theta}^*)\}$. Taking the ratio of these two approximations produces the approximation

$$\hat{E}_n[g] = (\sigma^*/\sigma)\exp\{n(L^*(\hat{\theta}^*) - L(\hat{\theta}))\} \tag{2.3}$$

to $E_n[g]$. We will refer to approximation (2.3) as the Laplace approximation.

A similar approximation applies in the multiparameter case: Set

$$\hat{E}_n[g] = \left(\frac{\det \Sigma^*}{\det \Sigma}\right)^{1/2} \exp\{n(L^*(\hat{\theta}^*) - L(\hat{\theta}))\},$$

where $\hat{\theta}^*$ and $\hat{\theta}$ maximize $L^*$ and $L$, respectively, and $\Sigma^*$ and $\Sigma$ are minus the inverse Hessians of $L^*$ and $L$ at $\hat{\theta}^*$ and $\hat{\theta}$, respectively.

The asymptotic accuracy of the Laplace approximation (2.3) is investigated in the Appendix. To summarize the results: The errors in the approximations to the two integrals in (2.1) are of order $O(n^{-1})$. The leading terms in the two errors are identical, however, and cancel when the ratio is taken. As a result, the error in (2.3) is of order $O(n^{-2})$; that is,

$$E_n[g] = \hat{E}_n[g](1 + O(n^{-2})). \tag{2.4}$$

Having obtained an approximation for the posterior mean of $g$, we would next like to approximate the posterior variance. The simplest way to obtain such an approximation is to set

$$\hat{V}_n[g] = \hat{E}_n[g^2] - \hat{E}_n[g]^2. \tag{2.5}$$

That is, we use (2.3) to approximate the posterior means of $g$ and $g^2$, and then we insert these approximations into the standard computational formula for the variance.

By (2.4) we might expect that (2.5) will have an absolute error of order $O(n^{-2})$ or a relative error of order $O(n^{-1})$ (since $V_n[g] \sim \sigma/n$). Another cancellation occurs, however: From the expression (A.2) given in the Appendix for the error of (2.3), we can see that the leading error terms in $\hat{E}_n[g^2]$ as an approximation to $E_n[g^2]$ and in $\hat{E}_n[g]^2$ as an approximation to $E_n[g]^2$ are identical. As a result, they cancel in computing (2.5). So $\hat{V}_n[g]$ has an absolute error of order $O(n^{-3})$ or a relative error of order $O(n^{-2})$; that is,

$$\hat{V}_n[g] = V_n[g](1 + O(n^{-2})). \tag{2.6}$$

A similar argument shows that if we approximate the posterior covariance $C_n[g, h]$ of two positive functions $g(\theta)$ and $h(\theta)$ by

$$\hat{C}_n[g, h] = \hat{E}_n[gh] - \hat{E}_n[g]\hat{E}_n[h], \tag{2.7}$$

then

$$C_n[g, h] = \hat{C}_n[g, h] + O(n^{-3}) \tag{2.8}$$

as well.

As a practical point, it is worth mentioning that approximation (2.5) should be used with caution if $n$ is very large, since it involves the computation of a small number as the difference between two large numbers. If computations are done with sufficient precision, however, then for most practical sample sizes this will not cause any problems. At the other end of the spectrum, if $n$ is very small, then it is possible for the variance approximation (2.5) to be negative and for a covariance matrix computed from (2.7) not to be positive semidefinite. This should be checked in any application, but in most cases it does not seem to be a problem even for moderate sample sizes. More work is needed to see if modified variance approximations can be obtained that are guaranteed to be positive. A second point concerns the maximizations required for (2.3), (2.5), and (2.7): Once $\hat{\theta}$, the posterior mode, has been determined, it can be used as a starting value for a numerical search for $\hat{\theta}^*$, the maximum of $L^*$. Generally, the number of iterations needed to find $\hat{\theta}^*$ from $\hat{\theta}$ will be quite small. In fact, since the asymptotic statement of (2.4) only depends on the leading asymptotic term in $\hat{\theta}^* - \hat{\theta}$, it remains valid if we replace $\hat{\theta}^*$ by a single Newton step from $\hat{\theta}$ towards $\hat{\theta}^*$. For (2.6) and (2.8) to remain valid, two Newton steps are sufficient.

In concluding this section, a comment on the assumed positivity of $g$ is appropriate. This assumption is needed to insure that the numerator and denominator integrands in (2.1) are similar in shape. This similarity in shape, in turn, is responsible for the cancellation of error terms in the approximation to the ratio (2.1). Thus for the approximation to be accurate for a function $g$ taking both positive and negative values, the posterior distribution of $g$ must be concentrated almost entirely to one side of the origin. If this is not the case, then this approach is not directly applicable. Work on extensions that will retain the computational simplicity of (2.3) and (2.5) but are applicable, for example, for computing posterior moments of regression parameters is currently in progress.

## 3. PREDICTIVE DISTRIBUTIONS

Consider approximating the predictive density

$$f_n(z) = f(z \mid x^{(n)}) = E_n[f(z \mid \theta)]$$

at specific values of $z$. Direct application of (2.3) produces the approximation

$$\hat{f}_n(z) = \hat{E}_n[f(z \mid \theta)],$$

which, by (2.4), has an error of order $O(n^{-2})$. This is similar to the approximation considered by Leonard (1982) in his comment on the paper by Lejeune and Faulkenberry (1982), who considered predictive densities from the frequentist point of view.

With its error of order $O(n^{-2})$, the approximation $\hat{f}_n(z)$ has a lower-order error than Dunsmore's (1976) modification of the simple approximation $f(z \mid \hat{\theta})$, where $\hat{\theta}$ is the maximum likelihood estimator (MLE). Dunsmore's approximation includes some but not all terms of order $O(n^{-1})$. In particular, he drops the integral of the second term in the integrand of his equation (1); Lindley's (1980) equation (7) shows that this integral is generally of order $O(n^{-1})$ but no smaller.

## 4. MARGINAL POSTERIOR DENSITIES

Laplace's method can also be used to approximate marginal posterior densities of individual parameters in multiparameter settings. The resulting approximation, first suggested by Leonard (1982), provides a useful alternative to generally more time-consuming numerical or Monte Carlo integration techniques. In its use of Laplace's method to integrate out one or more variables from a multivariate function to obtain a density, this approach is similar to the saddle-point method introduced by Daniels (1954) and studied further, for example, in Barndorff-Nielsen and Cox (1979) and Daniels (1980).

To obtain the approximation, set $\boldsymbol{\theta} = (\theta_1, \ldots, \theta_p) = (\theta_1, \boldsymbol{\theta}_2)$; that is, partition the $p$ vector $\boldsymbol{\theta}$ into its first component and the $(p-1)$ vector of the remaining components. Suppose $\hat{\theta} = (\hat{\theta}_1, \hat{\boldsymbol{\theta}}_2)$ maximizes $\pi e^{\mathcal{L}}$ (i.e., $\hat{\boldsymbol{\theta}}$ is the posterior mode), and let $\Sigma$ be minus the inverse of the Hessian of $(\mathcal{L} + \log \pi)/n$ at $\hat{\boldsymbol{\theta}}$; thus $\Sigma$ is a $p \times p$ matrix. For a given $\theta_1$, let the $(p-1)$ vector $\hat{\boldsymbol{\theta}}_2^* = \hat{\boldsymbol{\theta}}_2^*(\theta_1)$ maximize the function $h(\cdot) = \pi(\theta_1, \cdot)e^{\mathcal{L}(\theta_1, \cdot)}$, the function $\pi e^{\mathcal{L}}$ with $\theta_1$ held fixed, and let $\Sigma^* = \Sigma^*(\theta_1)$ be minus the inverse of the Hessian of $(\log h(\cdot))/n$, a $(p-1) \times (p-1)$ matrix. Applying Laplace's method to the integrals in the numerator and denominator of the expression

$$\pi_1(\theta_1 \mid x^{(n)}) = \pi_{n,1}(\theta_1) = \frac{\int \pi(\theta_1, \boldsymbol{\theta}_2)e^{\mathcal{L}(\theta_1, \boldsymbol{\theta}_2)}d\boldsymbol{\theta}_2}{\int \pi(\boldsymbol{\theta})e^{\mathcal{L}(\boldsymbol{\theta})}d\boldsymbol{\theta}},$$

for the marginal posterior density of $\theta_1$ we obtain the approximation

$$\hat{\pi}_{n,1}(\theta_1) = \left(\frac{\det \Sigma^*(\theta_1)}{2\pi n \det \Sigma}\right)^{1/2} \frac{\pi(\theta_1, \hat{\boldsymbol{\theta}}_2^*)e^{\mathcal{L}(\theta_1, \hat{\boldsymbol{\theta}}_2^*)}}{\pi(\hat{\boldsymbol{\theta}})e^{\mathcal{L}(\hat{\boldsymbol{\theta}})}}. \quad (4.1)$$

Using calculations similar to those presented in the Appendix, it is possible to show that

$$\pi_{n,1}(\theta_1) = \hat{\pi}_{n,1}(\theta_1)(1 + O_{\theta_1}(n^{-1})), \quad (4.2)$$

where $O_{\theta_1}(n^{-1})$ is of order $O(n^{-1})$ but depends on $\boldsymbol{\theta}_1$. If $\boldsymbol{\theta}_0$ is the true parameter vector, then under regularity conditions similar to those given in Walker (1969), it is possible to show that the error term $O_{\theta_1}(n^{-1})$ will be uniformly of order $O(n^{-1})$ for $\theta_1$ in some fixed neighborhood of $\theta_{0,1}$, the first component of $\boldsymbol{\theta}_0$. By contrast, the absolute error of the usual normal approximation is only of order $O(n^{-1/2})$. Moreover, the relative error of the normal approximation, or any Edgeworth-type approximation for that matter, usually only tends to zero on neighborhoods that shrink toward $\theta_{0,1}$ at rate $n^{-1/2}$.

The main reason that the error in (4.2) is as large as $O(n^{-1})$ is that the dimensionalities of the two integrals in the numerator and the denominator of $\pi_{n,1}$ are different. In fact, most of this error is due to the constant of integration; the error in the approximation of the functional form of $\pi_{n,1}(\theta_1)$ is only of order $O(n^{-3/2})$ in $n^{-1/2}$ neighborhoods of $\hat{\theta}_1$. To see this, fix $u$, let $\theta_1 = \hat{\theta}_1 + n^{-1/2}u$, and note that the error $O_{\theta_1}(n^{-1})$ is usually a smooth function of $\theta_1$. Thus we can write

$$O_{\theta_1}(n^{-1}) = O_{\hat{\theta}_1}(n^{-1}) + (u/n^{-1/2})O_u(n^{-1})$$

and

$$\pi_{n,1}(\theta_1) = \hat{\pi}_{n,1}(\theta)(1 + O_{\hat{\theta}_1}(n^{-1}))(1 + O_u(n^{-3/2})). \quad (4.3)$$

Since the term $O_{\hat{\theta}_1}(n^{-1})$ does not depend on $u$, it is an error in the constant of integration, and the error in the approximation $\hat{\pi}_{n,1}$ to the functional form of $\pi_{n,1}$ is of order $O(n^{-3/2})$, as claimed.

A similar result was pointed out by Daniels (1956) for the saddle-point approximation. As in that case, the main implication is that an approximate marginal posterior calculated using (4.1) should be renormalized by numerical integration to integrate to one.

To appreciate just how accurately (4.1) can capture the functional form of $\pi_{n,1}(\theta_1)$, consider, for example, the normal-gamma conjugate distribution for normal data with unknown mean and precision. Thus the joint posterior for the mean $m$ and the precision $r$ is of the form

$$\pi(m, r) \propto r^{\alpha - 1/2}e^{-r(\beta + \tau(m-\mu)^2/2)}$$

for some $\alpha$, $\beta$, $\mu$, and $\tau$. Leonard (1982) pointed out that (4.1) is remarkably accurate in this case. In fact, a simple calculation shows that (4.1) produces the correct functional forms!

Another joint distribution for which (4.1) produces the exact functional forms of the marginals is the Dirichlet distribution. It would be interesting to obtain a characterization of all joint distributions for which this occurs. A similar phenomenon occurs for the saddle-point approximation; that is, there are certain distributions for which that approximation produces the exact functional forms. For the saddle-point approximation, Daniels (1980) has obtained a characterization of all cases in which the approximation produces exact results. It may be possible to modify Daniels's approach to characterize the joint distributions for which (4.1) produces exact functional forms.

## 5. AN APPLICATION

As an example we consider a three-parameter model used by Turnbull, Brown, and Hu (1974) to describe data from the

Stanford heart-transplant program and referred to by them as the Pareto model. This model, described in section 4.3 of their paper, views individual patients in the nontransplant group as having exponential lifetimes with mean $\phi$, where $\phi$ is itself a random variable drawn independently for each patient from a gamma distribution with density proportional to $\phi^{p-1}e^{-\lambda\phi}$. Patients in the transplant group have a similar distribution but with $\tau\phi$ in place of $\phi$ for the residual lifetime after the transplant. The resulting likelihood function of the three parameters $\tau, \lambda, p$ is

$$\prod_{i=1}^{n} \frac{p\lambda^p}{(\lambda + x_i)^{p+1}} \prod_{i=n+1}^{N} \left(\frac{\lambda}{\lambda + x_i}\right)^p \times \prod_{j=1}^{m} \frac{\tau p\lambda^p}{(\lambda + y_j + \tau z_j)^{p+1}} \prod_{j=m+1}^{M} \left(\frac{\lambda}{\lambda + y_j + \tau z_j}\right)^p,$$

where the $x_i$ are the survival times in days of the $N = 30$ nontransplant patients, $n = 26$ of whom died, and $y_j$, $z_j$ are the times to transplant and survival times after transplant, respectively, of the $M = 52$ transplant patients, $m = 34$ of whom died.

Naylor and Smith (1982) used this model with an improper uniform prior on the parameters $\tau, \lambda, p$ to illustrate their computational approach based on Gauss–Hermite quadrature, and we use the same improper prior distribution for the present illustration. Naylor and Smith point out the possibility of integrating out the parameter $p$ analytically, but, following their example, we have chosen not to do this and to apply the Laplace approximations directly.

Table 1 lists the posterior means and standard deviations computed by the Laplace approximation method and by Naylor and Smith using Gauss–Hermite integration applied to an orthogonalized reparameterization. As can be seen from this table, the largest relative difference between any of the Laplace approximations and the results of Naylor and Smith is about 4%.

In approximating the marginal densities, we selected a set of 60 equally spaced points for each parameter and then at each point computed approximations to the marginal densities by formula (4.1). A simple rectangular integration of these approximate densities produced integrals of approximately 1.2 in all three cases; thus renormalization was necessary. We then obtained plots of spline interpolations of the renormalized densities. We used 60 points for increased accuracy; however, 30 points produced identical pictures. In performing the maximizations for the individual grid points, we proceeded outward from the MLE's, using each current set of optimal values as the starting values for the next maximization.

As a basis for comparison, for each of the 60 grid points selected for a given parameter we orthogonalized the other two parameters using the $\Sigma$ matrix computed for the Laplace approximation at that point. We then integrated with respect to each of the two orthogonalized parameters using a 20-point Gauss–Hermite quadrature. The results were renormalized using a rectangular integration formula and plots were obtained. In all three cases the resulting plots were indistinguishable from the renormalized Laplace approximations. Figure 1 shows the results for the marginal density of $p$. The solid line is the superposition of the renormalized Laplace approximation and the Gauss–Hermite calculations; the broken line is the asymptotic normal approximation.

As a final note on this example we mention the computing time requirements. On the University of Minnesota's Cyber 630 computer, computing the Laplace approximations to the moments took about .15 seconds of central processing unit (CPU) time; the Gauss–Hermite method took about 3 seconds. For marginals, the Laplace calculations took about 3 seconds and the Gauss–Hermite calculations about 60 seconds of CPU time.

*Table 1. Posterior Means and Standard Deviations for the Pareto Model*

| Method | Posterior Means | | | Posterior Standard Deviations | | |
|---|---|---|---|---|---|---|
| | $\tau$ | $\lambda$ | $p$ | $\tau$ | $\lambda$ | $p$ |
| Laplace | 1.044 | 32.11 | .4926 | .4944 | 16.09 | .1381 |
| Naylor and Smith | 1.04 | 32.5 | .50 | .47 | 16.2 | .14 |

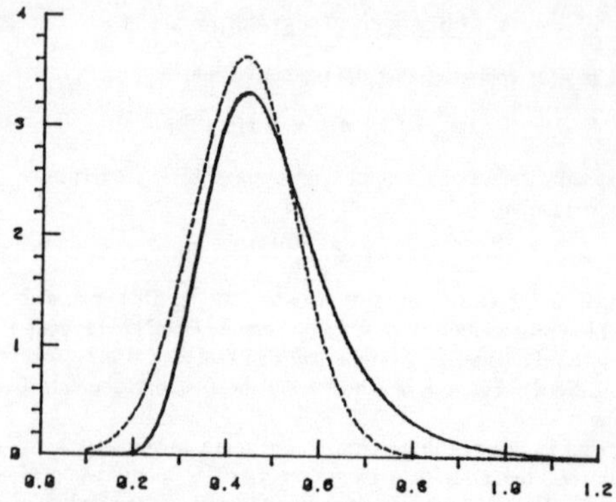


*Figure 1. Marginal Posterior Densities for p. —: Laplace and 20-point adaptive Gauss–Hermite approximations; ---: asymptotic normal approximation.*

## 6. CONCLUDING REMARKS

Several open questions remain to be investigated. The most important is to obtain approximations for moments of quantities taking on both positive and negative values, such as regression coefficients. Another is to determine the exact conditions under which the approximate marginalization approach produces exact results. It would also be useful to determine whether the approximations proposed here remain accurate when numerical derivatives are used in place of analytic ones in cases where closed-form derivatives are not available.

The approximations of this article may also prove helpful in certain theoretical problems such as developing tractable Bayesian approaches to log-linear models and to experimental design for nonlinear models. An extension that is currently being explored is to multimodal posteriors such as the poly-$t$ distributions of Dickey (1968) and Dreze (1977).

In conclusion, we would like to emphasize that we do not

think of these approximations as replacements for exact calculations in situations where extremely accurate results are needed. Instead, we consider them to be simple first approximations that are easy to obtain, are often sufficiently accurate in their own right, and generally provide good starting points for exact computations, should these be required.

## APPENDIX: ASYMPTOTIC ERRORS OF THE LAPLACE APPROXIMATIONS

Laplace's method for integrals as described, for example, in De Bruijn (1961) provides an approximation for integrals of the form $\int e^{nL(\theta)}d\theta$ when $n$ is large. The idea is that if $L$ has a unique maximum at $\hat{\theta}$, then for large $n$ the value of this integral depends only on the behavior of the function $L$ near its maximum. Thus if we set $\sigma^2 = -1/L''(\hat{\theta})$, then we can replace $L(\theta)$ by $L(\hat{\theta}) - (\theta - \hat{\theta})^2/(2\sigma^2)$. This produces the approximation

$$\int e^{nL(\theta)}d\theta \approx e^{nL(\hat{\theta})}\int \exp\{-[n(\theta - \hat{\theta})^2/2\sigma^2]\}d\theta = \sqrt{2\pi}\sigma n^{-1/2}\exp\{nL(\hat{\theta})\}.$$

By expanding $n(L(\theta) - L(\hat{\theta}) + (\theta - \hat{\theta})^2/(2\sigma^2))$ about $\hat{\theta}$ and $e^x$ about zero, it is possible to obtain the more refined result

$$\int e^{nL(\theta)}d\theta = \sqrt{2\pi}\sigma n^{-1/2}e^{nL(\hat{\theta})}\left(1 + \frac{a}{n} + \frac{b}{n^2} + O(n^{-3})\right), \quad \text{(A.1)}$$

where, setting $L_k = (d/d\theta)^k L(\hat{\theta})$, the constants $a$ and $b$ are given by

$$a = \frac{1}{8}\sigma^4 L_4 + \frac{5}{24}\sigma^6 L_3^2$$

and

$$b = \frac{1}{48}\sigma^6 L_6 + \frac{35}{384}\sigma^8 L_4^2 + \frac{7}{48}\sigma^8 L_3 L_5 + \frac{35}{64}\sigma^{10}L_3^2 L_4 + \frac{385}{1{,}152}\sigma^{12}L_3^4.$$

Result (A.1) remains valid if $L$ is replaced by a sufficiently well-behaved sequence $L^{(n)}$ of functions. In this case the coefficients $a$ and $b$ may depend on $n$, but this dependence will be suppressed. If $a$ and $b$ do indeed depend on $n$, we will assume regularity conditions for the sequence $L^{(n)}$ that insure that $a$ and $b$ are bounded in $n$.

The posterior expectation $E_n[g]$ of a positive function $g$ is a ratio of two integrals of the form $\int \exp\{nL^*(\theta)\}d\theta / \int \exp\{nL(\theta)\}d\theta$, where the difference $L^* - L$ is of order $O(n^{-1})$. Thus if we write $L^* = G/n + L$, with $G = \log g$, then

$$\frac{\int e^{nL^*(\theta)}d\theta}{\int e^{nL(\theta)}d\theta} = \frac{\sigma^* e^{nL^*(\hat{\theta}^*)}}{\sigma e^{nL(\hat{\theta})}}\left[1 + \frac{c}{n^2} + O(n^{-3})\right], \quad \text{(A.2)}$$

where $\hat{\theta}^*$ maximizes $L^*$, $\sigma^{*2} = -1/L^{*''}(\hat{\theta}^*)$,

$$c = G_1 d_1 + G_2 d_2 + G_3 d_3 + G_4 d_4,$$

and the $d_i$'s are given by

$$d_1 = \frac{1}{8}\sigma^6 L_5 + \frac{1}{4}\sigma^8 L_3 L_4 + \frac{5}{12}\sigma^8 L_3 L_4 + \frac{5}{8}\sigma^{10}L_3^3$$

$$d_2 = \frac{1}{4}\sigma^6 L_4 + \frac{5}{8}\sigma^8 L_3^2$$

$$d_3 = \frac{5}{12}\sigma^6 L_3$$

$$d_4 = \frac{1}{8}\sigma^4.$$

To see this, apply (A.1) to $\int \exp\{nL^*(\theta)\}d\theta$ and $\int \exp\{nL(\theta)\}d\theta$ to obtain

$$\int e^{nL^*(\theta)}d\theta \Big/ \int e^{nL(\theta)}d\theta = \frac{\sigma^*}{\sigma}\exp\{n(L^*(\hat{\theta}^*) - L(\hat{\theta}))\}\frac{\left(1 + \frac{a^*}{n} + \frac{b^*}{n^2} + O(n^{-3})\right)}{\left(1 + \frac{a}{n} + \frac{b}{n^2} + O(n^{-3})\right)}$$

$$= \frac{\sigma^*}{\sigma}\exp\{n(L^*(\hat{\theta}^*) - L(\hat{\theta}))\} \times \left(1 + \frac{a^* - a}{n} + \frac{b^* - b - a(a^* - a)}{n^2} + O(n^{-3})\right),$$

again suppressing any dependence of $a^*$, $b^*$ and $a$, $b$ on $n$.

Then observe that $\hat{\theta}^*$ solves $L^{*'}(\theta) = 0$ and $\hat{\theta}$ solves $L'(\theta) = 0$; thus

$$\begin{aligned} 0 = L^{*'}(\hat{\theta}^*) &= L'(\hat{\theta}^*) + (1/n)G'(\hat{\theta}^*) \\ &\approx L'(\hat{\theta}) + (\hat{\theta}^* - \hat{\theta})L''(\hat{\theta}) + (1/n)G'(\hat{\theta}) \\ &= -(\hat{\theta}^* - \hat{\theta})/\sigma^2 + (1/n)G_1. \end{aligned}$$

So $\hat{\theta}^* - \hat{\theta} = (1/n)G_1\sigma^2 + O(n^{-2})$. Together with the fact that $L^*(\theta) - L(\theta) = (1/n)G(\theta) = O(n^{-1})$ for any $\theta$, this implies that $a^* - a$ and $b^* - b$ are both of order $O(n^{-1})$.

[*Received June 1984. Revised September 1985.*]

## REFERENCES


Barndorff-Nielsen, O., and Cox, D. R. (1979), "Edgeworth and Saddle-point Approximations With Statistical Applications," *Journal of the Royal Statistical Society*, Ser. B, 41, 270–312.

Daniels, H. E. (1954), "Saddle Point Approximations in Statistics," *Annals of Mathematical Statistics*, 25, 631–650.

——— (1956), "The Approximate Distribution of Serial Correlation Coefficients," *Biometrika*, 67, 335–349.

——— (1980), "Exact Saddlepoint Approximations," *Biometrika*, 67, 59–63.

De Bruijn, N. G. (1961), *Asymptotic Methods in Analysis*, Amsterdam: North-Holland.

Dickey, J. (1968), "Three Multidimensional Integral Identities With Bayesian Applications," *Annals of Mathematical Statistics*, 39, 1615–1627.

Dreze, J. (1977), "Bayesian Regression Analysis Using Poly-*t* Densities," *Journal of Econometrics*, 6, 329–354.

Dunsmore, T. R. (1976), "Asymptotic Prediction Analysis," *Biometrika*, 63, 627–630.

Kloek, T., and Van Dijk, H. K. (1978), "Bayesian Estimation of Equation System Parameters: An Application of Integration by Monte Carlo," *Econometrica*, 46, 1–19.

Lejeune, M., and Faulkenberry, G. D. (1982), "A Simple Predictive Density Function," *Journal of the American Statistical Association*, 77, 654–657.

Leonard, T. (1982), Comment on "A Simple Predictive Density Function," by M. Lejeune and G. D. Faulkenberry, *Journal of the American Statistical Association*, 77, 657–658.

Lindley, D. V. (1980), "Approximate Bayesian Methods," in *Bayesian Statistics*, eds. J. M. Bernardo, M. H. Degroot, D. V. Lindley, and A. M. F. Smith, Valencia, Spain: University Press.

Mosteller, F., and Wallace, D. L. (1964), *Inference and Disputed Authorship: The Federalist*, Reading, MA: Addison-Wesley.

Naylor, J. C., and Smith, A. F. M. (1982), "Applications of a Method for the Efficient Computation of Posterior Distributions," *Applied Statistics*, 31, 214–225.

——— (1983), "A Contamination Model in Clinical Chemistry: An Illustration of a Method for the Efficient Computation of Posterior Distributions," *The Statistician*, 32, 82–87.

Turnbull, B. W., Brown, B. W., Jr., and Hu, M. (1974), "Survivorship Analysis of Heart Transplant Data," *Journal of the American Statistical Association*, 69, 74–80.

Walker, A. M. (1969), "On the Asymptotic Behaviour of Posterior Distributions," *Journal of the Royal Statistical Society*, Ser. B, 31, 80–88.

Zellner, A., and Rossi, P. E. (1982), "Bayesian Analysis of Dichotomous Quantal Response Models," technical report, University of Chicago, Graduate School Business.

# [8]

*J. R. Statist. Soc.* B (1979),
**41**, *No.* 2, *pp.* 184–189

## On Asymptotic Posterior Normality for Stochastic Processes

By C. C. Heyde and I. M. Johnstone

*CSIRO, Canberra* *Australian National University*

[Received August 1978. Revised October 1978]

Summary

Asymptotic normality of the posterior distribution of a parameter in a stochastic process is shown to hold under conditions which do little more than ensure consistency of a maximum likelihood estimator. Much more stringent conditions are required to ensure asymptotic normality of the MLE. This contrast, which has implications of considerable significance, does not emerge in the classical context of independent and identically distributed observations.

*Keywords*: PARAMETER ESTIMATION; STOCHASTIC PROCESSES; BAYESIAN METHODS; MAXIMUM LIKELIHOOD ESTIMATOR; ASYMPTOTIC NORMALITY

### 1. Introduction

In the classical context of independent and identically distributed observations it is well known that, under mild conditions, the maximum likelihood estimator (MLE) is consistent and, with appropriate normalization, asymptotically normally distributed. This "regular" behaviour is much less common when one moves to the more general context of estimation for stochastic processes. Indeed, quite stringent conditions are required to ensure that the MLE, when appropriately normalized, has an asymptotically normal distribution; e.g. Basawa *et al.* (1976), Heyde (1978).

Interestingly, a Bayesian approach to the problem has much greater universality. The conditions for asymptotic normality of the posterior distribution, appropriately scaled, are very similar to those which ensure consistency of the MLE and are much weaker than those required for asymptotic normality of the MLE. This contrast does not appear in the i.i.d. case, where the conditions for asymptotic normality of the MLE simplify to (essentially) a finite variance requirement and hence it appears to have escaped the attention which it deserves. Our methodology is similar to that of Walker (1969) who has discussed the i.i.d. case but the conditions have been simplified.

A Bayesian approach to some questions of "hypothesis testing" in a large sample situation also has an important advantage. It can provide, via the posterior distribution, a single categorical answer in each application while Neyman–Pearson based methodology can, in many cases, lead to complex questions of optimality of procedure in which categorical answers do not seem possible; e.g. Basawa and Scott (1976, 1977), Feigin (1978), Sweeting (1978).

### 2. The Setting and Results

Let $(X_1, X_2, \ldots, X_n)$ be a sample of consecutive observations from a stochastic process whose distribution depends on a single unknown parameter $\theta \in \Theta$, $\Theta$ denoting an interval. We suppose that $(X_1, \ldots, X_n)$ possesses a density $P_n(X_1, \ldots, X_n \mid \theta)$ with respect to a $\sigma$-finite measure $\mu_n$. Let $\hat{\theta}_n = \hat{\theta}_n(X_1, \ldots, X_n)$ be an MLE—a value of $\theta$ maximizing the likelihood function $P_n(X_1, \ldots, X_n \mid \theta)$ (uniqueness of $\hat{\theta}_n$ is not essential).

We put $L_n(\theta) = \log P_n(X_1, \ldots, X_n \mid \theta)$ and use $P_\theta$ and $E_\theta$ to denote probabilities and expectations computed under the assumption that $\theta$ is the true parameter value.

Let $\theta$ have an absolutely continuous prior distribution over $\Theta$, with density $\pi(\theta)$ with respect to Lebesgue measure. Then $\theta$ has an absolutely continuous posterior distribution with density

$$\pi_n(\theta \mid X_1, \ldots, X_n) = \pi(\theta)\, P_n(X_1, \ldots, X_n \mid \theta)/P_n(X_1, \ldots, X_n),$$

where $P_n(X_1, \ldots, X_n) = \int_\Theta \pi(\theta)\, P_n(X_1, \ldots, X_n \mid \theta)\, d\theta$.

Suppose that $\theta_0$ is an interior point of $\Theta$. We impose the following regularity conditions throughout:

(A1) The prior density $\pi(\theta)$ is continuous and positive at $\theta_0$.

(A2) $L_n(\theta)$ is a.s. twice continuously differentiable with respect to $\theta$ in some neighbourhood of $\theta_0 \subset \Theta$, $N_0$ say.

(A3) $[-L''_n(\theta_0)] \to \infty$ a.s. $(P_{\theta_0})$ as $n \to \infty$.

(A4) For any $\delta > 0$ for which $N_0(\delta) = (\theta_0 - \delta, \theta_0 + \delta) \subset \Theta$ there exists $k(\delta) > 0$ and $h(\delta) \downarrow 0$ as $\delta \downarrow 0$ satisfying

$$\liminf_{n\to\infty} P_{\theta_0}\{\sup_{\theta \notin \Theta \setminus N_0(\delta)} [-L''_n(\theta_0)]^{-1}[L_n(\theta) - L_n(\theta_0)] < -k(\delta)\} \geqslant 1 - h(\delta).$$

(A5) Given $\varepsilon > 0$, there exists $\delta = \delta(\varepsilon)$ satisfying

$$\liminf_{n\to\infty} P_{\theta_0}\{\sup_{\theta \in N_0(\delta)} |[L''_n(\hat\theta_n)]^{-1}(L''_n(\theta) - L''_n(\hat\theta_n))| < \varepsilon\} \geqslant 1 - \varepsilon.$$

Conditions (A1) and (A2) require no comment while Condition (A3) essentially ensures that the process does not degenerate. Condition (A4) is little more than a simple sufficient condition for the consistency of $\hat\theta_n$ for $\theta_0$ while Condition (A5) is a continuity condition for $L''_n$. The Conditions A2–A5 are, in fact, implied by familiar classical conditions of Wald (1948) for consistency of the MLE, provided that it is unique. This can be straightforwardly checked via the conditions $A$ in Basawa *et al.* (1976).

We are now in a position to establish the following theorem, the proof of which is deferred until the following section.

*Theorem.* Suppose that conditions $A$ hold. Let $\sigma_n$ be the positive square root of $\{-L''_n(\hat\theta_n)\}^{-1}$ whenever this exists (which will be true with probability tending to unity as $n \to \infty$ by (A3) and (A5)). Then, if $-\infty \leqslant b < a \leqslant \infty$, the posterior probability that $\hat\theta_n + b\sigma_n < \theta < \hat\theta_n + a\sigma_n$, namely

$$\int_{\hat\theta_n + b\sigma_n}^{\hat\theta_n + a\sigma_n} \pi_n(\theta \mid X_1, \ldots, X_n)\, d\theta,$$

tends in $P_{\theta_0}$ probability to

$$(2\pi)^{-\frac{1}{2}} \int_a^b e^{-\frac{1}{2}u^2}\, du$$

as $n \to \infty$.

This theorem provides the Bayesian analogue of confidence intervals. These are, of course, typically approached via asymptotic normality results for $\hat\theta_n$ in the Neyman–Pearson theory. Consequently, the assumptions underlying the result of the theorem should be contrasted with those necessary to ensure the asymptotic normality of $\hat\theta_n$, when appropriately normed, in a non-Bayesian situation.

The usual approach to asymptotic normality for $\hat\theta_n$ is via martingale theory and complicated regularity conditions are intrinsic (e.g. Heyde and Feigin (1975), Basawa *et al.* (1976), Heyde (1978). Indeed, under mild differentiability and integrability conditions,

$$\left\{L'_n(\theta) = \sum_{i=1}^{n} u_i(\theta), F_n, n \geqslant 1\right\},$$

the $u$'s being defined by differencing, is a martingale, $F_k$ denoting the $\sigma$-field generated by $X_1, \ldots, X_k$. Asymptotic normality of $\hat\theta_n - \theta$, when suitably normalized, holds under minor additional conditions if and only if a corresponding asymptotic result holds for $L_n'(\theta)$. This, however, involves additional stringent assumptions on the increments $u_i(\theta)$ of the $L_n'(\theta)$ process. Typical are conditions such as

$$s_n^{-2} \sum_1^n E_\theta(u_i^2(\theta) \mid F_{i-1}) \to \eta^2(\theta) \quad \text{in probability,}$$

as $n \to \infty$ for some r.v. $\eta^2 > 0$ a.s. and

$$s_n^{-2} \sum_1^n E_\theta(u_i^2(\theta)\, I(|u_i(\theta)| > \varepsilon s_n)) \to 0$$

as $n \to \infty$ for any $\varepsilon > 0$. Here $s_n^2 = \Sigma_1^n E_\theta u_i^2(\theta)$. Requirements of this kind are indeed stringent by comparison with their rather robust Bayesian counterparts. There are not even any formal requirements concerning existence of moments in the Bayesian formulation.

The theorem is given for the single parameter case to facilitate comparison with corresponding results based on the MLE. These are not yet available in such general form in the multiparameter case. There is, however, no special difficulty in extending the theorem to the multiparameter case and details are provided in Johnstone (1978).

The scope of the theorem may also be extended in another direction, namely to increase somewhat the family of distributions to which the results are applicable. Thus, using a formulation along the lines of Dawid (1970) it is possible to deal with some distributions whose range depends on the parameter. There is, however, a significant cost in terms of formal complications of the regularity conditions.

## 3. Proof of the Theorem

Write

$$P_n(X_1, \ldots, X_n \mid \theta) = P_n(X_1, \ldots, X_n \mid \hat\theta_n) \exp\{L_n(\theta) - L(\hat\theta_n)\}. \tag{1}$$

Whenever $\theta, \hat\theta_n \in N_0$ we have by Taylor expansion that

$$L_n(\theta) = L_n(\hat\theta_n) + \tfrac{1}{2}(\theta - \hat\theta_n)^2 L_n''(\theta_n^*) \tag{2}$$

for $\theta_n^* = \theta + \gamma(\hat\theta_n - \theta)$ with $0 < \gamma < 1$. Thus, letting

$$R_n = R_n(\theta, X_1, \ldots, X_n) = \sigma_n^2\{L_n''(\theta_n^*) - L_n''(\hat\theta_n)\},$$

we have

$$-(1 - R_n)/\sigma_n^2 = L_n''(\theta_n^*)$$

so that, for $\theta, \hat\theta_n \in N_0$,

$$P_n(X_1, \ldots, X_n \mid \theta) = P_n(X_1, \ldots, X_n \mid \hat\theta_n) \exp[-\{(\theta - \hat\theta_n)^2/2\sigma_n^2\}(1 - R_n)]. \tag{3}$$

Given $\varepsilon > 0$ we choose $\delta = \delta(\varepsilon)$ so small that $N_0(\delta) \subset N_0$ and

$$(1 - \varepsilon) < \inf_{\theta \in N_0(\delta)} \pi(\theta)/\pi(\theta_0) \leqslant \sup_{\theta \in N_0(\delta)} \pi(\theta)/\pi(\theta_0) < 1 + \varepsilon, \tag{4}$$

this last condition being justified by (A1). Now we split the integral defining $P_n(X_1, \ldots, X_n)$ into $I_1^{(n)}(\delta)$ and $I_2^{(n)}(\delta)$, taken over $\Theta \setminus N_0(\delta)$ and $N_0(\delta)$ respectively. Then, using (1),

$$I_1^{(n)}(\delta) = P_n(X_1, \ldots, X_n \mid \hat\theta_n) \int_{\Theta \setminus N_0(\delta)} \pi(\theta) \exp\{L_n(\theta) - L_n(\hat\theta_n)\}\, d\theta,$$

so that, conditional on $L_n''(\hat{\theta}_n)<0$ and

$$A_n = A_n(\varepsilon) = [\sup_{\Theta\setminus N_0(\delta)} \{-L_n''(\theta_0)\}^{-1}\{L_n(\theta)-L_n(\theta_0)\} < -k(\delta)],$$

$$[P_n(X_1, \ldots, X_n \mid \hat{\theta}_n)\sigma_n]^{-1} I_1^{(n)}(\delta)$$
$$= \exp\{L_n(\theta_0)-L_n(\hat{\theta}_n)\}\{-L_n''(\hat{\theta}_n)\}^{\frac{1}{2}} \times \int_{\Theta\setminus N_0(\delta)} \pi(\theta)\exp\{L_n(\theta)-L_n(\theta_0)\}\,d\theta$$
$$\leqslant \{-L_n''(\hat{\theta}_n)\}^{\frac{1}{2}} \exp[-k(\delta)\{-L_n''(\theta_0)\}]$$

by the definition of $\hat{\theta}_n$. Now

$$\{-L_n''(\theta_0)\}^{\frac{1}{2}} \exp[-k(\delta)\{-L(_n''\theta_0)\}] \to 0, \quad \text{a.s.}$$

as $n\to\infty$ in view of (A3) and hence, by virtue of (A3), (A5) and $P_{\theta_0}(L_n''(\hat{\theta}_n)<0)\to 1$ as $n\to\infty$, it follows that conditional on $A_n$ and

$$B_n = B_n(\varepsilon) = \{\sup_{\theta\in N_0(\delta)} |[L_n''(\hat{\theta}_n)]^{-1}(L_n''(\theta)-L_n''(\hat{\theta}_n))| < \varepsilon\},$$

we have

$$\{P_n(X_1, \ldots, X_n \mid \hat{\theta}_n)\sigma_n\}^{-1} I_1^{(n)}(\delta) \to 0, \tag{5}$$

in probability.

To deal with $I_2^{(n)}(\delta)$ we first suppose that $\hat{\theta}_n \in N_0(\delta) \subset N_0$ and use (3) to write

$$\{P_n(X_1, \ldots, X_n \mid \hat{\theta}_n)\pi(\theta_0)\}^{-1} I_1^{(n)}(\delta) = \int_{N_0(\delta)} \{\pi(\theta)/\pi(\theta_0)\}\exp[-\{(\theta-\hat{\theta}_n)^2/2\sigma_n^2\}(1-R_n)]\,d\theta.$$

Then, putting

$$I_3^{(n)}(\delta) = \int_{N_0(\delta)} \exp[-\{(\theta-\hat{\theta}_n)^2/2\sigma_n^2\}(1-R_n)]\,d\theta,$$

we have from (4) that

$$(1-\varepsilon)\, I_3^{(n)}(\delta) < \{P_n(X_1, \ldots, X_n \mid \hat{\theta}_n)\pi(\theta_0)\}^{-1} I_2^{(n)}(\delta) < (1+\varepsilon)\, I_3^{(n)}(\delta). \tag{6}$$

If $\sup_{N_0(\delta)} |R_n| < \varepsilon < 1$,

$$\int_{N_0(\delta)} \exp\{-(\theta-\hat{\theta}_n)^2(1+\varepsilon)/2\sigma_n^2\}\,d\theta < I_3^{(n)}(\delta) < \int_{N_0(\delta)} \exp\{-(\theta-\hat{\theta}_n)^2(1-\varepsilon)/2\sigma_n^2\}\,d\theta$$

whilst for $\eta = +\varepsilon$ or $-\varepsilon$,

$$\int_{N_0(\delta)} \exp\{-(\theta-\hat{\theta}_n)^2(1+\eta)/2\sigma_n^2\}\,d\theta$$
$$= (2\pi)^{\frac{1}{2}}\sigma_n(1+\eta)^{-\frac{1}{2}}[\Phi\{\sigma_n^{-1}(\theta_0+\delta-\hat{\theta}_n)(1+\eta)^{\frac{1}{2}}\} - \Phi\{\sigma_n^{-1}(\theta_0-\delta-\hat{\theta}_n)(1+\eta)^{\frac{1}{2}}\}], \tag{7}$$

where

$$\Phi(y) = (2\pi)^{-\frac{1}{2}} \int_{-\infty}^{y} e^{-\frac{1}{2}u^2}\,du.$$

Conditions (A3) and (A5) imply that $\sigma_n^{-1}\to\infty$, in probability, and, since $\hat{\theta}_n$ is consistent, the factor in curly brackets in (6) converges to unity in probability. Thus, using (A5) to secure the appropriate bound on $R_n$, it follows that

$$P_{\theta_0}((2\pi)^{\frac{1}{2}}(1+\varepsilon)^{-\frac{1}{2}} < \sigma_n^{-1} I_3^{(n)}(\delta) < (2\pi)^{\frac{1}{2}}(1-\varepsilon)^{-\frac{1}{2}} \mid A_n \cap B_n) \to 1 \tag{8}$$

as $n\to\infty$. The conditions (5), (6) and (8) then give

$$P_{\theta_0}(C_n \mid A_n \cap B_n) \to 1$$

as $n\to\infty$, where

$$C_n = [(1+\varepsilon)^{\frac{1}{2}}(1-\varepsilon) < \{P_n(X_1,\ldots,X_n \mid \hat{\theta}_n)\,\pi(\theta_0)\,(2\pi)^{\frac{1}{2}}\,\sigma_n\}^{-1} P_n(X_1,\ldots,X_n) < (1+\varepsilon)(1-\varepsilon)^{-\frac{1}{2}}].$$

But, by virtue of (A4) and (A5), and using a bar to denote complementation,

$$\begin{aligned}\liminf_{n\to\infty} P_{\theta_0}(A_n \cap B_n) &= 1 - \limsup_{n\to\infty} P_{\theta_0}(\bar{A}_n \cup \bar{B}_n)\\ &\geqslant 1 - \limsup_{n\to\infty} P_{\theta_0}(\bar{A}_n) - \limsup_{n\to\infty} P_{\theta_0}(\bar{B}_n) \geqslant 1 - h(\delta(\varepsilon)) - \varepsilon,\end{aligned}$$

and hence,

$$\begin{aligned}\liminf_{n\to\infty} P_{\theta_0}(C_n) &\geqslant \liminf_{n\to\infty} P_{\theta_0}(C_n \cap A_n \cap B_n)\\ &= \liminf_{n\to\infty} P_{\theta_0}(C_n \mid A_n \cap B_n)\, P_{\theta_0}(A_n \cap B_n)\\ &\geqslant 1 - h(\delta(\varepsilon)) - \varepsilon. \end{aligned} \tag{9}$$

The results (9) holds for all $\varepsilon$ (and hence $h(\delta)$) small enough and thus,

$$\{P_n(X_1,\ldots,X_n \mid \hat{\theta}_n)\,\sigma_n\}^{-1} P_n(X_1,\ldots,X_n) \to (2\pi)^{\frac{1}{2}}\,\pi(\theta_0), \tag{10}$$

in probability.

To complete the proof, we consider

$$I_2' = \int_{\hat{\theta}_n + b\sigma_n}^{\hat{\theta}_n + a\sigma_n} \pi(\theta)\, P_n(X_1,\ldots,X_n \mid \theta)\, d\theta,$$

which is $I_2^{(n)}(\delta)$ with $N_0(\delta)$ replaced by $(\hat{\theta}_n + b\sigma_n, \hat{\theta}_n + a\sigma_n)$. Similarly, let $I_3'$ be $I_3^{(n)}(\delta)$ with the limits of integration changed to $(\hat{\theta}_n + b\sigma_n, \hat{\theta}_n + a\sigma_n)$.

We use (A1) and the argument up to (6) so that if $\varepsilon > 0$ is given, there exists $\delta$ such that if $(\hat{\theta}_n + b\sigma_n, \hat{\theta}_n + a\sigma_n) \subset N_0(\delta)$,

$$(1-\varepsilon)\, I_3' < \{P_n(X_1,\ldots,X_n \mid \hat{\theta}_n)\,\pi(\theta_0)\}^{-1} I_2' < (1+\varepsilon)\, I_3',$$

while the right-hand side corresponding to (7) becomes

$$(2\pi)^{\frac{1}{2}}\,\sigma_n (1+\eta)^{-\frac{1}{2}} [\Phi\{a(1+\eta)^{\frac{1}{2}}\} - \Phi\{b(1+\eta)^{\frac{1}{2}}\}].$$

Clearly

$$\lim_{n\to\infty} P_{\theta_0}\{(\hat{\theta}_n + b\sigma_n, \hat{\theta}_n + a\sigma_n) \subset N_0(\delta)\} = 1$$

and our previous argument can be paralleled to obtain

$$\{P_n(X_1,\ldots,X_n \mid \hat{\theta}_n)\,\sigma_n\}^{-1} I_2' \to (2\pi)^{\frac{1}{2}}\,\pi(\theta_0)\,\{\Phi(a) - \Phi(b)\}, \tag{11}$$

in probability. Since

$$\int_{\hat{\theta}_n + b\sigma_n}^{\hat{\theta}_n + a\sigma_n} \pi_n(\theta \mid X_1,\ldots,X_n)\, d\theta = I_2' / P_n(X_1,\ldots,X_n),$$

the result of the theorem follows from (10) and (11).

## REFERENCES

BASAWA, I. V., FEIGIN, P. D. and HEYDE, C. C. (1976). Asymptotic properties of maximum likelihood estimators for stochastic processes. *Sankhyā* A, **38**, 259–270.

BASAWA, I. V. and SCOTT, D. J. (1976). Efficient tests for branching processes. *Biometrika*, **63**, 531–536.

—— (1977). Efficient tests for stochastic processes. *Sankhyā* A, **39**, 21–31.

DAWID, A. P. (1970). On the limiting normality of posterior distributions. *Proc. Camb. Phil. Soc.*, **67**, 625–633.

FEIGIN, P. D. (1978). The efficiency criteria problem for stochastic processes. *Stoch. Proc. Appl.*, **6**, 115–127

HEYDE, C. C. (1978). On an optimal property of the maximum likelihood estimator of a parameter from a stochastic process. *Stoch. Proc. Appl.* **8**, 1–9.

HEYDE, C. C. and FEIGIN, P. D. (1975). On efficiency and exponential families in stochastic process estimation. In *Statistical Distributions in Scientific Work* (G. P. Patil, S. Kotz and J. K. Ord, eds), Vol. 1, pp. 227–240. Dordrecht: Reidel.

JOHNSTONE, I. M. (1978). M.Sc. Thesis, Australian National University.

SWEETING, T. J. (1978). On efficient tests for branching processes. *Biometrika*, **65**, 123–127.

WALD, A. (1948). Asymptotic properties of the maximum likelihood estimate of an unknown parameter of a discrete stochastic process. *Ann. Math. Statist.*, **19**, 40–46.

WALKER, A. M. (1969). On the asymptotic behaviour of posterior distributions. *J. R. Statist. Soc.* B, **31**, 80–88.

# Part III
# Computation Via Monte Carlo Methods

# [9]



# Stochastic Relaxation, Gibbs Distributions, and the Bayesian Restoration of Images

STUART GEMAN AND DONALD GEMAN

***Abstract*—We make an analogy between images and statistical mechanics systems. Pixel gray levels and the presence and orientation of edges are viewed as states of atoms or molecules in a lattice-like physical system. The assignment of an energy function in the physical system determines its Gibbs distribution. Because of the Gibbs distribution, Markov random field (MRF) equivalence, this assignment also determines an MRF image model. The energy function is a more convenient and natural mechanism for embodying picture attributes than are the local characteristics of the MRF. For a range of degradation mechanisms, including blurring, nonlinear deformations, and multiplicative or additive noise, the posterior distribution is an MRF with a structure akin to the image model. By the analogy, the posterior distribution defines another (imaginary) physical system. Gradual temperature reduction in the physical system isolates low energy states ("annealing"), or what is the same thing, the most probable states under the Gibbs distribution. The analogous operation under the posterior distribution yields the maximum *a posteriori* (MAP) estimate of the image given the degraded observations. The result is a highly parallel "relaxation" algorithm for MAP estimation. We establish convergence properties of the algorithm and we experiment with some simple pictures, for which good restorations are obtained at low signal-to-noise ratios.**

***Index Terms*—Annealing, Gibbs distribution, image restoration, line process, MAP estimate, Markov random field, relaxation, scene modeling, spatial degradation.**

## I. Introduction

The restoration of degraded images is a branch of digital picture processing, closely related to image segmentation and boundary finding, and extensively studied for its evident practical importance as well as theoretical interest. An analysis of the major applications and procedures (model-based and otherwise) through approximately 1980 may be found in [47]. There are numerous existing models (see [34]) and algorithms and the field is currently very active. Here we adopt a Bayesian approach, and introduce a "hierarchical," stochastic model for the original image, based on the *Gibbs distribution*, and a new restoration algorithm, based on stochastic relaxation and *annealing*, for computing the maximum *a posteriori* (MAP) estimate of the original image given the degraded image. This algorithm is highly parallel and exploits the equivalence between Gibbs distributions and *Markov random fields* (MRF).

Manuscript received October 7, 1983; revised June 11, 1984. This work was supported in part by ARO Contract DAAG-29-80-K-0006 and in part by the National Science Foundation under Grants MCS-83-06507 and MCS-80-02940.

S. Geman is with the Division of Applied Mathematics, Brown University, Providence, RI 02912.

D. Geman is with the Department of Mathematics and Statistics, University of Massachusetts, Amherst, MA 01003.

The essence of our approach to restoration is a stochastic relaxation algorithm which generates a sequence of images that converges in an appropriate sense to the MAP estimate. This sequence evolves by *local* (and potentially *parallel*) changes in pixel gray levels and in locations and orientations of boundary elements. Deterministic, iterative-improvement methods generate a sequence of images that monotonically increase the posterior distribution (our "objective function"). In contrast, stochastic relaxation permits changes that *decrease* the posterior distribution as well. These are made on a *random* basis, the effect of which is to avoid convergence to *local maxima*. This should not be confused with "probabilistic relaxation" ("relaxation labeling"), which is deterministic; see Section X.

The stochastic relaxation algorithm can be informally described as follows.

1) A local change is made in the image based upon the current values of pixels and boundary elements in the immediate "neighborhood." This change is *random*, and is generated by sampling from a local conditional probability distribution.

2) The local conditional distributions are dependent on a global control parameter $T$ called "temperature." At *low* temperatures the local conditional distributions concentrate on states that *increase* the objective function, whereas at high temperatures the distribution is essentially uniform. The limiting cases, $T = 0$ and $T = \infty$, correspond respectively to greedy algorithms (such as gradient ascent) and undirected (i.e., "purely random") changes. (High temperatures induce a loose coupling between neighboring pixels and a chaotic appearance to the image. At low temperatures the coupling is tighter and the images appear more regular.)

3) Our image restorations avoid local maxima by beginning at high temperatures where many of the stochastic changes will actually decrease the objective function. As the relaxation proceeds, temperature is gradually lowered and the process behaves increasingly like iterative improvement. (This gradual reduction of temperature simulates "annealing," a procedure by which certain chemical systems can be driven to their low energy, highly regular, states.)

Our "annealing theorem" prescribes a schedule for lowering temperature which guarantees convergence to the global maxima of the posterior distribution. In practice, this schedule may be too slow for application, and we use it only as a guide in choosing the functional form of the temperature-time dependence. Readers familiar with Monte Carlo methods in statistical physics will recognize our stochastic relaxation algorithm as a "heat bath" version of the *Metropolis algorithm* [42]. The idea of introducing temperature and simulating an-

0162-8828/84/1100-0721$01.00 

nealing is due to Černý [8] and Kirkpatrick *et al.* [40], both of whom used it for combinatorial optimization, including the traveling salesman problem. Kirkpatrick also applied it to computer design.

Since our approach is Bayesian it is model-based, with the "model" captured by the prior distribution. Our models are "hierarchical," by which we mean layered processes reflecting the type and degree of *a priori* knowledge about the class of images under study. In this paper, we regard the original image as a pair X = (F, L) where F is the matrix of observable pixel intensities and L denotes a (dual) matrix of unobservable edge elements. Thus the usual gray levels are considered a marginal process. We refer to F as the *intensity process* and L as the *line process.* In future work we shall expand this model by adjoining other, mainly geometric, attribute processes.

The degradation model allows for noise, blurring, and some nonlinearities, and hence is characteristic of most photochemical and photoelectric systems. More specifically, the degraded image G is of the form $\phi(H(F)) \odot N$, where $H$ is the blurring matrix, $\phi$ is a possibly nonlinear (memoryless) transformation, N is an independent noise field, and $\odot$ denotes any suitably invertible operation, such as addition or multiplication. Surprisingly, these nonlinearities do not affect the computational burden.

To pin things down, let us briefly discuss the Markovian nature of the intensity process; similar remarks apply to the line process, the pair (F, L), and the distribution of (F, L) conditional on the "data" G. Of course, all of this will be discussed in detail in the main body of the paper.

Let $Z_m = \{(i, j): 1 \leq i, j \leq m\}$ denote the $m \times m$ integer lattice; then $F = \{F_{i,j}\}$, $(i, j) \in Z_m$, denotes the gray levels of the original, digitized image. Lowercase letters will denote the values assumed by these (random) variables; thus, for example, $\{F = f\}$ stands for $\{F_{i,j} = f_{i,j}, (i, j) \in Z_m\}$. We regard F as a sample realization of a random field, usually isotropic and homogeneous, and with significant correlations well beyond nearest neighbors. Specifically, we model F as an MRF, or, what is the same (see Section IV), we assume that the probability law of F is a Gibbs *distribution.* Given a *neighborhood system* $\mathcal{G} = \{\mathcal{G}_{i,j}, (i, j) \in Z_m\}$, where $\mathcal{G}_{i,j} \subseteq Z_m$ denotes the neighbors of $(i, j)$, an MRF over $(Z_m, \mathcal{G})$ is a stochastic process indexed by $Z_m$ for which, for every $(i, j)$ and every $f$,

$$P(F_{i,j} = f_{i,j} | F_{k,l} = f_{k,l}, (k, l) \neq (i, j)) = P(F_{i,j} = f_{i,j} | F_{k,l} = f_{k,l}, (k, l) \in \mathcal{G}_{i,j}). \tag{1.1}$$

The MRF-Gibbs equivalence provides an explicit formula for the *joint* probability distribution $P(F = f)$ in terms of an *energy function*, the choice of which, together with $\mathcal{G}$, supplies a powerful mechanism for modeling spatial continuity and other scene features.

The relaxation algorithm is designed to maximize the conditional probability distribution of (F, L) given the data $G = g$, i.e., find the mode of the *posterior distribution* $P(X = x | G = g)$. This form of Bayesian estimation is known as *maximum a posteriori* or MAP estimation, or sometimes as *penalized maximum likelihood* because one seeks to maximize $\log P(G = g | X = x) + \log P(X = x)$ as a function of $x$; the second term is the "penalty term." MAP estimation has been successfully employed in special settings (see, e.g., Hunt [31] and Hansen and Elliott [25]) and we share the opinion of many that the MAP formulation (and a Bayesian approach in general; see also [24], [43], [45]) is well-suited to restoration, particularly for handling general forms of spatial degradation. Moreover, the distribution of G itself need not be known, which is fortunate due to its usual complexity. On the other hand, MAP estimation clearly presents a formidable computational problem. The number of possible intensity images is $L^{m^2}$, where $L$ denotes the number of allowable gray levels, which rules out any direct search, even for small ($m = 64$), binary ($L = 2$) scenes. Consequently, one is usually obliged to make simplifying assumptions about the image and degradation models as well as compromises at the computational stage. Here, the computational problem is overcome by exploiting the pivotal observation that the posterior distribution is again Gibbsian with approximately the same neighborhood system as the original image, together with a sampling method which we call the *Gibbs Sampler.* Indeed, our principal theoretical contribution is a general, practical, and mathematically coherent approach for investigating MRF's by sampling (Theorem A), and by computing modes (Theorem B) and expectations (Theorem C).

The Gibbs Sampler generates realizations from a given MRF by a "relaxation" technique akin to site-replacement algorithms in statistical physics, such as "spin-flip" and "exchange" systems. The prototype is due to Metropolis *et al.* [42]; see also [7], [18], and Section X. Cross and Jain [12] use one of these algorithms invented for studying binary alloys. ("Relaxation labeling" in the sense of [13], [30], [46], [47] is different; see Section X.) The Markov property (1.1) permits parallel updating of the line and pixel sites, each of which is "refreshed" according to a simple recipe determined by the governing distribution. Thus, both parts of the MRF-Gibbs equivalence are exploited, for computing and modeling, respectively. Moreover, minimum mean-square error (MMSE) estimation is also feasible by using the (temporal) ergodicity of the relaxation chain to compute *means* w.r.t. the posterior distribution. However, we shall not pursue this approach.

We have used a comparatively slow, raster scan-serial version of the Gibbs Sampler to generate images and restorations (see Section XIII). But the algorithm is parallel; it could be executed in essentially one-half the time with two processors running simultaneously, or in one-third the time with three, and so on. The full parallel potential is realized by assigning one (simple) processor to each site of the intensity process and to each site of the line process. Whatever the number of processors, parallel implementation is made feasible by a small communications requirement among processors. The communications burden is related to the neighborhood size of the graph associated with the image model, and herein lies much of the power of the hierarchical structure: although the field model X = (F, L) has a local graph structure, the *marginal* distribution on the observable intensity process F has a *completely connected graph.* The introduction of a hierarchy dramatically expands the richness of the model of the observed process while only moderately adding to the computational burden. We shall return to these points in Sections IV and XI.

The MAP algorithm depends on an *annealing schedule*, which refers to the (sufficiently) slow decrease of a ("control") parameter $T$ that corresponds to *temperature* in a physical system. As $T$ decreases, samples from the posterior distribution are forced towards the minimal energy configurations; these correspond to the mode(s) of the distribution. Theorem B makes this precise, and is, to our knowledge, the first theoretical result of this nature. Roughly speaking, it says that if the temperature $T(k)$ employed in executing the $k$th site replacement (i.e., the $k$th image in the iteration scheme) satisfies the bound

$$T(k) \geqslant \frac{c}{\log(1+k)}$$

for every $k$, where $c$ is a constant independent of $k$, then with probability converging to one (as $k \to \infty$), the configurations generated by the algorithm will be those of minimal energy. Put another way, the algorithm generates a Markov chain which converges *in distribution* to the uniform measure over the minimal energy configurations. (It should be emphasized that *pointwise* convergence, i.e., convergence *with probability one*, is in general not possible.) These issues are discussed in Section XII, and the algorithm is demonstrated in Section XIII on a variety of degraded images. We also discuss the nature of the constant $c$ in regard to practical convergence rates. Basically, we believe that the logarithmic rate is best possible. However, the best (i.e., smallest) value of $c$ that we have obtained to date (see the Appendix) is far too large for computational value and our restorations are actually performed with small values of $c$. As yet, we do not know how to bring the theory in line with experimental results in this regard.

The role of the Gibbs (or Boltzmann) distribution, and other notions from statistical physics, in the construction of "expert systems" is expanding. To begin with, we refer the reader to [21] for the original formulation of our computational method and of a general approach to expert systems based on maximum entropy extensions. As previously mentioned, Černý [8] and Kirkpatrick *et al.* [40] introduced annealing into combinatorial optimization. Other examples include the work of Cheeseman [9] on maximum entropy and diagnosis and of Hinton and Sejnowski [29] on neural modeling of inference and learning.

This paper is organized as follows. The degradation model is described in the next section, and the undegraded image models are presented in Section IV after preliminary material on graphs and neighborhood systems in Section III. In particular, Section IV contains the definitions of MRF's, Gibbs distributions, and the equivalence theorem. Due to the plethora of Markovian models in the literature, we pause in Section V to compare ours to others, and in Section VI to explain some connections with maximum entropy methods. In Section VII we raise the issues of parameter estimation and model selection, and indicate why we are avoiding the former for the time being. The posterior distribution is computed in Section VIII and the corresponding optimization problem is addressed in Section IX. The concept of stochastic relaxation is reviewed in Section X, including its origins in physics. Sections XI and XII are devoted to the Gibbs Sampler, dealing, respectively, with its mechanical and mathematical workings. Our experimental results appear in Section XIII, followed by concluding remarks.

## II. Degraded Image Model

We follow the standard modeling of the (intensity) image formation and recording processes, and refer the reader to [31] or [47] for better accounts of the physical mechanisms.

Let $H$ denote the "blurring matrix" corresponding to a shift-invariant point-spread function. The formation of F gives rise to a blurred image $H(\mathrm{F})$ which is recorded by a sensor. The latter often involves a nonlinear transformation of $H(\mathrm{F})$, denoted here by $\phi$, in addition to random sensor noise $\mathrm{N} = \{\eta_{i,j}\}$, which we assume to consist of independent, and for definiteness, Gaussian variables with mean $\mu$ and standard deviation $\sigma$.

Our methods apply to essentially arbitrary noise processes $\mathrm{N} = \{\eta_{i,j}\}$, discrete or continuous. However, computational feasibility requires that the description of N as an MRF (this can always be done; see Section IV) has an associated graph structure that is approximately "local"; the same requirement is applied to the image process $\mathrm{X} = (\mathrm{F}, \mathrm{L})$. For clarity, we forgo full generality and focus on the traditional Gaussian white noise case. Extension to a general noise process is mostly a matter of notation.

The degraded image is then a function of $\phi(H(\mathrm{F}))$ and N, say $\psi(\phi(H(\mathrm{F})), \mathrm{N})$, for example, addition or multiplication. (To compute the posterior distribution, we only need to assume that $b \to \psi(a, b)$ is invertible for each $a$.) For notational ease, we will write

$$\mathrm{G} = \phi(H(\mathrm{F})) \odot \mathrm{N}. \tag{2.1}$$

At the pixel level, for each $(i,j) \in Z_m$,

$$G_{i,j} = \phi\left(\sum_{(k,l)} H(i-k, j-l) F_{k,l}\right) \odot \eta_{i,j}. \tag{2.2}$$

The mathematical results require an additional assumption, namely, that F and N be independent as stochastic processes (and likewise for L and N) and we assume this henceforth. This is customary, although we recognize the limitation in certain contexts, e.g., for nuclear scan pictures.

For computational purposes, the degree of locality of F should be approximately preserved by (2.1), so that the neighborhood systems for the prior and posterior distributions on (F, L) are comparable. This is achieved when $H$ is a simple convolution over a small window. For instance, take

$$H(k,l) = \begin{cases} \frac{1}{2}, & k=0, l=0 \\ \frac{1}{16}, & |k|, |l| \leqslant 1, (k,l) \neq (0,0) \end{cases} \tag{2.3}$$

so that the intensity at $(i,j)$ is weighted equally with the average of the eight nearest neighbors. The function $\phi$ is unrestricted, bearing in mind that the true noise level depends on $\phi$, $\odot$, and $\sigma$. Typically, $\phi$ is logarithmic (film) or algebraic (TV).

An important special case, which occurs in two-dimensional (2-D) signal theory, is the segmentation of noisy images into

724 IEEE TRANSACTIONS ON PATTERN ANALYSIS AND MACHINE INTELLIGENCE, VOL. PAMI-6, NO. 6, NOVEMBER 1984

coherent regions. The usual model is

$$G = F + N \tag{2.4}$$

where N is white noise and the number of intensity levels is small. This is the model entertained by Hansen and Elliott [25] for simple, binary MRF's F, and by many other workers with varying assumptions about F; see [14], [16], [17]. In this case, namely (2.4), we can extract simple images under extremely low signal-to-noise ratios.

The full degraded image is (G, L); that is, the "line process" is not transformed.

## III. Graphs and Neighborhoods

Here and in Section IV we present the general theory of MRF's on graphs, focusing on the aspects and examples which figure in the experimental restorations. The level of abstraction is warranted by the variety of MRF's, graphs, and probability distributions simultaneously under discussion.

Let $S = \{s_1, s_2, \cdots, s_N\}$ be a set of *sites* and let $\mathcal{G} = \{\mathcal{G}_s, s \in S\}$ be a *neighborhood system* for $S$, meaning any collection of subsets of $S$ for which 1) $s \notin \mathcal{G}_s$ and 2) $s \in \mathcal{G}_r \Leftrightarrow r \in \mathcal{G}_s$. Obviously, $\mathcal{G}_s$ is the set of *neighbors* of $s$ and the pair $\{S, \mathcal{G}\}$ is a graph in the usual way. A subset $C \subseteq S$ is a *clique* if every pair of distinct sites in $C$ are neighbors; $\mathcal{C}$ denotes the set of cliques.

The special cases below are especially relevant.

Case 1: $S = Z_m$. This is the set of pixel sites for the intensity process F; $\{s_1, s_2, \cdots, s_N\}$, $N = m^2$, is any ordering of the lattice points. We are interested in homogeneous neighborhood systems of the form

$$\mathcal{G} = \mathcal{F}_c = \{\mathcal{F}_{i,j}, (i,j) \in Z_m\}; \ \mathcal{F}_{i,j}$$
$$= \{(k,l) \in Z_m : 0 < (k-i)^2 + (l-j)^2 \leq c\}.$$

Notice that sites at or near the boundary have fewer neighbors than interior ones; this is the so-called "free boundary" and is more natural for picture processing than torodial lattices and other periodic boundaries. Fig. 1(a), (b), (c) shows the (interior) neighborhood configurations for $c = 1, 2, 8$; $c = 1$ is the first-order or nearest-neighbor system common in physics, in which $\mathcal{F}_{i,j} = \{(i, j-1), (i, j+1), (i-1, j), (i+1, j)\}$, with adjustments at the boundaries. In each case, $(i, j)$ is at the center, and the symbol $\circ$ stands for a neighboring pixel. The cliques for $c = 1$ are all subsets of $Z_m$ of the form $\{(i, j)\}$, $\{(i, j), (i, j+1)\}$ or $\{(i, j), (i+1, j)\}$, shown in Fig. 1(d). For $c = 2$, we have the cliques in Fig. 1(d) as well as those in Fig. 1(e). Obviously, the number of clique types grows rapidly with $c$. However, only small cliques appear in the model for F actually employed in this paper; indeed, the degree of progress with only *pair* interactions is somewhat surprising. Nonetheless, more complex images will likely necessitate more complex energies. Our experiments (see Section XIII) suggest that much of this additional complexity can be accommodated while maintaining modest neighborhood sizes by further developing the hierarchy.

*Case 2:* $S = D_m$, the "dual" $m \times m$ lattice. Think of these sites as placed midway between each vertical or horizontal pair of pixels, and as representing the possible locations of "edge

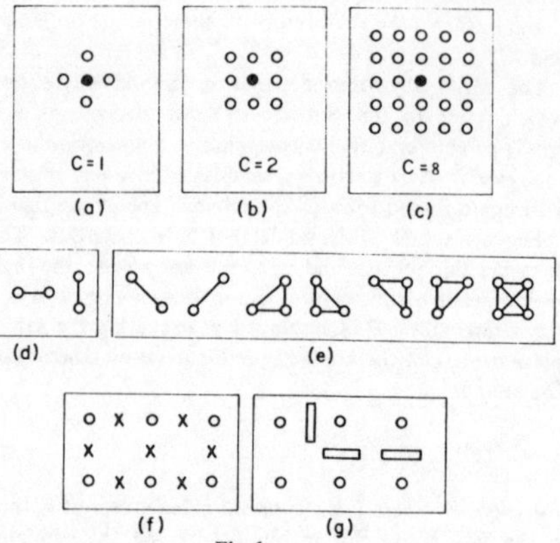


Fig. 1.

elements." Shown in Fig. 1(f) are six pixel sites together with seven line sites denoted by an X. The six surrounding X's are the neighbors of the middle X for the neighborhood system we denote by $\mathcal{L} = \{\mathcal{L}_d, d \in D_m\}$. Fig. 1(g) is a segment of a realization of a binary line process for which, at each line site, there may or may not be an edge element. We also consider line processes with more than two levels, corresponding to edge elements with varying orientations.

*Case 3:* $S = Z_m \cup D_m$. This is the setup for the field (F, L). $Z_m$ has neighborhood system $\mathcal{F}_1$ (nearest-neighbor lattice) and $D_m$ has the above-described system. The pixel neighbors of sites in $D_m$ are the two pixels on each side, and hence each (interior) pixel has four line site neighbors.

## IV. Markov Random Fields and Gibbs Distributions

We now describe a class of stochastic processes that includes both the prior and posterior distribution on the original image. In general, this class of processes (namely, MRF's) is neither homogeneous nor isotropic, assuming the index set $S$ has enough geometric structure to even *define* a suitable family of translations and rotations. However, the *particular* models we choose for prior distributions on the original image are in fact both homogeneous and isotropic in an appropriate sense. (This is not the case for the *posterior* distribution.) We refer the reader to Section XIII for a precise description of the prior models employed in our experiments, and in particular for specific examples of the role of the line elements.

As in Section III, $\{S, G\}$ denotes an arbitrary graph. Let $X = \{X_s, s \in S\}$ denote *any* family of random variables indexed by $S$. For simplicity, we can assume a common state space, say $\Lambda \doteq \{0, 1, 2, \cdots, L-1\}$, so that $X_s \in \Lambda$ for all $s$; the extension to site-dependent state spaces, appropriate when $S$ consists of both line and pixel sites, is entirely straightforward (although not merely a notational matter due to the "positivity condition" below). Let $\Omega$ be the set of all possible *configurations:*

$$\Omega = \{\omega = (x_{s_1}, \cdots, x_{s_N}) : x_{s_i} \in \Lambda, 1 \leq i \leq N\}.$$

As usual, the event $\{X_{s_1} = x_{s_1}, \cdots, X_{s_N} = x_{s_N}\}$ is abbreviated $\{X = \omega\}$.

X is an MRF with respect to $\mathcal{G}$ if

$$P(X = \omega) > 0 \quad \text{for all } \omega \in \Omega; \tag{4.1}$$

$$P(X_s = x_s | X_r = x_r, r \neq s) = P(X_s = x_s | X_r = x_r, r \in \mathcal{G}_s) \tag{4.2}$$

for every $s \in S$ and $(x_{s_1}, \cdots, x_{s_N}) \in \Omega$. Technically, what is meant here is that the pair $\{X, P\}$ satisfies (4.1) and (4.2) relative to some probability measure on $\Omega$. The collection of functions on the left-hand side of (4.2) is called the *local characteristics* of the MRF and it turns out that the (joint) probability distribution $P(X = \omega)$ of *any* process satisfying (4.1) is *uniquely* determined by these conditional probabilities; see, e.g., [6, p. 195].

The concept of an MRF is essentially due to Dobrushin [15] and is one way of extending Markovian dependence from 1-D to a general setting; there are, of course, many others, some of which will be reviewed in Section V.

Notice that *any* X satisfying (4.1) is an MRF if the neighborhoods are large enough to encompass the dependencies. The utility of the concept, at least in regard to image modeling, is that priors are available with neighborhoods that are small enough to ensure feasible computational loads and yet still rich enough to model and restore interesting classes of images (and textures: [12]).

Ordinary 1-D Markov chains are MRF's relative to the nearest-neighbor system on $S = \{1, 2, \cdots, N\}$ (i.e., $\mathcal{G}_1 = \{2\}$, $\mathcal{G}_i = \{i-1, i+1\}$ $2 \leq i \leq N-1$, $\mathcal{G}_N = \{N-1\}$) if we assume all positive transitions and the chain is started in equilibrium. In other words, the "one-sided" Markov property

$$P(X_k = x_k | X_j = x_j, j \leq k-1) = P(X_k = x_k | X_{k-1} = x_{k-1})$$

and the "two-sided" Markov property

$$P(X_k = x_k | X_j = x_j, j \neq k) = P(X_k = x_k | X_j = x_j, j \in \mathcal{G}_k)$$

are equivalent. Similarly for an $r$th order Markov process on the line with respect to the $r$ nearest neighbors on one side and on both sides. (This appears to be doubted in [1] but follows, eventually, from straightforward calculations or immediately from the Gibbs connection.)

Gibbs models were introduced into image modeling by Hassner and Sklansky [28], although the treatment there is mostly expository and limited to the binary case.

A *Gibbs distribution* relative to $\{S, \mathcal{G}\}$ is a probability measure $\pi$ on $\Omega$ with the following representation:

$$\pi(\omega) = \frac{1}{Z} e^{-U(\omega)/T} \tag{4.3}$$

where $Z$ and $T$ are constants and $U$, called the *energy function*, is of the form

$$U(\omega) = \sum_{C \in \mathcal{C}} V_C(\omega). \tag{4.4}$$

Recall that $\mathcal{C}$ denotes the set of cliques for $\mathcal{G}$. Each $V_C$ is a function on $\Omega$ with the property that $V_C(\omega)$ depends only on those coordinates $x_s$ of $\omega$ for which $s \in C$. Such a family $\{V_C, C \in \mathcal{C}\}$ is called a *potential.* $Z$ is the normalizing constant:

$$Z \doteq \sum_{\omega} e^{-U(\omega)/T} \tag{4.5}$$

and is called the *partition function.* Finally, $T$ stands for "temperature"; for our purposes, $T$ controls the degree of "peaking" in the "density" $\pi$. Choosing $T$ "small" exaggerates the mode(s), making them easier to find by sampling; this is the principle of annealing, and will be applied to the posterior distribution $\pi(f, l) = P(F = f, L = l | G = g)$ in order to find the MAP estimate. Of course, we will show that $\pi(f, l)$ is Gibbsian and identify the energy and neighborhood system in terms of those for the priors. The *choice* of the prior distributions, i.e., of the particular functions $V_C$ for the image model $\pi(\omega) = P(X = \omega)$, will be discussed later on; see Section VII for some general remarks and Section XIII for the particular models employed in our experiments.

The terminology obviously comes from statistical physics, wherein such measures are "equilibrium states" for physical systems, such as ferromagnets, ideal gases, and binary alloys. The $V_C$ functions represent contributions to the total energy from external fields (singleton cliques), pair interactions (doubletons), and so forth. Most of the interest there, and in the mathematical literature, centers on the case in which $S$ is an *infinite*, 2-D or 3-D lattice; singularities in $Z$ may then occur at certain ("critical") temperatures and are associated with "phase transitions."

Typically, several free parameters are involved in the specification of $U$, and $Z$ is then a function of those parameters—notoriously intractable. For more information see [3], [5], [6], [23], [32], and [39].

The best-known of these lattice systems is the Ising model, invented in 1925 by E. Ising [33] to help explain ferromagnetism. Here, $S = Z_m$ and $\mathcal{G} = \mathcal{F}_1$, the nearest-neighbor system. The most general form of $U$ is then

$$U(\omega) = \sum V_{\{i,j\}}(x_{i,j}) + \sum V_{\{(i,j),(i+1,j)\}}(x_{i,j}, x_{i+1,j}) + \sum V_{\{(i,j),(i,j+1)\}}(x_{i,j}, x_{i,j+1}) \tag{4.6}$$

where the sums extend over all $(i, j) \in Z_m$ for which the indicated cliques make sense. The Ising model is the special case of (4.6) in which X is binary ($L = 2$), homogeneous (= strictly stationary), and isotropic (= rotationally invarient):

$$U(\omega) = \alpha \sum x_{i,j} + \beta \left(\sum x_{i,j}x_{i+1,j} + \sum x_{i,j}x_{i,j+1}\right) \tag{4.7}$$

for some parameters $\alpha$ and $\beta$, which measure, respectively, the external field and bonding strengths.

Returning to the general formulation, recall that the local characteristics

$$\pi(x_s | x_r, r \neq s) = \frac{\pi(\omega)}{\sum_{x_s \in \Lambda} \pi(\omega)} \qquad s \in S, \omega \in \Omega$$

uniquely determine $\pi$ for any probability measure $\pi$ on $\Omega$, $\pi(\omega) > 0$ for all $\omega$. The difficulty with the MRF formulation *by itself* is that

i) the *joint* distribution of the $X_s$ is not apparent;

ii) it is extremely difficult to spot local characteristics, i.e., to determine when a given set of functions $\psi(x_s|x_r, r \neq s)$, $s \in S$, $(x_{s_1}, \cdots, x_{s_N}) \in \Omega$, are conditional probabilities for some (necessarily unique) distribution on $\Omega$.

For example, Chellappa and Kashyap [10] allude to i) as a disadvantage of the "conditional Markov" models. See also the discussion in [6]. In fact, these apparent limitations to the MRF formulation have been noted by a number of authors, many of whom were obviously not aware of the following theorem.

*Theorem: Let $\mathcal{G}$ be a neighborhood system. Then X is an MRF with respect to $\mathcal{G}$ if and only if $\pi(\omega) = P(X = \omega)$ is a Gibbs distribution with respect to $\mathcal{G}$.*

Among other benefits, this equivalence provides us with a simple, practical way of specifying MRF's, namely by specifying potentials, which is easy, instead of local characteristics, which is nearly impossible. In fact, with some experience, one can choose $U$'s in accordance with the desired *local* behavior, at least at the intensity level. In short, the modeling and consistency problems of i) and ii) are eliminated.

Proofs may be found in many places now; see, e.g., [39] and the references therein, or the approach via the Hammersley-Clifford expansion in [6]. An influential discussion of this correspondence appears in Spitzer's work, e.g., [48]. Explicit formulas exist for obtaining $U$ from the local characteristics. Conversely, the local characteristics of $\pi$ are obtained in a straightforward way from the potentials: use the defining ratios and make the allowable cancellations. Fix $s \in S$, $\omega = (x_{s_1}, \cdots, x_{s_N}) \in \Omega$, and let $\omega^x$ denote the configuration which is $x$ at site $s$ and agrees with $\omega$ everywhere else. Then if $\pi(\omega) = P(X = \omega)$ is Gibbsian,

$$P(X_s = x_s | X_r = x_r, r \neq s) = Z_s^{-1} \exp - \frac{1}{T} \sum_{C: s \in C} V_C(\omega) \tag{4.8}$$

$$Z_s \doteq \sum_{x \in \Lambda} \exp - \frac{1}{T} \sum_{C: s \in C} V_C(\omega^x). \tag{4.9}$$

Notice that the right-hand side of (4.8) only depends on $x_s$ and on $x_r$, $r \in \mathcal{G}_s$, since any site in a clique containing $s$ must be a neighbor of $s$. These formulas will be used repeatedly to program the Gibbs Sampler for local site replacements.

For the Ising model, the conditional probability that $X_{i,j} = x_{i,j}$, given the states at $S \backslash \{i, j\}$, or equivalently, just the four nearest neighbors, reduces to

$$\frac{e^{-x_{ij}(\alpha + \beta v_{i,j})}}{1 + e^{-(\alpha + \beta v_{i,j})}}$$

where $v_{i,j} = x_{i,j-1} + x_{i-1,j} + x_{i,j+1} + x_{i+1,j}$. This is also known as the autologistic model and has been used for texture modeling in [12]. More generally, if the local characteristics are given by an exponential family and if $V_C(\omega) \equiv 0$ for $|C| > 2$, then the pair potentials always "factor" into a product of two like terms; see [6].

We conclude with some further discussion of a remark made in Section I: that the hierarchical structure introduced with the line process L expands the graph structure of the *marginal* distribution of the intensity process F. Consider first an arbitrary MRF X with respect to a graph $\{S, \mathcal{G}\}$. Fix $r \in S$ and let $\hat{X} = \{X_s, s \in S, s \neq r\}$. The marginal distribution $\hat{P}$ of $\hat{X}$ is derived from the distribution $P$ of X by summing over the range of $X_r$. Use the Gibbs representation for $P$ and perform this summation: the resulting expression for $\hat{P}$ can be put in the Gibbs form, and from this the neighborhood system on $\hat{S} \doteq S \backslash \{r\}$ can be inferred. The conclusion of this exercise is that $s_1, s_2 \in \hat{S}$ are, in general, neighbors if either i) they were neighbors in $S$ under $\mathcal{G}$ or ii) each is a neighbor of $r \in S$ under $\mathcal{G}$. Now let X = (F, L), with neighborhood system defined at the end of Section III. Successive summations of the distribution of X over the ranges of the elements of L yields the marginal distribution of the observable intensity process F. Each summation leaves a graph structure associated with the marginal distribution of the remaining variables, and this can be related to the original neighborhood system by following the preceding discussion of the general case. It is easily seen that when all of the summations are performed, the remaining graph is completely connected; under the marginal distribution of F, all sites are neighbors. This calculation suggests that significant long-range interactions can be introduced through the development of hierarchical structures without sacrificing the computational advantages of local neighborhood systems.

## V. Related Markov Image Models

The use of neighborhoods is, of course, pervasive in the literature: they offer a geometric framework for the clustering of pixel intensities and for many types of statistical models. In particular, the Markov property is a natural way to formalize these notions. The result is a somewhat bewildering array of Markov-type image models and it seems worthwhile to puase to relate these to MRF's. The process under consideration is $F = \{F_{i,j}, (i, j) \in Z_m\}$, the gray levels, or really any pixel attribute.

An early work in this direction is Abend, Harley and Kanal [1] about pattern classification. Among many novel ideas, there is the notion of a *Markov mesh* (MM) process, in which the Markovian dependence is *causal:* generally, one assumes that, for all $(i, j)$ and $f$,

$$\begin{aligned} P(F_{i,j} = f_{i,j} | F_{k,l} = f_{k,l}, (k, l) \in A_{i,j}) \\ = P(F_{i,j} = f_{i,j} | F_{k,l} = f_{k,l}, (k, l) \in B_{i,j}) \end{aligned} \tag{5.1}$$

where $B_{i,j} \subseteq A_{i,j} \subseteq \{(k, l): k < i \text{ or } l < j\}$. A common example is $B_{i,j} = \{(i-1, j), (i-1, j-1), (i, j-1)\}$. Besag [6], Kanal [37], and Pickard [44] also discuss such "unilateral" processes, which are usually a subclass of MRF's, although the resulting (bilateral) neighborhoods can be irregular. Anyway, for MM models the emphasis is on the causal, iterative aspects, including a recursive representation for the joint probabilities. Incidentally, a Gibbs type description of $r$th order Markov chains is given in [1]; of course, the full Gibbs-MRF equivalence is not perceived and was not for about five years. Derin *et al.* [14] model F as an MM process and use recursive Bayes smoothing to recover F from a noisy version F + $N$; the algorithms exploit the causality to maximize the univariate posterior distribution at each pixel based on the data over a strip containing it, and are very effective at low $S/N$ ratios for some simple images.

Motivated by a paper of Lévy [41], Woods [51] defined "$P$-Markov" processes for the resolution of wavenumber spectra. The definition involves two spatial regions separated by a "boundary" of width $P$, and correspond to the past, future, and present in 1-D. Woods also considers a family of "wide-sense" Markov fields of the form

$$F_{i,j} = \sum_{(k,l)\in W_p} \theta_{k,l} F_{i-k,j-l} + U_{i,j} \tag{5.2}$$

where $W_p = \{(k, l): 0 < k^2 + l^2 \leqslant P\}$, $\theta_{k,l}$ are the MMSE coefficients for projecting $F_{i,j}$ on $\{F_{k,l}, (k, l) \in (i, j) + W_p\}$, and $\{U_{i,j}\}$ is the error, generally nonwhite. The main theoretical result is that if $\{U_{i,j}\}$ is homogeneous, Gaussian, and satisfies a few other assumptions, then $\mathrm{F}$ is Gaussian, $P$-Markov and vice-versa. In general, there are consistency problems and the $P$-Markov property is hard to verify. In the nearest-neighbor case, one gets a Gaussian MRF.

Other "wide-sense" Markov processes appear in Jain and Angel [35] and Stuller and Kurz [49]. The assumptions in [35] are a nearest-neighbor system, white noise, and no blur; restoration is achieved by recursively filtering the rows $\{F_{i,j}\}_{j=1}^m$, which form a vector-valued, second-order Markov chain, to find the optimal interpolator of each row. In [49], causality is introduced and earlier work is generalized by considering an arbitrary "scanning pattern."

The "spatial interaction models" in Chellappa and Kashyap [10], [38] satisfy (5.2) for general coefficients and $W$'s. The model is causal if $W$ lies in the third quadrant. The authors consider "simultaneous autoregressive" (SAR) models, wherein the noise is white, and "conditional Markov" (CM) models, wherein the "bilateral" Markov property holds (i.e., (1.1) with $\mathcal{F}_{i,j} = (i, j) + W$) in addition to (5.2), and the noise is nonwhite. Thus, the CM models are MRF's, although in [10], [38] the boundary of $Z_m$ is periodic, and hence boundary conditions must be adjoined to (5.2). Given any (homogeneous) SAR process there exists a unique CM process with the same spectral density, although different neighborhood structure. The converse holds in the Gaussian case but is generally false (see the discussion in Besag [6]). MMSE restoration of blurred images with additive Gaussian noise is discussed in [10]; the original image is SAR or CM, usually Gaussian.

Finally, Hansen and Elliott [25] and Elliott *et al.* [17] design MAP algorithms for the segmentation of remotely sensed data with high levels of additive noise. The image model is a nearest-neighbor, binary MRF. However, the autologistic form of the joint distribution is not recognized due to the lack of the Gibbs formulation. The conditional probabilities are approximated by the product of four 1-D transitions, and segmentation is performed by dynamic programming, first for each row and then for the entire images. More recent work in Elliott *et al.* [16] is along the same lines, namely MAP estimation, via dynamic programming, of very noisy but simple images; the major differences are the use of the Gibbs formulation and improvements in the algorithms. Similar work, applied to boundary finding, can be found in Cooper and Sung [11], who use a Markov boundary model and a deterministic relaxation scheme.

## VI. Maximum Entropy Restoration

There are several contact points. The Gibbs distribution can be derived (directly from physical principles in statistical mechanics) by maximizing entropy: basically, it has maximal entropy among all probability measures (equilibrium states) on $\Omega$ with the same *average* energy. Thus it is no accident that, like maximum entropy (ME) methods, ours are well-suited to nonlinear problems; see [50]. Moreover, based on the success of ME restoration (along the lines suggested by Jaynes [36]) for recovering randomly pulsed objects (cf. Frieden [19]), we intend in the future to analyze such data (e.g., starfield photographs) by our methods.

We should also like to mention the interesting observation of Trussell [50] that conventional ME restoration is a special case of MAP estimation in which the prior distribution on $\mathrm{F}$ is

$$P(\mathrm{F} = f) = \exp\left(-\beta \sum f_{i,j} \log f_{i,j}\right) \Big/ (\text{normalizing constant}).$$

By "conventional ME," we refer to maximizing the entropy $\Sigma f_{i,j} \log f_{i,j}$ subject to $\Sigma \eta_{i,j}^2 = \text{constant}$ ($\eta_{i,j}$ is here again the noise process); see [2]. Other ME methods (e.g., [19]) do not appear to be MAP-related.

## VII. Model Selection and Parameter Estimation

The quality of the restoration will clearly depend on choices made at the modeling stage, in our case about specific energy types, attribute processes, and parameters. Cross and Jain [12] use maximum likelihood estimation in the context of Besag's [6] "coding scheme," as well as standard goodness-of-fit tests, for matching realizations of autobinomial MRF's to real textures. Kashyap and Chellappa [38] introduce some new methods for parameter estimation and the choice of neighborhoods for the SAR and CM models, mostly in the Gaussian case. These are but two examples.

For uncorrupted, simple MRF's, the coding methods do finesse the problem of the partition function. However, for more complex models and for corrupted data, we feel that the coding methods are ultimately inadequate due to the complexity of the distribution of $\mathrm{G}$. This view seems to be shared by other authors, although in different contexts. Of course, for MRF's, the obstacles facing conventional statistical inference due to $Z$ have often been noted. Even for the Ising model, analytical results are rare; a famous exception is Onsager's work on the correlational structure.

At any rate, we have developed a new method [20] for estimating clique parameters from the "noisy" data, and this will be implemented in a forthcoming paper. For now, we are obliged to choose the parameters on an ad hoc basis (which is common), but hasten to add that the quality of restoration does not seem to have been adversely affected, probably due to the relative simplicity of the MRF's we actually use for the line and intensity processes; see Section XIII.

One should also address the *general* choice of $\pi$ and $\mathcal{G}$. This is really quite different than parameter estimation and somewhat related to "image understanding": how does one incorporate "real-world knowledge" into the modeling process? In image interpretation systems, various semantical and hierarchical models have been proposed (see, e.g., [26]). We have begun our study of hierarchical Gibbs models in this paper. A *general theory* of interactive, self-adjusting models that is practical and mathematically coherent may lie far ahead.

## VIII. Posterior Distribution

We now turn to the posterior distribution $P(\mathrm{F} = f, \mathrm{L} = l \mid \mathrm{G} = g)$ of the original image given the "data" $g$. In this section we take $S = Z_m \cup D_m$, the collection of pixel and line sites, with some neighborhood system $\mathcal{G} = \{\mathcal{G}_s, s \in S\}$; an example of such a "mixed" graph was given in Section III. The configuration space is the set of all pairs $\omega = (f, l)$ where the components of $f$ assume values among the allowable gray levels and those of $l$ among the (coded) line states.

We assume that $\mathrm{X}$ is an MRF relative to $\{S, \mathcal{G}\}$ with corresponding energy function $U$ and potentials $\{V_C\}$:

$$P(\mathrm{F} = f, \mathrm{L} = l) = e^{-U(f,l)/T}/Z$$

$$U(f, l) = \sum_C V_C(f, l).$$

For convenience, take $T = 1$

Recall that $\mathrm{G} = \phi(H(\mathrm{F})) \odot \mathrm{N}$, where N is white Gaussian noise with mean $\mu$ and variance $\sigma^2$ and is independent of X.

We emphasize that what follows is easily extended to processes N that are more general MRF's, although we still require that N be independent of X. The operation $\odot$ is assumed invertible and we will write $\mathrm{N} = \Phi(\mathrm{G}, \phi(H(\mathrm{F}))) = \{\Phi_s, s \in Z_m\}$ to indicate this inverse.

Let $\mathcal{H}_s$, $s \in Z_m$, denote the pixels which affect the blurred image $H(\mathrm{F})$ at $s$. For instance, for the $H$ in (2.3), $\mathcal{H}_s$ is the $3 \times 3$ square centered at $s$. Observe that $\Phi_s$, $s \in Z_m$, depends only on $g_s$ and $\{f_t, t \in \mathcal{H}_s\}$. By the shift-invariance of $H$, $\mathcal{H}_{r+s} = s + \mathcal{H}_r$ where $\mathcal{H}_r \subseteq Z_m$, $s + r \in Z_m$, and $s + \mathcal{H}_r$ is understood to be intersected with $Z_m$, if necessary. In addition, we will assume that $\{\mathcal{H}_s\}$ is "symmetric" in that $r \in \mathcal{H}_0 \Rightarrow -r \in \mathcal{H}_0$. Then the collection $\{\mathcal{H}_s \backslash \{s\}, s \in Z_m\}$ is a neighborhood system over $Z_m$. Let $\mathcal{H}^2$ denote the second-order system, i.e.,

$$\mathcal{H}_s^2 = \bigcup_{r \in \mathcal{H}_s} \mathcal{H}_r, \quad s \in Z_m.$$

Then it is not hard to see that $\{\mathcal{H}_s^2 \backslash \{s\}, s \in Z_m\}$ is also a neighborhood system. Finally, set $\mathcal{G}^P = \{\mathcal{G}_s^P, s \in S\}$ where

$$\mathcal{G}_s^P = \begin{cases} \mathcal{G}_s, & s \in D_m \\ \mathcal{G}_s \cup \mathcal{H}_s^2 \backslash \{s\}, & s \in Z_m. \end{cases} \tag{8.1}$$

The "$P$" stands for "posterior"; some thought shows that $\mathcal{G}^P$ is a neighborhood system on $S$.

Let $\mu \in \mathrm{R}^M (M = N^2)$ have all components $= \mu$ and let $\|\cdot\|$ denote the usual norm in $\mathrm{R}^M$: $\|V\|^2 = \Sigma_1^M V_i^2$.

*Theorem: For each $g$ fixed, $P(\mathrm{X} = \omega \mid \mathrm{G} = g)$ is a Gibbs distribution over $\{S, \mathcal{G}^P\}$ with energy function*

$$U^P(f, l) = U(f, l) + \|\mu - \Phi(g, \phi(H(f)))\|^2/2\sigma^2. \tag{8.2}$$

*Proof:* Using standard results about "regular conditional expectations," we can and do assume that

$$P(\mathrm{X} = \omega \mid \mathrm{G} = g) = \frac{P(\mathrm{G} = g \mid \mathrm{X} = \omega)\, P(\mathrm{X} = \omega)}{P(\mathrm{G} = g)} \tag{8.3}$$

for all $\omega = (f, l)$, for each $g$.

Since $P(\mathrm{G} = g)$ is a constant and $P(\mathrm{X} = \omega) = e^{-U(\omega)}/Z$, the key term is

$$\begin{aligned} P(\mathrm{G} = g \mid \mathrm{X} = \omega) &= P(\phi(H(\mathrm{F})) \odot \mathrm{N} = g \mid \mathrm{F} = f, \mathrm{L} = l) \\ &= P(\mathrm{N} = \Phi(g, \phi(H(f))) \mid \mathrm{F} = f, \mathrm{L} = l) \\ &= P(\mathrm{N} = \Phi(g, \phi(H(f)))) \end{aligned}$$

(since N is independent of F and L)

$$= (2\pi\sigma^2)^{-M/2} \exp - \left(\frac{1}{2\sigma^2}\right) \|\mu - \Phi\|^2.$$

We will write $\Phi$ for $\Phi(g, \phi(H(f)))$. Collecting constants we have, from (8.3),

$$P(\mathrm{X} = \omega \mid \mathrm{G} = g) = e^{-U^P(\omega)}/Z^P$$

for $U^P$ as in (8.2); $Z^P$ is the usual normalizing constant (which will depend on $g$). It remains to determine the neighborhood structure.

Intuitively, the line sites should have the *same* neighbors whereas the neighbors $\mathcal{G}_s$ of a pixel site $s \in Z_m$ should be augmented in accordance with the blurring mechanism.

Take $s \in D_m$. The local characteristics at $s$ for the posterior distribution are, by (8.2),

$$\begin{aligned} P(L_s = l_s \mid L_r = l_r, r \neq s, r \in D_m, \mathrm{F} = f, \mathrm{G} = g) \\ = \frac{e^{-U^P(f,l)}}{\sum_{l_s} e^{-U^P(f,l)}} = \frac{e^{-U(f,l)}}{\sum_{l_s} e^{-U(f,l)}} \end{aligned}$$

where the sum extends over all possible values of $L_s$. Hence $\mathcal{G}_s^P = \mathcal{G}_s$.

For $s \in Z_m$, the term in (8.2) involving $\Phi$ does not cancel out. Now $\Phi(g, \phi(H(f))) = \{\Phi_s, s \in Z_m\}$ and let us denote the dependencies in $\Phi_s$ by writing $\Phi_s = \Phi_s(g_s; f_t, t \in \mathcal{H}_s)$. Then

$$\begin{aligned} P(F_s = f_s \mid F_r = f_r, r \neq s, r \in Z_m, \mathrm{L} = l, \mathrm{G} = g) \\ = \frac{e^{-U^P(f,l)}}{\sum_{f_s} e^{-U^P(f,l)}}; \quad U^P(f, l) \\ = U(f, l) + \sum_{r \in Z_m} (\Phi_r - \mu)^2/2\sigma^2. \end{aligned} \tag{8.4}$$

Decompose $U^P$ as follows:

$$\begin{aligned} U^P(f, l) = & \sum_{C: s \in C} V_C(f, l) \\ & + (2\sigma^2)^{-1} \sum_{r: s \in \mathcal{H}_r} (\Phi_r(g_r; f_t, t \in \mathcal{H}_r) - \mu)^2 \\ & + \sum_{C: s \notin C} V_C(f, l) \\ & + (2\sigma^2)^{-1} \sum_{r: s \notin \mathcal{H}_r} (\Phi_r(g_r; f_t, t \in \mathcal{H}_r) - \mu)^2. \end{aligned}$$

Since the last two terms do not involve $f_s$ (remember that $V_C$ only depends on the sites in $C$), the ratio in (8.4) depends only on the first two terms above. The first term depends only on coordinates of $(f, l)$ for sites in $\mathcal{G}_s(s \in C \Rightarrow C \subseteq \mathcal{G}_s)$ and the second term only on sites in

$$\bigcup_{r: s \in \mathcal{H}_r} \mathcal{H}_r = \bigcup_{r \in \mathcal{H}_s} \mathcal{H}_r \doteq \mathcal{H}_s^2.$$

Hence, $\mathcal{G}_s^P = \mathcal{G}_s \cup \mathcal{H}_s^2 \backslash \{s\}$, as asserted in the theorem. □

## IX. The Computational Problem

The posterior distribution $P(X = \omega | g)$ is a powerful tool for image analysis; in principle, we can construct the optimal (Bayesian) estimator for the original image, examine images sampled from $P(X = \omega | g)$, estimate parameters, design near-optimal statistical tests for the presence or absence of special objects, and so forth. But a conventional approach to any of these involves prohibitive computations. Specifically, our job here is to find the value(s) of $\omega$ which maximize the posterior distribution for a fixed $g$, i.e., *minimize*

$$U(f, l) + \| \mu - \Phi(g, \phi(H(f))) \|^2 / 2\sigma^2, (f, l) \in \Omega \tag{9.1}$$

where (see Section VIII) $\Phi$ is defined by $\phi(H(f)) \odot \Phi = g$. Even without L, the size of $\Omega$ is at least $2^{4000}$, corresponding to a binary image on a small (64 × 64) lattice. Hence, the identification of even near-optimal solutions is extremely difficult for such a relatively complex function.

In Sections XI and XII we will describe our stochastic relaxation method for this kind of optimization. The same method works for sampling and for computing expectations (and hence forming likelihood ratios), as will be explained in Section XI. The algorithm is highly parallel, but our current implementation is serial: it uses a single processor. The restoration of more complex images than those in Section XIII, probably involving more levels in the hierarchy, may necessitate *some* parallel processing.

## X. Stochastic Relaxation

There are many types of "relaxation," two of them being the type used in statistical physics and the type developed in image processing called "relaxation labeling" (RL), or sometimes "probabilistic relaxation." Basically, ours is of the former class, referred to here as SR, although there are some common features with RL.

The "Metropolis algorithm" (Metropolis *et al.* [42]) and others like it [7], [18] were invented to study the equilibrium properties, especially ensemble averages, time-evolution, and low-temperature behavior, of very large systems of essentially identical, interacting components, such as molecules in a gas or atoms in binary alloys.

Let $\Omega$ denote the possible configurations of the system; for example, $\omega \in \Omega$ might be the molecular positions or site configuration. If the system is in thermal equilibrium with its surroundings, then the probability (or "Boltzmann factor") of $\omega$ is given by

$$\pi(\omega) = e^{-\beta \mathcal{E}(\omega)} \Big/ \sum_{\omega} e^{-\beta \mathcal{E}(\omega)}, \quad \omega \in \Omega$$

where $\mathcal{E}(\omega)$ is the potential energy of $\omega$ and $\beta = 1/KT$ where $K$ is Boltzmann's constant and $T$ is absolute temperature. We have already seen an example in the Ising model (4.7). Usually, one needs to compute ensemble averages of the form

$$\langle Y \rangle = \int_{\Omega} Y(\omega)\, d\pi(\omega) = \frac{\sum_{\omega} Y(\omega)\, e^{-\beta \mathcal{E}(\omega)}}{\sum_{\omega} e^{-\beta \mathcal{E}(\omega)}}$$

where $Y$ is some variable of interest. This cannot be done analytically. In the usual Monte Carlo method, one restricts the sums above to a *sample* of $\omega$'s drawn uniformly from $\Omega$. This, however, breaks down in the situation above: the exponential factor puts most of the mass of $\pi$ over a very small part of $\Omega$, and hence one tends to choose samples of very low probability. The idea in [42] is to choose the samples from $\pi$ instead of uniformly and then weight the samples evenly instead of by $d\pi$. In other words, one obtains $\omega_1, \omega_2, \cdots, \omega_R$ from $\pi$ and $\langle Y \rangle$ is approximated by the usual ergodic averages:

$$\langle Y \rangle \approx \frac{1}{R} \sum_{r=1}^{R} Y(\omega_r). \tag{10.1}$$

Briefly, the sampling algorithm in [42] is as follows. Given the state of the system at "time" $t$, say $X(t)$, one randomly chooses another configuration $\eta$ and computes the energy change $\Delta\mathcal{E} = \mathcal{E}(\eta) - \mathcal{E}(X(t))$ and the quantity

$$q = \frac{\pi(\eta)}{\pi(X(t))} = e^{-\beta \Delta \mathcal{E}}. \tag{10.2}$$

If $q > 1$, the move to $\eta$ is allowed and $X(t+1) = \eta$, whereas if $q \leqslant 1$, the transition is made *with probability q*. Thus we choose $0 \leqslant \xi \leqslant 1$ uniformly and set $X(t+1) = \eta$ if $\xi \leqslant q$ and $X(t+1) = X(t)$ if $\xi > q$. (A "parallel processing variant" of this for simulating certain binary MRF's is given by Berger and Bonomi [4].)

In binary, "single-flip" studies, $\eta = X(t)$ except at one site, whereas in "spin-exchange" [18] systems, a pair of neighboring sites is selected. In either case, the "flip" or "exchange" is made with probability $q/(1+q)$, where $q$ is given in (10.2). In special cases, the single-flip system is equivalent to our Gibbs Sampler. The exchange algorithm in Cross and Jain [12] is motivated by work on the evolution of binary alloys. The samples generated are used for visual inspection and statistical testing, comparing the real and simulated textures. The model is an autobinomial MRF; see [6] or [12]. The algorithm is not suitable (nor intended) for restoration: for one thing, the intensity histogram is constant throughout the iteration process. This is necessarily the case with exchange systems which depend heavily on the initial configuration.

The algorithm in Hassner and Sklansky [28] is apparently a modification of one in Bortz *et al.* [7]. Another application of these ideas outside statistical mechanics appears in Hinton and Sejnowski [29], a paper about neural modeling but a spiritual cousin of ours. In particular, the parallel nature of these algorithms is emphasized.

The essence of every SR scheme is that changes $(\omega \rightarrow \eta)$ which *increase* energy, i.e., *lower* probability, are permitted.

By contrast, deterministic algorithms only allow jumps to states of lower energy and invariably get "stuck" in *local* minima. To get to samples from $\pi$, we must occasionally "backtrack."

All of these algorithms can be cast in a general theory involving Markov chains with state space $\Omega$. See Hammersley and Handscomb [27] for a readable treatment. The goal is an irreducible, aperiodic chain with equilibrium measure $\pi$. If $\omega_1, \omega_2, \cdots, \omega_R$ is a realization of such a chain, then standard results yield (10.1), in fact at a rate $O(R^{-1/2})$ as $R \to \infty$. In this setup an auxiliary transition matrix is used to go from $\omega$ to $\eta$, and the general replacement recipe involves the same ratio $\pi(\eta)/\pi(\omega)$. The Markovian properties of the Gibbs Sampler will be described in the following sections.

Chemical annealing is a method for determining the low energy states of a material by a gradual lowering of temperature. The process is delicate: if $T$ is lowered too rapidly and insufficient time is spent at temperatures near the freezing point, then the process may bog down in nonequilibrium states, corresponding to flaws in the material, etc. In *simulated* annealing, Kirkpatrick *et al.* [40] identify the solution of an optimal (computer) design problem with the ground state of an imaginary physical system, and then employ the Metropolis algorithm to reach "steady-state" at each of a decreasing sequence of temperatures $\{T_n\}$. This sequence, and the time spent at each temperature, is called an "annealing schedule." In [40], this is done on an ad hoc basis using guidelines developed for chemical annealing. Here, we prove the existence of annealing schedules which guarantee convergence to minimum energy states (see Section XII for formal definitions), and we identify the *rate* of decrease relative to the number of full sweeps.

Turning to RL, there are many similarities with SR, both in purpose and, at least abstractly, in method. RL was designed for the assignment of numeric or symbolic labels to objects in a visual system, such as intensity levels to pixels or geometric labels to cube edges, in order to achieve a "global interpretation" that is consistent with the context and certain "local constraints." Ideally, the process evolves by a series of *local* changes, which are intended to be simple, homogeneous, and performed in parallel. The local constraints are usually so-called "compatibility functions," which are much like statistical correlations, and often defined in reference to a graph. We refer the reader to Davis and Rosenfeld [13] for an expository treatment, to Rosenfeld *et al.* [46] for the origins, to Hummel and Zucker [30] for recent work on the logical and mathematical foundations, and to Rosenfeld and Kak [47] for applications to iterative segmentation.

But there are also fundamental differences. First, most variants of RL are rather ad hoc and heuristic. Second, and more importantly, RL is essentially a *nonstochastic* process, both in the interaction model and in the updating algorithms. (Indeed, various probabilistic analogies are often avoided as misleading; see [30], for example.) There is nothing in RL corresponding to an equilibrium measure or even a joint probability law over configurations, whereas there is no analogue in SR of the all-important, iterative updating *formulas* and corresponding sequence of "probability estimates" for various hypotheses involving pixel or object classification.

In summary, there are shared goals and shared features (locality, parallelism, etc.) but SR and RL are quite distinct, at least as practiced in the references made here.

## XI. Gibbs Sampler: General Description

We return to the general notation of Section IV: $\backslash = \{X_s, s \in S\}$ is an MRF over a graph $\{\mathcal{G}_s, s \in S\}$ with state spaces $\Lambda_s$, configuration space $\Omega = \prod_s \Lambda_s$, and Gibbs distribution $\pi(\omega) = e^{-U(\omega)/T}/Z$, $\omega \in \Omega$.

The general computational problems are

A) sample from the distribution $\pi$;

B) minimize $U$ over $\Omega$;

C) compute expected values.

Of course, we are most concerned with B), which corresponds to MAP estimation when $\pi$ is the posterior distribution. The most basic problem is A), however, because A) together with annealing yields B) and A) together with the ergodic theorem yields C). We will state three theorems corresponding to A), B), and C) above. Theorem C is not used here and will be proven elsewhere; we state it because of its potential importance to other methods of restoration and to hypothesis testing.

Let us imagine a simple processor placed at each site $s$ of the graph. The connectivity relation among the processors is determined by the bonds: the processor at $s$ is connected to each processor for the sites in $\mathcal{G}_s$. In the cases of interest here (and elsewhere) the number of sites $N$ is very large. However, the size of the neighborhoods, and thus the number of connections to a given processor, is modest, only eight in our experiments, including line, pixel and mixed bonds.

The state of the machine evolves by discrete changes and it is therefore convenient to discretize time, say $t = 1, 2, 3, \cdots$. At time $t$, the state of the processor at site $s$ is a random variable $X_s(t)$ with values in $\Lambda_s$. The total configuration is $X(t) \doteq (X_{s_1}(t), X_{s_2}(t), \cdots, X_{s_N}(t))$, which evolves due to state changes of the individual processors. The starting configuration, $X(0)$, is arbitrary. At each epoch, only *one* site undergoes a (possible) change, so that $X(t-1)$ and $X(t)$ can differ in at most one coordinate. Let $n_1, n_2, \cdots$ be the sequence in which the sites are "visited" for replacement; thus, $n_t \in S$ and $X_{s_i}(t) = X_{s_i}(t-1)$, $i \neq n_t$. Each processor is programmed to follow the same algorithm: at time $t$, a sample is drawn *from the local characteristics* of $\pi$ for $s = n_t$ and $\omega = X(t-1)$. In other words, we choose a state $x \in \Lambda_{n_t}$ from the conditional distribution of $X_{n_t}$ given the observed states of the neighboring sites $X_r(t-1)$, $r \in \mathcal{G}_{n_t}$. The new configuration $X(t)$ has $X_{n_t}(t) = x$ and $X_s(t) = X_s(t-1)$, $s \neq n_t$.

These are *local* computations, and *identical* in nature when $\pi$ is homogeneous. Moreover, the actual calculation is *trivial* since the local characteristics are generally very simple. These conditional probabilities were discussed in Section IV and we refer the reader again to formulas (4.8) and (4.9). Notice that $Z$ does not appear.

Given an initial configuration $X(0)$, we thus obtain a sequence $X(1), X(2), X(3), \cdots$ of configurations whose convergence properties will be described in Section XII. The limits obtained do not depend on $X(0)$. The sequence $(n_t)$ we actually use is simply the one corresponding to a raster scan, i.e., repeatedly visiting all the sites in some "natural" fixed order. Of course, in this case one does not actually need a processor at each site. But the theorems are valid for very general (not necessarily periodic) sequences $(n_t)$ allowing for *asychronous* schemes in which each processor could be driven *by its own clock*. Let us briefly discuss such a parallel implementation of the Gibbs Sampler and its advantage over the serial version.

Computation is parallel in the sense that it is realized by simple and alike units operating largely independently. Units are dependent only to the extent that each must transmit its current state to its neighbors. Most importantly, the amount of time required for one complete update of the entire system is *independent of the number of sites.* In the raster version, we simply "move" a processor from site to site. Upon arriving at a site, this processor must first load the local neighborhood relations and state values, perform the replacement, and move on. The time required to refresh $S$ grows linearly with $N = |S|$. Thus, for example, for the purposes at hand, the parallel procedure is potentially at least $10^4$ times faster than the raster version we used, and which required considerable CPU time on a VAX 780. Of course, we recognize that the fully parallel version will require extremely sophisticated new hardware, although we understand that small prototypes of similar machines are underway at several places.

A more modest degree of parallelism can be simply implemented. Since the convergence theorems are independent of the details of the site replacement scheme $n_1, n_2, \cdots$ the graph associated with the MRF X can be divided into collections of sites with each collection assigned to an independently running (asynchronous) processor. Each such processor would execute a raster scan updating of its assigned sites. Communication requirements will be small if the division of the graph respects the natural topology of the scene, provided, of course, that the neighborhood systems are reasonably local. Such an implementation, with five or ten micro- or minicomputers, represents a straightforward application of available technology.

## XII. Gibbs Sampler: Mathematical Foundations

As in Section XI, $(n_t)$, $t = 1, 2, \cdots$, is the sequence in which the sites are visited for updating, and $X_s(t)$ denotes the state of site $s$ after $t$ replacement opportunities, of which only those for which $n_\tau = s$, $1 \leqslant \tau \leqslant t$, involve site $s$. For simplicity, we will assume a common state space $\Lambda_s \equiv \Lambda = \{0, 1, \cdots, L - 1\}$, and as usual that $0 < \pi(\omega) < 1$ for all $\omega \in \Omega$ or, what is the same, that $\sup_\omega |U(\omega)| < \infty$. The initial configuration is $X(0)$.

We now investigate the statistical properties of the random process $\{X(t), t = 0, 1, 2, \cdots\}$. The evolution $X(t - 1) \to X(t)$ of the system was explained in Section XI. In mathematical terms,

$$\begin{aligned} &P(X_s(t) = x_s, s \in S) \\ &\quad = \pi(X_{n_t} = x_{n_t} | X_s = x_s, s \neq n_t) P(X_s(t - 1) \\ &\quad = x_s, s \neq n_t) \end{aligned} \tag{12.1}$$

where, of course, $\pi = e^{-U/T}/Z$ is the Gibbs measure which drives the process. Our first result states that the distribution of $X(t)$ converges to $\pi$ as $t \to \infty$ regardless of $X(0)$. The only assumption is that we continue to visit every site, obviously a necessary condition for convergence.

*Theorem A (Relaxation): Assume that for each $s \in S$, the sequence $\{n_t, t \geqslant 1\}$ contains $s$ infinitely often. Then for every starting configuration $\eta \in \Omega$ and every $\omega \in \Omega$,*

$$\lim_{t \to \infty} P(X(t) = \omega | X(0) = \eta) = \pi(\omega). \tag{12.2}$$

The proof appears in the Appendix, along with that of Theorem B. Like the Metropolis algorithm, the Gibbs Sampler produces a Markov chain $\{X(t), t = 0, 1, 2, \cdots\}$ with $\pi$ as equilibrium distribution. The only complication is that the transition probabilities associated with the Gibbs Sampler are nonstationary, and their matrix representations do not commute. This precludes the usual algebraic treatment. These issues are discussed in more detail at the beginning of the Appendix.

We now turn to annealing. Hitherto the temperature has been fixed. Theorem B is an "annealing schedule" or rate of temperature decrease which forces the system into the lowest energy states. The necessary programming modification in the relaxation process is trivial, and the *local* nature of the calculations is preserved.

Let us indicate the dependence of $\pi$ on $T$ by writing $\pi_T$, and let $T(t)$ denote the temperature at stage $t$. The annealing procedure generates a different process $\{X(t), t = 1, 2, \cdots\}$ such that

$$\begin{aligned} &P(X_s(t) = x_s, s \in S) \\ &\quad = \pi_{T(t)}(X_{n_t} = x_{n_t} | X_s = x_s, s \neq n_t) \\ &\quad \cdot P(X_s(t - 1) = x_s, s \neq n_t). \end{aligned} \tag{12.3}$$

Let

$$\Omega_0 = \{\omega \in \Omega : U(\omega) = \min_\eta U(\eta)\}, \tag{12.4}$$

and let $\pi_0$ be the uniform distribution on $\Omega_0$. Finally, define

$$\begin{aligned} U^* &= \max_\omega U(\omega), \\ U_* &= \min_\omega U(\omega), \\ \Delta &= U^* - U_*. \end{aligned} \tag{12.5}$$

*Theorem B (Annealing): Assume that there exists an integer $\tau \geqslant N$ such that for every $t = 0, 1, 2, \cdots$ we have*

$$S \subseteq \{n_{t+1}, n_{t+2}, \cdots, n_{t+\tau}\}.$$

*Let $T(t)$ be any decreasing sequence of temperatures for which*

a) $T(t) \to 0$ as $t \to \infty$;
b) $T(t) \geqslant N\Delta/\log t$
for all $t \geqslant t_0$ for some integer $t_0 \geqslant 2$.

*Then for any starting configuration $\eta \in \Omega$ and for every $\omega \in \Omega$,*

$$\lim_{t \to \infty} P(X(t) = \omega | X(0) = \eta) = \pi_0(\omega). \tag{12.6}$$

The first condition is that the individual "clocks" do not slow to an arbitrarily low frequency as the system evolves, and imposes no limitations in practice. For raster replacement,

$\tau = N$. The major practical weakness is b); we cannot truly follow the "schedule" $N\Delta/\log t$. For example, with $N = 20{,}000$ and $\Delta = 1$, it would take $e^{40{,}000}$ site visits to reach $T = 0.5$. We single out this temperature because we have obtained good results by making $T$ decrease from approximately $T = 4$ to $T = 0.5$ over 300–1000 sweeps ($= 300N - 1000N$ replacements), using a schedule of the form $C/\log(1 + k)$, where $k$ is the number of full sweeps. (Notice that the condition in b) is then satisfied provided $C$ is sufficiently large.) Apparently, the bound in b) is far from optimal, at least as concerns the constant $N\Delta$. (In fact, the proof of Theorem B does establish something stronger, namely that $\Delta$ can be taken as the largest absolute difference in energies associated with pairs $\omega$ and $\omega^*$ which differ at only one coordinate. But this improvement still leaves $N\Delta$ too large for actual practice.) On the other hand, the logarithmic rate is not too surprising in view of the widespread experience of chemists that $T$ must be lowered very slowly, particularly near the freezing point. Otherwise one encounters undesirable physical embodiments of *local* energy minima.

Concerning ergodicity, in statistical physics one attempts to predict the observable quantities of a system in equilibrium; these are the "time averages" of functions on $\Omega$. Under the "ergodic hypothesis," one assuumes that (10.1) is in force, so that time averages approach the corresponding "phase averages" or expected values. The analog for our system is the assertion that, in some suitable sense,

$$\lim_{n\to\infty} \frac{1}{n} \sum_{t=1}^{n} Y(X(t)) = \int_{\Omega} Y(\omega)\, d\pi(\omega). \tag{12.7}$$

(Here again $T$ is fixed.) As we have already stated, a direct calculation of the righthand side of (12.7), namely,

$$\sum_{\omega} Y(\omega)\, e^{-U(\omega)/T} \Big/ \sum_{\omega} e^{-U(\omega)/T}$$

is impossible in general. The left-hand side of (12.7) suggests that we use the Gibbs Sampler and compute a time average of the function $Y$. For most physical systems, the ergodic hypothesis is just that—a *hypothesis*—which can rarely be verified in practice. Fortunately, for our system it is not too difficult to directly establish ergodicity.

*Theorem C (Ergodicity): Assume that there exists a $\tau$ such that $S \subseteq \{n_{t+1}, \cdots, n_{t+\tau}\}$ for all $t$. Then for every function $Y$ on $\Omega$ and for every starting configuration $\eta \in \Omega$, (12.7) holds with probability one.*

## XIII. Experimental Results

There are three groups of pictures. Each contains an original image, several degraded versions, and the corresponding restorations, usually at two stages of the annealing process to illustrate its evolution. The degradations are formed from combinations of

i) $\phi$ absent or $\phi(x) = \sqrt{x}$;
ii) multiplicative or additive noise;
iii) signal-to-noise levels.

The signal-to-noise ratios are all very low. For blurring, we always took the convolution $H$ in (2.3). The restorations are all MAP estimates generated by the serial Gibbs Sampler with annealing schedule

$$T(k) = \frac{C}{\log(1 + k)}, \qquad 1 \leqslant k \leqslant K$$

where $T(k)$ is the temperature during the $k$th *iteration* (= full sweep of $S$), so that $K$ is the total number of iterations. In each case, $C = 3.0$ or $C = 4.0$. No pre- or postfiltering, nor anything else was done. The models for the intensity and line processes were kept as simple as possible; indeed, only cliques of size two appear in the intensity model.

*Group 1:* The original image [Fig. 2(a)] is a sample of an MRF on $Z_{128}$ with $L = 5$ intensities and the eight-neighbor system (Fig. 1, $c = 2$). The potentials $V_C = 0$ unless $C = \{r, s\}$, in which case

$$V_C(f) = \begin{cases} \tfrac{1}{3}, & f_s = f_r \\ -\tfrac{1}{3}, & f_s \neq f_r. \end{cases}$$

Two hundred iterations (at $T \equiv 1$) were made to generate Fig. 2(a).

The first degraded version is Fig. 2(b), which is simply Fig. 2(a) plus Gaussian noise with $\sigma = 1.5$ *relative to* gray levels $f$, $1 \leqslant f \leqslant 5$. Fig. 2(c) is the restoration of Fig. 2(b) with $K = 25$ iterations only, i.e., early in the annealing process. In Fig. 2(d), $K = 300$.

The second degraded image [Fig. 3(b)] uses the model

$$G = H(F)^{1/2} \cdot N \tag{13.1}$$

where $\mu = 1$ and $\sigma = 0.1$, again relative to intensities $1 \leqslant f \leqslant 5$. Fig. 3(c) and 3(d) shows the restorations of Fig. 3(b) with $K = 25$ and $K = 300$, respectively.

*Group 2:* Fig. 4(a) is "hand-drawn." The lattice size is $64 \times 64$ and there are three gray levels. Gaussian noise ($\mu = 0$, $\sigma = 0.7$) was added to produce Fig. 4(b). We tried two types of restoration on Fig. 4(b). First, we used the "blob process" which generated Fig. 2(a) for the F-model. There was no line process and $K = 1000$. Obviously these are flaws; see Fig. 4(c).

A line process L was then adjoined to F for the original image model, and the corresponding restoration after 1000 iterations is shown in Fig. 4(d). L itself was described in Case 2 of Section III and the neighborhood system for (F, L) on $Z_{64} \cup D_{64}$ was discussed in Case 3 of Section III. The (prior) distribution on X = (F, L) was as follows. The range of F is $\{0, 1, 2\}$ ($L = 3$ intensities). The energy $U(f, l)$ consists of two terms, say $U(f|l) + U(l)$. To understand the interaction term $U(f|l)$, let $d$ denote a line site, say between pixels $r$ and $s$. If $L_d = 1$, i.e., an edge element is "present" at $d$, then the bond between $s$ and $r$ is "broken" and we set $V_{\{r,s\}}(f_r, f_s) = 0$ regardless of $f_r, f_s$; otherwise ($L_d = 0$) $V_{\{r,s\}}$ is as before except that $\pm\frac{1}{3}$ are replaced by $\pm 1$. As for $U(l)$, only cliques of size four are nonzero, of which there are six distinct types up to rotations. These are shown in Fig. 5(a) with their associated energy values.

Then we corrupted the hand-drawn figure using (13.1) with the same noise parameters as Fig. 3(b), obtaining Fig. 6(b), which is restored in Fig. 6(c) using the same prior on (F, L) as above and with $K = 1000$ iterations.

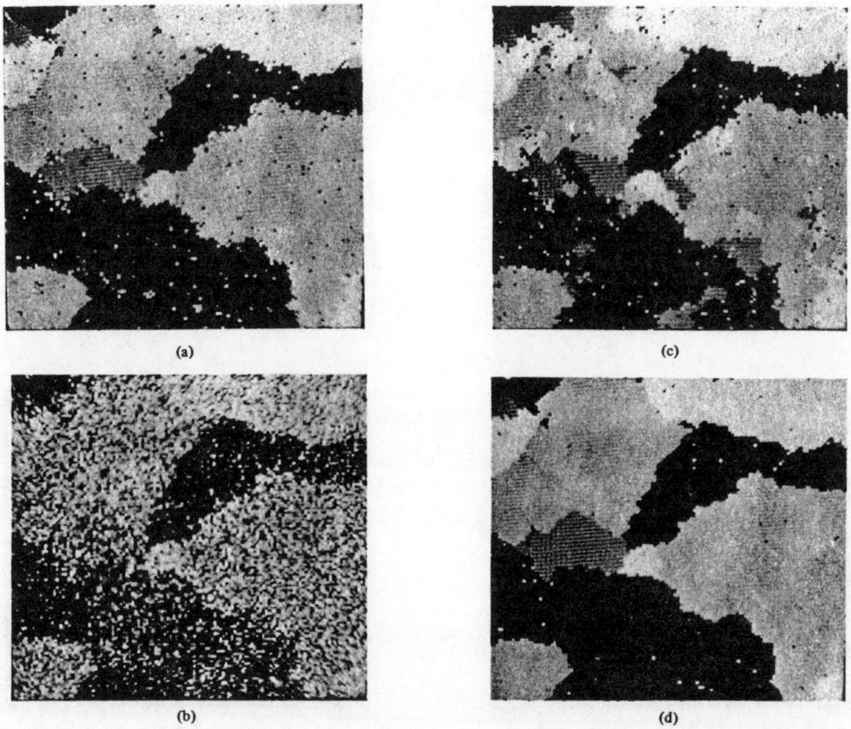

Fig. 2. (a) Original image: Sample from MRF. (b) Degraded image: Additive noise. (c) Restoration: 25 iterations. (d) Restoration: 300 iterations.

*Group 3:* The results in Group 2 suggest a boundary-finding algorithm for general shapes: allow the line process more directional freedom. Group 3 is an exercise in boundary finding at essentially 0 dB. Fig. 7(a) is a 64 × 64 segment of a roadside photograph that we obtained from the Visions Research Group at the University of Massachusetts. The levels are scaled so that the (existing) two peaks in the histogram occur at $f = 0$ and $f = 1$. We regard Fig. 7(a) as the *blurred image* $H(F)$. Noise is added in Fig. 7(b); the standard error is $\sigma = 0.5$ *relative to the two main gray levels* $f = 0, 1$.

Figs. 7(c) and 7(d) are "restorations" of Fig. 7(b) for $K = 100$ and $K = 1000$ iterations, respectively. The outcome of the line process is indicated by painting black any pixels to the left of or above a "broken bond." The two main regions, comprising the sign and the arrow, are perfectly circumscribed by a continuous sequence of line elements.

The model for $X$ is more complex than the one in Group 2. There are now four possible states for each line site corresponding to "off" ($l = 0$) and three directions, shown in Fig. 5(b). The $U(f|l)$ term is the same as before in that the pixel bond between $r$ and $s$ is broken whenever $l_d \neq 0$. The range of $F$ is $\{0, 1\}$ ($L = 2$).

Only cliques of size four are nonzero in $U(l)$, as before. However, there are now many combinations for $(l_{d_1}, l_{d_2}, l_{d_3}, l_{d_4})$ given such a clique $C = \{d_1, d_2, d_3, d_4\}$ of line sites, although the number is substantially reduced by assuming rotational invariance, which we do. Fig. 5(c) shows the convention we will use for the ordering and an example of the notation. The energies for the possible configurations ($l_{d_i}, 1 \leq i \leq 4$) range from 0 to 2.70. (Remember that high energies correspond to low probability, and that the exponential exaggerates differences.) We took $V(0, 0, 0, 0) = 0$ and $V(l_{d_i}, 1 \leq i \leq 4) = 2.70$ otherwise, except when exactly two of the $l_{d_i}$ are nonzero. Parallel segments [e.g., (1, 0, 1, 0)] receive energy 2.70; sharp turns [e.g., (0, 2, 1, 0)] and other "corner" types get 1.80; mild turns [e.g., (0, 2, 3, 0)] are 1.35; and continuations [e.g., (2, 0, 2, 0) or (0, 1, 3, 0)] are 0.90.

## XIV. Concluding Remarks

We have introduced some new theoretical and processing methods for image restoration. The models and estimates are noncausal and nonlinear, and do not represent extensions into two dimensions of one-dimensional filtering and smoothing

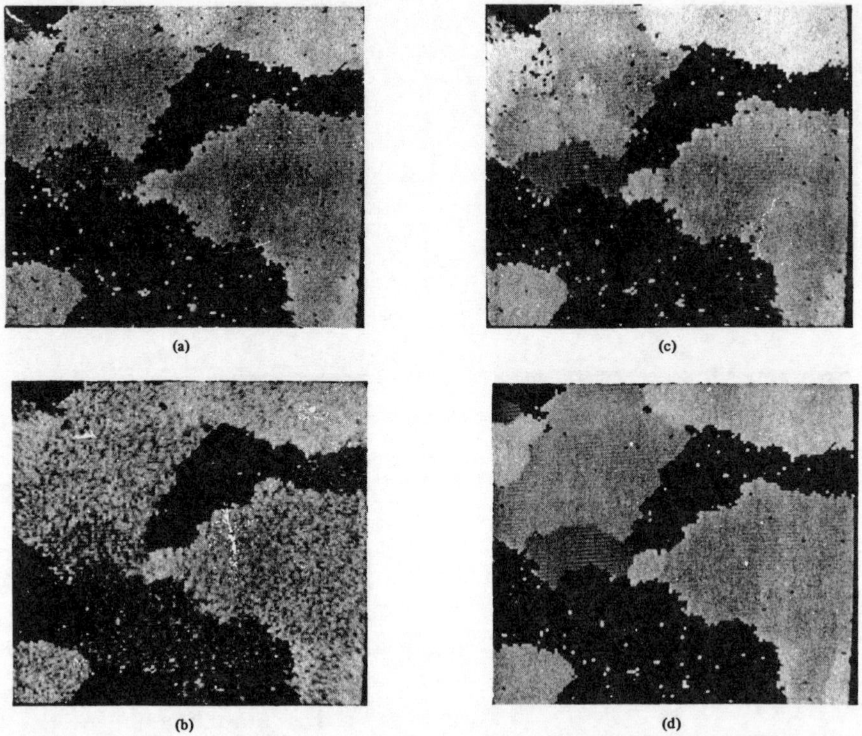

Fig. 3. (a) Original image: Sample from MRF. (b) Degraded image: Blur, nonlinear transformation, multiplicative noise. (c) Restoration: 25 iterations. (d) Restoration: 300 iterations.

algorithms. Rather, our work is largely inspired by the methods of statistical physics for investigating the time-evolution and equilibrium behavior of large, lattice-based systems.

There are, of course, *many* well-known and remarkable features of these massive, homogeneous physical systems. Among these is the evolution to minimal energy states, regardless of initial conditions. In our work posterior (Gibbs) distribution represents an *imaginary* physical system whose lowest energy states are exactly the MAP estimates of the original image given the degraded "data."

The approach is very flexible. The MRF–Gibbs class of models is tailor-made for representing the dependencies among the intensity levels of nearby pixels as well as for augmenting the usual, pixel-based process by other, unobservable attribute processes, such as our "line process," in order to bring exogenous information into the model. Moreover, the degradation model is almost unrestricted; in particular, we allow for deformations due to the image formation and recording processes. All that is required is that the posterior distribution have a "reasonable" neighborhood structure as a MRF, for in that case the computational load can be accommodated by appropriate variants (such as the Gibbs Sampler) of relaxation algorithms for dynamical systems.

## Appendix

## Proofs of Theorems

*Background and Notation*

Recall that $\Lambda = \{0, 1, 2, \cdots, L-1\}$ is the common state space, that $\eta, \eta', \omega$, etc. denote elements of the configuration space $\Omega = \Lambda^N$, and that the sites $S = \{s_1, s_2, \cdots, s_N\}$ are visited for updating in the order $\{n_1, n_2, \cdots\} \subset S$. The resulting stochastic process is $\{X(t), t = 0, 1, 2, \cdots\}$, where $X(0)$ is the initial configuration.

For Theorem A, the transitions are governed by the Gibbs distribution $\pi(\omega) = e^{-U(\omega)/T}/Z$ in accordance with (12.1), whereas, for Theorem B (annealing), we use $\pi_{T(t)}$ (see Section XII) for the transition $X(t-1) \to X(t)$ [see (12.3)].

Let us briefly discuss the process $\{X(t), t \geqslant 0\}$, restricting attention to constant temperature; the annealing case is essentially the same. To begin with, $\{X(t), t \geqslant 0\}$ is indeed a Markov chain; this is apparent from its construction. Fix $t$ and

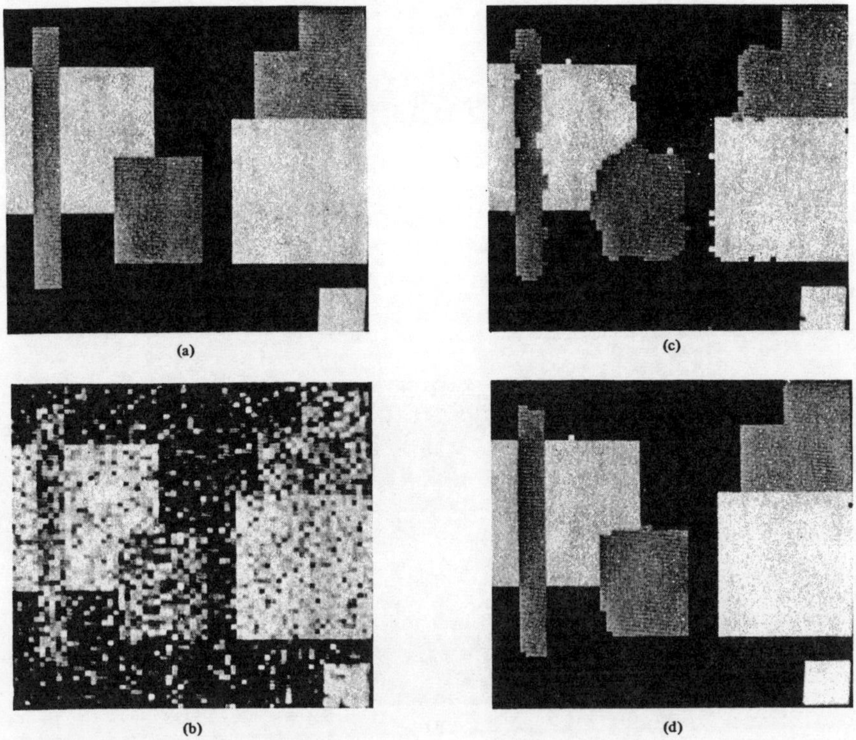


Fig. 4. (a) Original image: "Hand-drawn." (b) Degraded image: Additive noise. (c) Restoration: Without line process; 1000 iterations. (d) Restoration: Including line process; 1000 iterations.

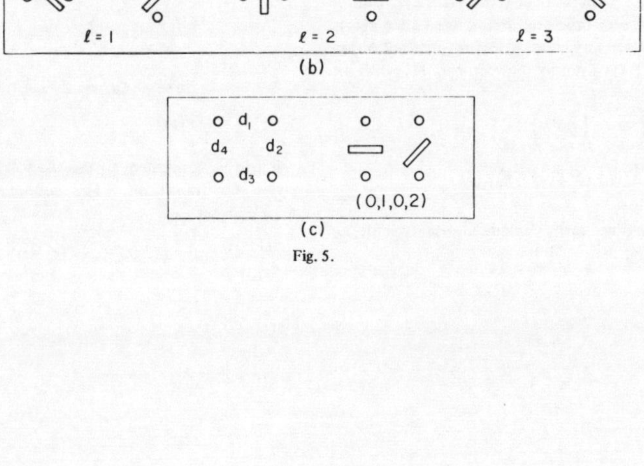


Fig. 5.

736 IEEE TRANSACTIONS ON PATTERN ANALYSIS AND MACHINE INTELLIGENCE, VOL. PAMI-6, NO. 6, NOVEMBER 1984

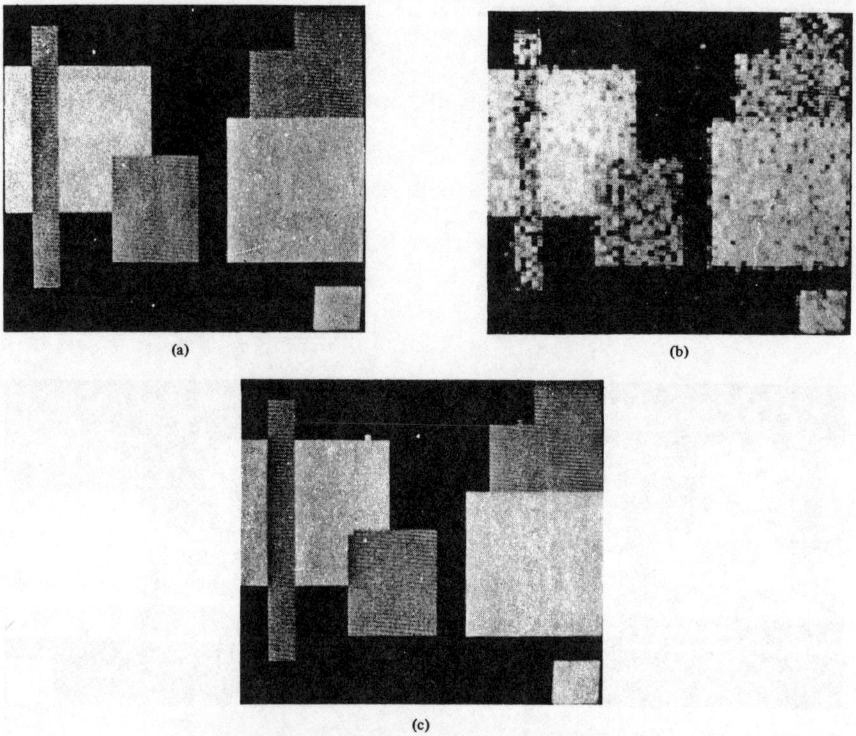

Fig. 6. (a) Original image: "Hand-drawn." (b) Degraded image: Blur, nonlinear transformation, multiplicative noise. (c) Restoration: including line process; 1000 iterations.

$\omega \in \Omega$. For any $x \in \Lambda$, let $\omega^x$ denote the configuration which is $x$ at site $n_t$ and agrees with $\omega$ elsewhere. The transition matrix *at time t* is

$$(M_t)_{\eta,\omega} = \begin{cases} \pi(X_{n_t} = x_{n_t} \mid X_s = x_s, s \neq n_t), \\ \qquad \text{if } \eta = \omega^x \text{ for some } x \in \Lambda \\ 0, \quad \text{otherwise} \end{cases}$$

where $(M_t)_{\eta,\omega}$ denotes the row $\eta$, column $\omega$ entry of $M_t$, and $\omega = (x_{s_1}, x_{s_2}, \cdots, x_{s_N})$. In particular, the chain is *nonstationary*, although clearly *aperiodic* and *irreducible* (since $\pi(\omega) > 0 \; \forall \omega$). Moreover, given any starting vector (distribution) $\mu_0$, the distribution of $X(t)$ is given by the vector $\mu_0 \prod_{j=1}^t M_j$, i.e.,

$$P_{\mu_0}(X(t) = \omega) = \left(\mu_0 \times \prod_{j=1}^{t} M_j\right)_\omega = \sum_\eta P(X(t) = \omega \mid X(0) = \eta)\, \mu_0(\eta).$$

Notice that $\pi$ is the (necessarily) unique invariant vector, i.e., for every $t = 1, 2, \cdots$,

$$\pi(\omega) = (\pi M_t)_\omega = \sum_\eta P(X(t) = \omega \mid X(0) = \eta)\, \pi(\eta). \tag{A.1}$$

To see this, fix $t$ and $\omega = \{x_s\}$, and write

$$\begin{aligned} (\pi M_t)_\omega &= \sum_\eta \pi(\eta)(M_t)_{\eta,\omega} \\ &= \sum_{x \in \Lambda} \pi(\omega^x)(M_t)_{\omega^x,\omega} \\ &= (M_t)_{\omega^{x'},\omega} \sum_{x \in \Lambda} \pi(\omega^x) \qquad (\text{for } any \;\; x' \in \Lambda) \\ &= \pi(X_{n_t} = x_{n_t} \mid X_s = x_s, s \neq n_t)\, \pi(X_s = x_s, s \neq n_t) \\ &= \pi(\omega). \end{aligned}$$

It will be convenient to use the following, semistandard notation for transitions. For nonnegative integers $r < t$ and $\omega, \eta \in \Omega$, set

$$P(t, \omega \mid r, \eta) = P(X(t) = \omega \mid X(r) = \eta)$$

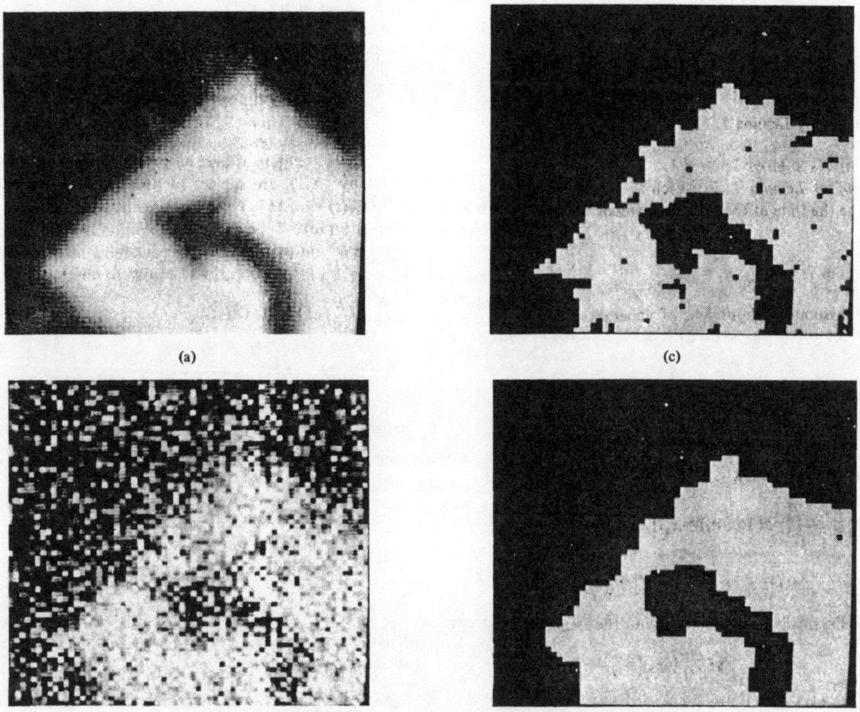

Fig. 7. (a) Blurred image (roadside scene). (b) Degraded image: Additive noise. (c) Restoration including line process; 100 iterations. (d) Restoration including line process; 1000 iterations.

and, for any *distribution* $\mu$ on $\Omega$, set

$$P(t, \omega | r, \mu) = \sum_{\eta} P(t, \omega | r, \eta)\, \mu(\eta).$$

Finally, $\|\mu - \nu\|$ denotes the $L^1$ distance between two distributions on $\Omega$:

$$\|\mu - \nu\| = \sum_{\omega} |\mu(\omega) - \nu(\omega)|.$$

Obviously, $\mu_n \to \mu (n \to \infty)$ in distribution (i.e., $\mu_n(\omega) \to \mu(\omega)$ $\forall \omega$) if and only if $\|\mu_n - \mu\| \to 0$, $n \to \infty$. (Remember that $\Omega$ is finite.)

*Proof of Theorem A:* Set $T_0 = 0$ and define $T_1 < T_2 < \cdots$ such that

$$S \subseteq \{n_{T_{k-1}+1}, n_{T_{k-1}+2}, \cdots, n_{T_k}\}, \qquad k = 1, 2, \cdots.$$

This is possible since every site is visited infinitely often. Clearly (at least) $k$ iterations or full sweeps have been completed by "time" $T_k$. In particular, $kN \leqslant T_k < \infty\ \forall k$. Let

$$K(t) = \sup \{k : T_k < t\}.$$

Obviously $K(t) \to \infty$ at $t \to \infty$. The proof of Theorem A is based on the following lemma, which also figures in the proof of the annealing theorem.

*Lemma 1: There exists a constant* $r, 0 \leqslant r < 1$, *such that for every* $t = 1, 2, \cdots$,

$$\sup_{\omega, \eta', \eta''} |P(X(t) = \omega | X(0) = \eta') - P(X(t) = \omega | X(0) = \eta'')| \leqslant r^{K(t)}.$$

Assume for now that the lemma is true. Since $\pi$ is an invariant vector for the chain:

$$\overline{\lim_{t \to \infty}} \sup_{\omega, \eta} |P(X(t) = \omega | X(0) = \eta) - \pi(\omega)|$$

$$= \overline{\lim_{t \to \infty}} \sup_{\omega, \eta} \left| \sum_{\eta'} \pi(\eta') \{P(X(t) = \omega | X(0) = \eta) - P(X(t) = \omega | X(0) = \eta')\} \right|$$

[by (A.1)]

$$\leqslant \overline{\lim_{t\to\infty}} \sup_{\omega,\eta',\eta''} |P(X(t)=\omega|X(0)=\eta') - P(X(t)=\omega|X(0)=\eta'')|$$
$$= 0, \text{ by Lemma 1.}$$

So it suffices to prove Lemma 1.

*Proof of Lemma 1:* For each $k = 1, 2, \cdots$ and $1 \leqslant i \leqslant N$, let $m_i$ be the time of the last replacement of site $s_i$ before $T_k + 1$, i.e.,

$$m_i = \sup\{t : t \leqslant T_k, n_t = s_i\}.$$

We can assume, without loss of generality, that $m_1 > m_2 > \cdots > m_N$; otherwise, relabel the sites. For any $\omega = (x_{s_1}, \cdots, x_{s_N})$ and $\omega'$,

$$\begin{aligned} P(X(T_k) = \omega|X(T_{k-1}) = \omega') &= P(X_{s_1}(m_1) = x_{s_1}, \cdots, X_{s_N}(m_N) = x_{s_N}|X(T_{k-1}) = \omega') \\ &= \prod_{j=1}^{N} P(X_{s_j}(m_j) = x_{s_j}|X_{s_{j+1}}(m_{j+1}) = x_{s_{j+1}}, \cdots, X_{s_N}(m_N) = x_{s_N}, X(T_{k-1}) = \omega'). \end{aligned}$$

Let $\delta$ be the smallest probability among the local characteristics:

$$\delta = \inf_{\substack{(x_{s_1},\cdots,x_{s_N})\in\Omega \\ 1\leqslant i\leqslant N}} \pi(X_{s_i} = x_{s_i}|X_{s_j} = x_{s_j}, j \neq i).$$

Then $0 < \delta < 1$ and a little reflection shows that every term in the product above is at least $\delta$. Hence,

$$\inf_{\substack{k=1,2,\cdots \\ \omega,\omega'}} P(X(T_k) = \omega|X(T_{k-1}) = \omega') \geqslant \delta^N. \tag{A.2}$$

Consider now the inequality asserted in Lemma 1. It is trivial for $t \leqslant T_1$ since in this case $K(t) = 0$. For $t > T_1$,

$$\begin{aligned} &\sup_{\omega,\eta',\eta''} |P(X(t)=\omega|X(0)=\eta') - P(X(t)=\omega|X(0)=\eta'')| \\ &= \sup_\omega \{\sup_\eta P(X(t)=\omega|X(0)=\eta) - \inf_\eta P(X(t)=\omega|X(0)=\eta)\} \\ &= \sup_\omega \Big\{\sup_\eta \sum_{\omega'} P(X(t)=\omega|X(T_1)=\omega')P(X(T_1)=\omega'|X(0)=\eta) \\ &\quad - \inf_\eta \sum_{\omega'} P(X(t)=\omega|X(T_1)=\omega')P(X(T_1)=\omega'|X(0)=\eta)\Big\} \\ &\doteq \sup_\omega Q(t,\omega). \end{aligned}$$

Certainly, for each $\omega \in \Omega$,

$$\sup_\eta \sum_{\omega'} P(X(t)=\omega|X(T_1)=\omega')P(X(T_1)=\omega'|X(0)=\eta) \leqslant \sup_\mu \sum_{\omega'} P(X(t)=\omega|X(T_1)=\omega')\mu(\omega')$$

where the supremum is over all probability measures $\mu$ on $\Omega$ which, by (A.2), are subject to $\mu(\omega') \geqslant \delta^N\ \forall\omega'$. Suppose $\omega' \to P(X(t)=\omega|X(T_1)=\omega')$ is maximized at $\omega' = \omega^*$ (which depends on $\omega$). Then the last supremum is attained by placing mass $\delta^N$ on each $\omega'$ and the remaining mass, namely, $1 - |\Omega|\delta^N = 1 - L^N\delta^N$, on $\omega^*$. The value so obtained is

$$(1-(L^N-1)\delta^N)P(X(t)=\omega|X(T_1)=\omega^*) + \delta^N \sum_{\omega'\neq\omega^*} P(X(t)=\omega|X(T_1)=\omega').$$

Similarly,

$$\begin{aligned} &\inf_\eta \sum_{\omega'} P(X(t)=\omega|X(T_1)=\omega')P(X(T_1)=\omega'|X(0)=\eta) \\ &\geqslant (1-(L^N-1)\delta^N)P(X(t)=\omega|X(T_1)=\omega_*) + \delta^N \sum_{\omega'\neq\omega_*} P(X(t)=\omega|X(T_1)=\omega_*) \end{aligned}$$

where $\omega' \to P(X(t)=\omega|X(T_1)=\omega')$ is *minimized* at $\omega_*$. It follows immediately that

$$Q(t,\omega) \leqslant (1-L^N\delta^N)\{P(X(t)=\omega|X(T_1)=\omega^*) - P(X(t)=\omega|X(T_1)=\omega_*)\},$$

and hence,

$$\begin{aligned} &\sup_{\omega,\eta',\eta''} |P(X(t)=\omega|X(0)=\eta') - P(X(t)=\omega|X(0)=\eta'')| \\ &\leqslant (1-L^N\delta^N) \sup_{\omega,\eta',\eta''} |P(X(t)=\omega|X(T_1)=\eta') - P(X(t)=\omega|X(T_1)=\eta'')|. \end{aligned}$$

Proceeding in this way, we obtain the bound

$$(1-L^N\delta^N)^{K(t)} \sup_{\omega,\eta',\eta''} |P(X(t)=\omega|X(T_{K(t)})=\eta') - P(X(t)=\omega|X(T_{K(t)})=\eta'')|$$

and the lemma now follows with $r = 1 - L^N\delta^N$. Notice that $r = 0$ corresponds to the (degenerate) case in which $\delta = L^{-1}$, i.e., all the local characteristics are uniform on $\Lambda$. Q.E.D.

*Proof of Theorem B:* We first state two lemmas.

*Lemma 2: For every* $t_0 = 0, 1, 2, \cdots$,

$$\lim_{t\to\infty} \sup_{\omega,\eta',\eta''} |P(X(t)=\omega|X(t_0)=\eta') - P(X(t)=\omega|X(t_0)=\eta'')| = 0.$$

*Lemma 3:*

$$\lim_{t_0 \to \infty} \sup_{t \geq t_0} \|P(t, \cdot \mid t_0, \pi_0) - \pi_0\| = 0.$$

Recall that $\pi_0$ is the uniform probability measure over the minimal energy states $\Omega_0 = \{\omega : U(\omega) = \min_\eta U(\eta)\}$.

First we show how these lemmas imply Theorem B, which states that $P(X(t) = \cdot \mid X(0) = \eta)$ converges to $\pi_0$ as $t \to \infty$. For any $\eta \in \Omega$,

$$\overline{\lim_{t \to \infty}} \|P(X(t) = \cdot \mid X(0) = \eta) - \pi_0\|$$

$$= \overline{\lim_{t_0 \to \infty}} \overline{\lim_{\substack{t \to \infty \\ t \geq t_0}}} \Big\| \sum_{\eta'} P(t, \cdot \mid t_0, \eta') \cdot P(t_0, \eta' \mid 0, \eta) - \pi_0 \Big\|$$

$$\leq \overline{\lim_{t_0 \to \infty}} \overline{\lim_{\substack{t \to \infty \\ t \geq t_0}}} \Big\| \sum_{\eta'} P(t, \cdot \mid t_0, \eta') \cdot P(t_0, \eta' \mid 0, \eta) - P(t, \cdot \mid t_0, \pi_0) \Big\| + \overline{\lim_{t_0 \to \infty}} \overline{\lim_{\substack{t \to \infty \\ t \geq t_0}}} \|P(t, \cdot \mid t_0, \pi_0) - \pi_0\|.$$

The last term is zero by Lemma 3. Furthermore, since $P(t_0, \cdot \mid 0, \eta)$ and $\pi_0$ have total mass 1, we have

$$\Big\| \sum_{\eta'} P(t, \cdot \mid t_0, \eta') P(t_0, \eta' \mid 0, \eta) - P(t, \cdot \mid t_0, \pi_0) \Big\|$$

$$= \sum_\omega \sup_{\eta''} \Big| \sum_{\eta'} (P(t, \omega \mid t_0, \eta') - P(t, \omega \mid t_0, \eta'')) \times (P(t_0, \eta' \mid 0, \eta) - \pi_0(\eta')) \Big|$$

$$\leq 2 \sum_\omega \sup_{\eta', \eta''} |P(t, \omega \mid t_0, \eta') - P(t, \omega \mid t_0, \eta'')|.$$

Finally, then,

$$\overline{\lim_{t \to \infty}} \|P(X(t) = \cdot \mid X(0) = \eta) - \pi_0\|$$

$$\leq 2 \sum_\omega \overline{\lim_{t_0 \to \infty}} \overline{\lim_{\substack{t \to \infty \\ t \geq t_0}}} \sup_{\eta', \eta''} |P(t, \omega \mid t_0, \eta') - P(t, \omega \mid t_0, \eta'')|$$

$$= 0 \quad \text{by Lemma 2.}$$

Q.E.D.

*Proof of Lemma 2:* We follow the proof of Lemma 1. Fix $t_0 = 0, 1, \cdots$ and define $T_k = t_0 + k\tau$, $k = 0, 1, 2, \cdots$. Recall that $S \subseteq \{n_{t+1}, \cdots, n_{t+\tau}\}$ for all $t$ by hypothesis, that $\pi_{T(t)}(\omega) = e^{-U(\omega)/T(t)}/Z$ and that $U^*$, $U_*$ are the maximum and minimum of $U(\omega)$, respectively, the range being $\Delta = U^* - U_*$. Let

$$\delta(t) = \inf_{\substack{1 \leq i \leq N \\ (x_{s_1}, \cdots, x_{s_N}) \in \Omega}} \pi_{T(t)}(X_{s_i} = x_{s_i} \mid X_{s_j} = x_{s_j}, j \neq i).$$

Observe that

$$\delta(t) \geq \frac{e^{-U^*/T(t)}}{Le^{-U_*/T(t)}} = \frac{1}{L} e^{-\Delta/T(t)}.$$

Now fix $k$ for the moment and define the $m_i$ as before:

$$m_i = \sup \{t : t \leq T_k, n_t = s_i\}, \quad 1 \leq i \leq N.$$

We again assume that $m_1 > m_2 > \cdots > m_N$. Then

$$P(X(T_k) = \omega \mid X(T_{k-1}) = \omega')$$

$$= P(X_{s_1}(m_1) = x_{s_1}, \cdots, X_{s_N}(m_N) = x_{s_N} \mid X(T_{k-1}) = \omega')$$

$$= \prod_{j=1}^N P(X_{s_j}(m_j) = x_{s_j} \mid X_{s_{j+1}}(m_{j+1}) = x_{s_{j+1}}, \cdots, X_{s_N}(m_N) = x_{s_N}, X(T_{k-1}) = \omega')$$

$$\geq \prod_{j=1}^N \delta(m_j) \quad \text{(using (12.3) and the definition of } \delta)$$

$$\geq L^{-N} \prod_{j=1}^N e^{-\Delta/T(m_j)}$$

$$\geq L^{-N} \exp - \left\{ \frac{\Delta N}{T(t_0 + k\tau)} \right\} \quad (\text{since } m_j \leq T_k = t_0 + k\tau, j = 1, 2, \cdots, N, \text{ and } T(\cdot) \text{ is decreasing})$$

$$\geq L^{-N} (t_0 + k\tau)^{-1}$$

wherever $t_0 + k\tau$ is sufficiently large. In fact, for a sufficiently small constant $C$, we can and do assume that

$$\inf_{\omega, \omega'} P(X(T_k) = \omega \mid X(T_{k-1}) = \omega') \geq \frac{CL^{-N}}{t_0 + k\tau} \tag{A.3}$$

for every $t_0 = 0, 1, 2, \cdots$ and $k = 1, 2, \cdots$, bearing in mind that $T_k$ depends on $t_0$.

For each $t > t_0$, define $K(t) = \sup \{k : T_k < t\}$ so that $K(t) \to \infty$ as $t \to \infty$. Fix $t > T_1$ and continue to follow the argument in Lemma 1, but using (A.3) in place of (A.2), obtaining

$$\sup_{\omega, \eta', \eta''} |P(X(t) = \omega \mid X(t_0) = \eta') - P(X(t) = \omega \mid X(t_0) = \eta'')|$$

$$\leq \prod_{k=1}^{K(t)} \left(1 - \frac{C}{t_0 + k\tau}\right).$$

Hence it will be sufficient to show that

$$\lim_{m \to \infty} \prod_{k=1}^m \left(1 - \frac{C}{t_0 + k\tau}\right) = 0 \tag{A.4}$$

for every $t_0$. However, (A.4) is a well-known consequence of the divergence of the series $\Sigma_k (t_0 + k\tau)^{-1}$ for all $t_0$, $\tau$. This completes the proof of Lemma 2.

*Proof of Lemma 3:* The probability measures $P(t, \cdot \mid t_0, \pi_0)$ figure prominently in the proof, and for notational ease we prefer to write $P_{t_0, t}(\cdot)$, so that for any $t \geq t_0 > 0$ we have

$$P_{t_0, t}(\omega) = \sum_\eta P(X(t) = \omega \mid X(t_0) = \eta) \, \pi_0(\eta).$$

To begin with, we claim that for any $t > t_0 \geq 0$,

$$\|P_{t_0, t} - \pi_{T(t)}\| \leq \|P_{t_0, t-1} - \pi_{T(t)}\|. \tag{A.5}$$

Assume for convenience that $n_t = s_1$. Then

$$\begin{aligned}
&\|P_{t_0,t} - \pi_{T(t)}\| \\
&= \sum_{(x_{s_1},\cdots,x_{s_N})} |\pi_{T(t)}(X_{s_1} = x_{s_1}|X_s = x_s, s \neq s_1) \\
&\quad \cdot P_{t_0,t-1}(X_s = x_s, s \neq s_1) \\
&\quad - \pi_{T(t)}(X_s = x_s, s \in S)| \\
&= \sum_{x_{s_2},\cdots,x_{s_N}} \Big\{ \sum_{x_{s_1}\in\Lambda} \pi_{T(t)}(X_{s_1} = x_{s_1}|X_s = x_s, s \neq s_1) \\
&\quad \times |P_{t_0,t-1}(X_s = x_s, s \neq s_1) \\
&\quad - \pi_{T(t)}(X_s = x_s, s \neq s_1)|\Big\} \\
&= \sum_{x_{s_2},\cdots,x_{s_N}} |P_{t_0,t-1}(X_s = x_s, s \neq s_1) \\
&\quad - \pi_{T(t)}(X_s = x_s, s \neq s_1)| \\
&= \sum_{x_{s_2},\cdots,x_{s_N}} \Big| \sum_{x_{s_1}} \{P_{t_0,t-1}(X_s = x_s, s \in S) \\
&\quad - \pi_{T(t)}(X_s = x_s, s \in S)\}\Big| \\
&\leq \sum_{(x_{s_1},\cdots,x_{s_N})\in\Omega} |P_{t_0,t-1}(X_s = x_s, s \in S) \\
&\quad - \pi_{T(t)}(X_s = x_s, s \in S)| \\
&= \|P_{t_0,t-1} - \pi_{T(t)}\|.
\end{aligned}$$

Observe that $\|\pi_0 - \pi_{T(t)}\| \to 0$ as $t \to \infty$. To see this, let $|\Omega_0|$ be the size of $\Omega_0$. Then

$$\begin{aligned}
\pi_{T(t)}(\omega) &= \frac{e^{-U(\omega)/T(t)}}{\sum_{\omega'\in\Omega_0} e^{-U(\omega')/T(t)} + \sum_{\omega'\in\Omega\setminus\Omega_0} e^{-U(\omega')/T(t)}} \\
&= \frac{e^{-(U(\omega)-U_*)/T(t)}}{|\Omega_0| + \sum_{\omega'\in\Omega\setminus\Omega_0} e^{-(U(\omega')-U_*)/T(t)}} \\
&\xrightarrow{t\to\infty} \begin{cases} 0, & \omega \notin \Omega_0 \\ \dfrac{1}{|\Omega_0|}, & \omega \in \Omega_0. \end{cases} \qquad \text{(A.6)}
\end{aligned}$$

Next, we claim that

$$\sum_{t=1}^{\infty} \|\pi_{T(t)} - \pi_{T(t+1)}\| < \infty. \qquad \text{(A.7)}$$

Since

$$\sum_{t=1}^{\infty} \|\pi_{T(t)} - \pi_{T(t+1)}\| = \sum_{\omega} \sum_{t=1}^{\infty} |\pi_{T(t)}(\omega) - \pi_{T(t+1)}(\omega)|$$

and since $\pi_{T(t)}(\omega) \to \pi_0(\omega)$ for every $\omega$, it will be enough to show that, for every $\omega$, $\pi_T(\omega)$ is monotone (increasing or decreasing) in $T$ for all $T$ sufficiently small. But this is clear from (A.6): if $\omega \notin \Omega_0$, then a little calculus shows that $\pi_T(\omega)$ is strictly increasing for $T \in (0, \epsilon)$ for some $\epsilon$, whereas if $\omega \in \Omega_0$, then $\pi_T(\omega)$ is strictly decreasing for all $T > 0$.

Lemma 3 can now be obtained from (A.5) and (A.7) in the following way. Fix $t > t_0 \geq 0$:

$$\begin{aligned}
&\|P_{t_0,t} - \pi_0\| \\
&\leq \|P_{t_0,t} - \pi_{T(t)}\| + \|\pi_{T(t)} - \pi_0\| \\
&\leq \|P_{t_0,t-1} - \pi_{T(t)}\| + \|\pi_{T(t)} - \pi_0\|, \qquad \text{by (A.5)} \\
&\leq \|P_{t_0,t-1} - \pi_{T(t-1)}\| + \|\pi_{T(t-1)} \\
&\quad - \pi_{T(t)}\| + \|\pi_{T(t)} - \pi_0\| \\
&\leq \|P_{t_0,t-2} - \pi_{T(t-1)}\| + \|\pi_{T(t-1)} \\
&\quad - \pi_{T(t)}\| + \|\pi_{T(t)} - \pi_0\| \\
&\leq \|P_{t_0,t-2} - \pi_{T(t-2)}\| + \|\pi_{T(t-2)} \\
&\quad - \pi_{T(t-1)}\| + \|\pi_{T(t-1)} - \pi_{T(t)}\| \\
&\quad + \|\pi_{T(t)} - \pi_0\|.
\end{aligned}$$

Proceeding in this way,

$$\|P_{t_0,t} - \pi_0\| \leq \|P_{t_0,t_0} - \pi_{T(t_0)}\| + \sum_{k=t_0}^{t-1} \|\pi_{T(k)} - \pi_{T(k+1)}\| + \|\pi_{T(t)} - \pi_0\|.$$

Since $P_{t_0,t_0} = \pi_0$ and $\|\pi_{T(t)} - \pi_0\| \to 0$ as $t \to \infty$, we have,

$$\begin{aligned}
&\overline{\lim_{t_0\to\infty}} \sup_{t \geq t_0} \|P_{t_0,t} - \pi_0\| \\
&\leq \overline{\lim_{t_0\to\infty}} \sup_{t>t_0} \sum_{k=t_0}^{t-1} \|\pi_{T(k)} - \pi_{T(k+1)}\| \\
&= \overline{\lim_{t_0\to\infty}} \sum_{k=t_0}^{\infty} \|\pi_{T(k)} - \pi_{T(k+1)}\| \\
&= 0 \quad \text{due to (A.7).}
\end{aligned}$$

Q.E.D.

## Acknowledgment

The authors would like to acknowledge their debt to U. Grenander for a flow of ideas; his work on pattern theory [23] prefigures much of what is here. They also thank D. E. McClure and S. Epstein for their sound advice and technical assistance, and V. Mirelli for introducing them to the practical side of image processing as well as arguing for MRF scene models.

## References

[1] K. Abend, T. J. Harley, and L. N. Kanal, "Classification of binary random patterns," *IEEE Trans. Inform. Theory*, vol. IT-11, pp. 538–544, 1965.

[2] H. C. Andrews and B. R. Hunt, *Digital Image Restoration*. Englewood Cliffs, NJ, Prentice-Hall, 1977.

[3] M. S. Bartlett, *The Statistical Analysis of Spatial Pattern*. London: Chapman and Hall, 1976.

[4] T. Berger and F. Bonomi, "Parallel updating of certain Markov random fields," preprint.

[5] J. Besag, "Nearest-neighbor systems and the auto-logistic model for binary data," *J. Royal Statist. Soc.*, series B, vol. 34, pp. 75–83, 1972.

[6] —, "Spatial interaction and the statistical analysis of lattice systems (with discussion)," *J. Royal Statist. Soc.*, series B, vol. 36, pp. 192–326, 1974.

[7] A. B. Bortz, M. H. Kalos, and J. L. Lebowitz, "A new algorithm

for Monte Carlo simulation of Ising spin systems," *J. Comp. Phys.*, vol. 17, pp. 10–18, 1975.

[8] V. Cerný, "A thermodynamical approach to the travelling salesman problem: an efficient simulation algorithm," preprint, Inst. Phys. & Biophys., Comenius Univ., Bratislava, 1982.

[9] P. Cheeseman, "A method of computing maximum entropy probability values for expert systems," preprint.

[10] R. Chellappa and R. L. Kashyap, "Digital image restoration using spatial interaction models," *IEEE Trans. Acoust., Speech, Signal Processing*, vol. ASSP-30, pp. 461–472, 1982.

[11] D. B. Cooper and F. P. Sung, "Multiple-window parallel adaptive boundary finding in computer vision," *IEEE Trans. Pattern Anal. Machine Intell.*, vol. PAMI-5, pp. 299–316, 1983.

[12] G. C. Cross and A. K. Jain, "Markov random field texture models," *IEEE Trans. Pattern Anal. Machine Intell.*, vol. PAMI-5, pp. 25–39, 1983.

[13] L. S. Davis and A. Rosenfeld, "Cooperating processes for low-level vision: A survey," 1980.

[14] H. Derin, H. Elliott, R. Christi, and D. Geman, "Bayes smoothing algorithms for segmentation of images modelled by Markov random fields," Univ. Massachusetts Tech. Rep., Aug. 1983.

[15] R. L. Dobrushin, "The description of a random field by means of conditional probabilities and conditions of its regularity," *Theory Prob. Appl.*, vol. 13, pp. 197–224, 1968.

[16] H. Elliott, H. Derin, R. Christi, and D. Geman, "Application of the Gibbs distribution to image segmentation," Univ. Massachusetts Tech. Rep., Aug. 1983.

[17] H. Elliott, F. R. Hansen, L. Srinivasan, and M. F. Tenorio, "Application of MAP estimation techniques to image segmentation," Univ. Massachusetts Tech. Rep., 1982.

[18] P. A. Flinn, "Monte Carlo calculation of phase separation in a 2-dimensional Ising system," *J. Statist. Phys.*, vol. 10, pp. 89–97, 1974.

[19] B. R. Frieden, "Restoring with maximum likelihood and maximum entropy," *J. Opt. Soc. Amer.*, vol. 62, pp. 511–518, 1972.

[20] D. Geman and S. Geman, "Parameter estimation for some Markov random fields," Brown Univ. Tech. Rep., Aug. 1983.

[21] S. Geman, "Stochastic relaxation methods for image restoration and expert systems," in *Proc. ARO Workshop: Unsupervised Image Analysis*, Brown Univ., 1983; to appear in *Automated Image Analysis: Theory and Experiments*, D. B. Cooper, R. L. Launer, and D. E. McClure, Eds. New York: Academic, 1984.

[22] U. Grenander, *Lectures in Pattern Theory*, Vols. I–III. New York: Springer-Verlag, 1981.

[23] D. Griffeath, "Introduction to random fields," in *Denumerable Markov Chains*, Kemeny, Knapp and Snell, Eds. New York: Springer-Verlag, 1976.

[24] A. Habibi, "Two-dimensional Bayesian estimate of images," *Proc. IEEE*, vol. 60, pp. 878–883, 1972.

[25] F. R. Hansen and H. Elliott, "Image segmentation using simple Markov field models," *Comput. Graphics Image Processing*, vol. 20, pp. 101–132, 1982.

[26] A. R. Hanson and E. M. Riseman, "Segmentation of natural scenes," in *Computer Visions Systems.* New York: Academic, 1978.

[27] J. M. Hammersley and D. C. Handscomb, *Monte Carlo Methods.* London: Methuen, 1964.

[28] M. Hassner and J. Sklansky, "The use of Markov random fields as models of texture," *Comput. Graphics Image Processing*, vol. 12, pp. 357–370, 1980.

[29] G. E. Hinton and T. J. Sejnowski, "Optimal perceptual inference," in *Proc. IEEE Conf. Comput. Vision Pattern Recognition*, 1983.

[30] R. A. Hummel and S. W. Zucker, "On the foundations of relaxation labeling processes," *IEEE Trans. Pattern Anal. Machine Intell.*, vol. PAMI-5, pp. 267–287, 1983.

[31] B. R. Hunt, "Bayesian methods in nonlinear digital image restoration," *IEEE Trans. Comput.*, vol. C-23, pp. 219–229, 1977.

[32] V. Isham, "An introduction to spatial point processes and Markov random fields," *Int. Statist. Rev.*, vol. 49, pp. 21–43, 1981.

[33] E. Ising, *Zeitschrift Physik*, vol. 31, p. 253, 1925.

[34] A. K. Jain, "Advances in mathematical models for image processing," *Proc. IEEE*, vol. 69, pp. 502–528, 1981.

[35] A. K. Jain and E. Angel, "Image restoration, modeling and reduction of dimensionality," *IEEE Trans. Comput.*, vol. C-23, pp. 470–476, 1974.

[36] E. T. Jaynes, "Prior probabilities," *IEEE Trans. Syst. Sci. Cybern.*, vol. SSC-4, pp. 227–241, 1968.

[37] L. N. Kanal, "Markov mesh models," in *Image Modeling.* New York: Academic, 1980.

[38] R. L. Kashyap and R. Chellappa, "Estimation and choice of neighbors in spatial interaction models of images," *IEEE Trans. Inform. Theory*, vol. IT-29, pp. 60–72, 1983.

[39] R. Kinderman and J. L. Snell, *Markov Random Fields and Their Applications.* Providence, RI: Amer. Math. Soc., 1980.

[40] S. Kirkpatrick, C. D. Gellatt, Jr., and M. P. Vecchi, "Optimization by simulated annealing," IBM Thomas J. Watson Research Center, Yorktown Heights, NY, 1982.

[41] P. A. Levy, "A special problem of Brownian motion and a general theory of Gaussian random functions," in *Proc. 3rd Berkeley Symp. Math. Statist. and Prob.*, vol. 2, 1956.

[42] N. Metropolis, A. W. Rosenbluth, M. N. Rosenbluth, A. H. Teller, and E. Teller, "Equations of state calculations by fast computing machines," *J. Chem. Phys.*, vol. 21, pp. 1087–1091, 1953.

[43] N. E. Nahi and T. Assefi, "Bayesian recursive image estimation," *IEEE Trans. Comput.*, vol. C-21, pp. 734–738, 1972.

[44] D. K. Pickard, "A curious binary lattice process," *J. Appl. Prob.*, vol. 14, pp. 717–731, 1977.

[45] W. H. Richardson, "Bayesian-based iterative method of image restoration," *J. Opt. Soc. Amer.*, vol. 62, pp. 55–59, 1972.

[46] A. Rosenfeld, R. A. Hummel, and S. W. Zucker, "Scene labeling by relaxation operations," *IEEE Trans. Syst., Man, Cybern.*, vol. SMC-6, pp. 420–433, 197.

[47] A. Rosenfeld and A. C. Kak, *Digital Picture Processing*, vols. 1, 2, 2nd ed. New York: Academic, 1982.

[48] F. Spitzer, "Markov random fields and Gibbs ensembles," *Amer. Math. Mon.*, vol. 78, pp. 142–154, 1971.

[49] J. A. Stuller and B. Kruz, "Two-dimensional Markov representations of sampled images," *IEEE Trans. Commun.*, vol. COM-24, pp. 1148–1152, 1976.

[50] H. J. Trussell, "The relationship between image restoration by the maximum a posteriori method and a maximum entropy method," *IEEE Trans. Acoust., Speech, Signal Processing*, vol. ASSP-28, pp. 114–117, 1980.

[51] J. W. Woods, "Two-dimensional discrete Markovian fields," *IEEE Trans. Inform. Theory*, vol. IT-18, pp. 232–240, 1972.

**Stuart Geman** received the B.A. degree in physics from the University of Michigan in 1971, the M.S. degree in physiology from Dartmouth College in 1973, and the Ph.D. degree in applied mathematics from the Massachusetts Institute of Technology in 1977.

Since 1977 he has been a member of the Division of Applied Mathematics at Brown University, Providence, RI, where he is currently an Associate Professor. His research interests include statistical inference, parallel computing, image processing, and stochastic processes.

Dr. Geman is an Associate Editor of *The Annals of Statistics* and is a recipient of the Presidential Young Investigator Award.

**Donald Geman**, for a photograph and biography, see this issue, p. 720.

# [10]

# The Calculation of Posterior Distributions by Data Augmentation

MARTIN A. TANNER and WING HUNG WONG*

The idea of data augmentation arises naturally in missing value problems, as exemplified by the standard ways of filling in missing cells in balanced two-way tables. Thus data augmentation refers to a scheme of augmenting the observed data so as to make it more easy to analyze. This device is used to great advantage by the EM algorithm (Dempster, Laird, and Rubin 1977) in solving maximum likelihood problems. In situations when the likelihood cannot be approximated closely by the normal likelihood, maximum likelihood estimates and the associated standard errors cannot be relied upon to make valid inferential statements. From the Bayesian point of view, one must now calculate the posterior distribution of parameters of interest. If data augmentation can be used in the calculation of the maximum likelihood estimate, then in the same cases one ought to be able to use it in the computation of the posterior distribution. It is the purpose of this article to explain how this can be done.

The basic idea is quite simple. The observed data $y$ is augmented by the quantity $z$, which is referred to as the latent data. It is assumed that if $y$ and $z$ are both known, then the problem is straightforward to analyze, that is, the augmented data posterior $p(\theta \mid y, z)$ can be calculated. But the posterior density that we want is $p(\theta \mid y)$, which may be difficult to calculate directly. If, however, one can generate multiple values of $z$ from the predictive distribution $p(z \mid y)$ (i.e., multiple imputations of $z$), then $p(\theta \mid y)$ can be approximately obtained as the average of $p(\theta \mid y, z)$ over the imputed $z$'s. However, $p(z \mid y)$ depends, in turn, on $p(\theta \mid y)$. Hence if $p(\theta \mid y)$ was known, it could be used to calculate $p(z \mid y)$. This mutual dependency between $p(\theta \mid y)$ and $p(z \mid y)$ leads to an iterative algorithm to calculate $p(\theta \mid y)$. Analytically, this algorithm is essentially the method of successive substitution for solving an operator fixed point equation. We exploit this fact to prove convergence under mild regularity conditions.

Typically, to implement the algorithm, one must be able to sample from two distributions, namely $p(\theta \mid y, z)$ and $p(z \mid \theta, y)$. In many cases, it is straightforward to sample from either distribution. In general, though, either sampling can be difficult, just as either the E or the M step can be difficult to implement in the EM algorithm. For $p(\theta \mid y, z)$ arising from parametric submodels of the multinomial, we develop a primitive but generally applicable way to approximately sample $\theta$. The idea is first to sample from the posterior distribution of the cell probabilities and then to project to the parametric surface that is specified by the submodel, giving more weight to those observations lying closer to the surface. This procedure should cover many of the common models for categorical data.

There are several examples given in this article. First, the algorithm is introduced and motivated in the context of a genetic linkage example. Second, we apply this algorithm to an example of inference from incomplete data regarding the correlation coefficient of the bivariate normal distribution. It is seen that the algorithm recovers the bimodal nature of the posterior distribution. Finally, the algorithm is used in the analysis of the traditional latent-class model as applied to data from the General Social Survey.

KEY WORDS: Bayesian inference; Monte Carlo sampling; Imputation; Correlation coefficient; Latent class analysis; Convergence results; Dirichlet sampling.

* Martin A. Tanner is Assistant Professor, Departments of Statistics and Human Oncology, University of Wisconsin, Madison, WI 53706. Wing Hung Wong is Associate Professor, Department of Statistics, University of Chicago, Chicago, IL 60637. This work was supported by National Science Foundation Research Grant MCS-8301459 and National Institutes of Health Grant R23 CA35464. The authors thank David Wallace for suggesting the covariance matrix example and Stephen Stigler for his helpful comments.

## 1. INTRODUCTION

This article introduces an iterative method for the computation of posterior distributions. The method applies whenever the data can be augmented in such a way that (a) it becomes easy to analyze the augmented data and (b) it is easy to generate the augmented data given the parameter. Let $y$ denote the observed data whose distribution depends on a parameter vector $\theta$. Suppose that there is a way to augment $y$ with latent data $z$ (unobserved) so that the augmented data, $x = (y, z)$, is straightforward to analyze [i.e., the augmented data posterior density, $p(\theta \mid x)$, is of known form]. The method consists of iterating the following two steps: (a) Given the current guess of the posterior distribution of $\theta$ given $y$, generate a sample of $m > 0$ latent data patterns from the predictive distribution of $z$ given $y$. (b) Update the posterior of $\theta$, given $y$, to be the mixture of the $m$ augmented data posteriors.

The sample size $m$ can change from iteration to iteration. If $m$ is always taken to be very large, then the algorithm can be interpreted as the method of successive substitution for solving a fixed point problem characterizing the true posterior distribution. The updated posterior at the end of the iterations can then be taken to be a close approximation of the true posterior distribution. When $m$ is small, however, we will need to pool over the latent data patterns generated near the end of the iterations to get a reasonable approximation to the true posterior distribution.

The plan of the article is as follows. In the remaining part of this introduction, we discuss data augmentation as a general tool for the analysis of data in complex models. At the same time, we will review relevant literature. In Section 2, we motivate and present the basic algorithm and illustrate the steps of the algorithm in the context of a simple example. In Section 3, we apply the method to the problem of inference on the covariance matrix of the multivariate normal distribution with missing values. In Section 4, we introduce the Dirichlet sampling procedure as a way to facilitate the approximate sampling from the posterior distribution in complex models of multinomial data. In Section 5, this procedure is applied to the study of social survey data modeled by a log-linear model with a latent variable. We also use this example to illustrate and discuss issues of identifiability in Bayesian modeling. In Section 6, we return to the study of the basic algorithm. We will discuss the uniqueness of the fixed point characterization that motivates the basic algorithm and will present convergence results for the algorithm. The reader who


Journal of the American Statistical Association
June 1987, Vol. 82, No. 398, Theory and Methods

is interested in applications, rather than theoretical details, may skip Section 6 without loss of continuity. In Section 7, variations of the basic algorithm will be presented and issues in its practical implementation will be discussed.

We now turn to the idea of data augmentation. In well-designed experiments, it often happens that, if not for the presence of missing values, the estimation of parameters will be straightforward. In currently popular terminology, the observed data are called the incomplete data. The complete data refer to the set of missing and observed values. Through the work of many authors, a large body of iterative techniques for maximum likelihood estimation from incomplete data has recently emerged, all of which exploit the simple structure of the complete data problem. This area is elegantly synthesized and further developed in the influential paper of Dempster, Laird, and Rubin (1977), in which references to earlier research can be found. Briefly, based on a current estimate of the parameter value, the method seeks to compute the expected value of the log-likelihood of the complete data and then maximizes the log-likelihood to obtain the updated parameter value. Dempster et al. called this approach the EM algorithm because of the expectation and maximization calculations involved. Although the details of the EM algorithm are not of direct interest for the present article, the aspect of Dempster et al. (1977) that is most important for our purpose is the impressive list of examples, which includes missing data problems, mixture problems, factor analysis, iteratively reweighted least squares, and many others. In each example, enough detail is presented to show how the EM algorithm can be applied. By these examples, the authors make it clear that even in cases that at first sight may not appear to be an incomplete data problem, one may sometimes still profit by artificially formulating it as such to facilitate the maximum likelihood estimation.

It seems that the potential usefulness of this problem formulation is still not fully appreciated by some practitioners, possibly because their problems appear to have little to do with missing values or incomplete data. For this reason, we will use the terms observed data (denoted by $y$) and augmented data (denoted by $x$), instead of incomplete data ($y$) and complete data ($x$). We will also use the term latent data ($z$) to denote the unobserved supplementary data needed for the augmentation of $y$ so that the augmented data, $x = (y, z)$, is straightforward to analyze.

In general, this data augmentation scheme is used for the calculation of maximum likelihood estimates or posterior modes. For making inferential statements, the validity of the normal approximation is assumed and the precision of the estimate is given by the observed Fisher information. In most cases, however, it is not possible to obtain the Fisher information directly from the basic EM calculations and one must do further calculations to obtain standard errors [see the discussion following Dempster et al. (1977); see also Louis (1982)]. Except in simple cases, it is difficult to obtain an indication to the validity of the normal approximation.

In the present article, we are interested in the entire likelihood or posterior distribution, not just the maximizer and the curvature at the maximizer. The method we propose exploits the simplicity of the posterior distribution of the parameter given the augmented data, just as the EM algorithm exploits the simplicity of maximum likelihood estimation given the complete data. Even in large sample situations, when the normal approximation is expected to be valid, it would still be comforting to note that the obtained posterior is consistent with the picture given by the maximum likelihood analysis. In small sample situations, the pitfalls of maximum likelihood estimation are well known, and the present method will provide a way of improving inference based on the entire posterior distribution (or the entire likelihood). The examples presented in this article will illustrate that a few steps of the iterative algorithm will provide a diagnostic for the adequacy of the normal approximation for the maximum likelihood estimate.

In practice, one is often interested in the marginal distribution of various parameters of interest. Even if one can evaluate the joint posterior distribution, obtaining the marginal distribution can be difficult and is a topic of current interest (Smith, Skene, Shaw, Naylor, and Dransfield 1985; Tierney and Kadane 1985; Zellner and Rossi 1984). In the data augmentation setup, one is faced with the additional complication that the posterior distribution given the observed data may not be expressible in closed form. Ideally, one would want to choose the augmentation such that the posterior given the augmented data can be sampled from with ease. In cases where this cannot be done, one would have to resort to approximate sampling methods. The Dirichlet sampling scheme discussed in Section 4 provides a simple approach for approximate sampling in the case of multinomial data. Moreover, the recent works on marginalization referred to previously may potentially be helpful in this regard.

We wish to draw the reader's attention to the concurrent and independent work of K. H. Li (1985a,b), who has devised an algorithm for doing multiple imputation of missing values that is very similar in its formal structure to our method. Whereas the main goal in the present article is to exploit the data augmentation formulation in the Bayesian inference of parameters, in Li's work, the initial focus, as well as sources of examples, have been the imputation of missing values. Thus the essential difference is that Li's method exploits the simplicity of the distribution of one component of the missing values given both the observed data and the remainder of the missing values, whereas our method relies on the simplicity of the posterior distribution of the parameter given the augmented data. Upon completion of both works, it was realized that when one identifies the unknown parameters as part of the missing values, then the two algorithms become essentially the same.

Our present results are, to a considerable extent, anticipated in the work of Rubin. In particular, the two key concepts of data augmentation and multiple imputation have been advocated and studied by Rubin in a series of papers on inference in the presence of incomplete data (Dempster et al. 1977; Rubin 1978, 1980).

## 2. THE BASIC ALGORITHM

The algorithm is motivated by the following simple representation of the desired posterior density:

$$p(\theta \mid y) = \int_Z p(\theta \mid z, y) p(z \mid y)\, dz, \qquad (2.1)$$

where $p(\theta \mid y)$ denotes the posterior density of the parameter $\theta$ given the data $y$, $p(z \mid y)$ denotes the predictive density of the latent data $z$ given $y$, and $p(\theta \mid z, y)$ denotes the conditional density of $\theta$ given the augmented data $x = (z, y)$. The predictive density of $z$ can, in turn, be related to the desired posterior density by

$$p(z \mid y) = \int_\Theta p(z \mid \phi, y) p(\phi \mid y)\, d\phi. \qquad (2.2)$$

In the above equations, the sample space for the latent data $z$ is denoted by $Z$ and the parameter space for $\theta$ is denoted by $\Theta$. (From this point on the range of integration will be omitted from the expressions, as it will be specified implicitly by the differentials $dz$ or $d\phi$.) Substituting (2.2) into (2.1) and interchanging the order of integration, we see that $p(\theta \mid y)$ must satisfy the integral equation

$$g(\theta) = \int K(\theta, \phi) g(\phi)\, d\phi,$$

$$\text{where } K(\theta, \phi) = \int p(\theta \mid z, y) p(z \mid \phi, y)\, dz. \qquad (2.3)$$

Let $T$ be the integral transformation that transforms any integrable function $f$ into another integrable function $Tf$ by the equation

$$Tf(\theta) = \int K(\theta, \phi) f(\phi)\, d\phi. \qquad (2.4)$$

The method of successive substitution for solving (2.3) thereby suggests an iterative method for the calculation of $p(\theta \mid y)$. Namely, start with any initial approximation $g_0(\theta)$ to $p(\theta \mid y)$, and successively calculate

$$g_{i+1}(\theta) = (Tg_i)(\theta). \qquad (2.5)$$

In Section 6 we will show that under mild conditions the $g_i$'s calculated this way will always converge to the desired posterior $p(\theta \mid y)$.

If the integral transform (2.5) can be calculated analytically, then the implementation of this method is straightforward. Unfortunately, this is seldom the case. In typical cases, the integration in (2.1), (2.2), and (2.5) is difficult to perform analytically. It is often possible, however, by the Monte Carlo method, to perform the integration. Equation (2.1) then motivates the following iterative scheme: Given the current approximation $g_i$ to $p(\theta \mid y)$,

(a) generate a sample $z^{(1)}, \ldots, z^{(m)}$ from the current approximation to the predictive density $p(z \mid y)$
(b) update the current approximation to $p(\theta \mid y)$ to be the mixture of conditional densities of $\theta$ given the augmented data patterns generated in (a), that is,

$$g_{i+1}(\theta) = m^{-1} \sum_{j=1}^{m} p(\theta \mid z^{(j)}, y).$$

In the above, we must either be able to calculate $p(\theta \mid z, y)$ for any augmented data $(z, y)$ or we must be able to sample numerically from this distribution. This is a prerequisite for the data augmentation scheme, and we will assume that it is true for the remainder of this discussion. Now consider step (a), that is, the generation of the latent data from $p(z \mid y)$. Given that the current approximation to $p(\theta \mid y)$ is $g_i(\theta)$, (2.2) then suggests that $z$ can be generated from the current predictive distribution in two steps:

(a1) generate $\theta$ from $g_i(\theta)$.
(a2) generate $z$ from $p(z \mid \phi, y)$, where $\phi$ is the value obtained in (a1).

Clearly, when $m$ is large, the two steps (a) and (b), where (a) may be implemented by (a1) and (a2), will provide a close approximation to one iteration of (2.5). Furthermore, as we will see in Sections 6 and 7, even when $m$ is as small as 1, the iteration is still "in the right direction" in the sense that the average of $p(\theta \mid x)$ over the augmented data patterns generated across iterations will converge to the $p(\theta \mid y)$. It is noted that $m$ need not be held fixed from iteration to iteration, and in Section 7 comments on how $m$ should be adaptively varied are presented.

Step (a) requires the generation of multiple values of the latent data $z$ by sampling from the conditional density of $z$ given $y$. This process is termed *multiple imputation* by Rubin (1980), who first introduced it as a method for handling nonresponse in sample surveys and in censuses. Thus step (a) can be referred to as the "imputation" step. Step (b) requires the computation (or sampling) of the posterior distribution of $\theta$ based on the augmented data sets. We will call this step the "posterior" step. The algorithm consists of iterating between the imputation and posterior steps.

The usefulness of the algorithm depends to a large extent on the ease of implementation of the imputation and posterior steps. In general, neither step is guaranteed to be easy. There is a parallel limitation on the EM algorithm; namely, that in general both the E and M steps may be difficult to implement. There remains, however, a rich class of problems, especially those connected with exponential families, for which there are natural ways to carry out these steps. This is illustrated by the examples here and the examples in Dempster et al. (1977).

### Linkage Example

To illustrate the basic algorithm, we consider an example that was presented in Rao (1973) and reexamined in Dempster et al. (1977) and Louis (1982). In particular, from a genetic linkage model, it is believed that 197 animals are distributed multinomially into four categories,

$y = (y_1, y_2, y_3, y_4) = (125, 18, 20, 34)$, with cell probabilities specified by

$$\left(\frac{1}{2} + \frac{\theta}{4}, \frac{(1-\theta)}{4}, \frac{(1-\theta)}{4}, \frac{\theta}{4}\right).$$

To illustrate the algorithm, $y$ is augmented by splitting the first cell into two cells, one of which having cell probability $\frac{1}{2}$, the other having cell probability $\theta/4$. Thus the augmented data set is given by $x = (x_1, x_2, x_3, x_4, x_5)$, where $x_1 + x_2 = 125$, $x_3 = y_2$, $x_4 = y_3$, and $x_5 = y_4$. The likelihood is of the form

$$p(y \mid \theta) \propto (2 + \theta)^{y_1}(1 - \theta)^{y_2+y_3}\theta^{y_4},$$

and the augmented likelihood is of the form

$$p(x \mid \theta) \propto \theta^{x_2+x_5}(1 - \theta)^{x_3+x_4}.$$

Thus, in this example, the augmented likelihood has a very simple form.

The implementation of our algorithm is then given as follows:

(a) *I Step* (*Imputation Step*).

(a1) Draw $\theta$ from the current estimate of $p(\theta \mid y)$.

(a2) Generate $x_2$ by drawing from the binomial distribution with parameters $(125, \theta/(\theta + 2))$.

Repeat steps (a1) and (a2) $m$ times.

(b) *P Step* (*Posterior Step*). Set the posterior density of $\theta$ equal to the mixture of beta distributions, mixed over the $m$ imputed values of $x_2$; that is,

$$p(\theta \mid y) = \frac{1}{m}\sum_{i=1}^{m} Be(\nu_1^{(i)}, \nu_2^{(i)})(\theta),$$

where

$$\nu_1^{(i)} = x_2^{(i)} + x_5 + 1, \quad \nu_2^{(i)} = x_3 + x_4 + 1,$$

and

$$Be(\nu_1, \nu_2)(\theta) = \frac{\Gamma(\nu_1 + \nu_2)}{\Gamma(\nu_1)\Gamma(\nu_2)}\theta^{\nu_1-1}(1 - \theta)^{\nu_2-1}.$$

In this step, the prior for $\theta$ is assumed to be uniform in (0, 1).

Figure 1 presents the posterior density estimates of $\theta$ for this example. In particular, the normal approximation with $\hat{\mu} = .63$ and $\hat{\sigma} = .05$ (solid line) is plotted along with the true posterior distribution (dotted line)

$$p(\theta \mid y) \propto (2 + \theta)^{y_1}(1 - \theta)^{y_2+y_3}\theta^{y_4},$$

and the estimated posterior (dashed line) obtained by plotting the mixture of the beta distributions at the final iteration in which $m = 1{,}600$. In the density scale, all three estimates are congruent. In the log-scale, however, even in this large sample situation a departure of the true posterior from the quadratic approximation at the mode is evident (Fig. 2).

Alternatively, we consider a second version of the data in which the sample size is reduced by a factor of 10, though the cell proportions are approximately unchanged; that is, $y = (13, 2, 2, 3)$. The resulting posterior density estimates are plotted in Figure 3. In this case, although the true posterior density and the estimated posterior density are congruent, the validity of the normal approximation may be in doubt, even when viewed on the density scale. An even more dramatic illustration is given in Figure 4, where $y = (14, 0, 1, 5)$. In cases with such a dramatic departure from normality, one or two iterations of our

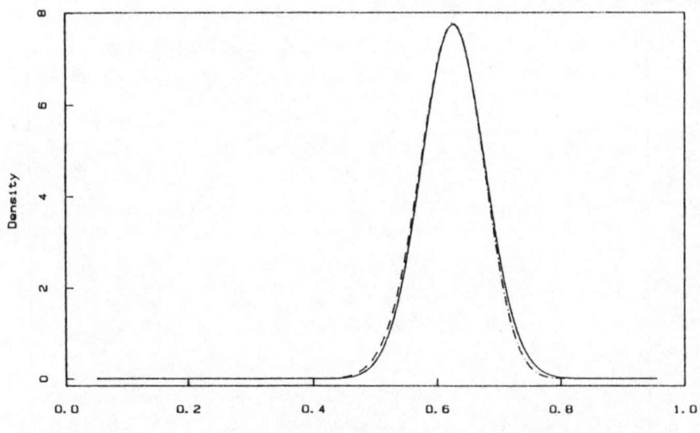


*Figure 1. Posterior Density of θ for Data (125, 18, 20, 34). The solid, dashed, and dotted lines represent the normal approximation, the estimated posterior distribution, and the true posterior, respectively. The dashed and dotted lines are superimposed.*

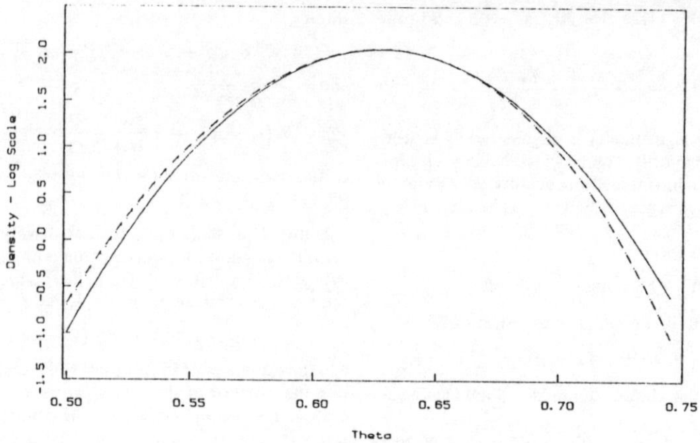


*Figure 2. Log-Posterior Density of θ (same data and legend as in Fig. 1).*

algorithm would indicate the inadequacy of the normal approximation.

## 3. FUNCTIONALS OF THE MULTIVARIATE NORMAL COVARIANCE MATRIX

In this section, the posterior distribution of the correlation coefficient from the bivariate normal distribution will be investigated. To illustrate, suppose that the data in Table 1 (Murray 1977) represent 12 observations from the bivariate normal distribution with $\mu_1 = \mu_2 = 0$, correlation coefficient $\rho$, and variances $\sigma_1^2$ and $\sigma_2^2$. Before proceeding to the formal analysis, we note that in the four pairs of observations, two pairs have correlation 1 and the remaining two pairs have correlation $-1$. Thus we can expect a nonunimodal posterior distribution for $\rho$ in this data set. In such a case, the maximum likelihood estimate and the associated standard error will clearly be misleading. Furthermore, we point out that the information regarding $\sigma_1^2$ and $\sigma_2^2$ in the eight incomplete observations cannot be ignored because information regarding $\sigma_1^2$ and $\sigma_2^2$ is of use in making inference regarding $\rho$.

The implementation of the algorithm in this problem is straightforward. Given the covariance matrix $\Sigma$, the unobserved data is generated as follows:

1. If $x_1$ is known, then generate the unobserved observation from

$$N\left(\rho \frac{\sigma_2}{\sigma_1} x_1, \sigma_2^2(1 - \rho^2)\right).$$

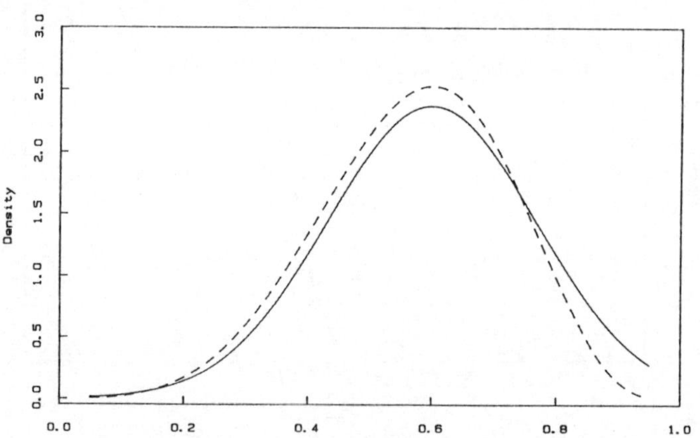


*Figure 3. Posterior Density of θ for Data (13, 2, 2, 3) (same legend as in Fig. 1). The dashed lines and dotted lines are superimposed.*

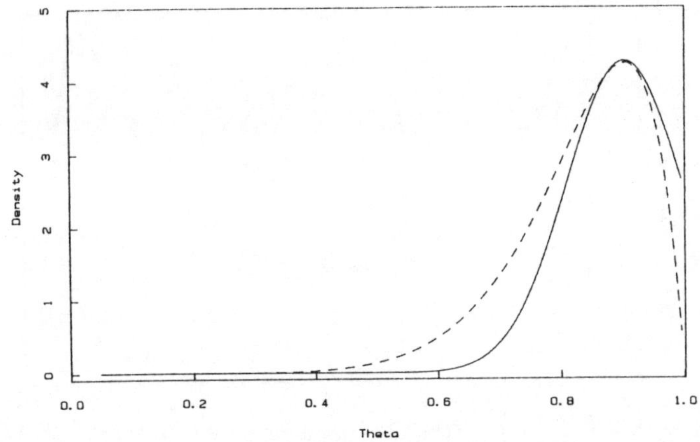


*Figure 4. Posterior Density of θ for Data (14, 0, 1, 5) (same legend as in Fig. 1). The dashed and dotted lines are superimposed.*

2. If $x_2$ is known, then generate the unobserved observation from

$$N\left(\rho \frac{\sigma_1}{\sigma_2} x_2, \sigma_1^2(1 - \rho^2)\right).$$

The covariance matrix $\Sigma$ is then generated from the current guess of the posterior distribution $p(\Sigma \mid y)$. At the first iteration, $\rho$ can be generated from $U[-1, 1]$ and $\sigma_1^2$ and $\sigma_2^2$ can be generated from weighted $\chi_7^2$ distributions. At succeeding iterations, the updated posterior $p(\Sigma \mid y)$ is a mixture of inverted Wishart distributions. This last point follows from the fact that $p(\Sigma \mid x)$ is an inverted Wishart distribution (Box and Tiao 1973, p. 428) when the prior of $\Sigma$ is given as

$$p(\Sigma) \propto |\Sigma|^{-(p+1)/2},$$

where $p$ is the dimension of the multivariate normal distribution. Thus, in the second step of the algorithm, we generate $m$ observations from this mixture of inverted Wishart distributions and compute the associated correlation coefficient for each observation.

Regarding the implementation of the algorithm, it is noted that the algorithm of Odell and Feiveson (1966) can be used to generate observations from the inverted Wishart distribution. The amount of computation in this algorithm is not extensive, since the computation is of order $p(p + 1)/2$, which does not depend on the sample size.

In Figure 5, we plot the histogram of the imputed correlation coefficients based on pooling the tenth through fifteenth iterations ($m = 6{,}400$). In addition, the true posterior of the correlation coefficient, which is proportional to $[(1 - \rho^2)^{4.5}]/[(1.25 - \rho^2)^8]$, is also plotted. As is evident from the plot, the estimated posterior distribution recovers the bimodal nature of the true distribution.

Finally, it is noted that the algorithm presented in this article can be used to examine the posterior distribution of any functional of the covariance matrix. For example, the posterior distribution of the largest eigenvalue of the covariance matrix (Tiao and Fienberg 1969) may be examined by simply computing the largest eigenvalue of each of the observations from the inverted Wishart distribution computed in the second step of the algorithm.

*Table 1. Twelve Observations From a Bivariate Normal Distribution*

| | | | | | | | | | | | |
|---|---|---|---|---|---|---|---|---|---|---|---|
| 1 | 1 | −1 | −1 | 2 | 2 | −2 | −2 | * | * | * | * |
| 1 | −1 | 1 | −1 | * | * | * | * | 2 | 2 | −2 | −2 |

* Value not observed (missing at random).

## 4. THE DIRICHLET SAMPLING PROCESS

In the linkage example of Section 2, the augmented posterior distribution $p(\theta \mid x)$ is a beta distribution. Thus it is a trivial matter to carry out the P step. In more complicated models, the sampling of $\theta$ from $p(\theta \mid x)$ may not be so simple. We now present a primitive but generally applicable procedure, based on a Dirichlet sampling process, which can be used to approximately sample from the posterior distribution of parametric models for multinomial data. In this section, we develop and illustrate the procedure using the linkage example. Further uses will be illustrated in Section 5.

In the linkage example, conditional on the augmented data, the distribution of the last four cell probabilities ($P_2$, $P_3$, $P_4$, $P_5$) is equal in distribution to that of ($\nu_2/2$, $\nu_3/2$, $\nu_4/2$, $\nu_5/2$), where ($\nu_2$, $\nu_3$, $\nu_4$, $\nu_5$) has the Dirichlet distribution

$$\frac{\Gamma(x_2 + x_3 + x_4 + x_5 + 4)}{\Gamma(x_2 + 1)\Gamma(x_3 + 1)\Gamma(x_4 + 1)\Gamma(x_5 + 1)} \nu_2^{x_2} \nu_3^{x_3} \nu_4^{x_4} \nu_5^{x_5},$$

$$\sum_{i=2}^{5} \nu_i = 1, \quad (4.1)$$

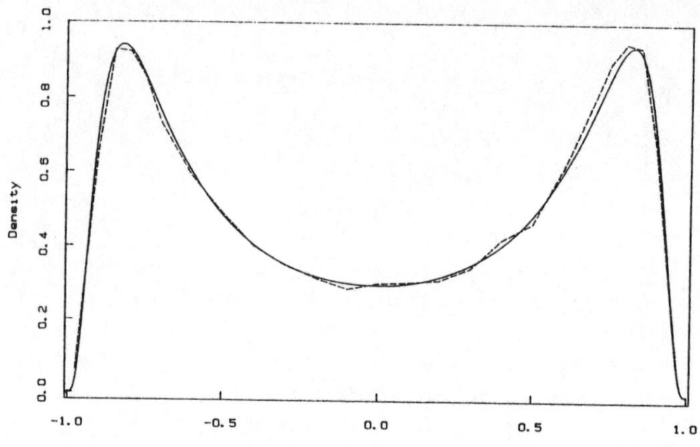


*Figure 5. Posterior Density of the Correlation Coefficient. The solid and dashed lines represent the true and estimated posterior, respectively.*

which will be denoted by $D(x_2, x_3, x_4; x_5)$. It is a trivial matter to generate observations from such a Dirichlet distribution. Our model, however, is not a saturated multinomial model. In fact, the linkage model specifies that $(P_2, P_3, P_4, P_5)$ must lie on a linear parametric curve,

$$C = \left\{\left(\frac{\theta}{4}, \frac{1}{4} - \frac{\theta}{4}, \frac{1}{4} - \frac{\theta}{4}, \frac{\theta}{4}\right): \theta \in [0, 1]\right\}.$$

The posterior distribution $p(\theta \mid x)$ will only induce a distribution of $(P_2, P_3, P_4, P_5)$ on the curve $C$. How is this induced distribution related to the Dirichlet distribution (4.1)? The answer is simple:

*Lemma.* The distribution induced by $p(\theta \mid x)$ on the curve $C$ is the same as the conditional distribution induced by the Dirichlet distribution (4.1) on $C$ (through the relationship $\mathbf{P} = \frac{1}{2}\boldsymbol{\nu}$).

*Proof.* To verify the lemma, it is sufficient to check that the ratio of the densities evaluated at any two points on $C$ is identical under either distribution.

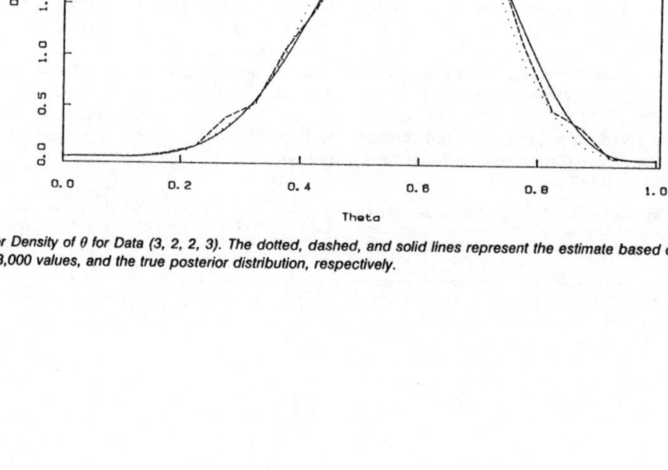


*Figure 6. Posterior Density of θ for Data (3, 2, 2, 3). The dotted, dashed, and solid lines represent the estimate based on 10,000 values, the estimate based on 3,000 values, and the true posterior distribution, respectively.*

This lemma suggests a simple two-stage algorithm: (a) generate observations from the Dirichlet distribution (4.1) and (b) accept only those points lying relatively close to the parametric curve $C$.

To be specific, observations are drawn from $D(x_2, x_3, x_4; x_5)$ and for each of these observations, we find the $\hat{\theta}$ that gives cell probabilities $(\hat{p}_2, \hat{p}_3, \hat{p}_4, \hat{p}_5)$ closest to the observed Dirichlet observation $(p_2, p_3, p_4, p_5)$. Given the functional dependencies of each of the probabilities on $\theta$: $P_2 = \theta/4$, $P_3 = 1/4 - \theta/4$, $P_4 = 1/4 - \theta/4$, and $P_5 = \theta/4$, the least squares solution yields $\hat{\theta} = 2(p_2 + p_5)$. The approximate posterior distribution for $\theta$ is then obtained by forming the histogram of those $\hat{\theta}$ values whose corresponding $(\hat{p}_2, \hat{p}_3, \hat{p}_4, \hat{p}_5)$ vector is within an $\varepsilon$-neighborhood of $(p_2, p_3, p_4, p_5)$, that is, such that

$$\left(\sum_{i=2}^{5} (p_i - \hat{p}_i)^2\right)^{1/2} < \varepsilon.$$

According to the above lemma, if $\varepsilon$ is sufficiently small, then the $\hat{\theta}$ values obtained in this way will have a distribution approximately equal to $p(\theta \mid x)$.

In practice, the value of $\varepsilon$ is selected by plotting a sequence of estimated posterior distributions of $\theta$ corresponding to a sequence of decreasing $\varepsilon$ values. The curves tend to converge as the value of $\varepsilon$ is decreased. The aforementioned procedure is generally applicable to parametric models for multinomial data if the cell probabilities are linear in $\theta$ *or* if the posterior distribution is relatively concentrated in comparison with the curvature of the parametric surface. Otherwise, the raw histogram of $\hat{\theta}$ must be multiplied by some adjustment factor.

To test the procedure in the linkage example, assume that the augmented data vector is given by (3, 2, 2, 3). To obtain the posterior distribution of $\theta$, we begin by drawing 10,000 observations from the Dirichlet distribution corresponding to this data vector. For each of these Dirichlet observations, the value of $\theta$ that gives the closest $(\hat{p}_2, \hat{p}_3, \hat{p}_4, \hat{p}_5)$ vector is found using least squares. The resulting histograms of the $\hat{\theta}$ values (using 10,000 initial values and 3,000 accepted values) and the true posterior distribution are presented in Figure 6. An examination of this figure reveals that the estimated distribution of $\theta$ based on the restricted set of $\hat{\theta}$ values is quite similar to the true distribution.

## 5. THE TRADITIONAL LATENT-CLASS MODEL

The data in Table 2 represent the responses of 3,181 participants in the 1972, 1973, and 1974 General Social Surveys, as presented in Haberman (1979). The participants in these surveys are cross-classified by the year of the survey and their responses to each of three questions regarding abortion. Thus the cell entry $n_{abcd}$ represents the number of subjects who in year $D = d$ give responses $a$ to question A, $b$ to question B, and $c$ to question C. Regarding question A, subjects are asked, "Please tell me whether or not you think it should be possible for a pregnant woman to obtain a legal abortion *if she is married and does not want any more children*." In question B, the italicized phrase is replaced with "if the family has a very low income and cannot afford any more children," and in question C it is replaced with "if she is not married and does not want to marry the man." For these data, Haberman (1979) considered several models, one of which is the traditional latent-class model. [See Goodman (1974a,b), Haberman (1979), or Clogg (1977) for an exposition of this model.] In this example, the traditional latent-class model assumes that the manifest variables ($A$, $B$, $C$, $D$) are conditionally independent, given a dichotomous latent variable ($X$). In other words, if the value of the dichotomous latent variable is known for a given participant, then knowledge of the response to a given question provides no further information regarding the responses to either of the other two questions. Haberman used the EM and scoring algorithms to obtain maximum likelihood estimates of the cell probabilities.

*Table 2. White Christian Subjects in the 1972–1974 General Social Surveys, Cross-Classified by Year of Survey and Responses to Three Questions on Abortion Attitudes*

| Year (D) | Response to A | Response to B | Response to C | Observed count |
|---|---|---|---|---|
| 1972 | Yes | Yes | Yes | 334 |
| | Yes | Yes | No | 34 |
| | Yes | No | Yes | 12 |
| | Yes | No | No | 15 |
| | No | Yes | Yes | 53 |
| | No | Yes | No | 63 |
| | No | No | Yes | 43 |
| | No | No | No | 501 |
| 1973 | Yes | Yes | Yes | 428 |
| | Yes | Yes | No | 29 |
| | Yes | No | Yes | 13 |
| | Yes | No | No | 17 |
| | No | Yes | Yes | 42 |
| | No | Yes | No | 53 |
| | No | No | Yes | 31 |
| | No | No | No | 453 |
| 1974 | Yes | Yes | Yes | 413 |
| | Yes | Yes | No | 29 |
| | Yes | No | Yes | 16 |
| | Yes | No | No | 18 |
| | No | Yes | Yes | 60 |
| | No | Yes | No | 57 |
| | No | No | Yes | 37 |
| | No | No | No | 430 |

*Source:* Haberman (1979, p. 559).

One parameter of interest associated with this model is the conditional probability of a response $a$ to question A, given that $X = 1$ (which will be denoted as $\pi_{a1}^{AX}$). In conjunction with $\pi_{a2}^{AX}$, the magnitude of this conditional probability indicates the accuracy of the response $a$ to question A in identifying the latent classification $X = 1$, since the ratio $\pi_{a1}^{AX}/\pi_{a2}^{AX}$ is the likelihood ratio for identifying $X$ based on an observation of $A$. In the present example, Haberman estimated $\pi_{11}^{AX}$ to be .892. The estimated standard error can also be obtained using the delta method, though Haberman did not include this value in his presentation.

To obtain the posterior distribution of $\pi_{11}^{AX}$, the IP algorithm is implemented as follows. In the initial iteration, the odds of being in the latent class $X = 2$ (which will be denoted as $\theta_{abcd.}$) is taken to be $\frac{1}{2}$ for all values of $a$, $b$, $c$, and $d$. The unobserved cell counts ($n_{abcdx}$) are imputed by noticing that conditional on both $\theta_{abcd.}$ and the observed cell counts $n_{abcd}$, the posterior distribution of $n_{abcd1}$ follows a binomial distribution with parameters $n_{abcd}$ and $1/(1 + \theta_{abcd.})$. The posterior distribution of $\pi_{11}^{AX}$ is then obtained by drawing from the mixture of augmented posterior distributions. In particular, for a given augmented data set, a vector of probabilities $\{P_{abcdx}\}$ is drawn from the Dirichlet distribution $D(n_{11111}, \ldots, n_{22231}; n_{22232})$ and some of the observations are discarded using the Euclidean distance criterion, as discussed in the previous section. The odds of being in the latent class $X = 2$ given that $A = a$, $B = b$, $C = c$, and $D = d$ is updated using the maximum likelihood estimate (under the conditional independence model)

$$\left(\frac{\sum_{b,c,d} p_{abcd2}}{\sum_{b,c,d} p_{abcd1}}\right)\left(\frac{\sum_{a,c,d} p_{abcd2}}{\sum_{a,c,d} p_{abcd1}}\right)\left(\frac{\sum_{a,b,d} p_{abcd2}}{\sum_{a,b,d} p_{abcd1}}\right)\left(\frac{\sum_{a,b,c} p_{abcd2}}{\sum_{a,b,c} p_{abcd1}}\right)\cdot\left(\frac{\sum_{a,b,c,d} p_{abcd1}}{\sum_{a,b,c,d} p_{abcd2}}\right)^3$$

and the algorithm cycles until convergence is achieved. For each augmented data set, the conditional probability of interest is calculated from the equation

$$\pi_{11}^{AX} = \frac{\sum_{b,c,d} p_{1bcd1}}{\sum_{a,b,c,d} p_{abcd1}}.$$

In Figures 7a and 7b, the estimated posterior distribution of $\pi_{11}^{AX}$ is presented, where the values from the fifteenth through the twentieth iteration are pooled ($m = 1{,}600$) to form the histogram in these figures. As can be seen from the figures, the posterior distribution appears to be bimodal, with one mode occurring at about .039 and the other mode occurring at about .886. The reason for this bimodality stems from the unidentifiability inherent in the problem. In the latent-class model, the data analyst has the choice of identifying a positive attitude toward abortion with the condition that $X = 1$ or with the condition that $X = 2$. The mode occurring at .039 occurs if one identifies a positive attitude with $X = 2$; the second mode occurs if a positive attitude is identified with $X = 1$. In this regard, it is important to note that the modes are well separated. Thus, for the present data set, the conditional probability is, in the Bayesian sense, *locally identifiable*.

Conditioning on the identification of a positive attitude toward abortion with $X = 1$, that is, examining the right mode, we find that our point estimate for $\pi_{11}^{AX}$ is close to the maximum likelihood estimate (.886 versus .892). (Such an identification is reasonable given the nature of the question.) In addition, there is little evidence of a departure of the normal approximation from the posterior distribution. Comparing the estimated density to the normal curve with matching mean and standard error (.009), an overall concordance is observed (Fig. 7b). A similar conclusion is reached by examining the corresponding rankit plot (Fig. 8). Regarding the lower mode (Figs. 7a and 9), some evidence against the normal approximation ($\hat{\mu} = .039$, $\hat{\sigma} = .006$) is noted. In particular, the posterior distribution is slightly skewed to the right.

## 6. THEORETICAL DEVELOPMENT

In this section, we return to the study of the algorithm motivated and outlined in Section 2. In previous examples, it was seen that the algorithm converged to the true posterior. The results in this section will explain *why* the al-

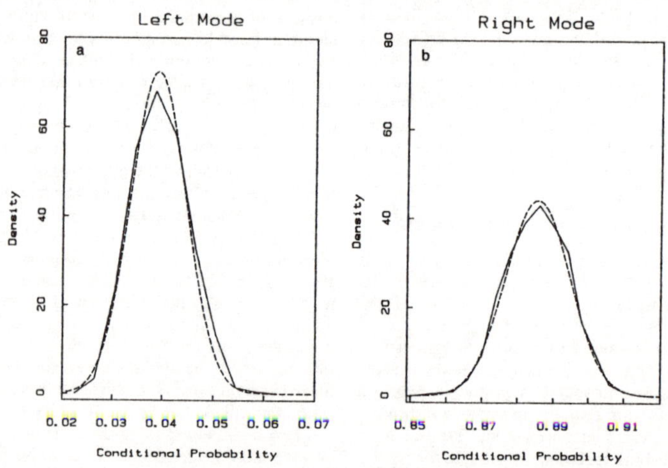


*Figure 7. Posterior Density of $\pi_{11}^{AX}$. The solid and dashed lines represent the estimated and true posterior density, respectively. (a) Left mode. (b) Right mode.*

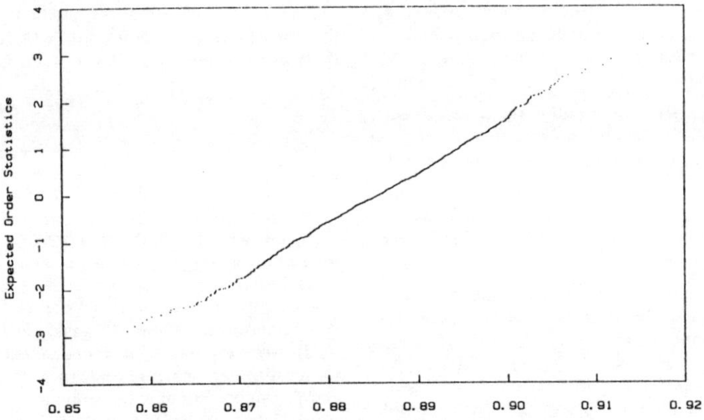


Figure 8. Rankit Plot for Right Mode.

gorithm should converge and at what rate it does so. For simplicity we will first assume that $\Theta$ is a connected subset of $R^p$. The theory is essentially the same for discrete $\Theta$, as discussed briefly at the end of this section. Let $L_1$ be the space of (Lebesque) integrable functions of $\theta \in \Theta$, and $\|f\| = \int |f(\theta)|\, d\theta$ for $f \in L_1$. Let $g_i(\theta)$, $K(\theta, \phi)$, and $T$ be defined as in (2.3)–(2.5). Clearly, $T$ is a bounded linear operator on $L_1$. Let us denote the true posterior density by $g_*(\theta)$. Then according to (2.3), $g_*$ is a fixed point under $T$; that is, $Tg_* = g_*$.

The main results of this section are, roughly, (a) $g_*$ is the only density that satisfies the fixed point equation and (b) for essentially any starting value, the iteration (2.5) converges linearly to $g_*$, that is, the deviation in the $L_1$ norm decreases at a geometric rate. These statements hold under some regularity conditions [Condition (C), given subsequently].

The first theorem shows that the $L_1$ distances from the true posterior are nonincreasing in the iterations.

*Theorem 1.* $\|g_{i+1} - g_*\| \leq \|g_i - g_*\|$.

*Proof.* The proof will make use of the following elementary facts: (a) $\int K(\theta, \phi)\, d\theta = 1$; thus if $f(\theta) \geq 0$ for all $\theta$, then $\|Tf\| = \|f\|$. (b) If $f(\theta) \geq g(\theta)$ for all $\theta$, then $Tf(\theta) \geq Tg(\theta)$ for all $\theta$. To prove the theorem, let $f = g_i - g_*$. Then

$$Tf = g_{i+1} - g_*,$$

$$\|Tf\| = \int |Tf(\theta)|\, d\theta \leq \int (T|f|)(\theta)\, d\theta = \|T|f|\| = \|f\|.$$

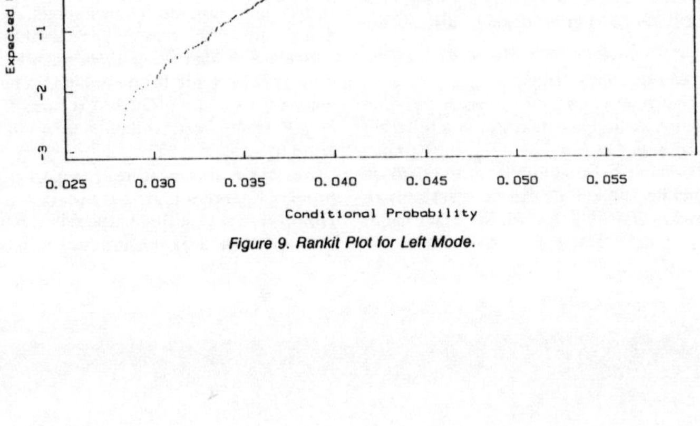


Figure 9. Rankit Plot for Left Mode.

Can the distances from the truth be strictly decreasing? Is $g_*$ the only density that satisfies the fixed point equation? To obtain positive results, we must impose some regularity conditions.

*Condition (C).* $K(\theta, \phi)$ is uniformly bounded and is equicontinuous in $\theta$. For any $\theta_0 \in \Theta$, there is an open neighborhood $U$ of $\theta_0$, so $K(\theta, \phi) > 0$ for all $\theta, \phi \in U$.

The second part of this condition says that if $\theta$ and $\phi$ are close, then it is possible to generate some latent data pattern $z$ from $p(z \mid \phi, y)$ such that $p(\theta \mid z, y)$ is nonzero, which is a reasonable condition.

*Lemma 1.* Under Condition (C), any density $g$ that is a fixed point of $T$ must be continuous and strictly positive.

*Proof.* By hypothesis, $g(\theta) \geq 0$, $g(\theta) = \int K(\theta, \phi)g(\phi)\, d\phi$. Hence $|g(\theta_1) - g(\theta)| \leq \int |K(\theta_1, \phi) - K(\theta, \phi)|\, g(\phi)\, d\phi$, which tends to 0 as $\theta_1 \to \theta$, by dominated convergence. This proves continuity of $g$. To prove positivity, consider $A = \{\theta \in \Theta : g(\theta) > 0\}$. If $A \neq \Theta$, then there must be a $\theta_0 \in \Theta$ that is also on the boundary of $A$. By Condition (C), there is a neighborhood $U$ of $\theta_0$ such that $K(\theta, \phi) > 0$ for all $\theta, \phi \in U$. Since $\theta_0$ is on the boundary we must have $g(\phi) > 0$ for some open subset of $U$. Hence $0 = g(\theta_0) \geq \int_U K(\theta_0, \phi)g(\phi)\, d\phi > 0$, a contradiction. Hence $A = \Theta$.

*Lemma 2.* Under Condition (C), if $f \in L_1$ is a function so that neither its positive part $f^+$ nor its negative part $f^-$ are identically 0, then $\|Tf\| < \|f\|$.

*Proof.* By connectedness of $\Theta$ and Condition (C), we must have support of $Tf^+ \supset$ support of $f^+$, and support of $Tf^- \supset$ support of $f^-$. Note that the inclusions are strict. It follows that

$$(\text{support of } Tf^+) \cap (\text{support of } Tf^-) \tag{6.1}$$

is nonempty. Now

$$|(Tf)(\theta)| = |Tf^+(\theta) - Tf^-(\theta)|,$$

$$(T|f|)(\theta) = Tf^+(\theta) + Tf^-(\theta).$$

Hence under (6.1) we must have

$$\int |(Tf)(\theta)|\, d\theta < \int (T|f|)(\theta)\, d\theta.$$

*Corollary.* Under (C), the distance of $g_i$ to $g_*$ is strictly decreasing.

Now we are ready to state and prove the main theorems. Theorem 2 guarantees the uniqueness of the solution to the fixed point equation. Theorem 3 gives the rate of convergence of the iteration (2.5) in terms of $L_1$ distances.

*Theorem 2.* Under Condition (C), the posterior density $g_*$ is the only density that satisfies $Tg = g$.

*Proof.* The fact that $g_*$ satisfies the fixed point equation was derived in Section 2. Suppose that $g_{**}$ is a different density satisfying $Tg = g$. Let $f = g_* - g_{**}$, then $f$ must be continuous by Lemma 1. In addition, since $\int f(\theta)\, d\theta = 0$ and $f \neq 0$, neither $f^+$ nor $f^-$ can be identically 0. Hence, by Lemma 2, $\|Tf\| < \|f\|$. But on the other hand, $Tf = Tg_* - Tg_{**} = g_* - g_{**} = f$, a contradiction.

*Theorem 3.* Suppose that Condition (C) holds and that the starting value $g_0$ satisfies $\sup_\theta (g_0(\theta)/g_*(\theta)) < \infty$. Then there exists a constant $\alpha$ $(0 < \alpha < 1)$, such that

$$\|g_{i+1} - g_*\| \leq \alpha^i \|g_0 - g_*\|.$$

*Proof.* The proof proceeds in five steps:

(a) For any $M > 0$, if $(g_0(\theta)/g_*(\theta)) < M$ for all $\theta \in \Theta$, then $(g_i(\theta)/g_*(\theta)) < M$ for all $i$, for all $\theta \in \Theta$.

(b) For any $M > 0$, the set $\{f \in L_1 : |f(\theta)/g_*(\theta)| < M$ for all $\theta\}$ is weakly sequentially compact in $L_1$.

(c) Let $f_i = g_i - g_*$ and let $\alpha = \sup_{i>1} (\|Tf_i\|/\|f_i\|)$. There exists a subsequence $\{f_{i'}\}$ such that $\|Tf_{i'}\|/\|f_{i'}\| \to \alpha$, and $f_{i'}$ converges to some $f_*$ weakly in $L_1$.

(d) Since the set $\{f_{i'}\}$ is bounded and equicontinuous, we must actually have $f_{i'}$ converges to $f_*$ strongly in $L_1$, and $f_*$ can be chosen to be continuous.

(e) Hence $\alpha = \lim(\|Tf_{i'}\|/\|f_{i'}\|) = \|Tf_*\|/\|f_*\|$. But $\int f_*(\theta)\, d\theta = 0$; hence by Lemma 2, $0 \leq \alpha < 1$. From this, the theorem follows directly.

It remains to establish statements (a)–(e). Statement (e) needs no proof, statement (a) follows from elementary manipulation, and statement (b) is a well-known property of $L_1$ spaces (see, e.g., Dunford and Schwartz 1958, p. 294). To prove (c), let $\{f_{i'}\}$ be a subsequence of $\{f_i\}$ such that $\|Tf_{i'}\|/\|f_{i'}\| \to \alpha$. Now by (a) and (b), $\{f_{i'}\}$ is weakly sequentially compact, so there must exist a further subsequence $\{f_{i''}\}$ of $\{f_{i'}\}$ convergent weakly in $L_1$. This establishes (c). Finally, (d) can be established by standard analytical arguments.

*Remark 1.* One of the conditions of Theorem 3 requires that $g_0(\theta)/g_*(\theta)$ be uniformly bounded. For a compact parameter space $\Theta$, this condition is automatic if Condition (C) holds, since under (C), $g_*$ is continuous and strictly positive. For an unbounded parameter space, we need to make sure that the decay of $g_0(\theta)$ when $|\theta| \to \infty$ is not slower than that of $g_*(\theta)$. This suggests using $g_0$ of bounded support.

*Remark 2.* Theorem 3 says that the convergence rate is linear. Unfortunately, the rate $\alpha$ is dependent on the initial value $g_0$. If $\Theta$ is compact, it can be shown that the supremum of $\alpha$ over all possible $g_0$ is still less than 1; that is, we get a linear rate independent of the starting values. If $\Theta$ is unbounded, however, $\alpha$ can be arbitrarily close to 1, depending on the starting value. This seems to be an intrinsic limit imposed by an unbounded parameter space and should not be regarded as a weakness of the method.

*Remark 3.* The whole theory can be developed in the same way for finite or countable $\Theta$. The simplest replacement for Condition (C) is to require $K(\theta, \phi) > 0$ for all $\theta, \phi \in \Theta$. Weaker conditions exist but they are cumbersome to state.

*Remark 4.* It is clear from properties (a) and (b) in the proof of Theorem 1 that $T$ is a Markov transition operator. However, a search through standard references, including Doob (1953), does not produce results directly suitable

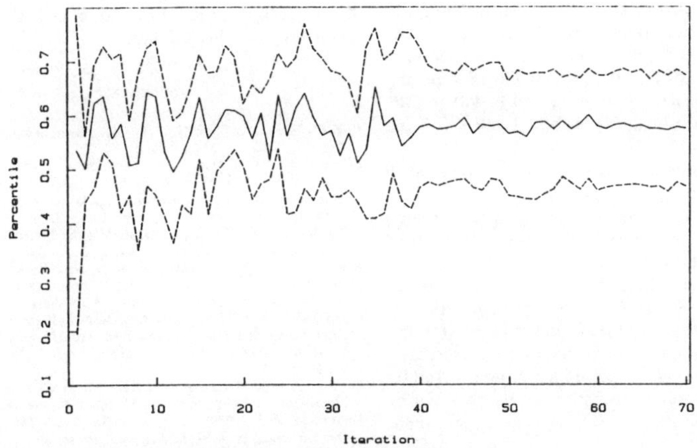


*Figure 10. Median and Upper and Lower Quartiles of θ Values Across Iterations. The upper dashed line, the solid line, and the lower dashed line represent the upper quartile, the median, and the lower quartile, respectively.*

for our use. Especially, the $L_1$ convergence rate in Theorem 3 seems to be new.

*Remark 5.* Similarly, there is a vast literature on fixed point operator equations and the method of successive substitution (see, e.g., Rall 1969, pp. 64–74). Again, we have not found results directly usable here.

## 7. PRACTICAL IMPLEMENTATION OF THE ALGORITHM

As indicated in the introduction, if the sample size $m$ is taken to be large in each iteration, then the algorithm can be interpreted as the method of successive substitution for solving a fixed point problem. In practice, however, it is inefficient to take $m$ large during the first few iterations when the estimated posterior distribution is far from the true distribution. Rather, it is suggested that $m$ initially be small and then increased with successive iterations. In addition, we have found it helpful to monitor the progress of the algorithm by examining selected percentiles of the estimated posterior distribution, for example, the 25%, 50%, and 75% percentiles.

To illustrate these ideas, let us return to the linkage

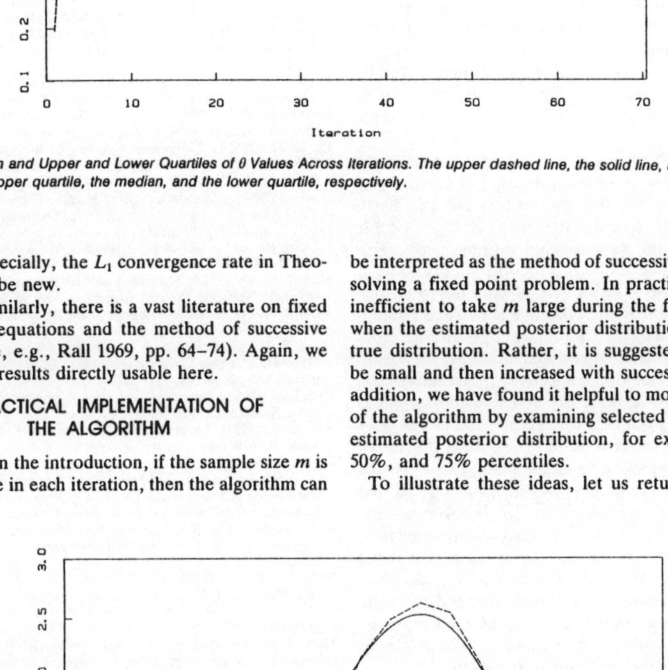


*Figure 11. The Posterior Density of θ. The dashed and solid lines represent the estimated and true posterior, respectively.*

example, where the observed data is taken to be (13, 2, 2, 3). At the initial iteration, $m$ is taken to be 20. The algorithm then runs through 40 iterations, at which point it appears (see Fig. 10) that the process has become stationary. The sample size is then increased to 400 and the algorithm proceeds through 20 further iterations. From Figure 10, we see that the effect of increasing $m$ has been to reduce substantially the system variability. The final 10 iterations are run with $m = 1{,}600$, and the estimated posterior distribution is then obtained by pooling the imputed theta values from the final iterations. Figure 11 is obtained by pooling the results of iterations 67–70.

For obvious reasons, the statistical fluctuations exhibited in iterations 20–40 cannot be reduced by further iterations without increasing the sample size $m$ (for the sample of augmented data). Typically, graphical displays, such as Figure 10, will give a good idea of how $m$ should be varied. A more formal procedure can be obtained by comparing the within-iteration variance to the between-iteration variance.

Another point illustrated in the linkage example is the possibility of pooling among iterations. For example, in iterations 20–40 we see that the process has stabilized. These samples are then pooled to form a combined sample of 400 to initialize the new iteration with $m = 400$. This pooled sample should not be regarded for all purposes as a random sample because the values from different iterations are dependent. If the process has reached equilibrium, however, then the histogram constructed from the sample will give the correct shape. Thus, for example, let $m$, $\bar{\theta}_m$, and $s$ denote, respectively, the sample size, mean, and standard deviation of the pooled sample. It then follows that $\bar{\theta}_m$ will be a consistent estimate (as $m \to \infty$) of the posterior mean of $\theta$, but the standard error of this estimate will typically be larger than $s/\sqrt{m}$. To see this, consider the extreme case in which $m = 1$, so that iteration $i$ produces only one value $\theta(i)$. In this case, $\theta(i)$ ($i = 1, 2, \ldots$) forms a Markov process with transition function equal to $K(\theta, \phi)$, as defined in (2.3). Under the regularity conditions of Section 6, this is an ergodic Markov process with an equilibrium distribution satisfying the fixed point equation given in (2.3). Hence $\bar{\theta}_m$ will converge to the mean of this equilibrium distribution, which is identical to the mean of the posterior distribution.

Finally, it is noted that the computation in Section 2.1 (10 iterations with $m = 1{,}600$) required 13 minutes on a VAX 750, whereas the computations in Section 3 (15 iterations with $m = 6{,}400$) and Section 5 (15 iterations with $m = 1{,}600$) required 23 minutes and 171 minutes, respectively, on a VAX 750.

[*Received April 1985. Revised October 1986.*]

## REFERENCES

Box, G. E. P., and Tiao, G. C. (1973), *Bayesian Inference in Statistical Analysis*, Reading, MA: Addison-Wesley.

Clogg, C. C. (1977), "Unrestricted and Restricted Maximum Likelihood Latent Structure Analysis: A Manual for Users," Working Paper 1977-09, Pennsylvania State University, Population Issues Research Center.

Dempster, A. P., Laird, N. M., and Rubin, D. B. (1977), "Maximum Likelihood From Incomplete Data Via the EM Algorithm," *Journal of the Royal Statistical Society*, Ser. B, 39, 1–38.

Doob, J. L. (1953), *Stochastic Processes*, New York: John Wiley.

Dunford, N., and Schwartz, J. T. (1958), *Linear Operators. Part I: General Theory*, New York: Interscience Publishers.

Goodman, L. A. (1974a), "The Analysis of Systems of Qualitative Variables When Some of the Variables Are Unobservable. Part I—A Modified Latent Structure Approach," *American Journal of Sociology*, 79, 1179–1259.

——— (1974b), "Exploratory Latent Structure Analysis Using Both Identifiable and Unidentifiable Models," *Biometrika*, 61, 215–231.

Haberman, S. J. (1979), *Analysis of Qualitative Data, Volume 2: New Developments*, New York: Academic Press.

Li, K. H. (1985a), "Hypothesis Testing in Multiple Imputation With Emphasis on Mixed-Up Frequencies in Contingency Tables," unpublished Ph.D. dissertation, University of Chicago, Dept. of Statistics.

——— (1985b), "Imputation Using Markov Chains," technical report, The Chinese University of Hong Kong, Dept. of Statistics.

Louis, T. A. (1982), "Finding the Observed Information Matrix When Using the EM Algorithm," *Journal of the Royal Statistical Society*, Ser. B, 44, 226–233.

Murray, G. D. (1977), Comment on "Maximum Likelihood From Incomplete Data Via the EM Algorithm," by A. P. Dempster, N. M. Laird, and D. B. Rubin, *Journal of the Royal Statistical Society*, Ser. B, 39, 27–28.

Odell, P. L., and Feiveson, A. H. (1966), "A Numerical Procedure to Generate a Sample Covariance Matrix," *Journal of the American Statistical Association*, 61, 199–203.

Rall, L. B. (1969), *Computational Solution of Nonlinear Operator Equations*, New York: John Wiley.

Rao, C. R. (1973), *Linear Statistical Inference and Its Applications*, New York: John Wiley.

Rubin, D. B. (1978), "Multiple Imputation in Sample Surveys. A Phenomenological Bayesian Approach to Nonresponse," in *Imputation and Editing of Faulty or Missing Survey Data*, Washington, DC: U.S. Department of Commerce, Social Security Administration.

——— (1980), "Handling Non-Response in Sample Surveys by Multiple Imputation," U.S. Department of Commerce Bureau of Census Monograph.

Smith, A. F. M., Skene, A. M., Shaw, J. E. H., Naylor, J. C., and Dransfield, M. (1985), "The Implementation of the Bayesian Paradigm," *Communications in Statistics, Part A—Theory and Methods*, 14, 1079–1102.

Tiao, G. C., and Fienberg, S. (1969), "Bayesian Estimation of Latent Roots and Vectors With Special Reference to the Bivariate Normal Distribution," *Biometrika*, 56, 97–104.

Tierney, L., and Kadane, J. B. (1986), "Accurate Approximations for Posterior Moments and Marginal Densities," *Journal of the American Statistical Association*, 81, 82–86.

Zellner, A., and Rossi, P. E. (1984), "Bayesian Analysis of Dichotomous Quantal Response Models," *Journal of Econometrics*, 25, 365–393.

# Comment

A. P. DEMPSTER*

The article by Tanner and Wong is a notable achievement because it combines theoretical elegance and insight with an important contribution to the emerging field of practical Bayesian statistics. I believe that Bayesian statistics is fundamentally a computational theory whereby the implications of a set of statistical data are understood by constructing a probability model and computing a set of relevant probabilities and expectations. There is little need for the traditional mathematical studies of properties of statistical techniques that dominate statistical journals. There is a great need for better understanding of the principles of model construction, criticism, and revision, and for studies of sensitivity of inferences to model change as a function of features of the data. Such studies depend critically on the availability of sophisticated Bayesian computing tools, especially for realistically complex data systems. The data augmentation idea seems destined to greatly facilitate the handling of complexity, so that situations that recently were computationally intractable now appear feasible. I look forward to much further development, both of related algorithms and of implementations for specific models, now that Tanner and Wong have shown us the way.

For example, hierarchical models with several levels of Gaussian randomness have a long history and many important applications. Standard practice in this area has been to replace the unknown hyperparameters (i.e., variance components) by point estimates, and then to report posterior distributions for interesting quantities, assuming that the point estimates are known true values. Standard methods for adjusting these posterior distributions to reflect increased variability due to inaccuracy of the point estimates of the hyperparameters do not exist. For a Bayesian such standard methods are necessarily dubious in principle, in the sense that second-level variability typically has relatively small sample size so that dependence on specific choices of hyperpriors is practically meaningful. The immediate task is to implement methods for constructing and computing with genuine hyperpriors, and then to study the results of applying the methods to data sets with specified features. Data augmentation methods (EM algorithms) have in recent years facilitated the implementation and study of methods based on maximum likelihood estimation of hyperparameters. I now expect parallel developments for logically more satisfying Bayesian methods.

I believe that Tanner and Wong might have helped readers unfamiliar with the problem of Bayesian computing by presenting an outline of the range of currently available methods and of their strengths and weaknesses in various settings. These include exact analytic expressions, analytic approximations and expansions, numerical integration procedures, and Monte Carlo methods, especially importance sampling. The basic idea of data augmentation exhibited in Tanner and Wong's formula (2.1) is to depend on numerical mixing via $p(z \mid y)$ of the analytically tractable complete data posteriors $p(\theta \mid z, y)$, where the relative contributions of each part to posterior variability depends on the fraction of missingness inherent in the adopted model. The numerical mixing operation could in principle draw on any of the approximate methods listed previously, and opportunities exist for developing algorithms along many different lines.

Tanner and Wong do not use any of the standard approximate methods, but instead adopt an ingenious iterative scheme based on successive substitution. The result is that the Tanner and Wong method has two striking parallels to the EM algorithm, namely, (a) dependence on the analytic simplicity of the likelihood function given augmented data, and (b) use of an iterative process with a monotone convergence property. Although the iterative scheme is mathematically very appealing, its use in the actual numerical problem may be somewhat unnatural, because Monte Carlo approximation is typically used for each successive iterate $g_i(\theta)$, whereas it seems to me that if Monte Carlo is needed it might more efficiently be applied directly to approximating $g(\theta)$.

A problem with direct use of Monte Carlo for numerical mixing is that available algorithms generally permit only sampling from approximate posteriors, whence importance sampling weights are required for the direct use of (2.1). Two desiderata for importance sampling are (a) it should be easy to compute the weights, and (b) the weights should be as constant as possible. The feasibility of (a) depends on the model. For the hierarchical Gaussian models mentioned previously, simulation of $z$ produces the likelihood $l(\theta)$ as a byproduct so that weights depending only on $\theta$ may be used. For other models, more variable weights based on $(\theta, z)$ pairs may be necessary. There would still be a role for successive iterates $g_i(\theta)$ for successive stages of sampling, in order to stabilize the importance sampling weights, but the iterates themselves would involve weights and so would differ from the Tanner and Wong iterates. My central point is that there are many options for using Monte Carlo in conjunction with the basic data augmentation principle (2.1), whence opportunities for studying and comparing them need to be pursued.

* A. P. Dempster is Professor, Department of Statistics, Harvard University, Cambridge, MA 02138. This work was facilitated in part by National Science Foundation Grant DMS-85-04332.


Journal of the American Statistical Association
June 1987, Vol. 82, No. 398, Theory and Methods

# Comment

## Simulation in Hierarchical Models

C. N. MORRIS*

### 1. SIMULATION FOR GENERAL HIERARCHICAL MODELS

Hierarchical Bayes models and related empirical Bayes models are suited ideally for analysis by Tanner and Wong's data augmentation algorithm. The following amplifies on the general approach and examines the case of normal means in particular.

In the usual three-stage hierarchical setting, assume known forms of densities: $p_1(y \mid \theta)$ for observed data $y$ (stage I in the hierarchy); $p_2(\theta \mid \alpha)$ for unknown parameters $\theta$ (stage II), given the unknown hyperparameter $\alpha$; and $p_3(\alpha)$ (stage III) for $\alpha$. The joint distribution is

$$p(y, \theta, \alpha) = p_1(y \mid \theta)p_2(\theta \mid \alpha)p_3(\alpha). \quad (1.1)$$

This can be rewritten in two alternative forms:

$$p(y, \theta, \alpha) = q_1(y)q_2(\theta \mid y)q_3(\alpha \mid \theta) \quad (1.2)$$

$$= r_1(y)r_2(\theta \mid \alpha, y)r_3(\alpha \mid y), \quad (1.3)$$

with the $q_i$ and $r_j$ densities derived from $p_1$, $p_2$, $p_3$. Of course $q_1(y) = r_1(y)$. Note the useful simplifying fact that the conditional probability of $\alpha$ given $\theta$ and $y$ does not depend on $y$ and hence is $q_3(\alpha \mid \theta)$.

Tanner and Wong obtain a sample $\theta^{(1)}, \ldots, \theta^{(m)}$ of size $m$ from $q_2(\theta \mid y)$, the posterior density, or a jointly distributed sample $(\theta^{(j)}, \alpha^{(j)})$ $(j = 1, \ldots, m)$ from

$$\bar{p}(\theta, \alpha \mid y) \equiv p(y, \theta, \alpha)/q_1(y). \quad (1.4)$$

Their method accomplishes the latter objective, and hence the former, by ignoring the $\alpha^{(j)}$. A sample of size $m$ is drawn from (1.4) by iterating two steps, the imputation and posterior steps. The imputation step (I step) imputes missing data, which in this model are taken to be values of the unknown hyperparameter $\alpha$. Given an initial sample $\theta^{(1)}, \ldots, \theta^{(m)}$, impute $\alpha^{(1)}, \ldots, \alpha^{(m)}$ independently:

$$\alpha^{(j)} \sim q_3(\alpha \mid \theta^{(j)}), \qquad j = 1, \ldots, m. \quad (1.5)$$

Then, given $\alpha^{(1)}, \ldots, \alpha^{(m)}$, the posterior step (P step) simulates a new sample $\theta^{(1)}, \ldots, \theta^{(m)}$ according to

$$\theta^{(j)} \sim r_2(\theta \mid \alpha^{(j)}, y) \quad (1.6)$$

independently for $j = 1, \ldots, m$. Iterating (1.5) and (1.6) many times, beginning with any starting point, produces a final sample $(\theta^{(j)}, \alpha^{(j)})$ $(j = 1, \ldots, m)$ from the desired distribution $\bar{p}(\theta, \alpha \mid y)$, even if the initial sample is not distributed as $\bar{p}$.

In some important hierarchical problems the P step is conducted easily whenever $p_2(\theta \mid \alpha)$ is a conjugate prior distribution relative to $p_1(y \mid \theta)$, because then $r_2(\theta \mid \alpha, y)$ takes the same form as $p_2(\theta \mid \alpha)$. For example, inferences about the usual parameters $\theta$ when $p_1$ is a normal, Poisson, gamma, binomial, or negative binomial distribution and $p_2$ is conjugate to $p_1$, only require normal, gamma, or beta variates to sample from $r_2$. The I step is simplified in all hierarchical models, because $q_3(\alpha \mid \theta)$ does not depend on $y$, and is simplified further when $p_3(\alpha)$ is chosen as conjugate to $p_2(\theta \mid \alpha)$. Thus simple forms for $q_3$ and $r_2$ often can be chosen for the iterations (1.5)–(1.6).

### 2. EXAMPLE: INFERENCES ABOUT NORMALLY DISTRIBUTED OBSERVATIONS

Suppose that $k$ population means $\theta = (\theta_1, \ldots, \theta_k)'$ are to be estimated after observing independent normally distributed sample means $y = (y_1, \ldots, y_k)'$ with

$$y_i \mid \theta_i \overset{\text{iid}}{\sim} N(\theta_i, V_i), \qquad i = 1, \ldots, k, \quad (2.1)$$

and $V_i = \text{var}(y_i \mid \theta_i)$ known. The conjugate prior distribution $p_2(\theta \mid \alpha)$ is assumed for each $\theta_i$ independently with $\alpha = A > 0$ unknown and

$$\theta_i \mid A \overset{\text{iid}}{\sim} N(0, A), \qquad i = 1, \ldots, k. \quad (2.2)$$

Further assume that $p_3(\alpha)$ has the conjugate (relative to $p_2$) form

$$p_3(A) = cA^{-1-q/2}\exp(-.5\lambda/A) \quad (2.3)$$

for known $q > -k/2$ and $\lambda \geq 0$. These choices lead to proper posterior densities, but proper prior densities require $q > 0$ and $\lambda > 0$, in which case $A$ is distributed as $\lambda/\chi^2_q$.

Let initial values $A^{(1)}, \ldots, A^{(m)}$ be given. The posterior distribution of $\theta$ given $(y, A)$ is normally distributed and the P step samples

$$\theta_i^{(j)} \overset{\text{iid}}{\sim} N((1 - B_i^{(j)})y_i, V_i(1 - B_i^{(j)})) \quad (2.4)$$

$(i = 1, \ldots, k; j = 1, \ldots, m)$ independently, with $B_i^{(j)} \equiv V_i/(V_i + A^{(j)})$ $(i = 1, \ldots, k; j = 1, \ldots, m)$. Next, given values $\theta^{(j)} = (\theta_1^{(j)}, \ldots, \theta_k^{(j)})$ from (2.4), (2.2), and (2.3) imply a reciprocal chi-squared distribution for

* C. N. Morris is Professor, Department of Mathematics, and Director, Center for Statistical Sciences, University of Texas, Austin, TX 78712. Support for this research was provided by National Science Foundation Research Grant DMS-8407876. This comment was written while the author was a Visiting Fellow at the University of Warwick, England. The author extends his appreciation to Don Rubin for key discussions on this topic.


Journal of the American Statistical Association
June 1987, Vol. 82, No. 398, Theory and Methods

$A$. The I step (1.5), therefore, samples new values $A^{(1)}, \ldots, A^{(m)}$ according to

$$A^{(j)} \sim \frac{\lambda + \|\theta^{(j)}\|^2}{\chi^2_{k+q}}, \quad j = 1, \ldots, m, \tag{2.5}$$

with $\chi^2_{k+q}$ sampled independently for each $j$, $\|\theta\|^2$ denoting sum of squares. The final iteration yields $m$ independent samples $(\theta_1^{(j)}, \ldots, \theta_k^{(j)}, A^{(j)})$ $(j = 1, \ldots, m)$, distributed as $\bar{p}(\theta, \alpha \mid y)$.

There can be ambiguities in the use of these data. Estimation of the posterior mean $(1 - B_i)y_i$, for example, is achieved by averaging either the $(1 - B_i^{(j)})y_i$ values or the $\theta_i^{(j)}$ values for $j = 1, \ldots, m$, but these values will not agree precisely.

The preceding can be expanded straightforwardly to encompass an unknown prior mean so that $\alpha = (\mu, A)$, $\mu = E\theta_i$, or to include regression forms $E\theta_i = \beta' x_i$ with $\alpha = (\beta_1, \ldots, \beta_r, A) = (\beta', A)$ and each $x_i$ a known vector, assuming a normal or a flat prior distribution is used for $\mu$ or $\beta$.

The simulations described here will be costly when substantial accuracy is required, because that usually necessitates large $m$ and many iterations. The number of iterations will be reduced dramatically, however, if a good starting approximation to the distribution $r_3(\alpha \mid y)$ is available. Assuming the model (2.1)–(2.3), a suggestion follows, based on Morris (1987).

Let $l(A)$ denote the logarithm of the modified posterior density $A \cdot r_3(A \mid y)$, and let $l'(A)$, $l''(A)$ be its first two derivatives.

(a) Find $\hat{A} > 0$ a (usually unique) value satisfying $l'(\hat{A}) = 0$, with

$$2l'(A) = -(k + q)A^{-1} + \lambda A^{-2} + \sum_1^k y_i^2/(V_i + A)^2. \tag{2.6}$$

(b) Define

$$p \equiv (-2\hat{A}^2 l''(\hat{A}))^{-1/2}. \tag{2.7}$$

(c) For $j = 1, \ldots, m$, let $u_j = (j - .5)/m$, and set

$$A^{(j)} = \hat{A}(u_j/(1 - u_j))^p. \tag{2.8}$$

The values (2.8) are approximations to the expected order statistics from the posterior density $r_3(A \mid y)$.

Tanner and Wong are to be congratulated for offering this approach to Bayesian simulation. Perhaps this note helps to expand further the usefulness of their method and to emphasize that their "missing data" concept can be used to include unknown parameters or latent data. Promising as their data augmentation method is, however, much still must be learned about affordable accuracy in large problems, about criteria for stopping the iterations (1.5)–(1.6), and whether simulation times are feasible in high-dimensional problems.

## ADDITIONAL REFERENCE

Morris, C. N. (1987), "Empirical Bayes Interval Estimation," Technical Report 42, The University of Texas at Austin, Center for Statistical Sciences.

# Comment

## A Noniterative Sampling/Importance Resampling Alternative to the Data Augmentation Algorithm for Creating a Few Imputations When Fractions of Missing Information Are Modest: The SIR Algorithm

DONALD B. RUBIN*

Tanner and Wong are to be congratulated for this fine addition (henceforth, TW) to modern statistical theory and practice, which makes heavy use of current computational capabilities to draw inferences using simulation techniques. I firmly believe (Rubin 1985, sec. 2.5) that given today's computing environments, obtaining inferences in applied problems using simulation is often highly desirable because of the resultant flexibility of matching models to data without tedious and tangential mathematical analysis.

I expect that TW will simulate a variety of interesting efforts involving inference via simulation. For instance, it is easy to imagine a major article or sequence of articles providing details of the data augmentation algorithm in the series of EM examples outlined in Dempster, Laird, and Rubin (1977), and explicated and extended in subsequent literature. Similarly, more technical articles on this algorithm concerning rates of convergence, choices of

* Donald B. Rubin is Professor, Department of Statistics, Harvard University, Cambridge, MA 02138. This comment was written while the author was visiting the University of Warwick, England, and was also partially supported by National Science Foundation Grants SES-8311428 and DMS-8504332.


Journal of the American Statistical Association
June 1987, Vol. 82, No. 398, Theory and Methods

number of replications at each iteration, methods for speeding convergence, and so forth certainly should be forthcoming.

## 1. RELEVANCE OF THE DATA-AUGMENTATION ALGORITHM TO THE CREATION OF MULTIPLY-IMPUTED DATA BASES

TW should also stimulate related work concerning the use of simulated complete data sets—multiply-imputed data sets—to draw inferences. I have been deeply involved in this topic in recent years [work now summarized in Rubin (1987)], although my efforts have been primarily focused on cases with only a few (e.g., $m = 2$–$10$) draws from the posterior predictive distribution of the missing values, $p(z \mid y)$ in TW's notation, where $z$ is the missing data and $y$ is the observed data. This focus has arisen from the desirability of supplementing an incomplete public-use data base with a few imputations for each missing value so that the resultant multiply-imputed public-use data base can be analyzed using standard complete-data methods applied to each of the $m$ data sets completed by imputation. Such complete-data analyses can be easily combined to form one posterior distribution that is conditional on the output of the $m$ complete-data analyses rather than on $y$ (Rubin 1987, chap. 3). Inferences based on this posterior distribution, which include an adjustment for small $m$, have been evaluated (Li 1985; Rubin 1987, chap. 4; Rubin and Schenker 1986), and they can be perfectly adequate even when $m$ is 2 or 3, provided that the fraction of missing information is modest, where the fraction of missing information about the parameter $\Theta$ is measured with normal posterior distributions, for example, by the eigenvalues of $\text{var}[E(\Theta \mid y, z) \mid y]$ relative to $\text{var}(\Theta \mid y)$.

Creating a multiply-imputed data set can be a major effort, however, depending on the model and pattern of missing data. Consequently, TW's work on the data augmentation algorithm initially appeared to be of great relevance to this work, even though the algorithm could be expensive to apply to large public-use data bases. After some reflection, however, I believe that in those situations where useful inferences can be drawn from a multiply-imputed data set with small $m$ (i.e., situations where the fraction of missing formation is modest), a few draws from $p(z \mid y)$ can often be easily created *noniteratively* using a method that may be of general interest to potential users of the data augmentation algorithm, especially considering that it can be applied even when the imputation step of the data augmentation algorithm is intractable.

## 2. A SAMPLING/IMPORTANCE RESAMPLING ALGORITHM FOR DRAWING A FEW IMPUTATIONS FROM THE POSTERIOR PREDICTIVE DISTRIBUTION OF THE MISSING VALUES

*Step 1*. Obtain a decent first-pass approximation to the joint posterior density of $(\Theta, z)$ say $h(\Theta, z \mid y) > 0$ for all possible $(\Theta, z)$, usually formulated as

$$h(\Theta, z \mid y) = h(\Theta \mid y)h(z \mid \Theta, y),$$

where $h(\Theta \mid y)$ is an approximate posterior density for $\Theta$, and $h(z \mid \Theta, y)$ is an approximate posterior predictive density for $z$ given $\Theta$.

*Step 2*. Draw $M$ values of $(\Theta, z)$ at random from $h(\Theta, z \mid y)$, where $M$ is large relative to $m$ = the final number of imputations desired for each missing value. Call these $(\Theta_j, z_j)$ $(j = 1, \ldots, M)$.

*Step 3*. Calculate the importance ratios for each $(\Theta_j, z_j)$,

$$r(\Theta_j, z_j \mid y) \propto p(y, z_j \mid \Theta_j)p(\Theta_j)/h(\Theta_j, z_j \mid y),$$

where the actual joint sampling density for $(y, z)$, $p(y, z \mid \Theta)$, is designed to be easy to evaluate (up to a multiplier depending only on $y$) by the construction of the missing data, $z$.

*Step 4*. Draw $m$ values of $z$ from the $z_j$ $(j = 1, \ldots, M)$ with probability proportional to $r_j = r(\Theta_j, z_j \mid y)$, thereby creating $m$ values of $z$, $z_l^*$ $(l = 1, \ldots, m)$; the associated values of $\Theta$, $\Theta_l^*$ $(l = 1, \ldots, m)$ are not needed for imputation. Methods for such drawing appear in the survey literature for pps sampling (e.g., Cochran 1977, chap. 9).

If $p(z \mid \Theta, y)$ is tractable in the sense that (a) the imputation step of the data augmentation algorithm is straightforward [i.e., if $h(z \mid \Theta, y) = p(z \mid \Theta, y)$] and (b) $p(z \mid \Theta, y)$ can be explicitly evaluated [i.e., if $p(y \mid \Theta) = p(y, z \mid \Theta)/p(z \mid \Theta, y)$, in addition to $p(y, z \mid \Theta)$, can be explicitly evaluated up to a multiplier depending only on $y$], the sampling/importance resampling (SIR) algorithm can be streamlined. In particular, in Step 2, only the $\Theta_j$ then need to be drawn, since the $r_j$ calculated in Step 3 and used in Step 4 do not depend on the $z_j$:

$$\begin{aligned} r(\Theta_j, z_j \mid y) & \\ &\propto p(y \mid \Theta_j)p(z_j \mid \Theta_j, y)p(\Theta_j)/[p(z_j \mid \Theta_j, y)h(\Theta_j \mid y)] \\ &\propto p(y \mid \Theta_j)p(\Theta_j)/h(\Theta_j \mid y). \end{aligned}$$

Having calculated the $r_j$ in Step 3 from the $\Theta_j$ $(j = 1, \ldots, M)$, then in Step 4, $m$ values of $\Theta_j$ are drawn with probability proportional to $r_j$, and for each $\Theta_l^*$, $z_l^*$ is drawn from $p(z \mid \Theta = \Theta_l^*, y)$, thereby creating the $m$ imputed values of $z$.

An important feature of the SIR algorithm is that only one set of $M$ (or $m$ in the streamlined version) imputations of $z$ is created. When a data base is large, the number of missing values can also be large, even with a small fraction of missing information, and then repeatedly passing through the database and creating sets of imputations for $z$ can be quite expensive. Furthermore, the process of drawing $M$ values of $(\Theta, z)$ (or $\Theta$ in the streamlined version) in Step 2 is designed to be relatively inexpensive by the choice of $h(\Theta, z \mid y)$ [or $h(\Theta \mid y)$ in the streamlined version].

The rationale for the SIR algorithm is based on the fact that as $M/m \to \infty$, the $m$ values $(z_l^*, \Theta_l^*)$ are drawn with probabilities given by

$$h(\Theta, z \mid y) \frac{r(\Theta, z \mid y)}{\iint h(\Theta, z \mid y) r(\Theta, z \mid y)\, d\Theta\, dz}$$

$$= \frac{p(y, z \mid \Theta) p(\Theta)}{\iint p(y, z \mid \Theta) p(\Theta)\, d\Theta\, dz}$$

$$= p(z, \Theta \mid y),$$

which implies that the $m$ imputations, $z_l^*$ ($l = 1, \ldots, m$), are independent draws from $p(z \mid y)$, as desired. This sampling/importance resampling technique for simulating a posterior distribution has been previously applied in small sample logistic regression problems (Rubin 1983).

The choice of an adequate ratio $M/m$ depends on the fraction of missing information, $\gamma$: smaller $\gamma$ implies satisfactory inferences from smaller $M/m$ for two reasons. First, $h(\Theta, z \mid y)$ should be a better approximation to $p(\Theta, z \mid y)$ with smaller $\gamma$, since $p(\Theta \mid y, z)$ is easy to find—a simple measure of the adequacy of the approximation is the variance of the log $r_j$. Second, with smaller $\gamma$, final inferences from a multiply-imputed data set will be proportionately more determined by $y$ than the imputed data, so there is less sensitivity to the imputed values. If $\gamma$ is large, very large ratios $M/m$ may be required for adequate performance of SIR, but in such cases a multiply-imputed data base with modest $m$ is of limited utility anyway. In common practical cases, I expect that $M/m = 20$ will often be more than adequate, especially considering that with modest $m$, accurately approximating the tails of $p(z \mid y)$ is of limited importance.

## 3. EXAMPLE: MISSING VALUES IN A NORMAL/CONDITIONALLY NORMAL BIVARIATE SAMPLE

The case of a bivariate normal sample with missing values on both variables is a classic example (e.g., Wilks 1932) of a missing-data problem without a general closed-form solution. This case, however, is easily handled by the EM, data-augmentation, and streamlined SIR algorithms. A slightly modified situation, which has no closed-form solution and cannot be directly handled by either the EM or data-augmentation algorithms, will be used to illustrate the general SIR algorithm.

Specifically, let $(w_1, w_2)$ be an iid sample from

$$w_1 \mid \Theta \sim N(\mu, \sigma^2), \tag{1}$$

$$w_2 \mid \Theta, w_1 \sim N(\alpha + \beta w_1 + \gamma w_1^2, \tau^2), \tag{2}$$

where

$$\Theta = (\mu, \log \sigma, \alpha, \beta, \gamma, \log \tau) \tag{3}$$

and

$$p(\Theta) \propto \text{constant}. \tag{4}$$

Suppose that a sample of $n$ units is taken, where $n_1$ units have only $w_1$ observed, $n_2$ units have only $w_2$ observed, and $n_{12}$ units have both $w_1$ and $w_2$ observed, where $N = n_1 + n_2 + n_{12}$, $n_{12} > n_1 + n_2 > n_2 > n_1$, and we assume that the missing data are missing at random (Rubin 1976). In TW's notation, $y$ consists of the $n_1$ observations of $w_1$, the $n_2$ observations of $w_2$, and the $n_{12}$ observations of $(w_1, w_2)$, and $z$ consists of the $n_1$ missing values of $w_2$ and the $n_2$ missing values of $w_1$. Notice that the fraction of missing information for each component of $\Theta$ is constructed to be modest, especially if $\tau^2$ is small relative to var$(w_2 \mid \Theta)$ or if $n_{12} \gg n_1 + n_2$.

The only specific issue in implementing the SIR algorithm in this example is Step 1: choosing a decent approximation $h(\Theta, z \mid y) = h(\Theta \mid y) h(z \mid \Theta, y)$. Note that the density $p(y, z \mid \Theta)$ is easily evaluated, since it is the product over the $n$ units of the two normal densities implied by (1) and (2), and from (3) and (4), this is the numerator of the importance ratio, $r(\Theta, z \mid y)$, used in Step 3.

Regarding $h(\Theta \mid y)$, it can be easily approximated by independent normal densities:

$$\log \sigma \mid y \sim N(\log s_1, [2(n_1 + n_{12} - 1)]^{-1}), \tag{5}$$

$$\mu \mid y \sim N(\overline{w}_1, s_1^2/(n_1 + n_{12})), \tag{6}$$

$$\log \tau \mid y \sim N(\log \hat{\tau}, [2(n_{12} - 3)]^{-1}, \tag{7}$$

and

$$(\alpha, \beta, \gamma) \mid y \sim N((\hat{\alpha}, \hat{\beta}, \hat{\gamma}), \hat{\tau}^2 C), \tag{8}$$

where $\overline{w}_1$ and $s_1^2$ are the mean and variance of the $(n_1 + n_{12})$ observations of $w_1$, and $(\hat{\alpha}, \hat{\beta}, \hat{\gamma}, \hat{\tau}^2, C)$ are the standard least squares summaries obtained by regressing $w_2$ on $(1, w_1, w_1^2)$ using the $n_{12}$ observations of $(w_1, w_2)$. Similarly, $h(z \mid \Theta, y)$ can be easily approximated by $n_1 + n_2$ independent normal densities:

$$n_1 \text{ missing } w_2 \mid \Theta, y \overset{\text{ind}}{\sim} N(\hat{\alpha} + \hat{\beta} w_1 + \hat{\gamma} w_1^2, \hat{\tau}^2), \tag{9}$$

and

$$n_2 \text{ missing } w_1 \mid \Theta, y \overset{\text{ind}}{\sim} N(a + b w_2 + c w_2^2, s_e^2), \tag{10}$$

where $(a, b, c, s_e^2)$ are the standard least squares summaries obtained by regressing $w_1$ on $(1, w_2, w_2^2)$ using the $n_{12}$ observations of $(w_1, w_2)$. Thus $h(\Theta, z \mid y)$ is the product of the $(n_1 + n_2 + 3)$ univariate normal densities specified by (5), (6), (7), (9), and (10), and the trivariate normal density specified by (8), and is therefore easy to draw from at Step 2, and easy to evaluate as the denominator of the importance ratios, $r(\Theta, z \mid y)$, in Step 3. Better approximations are available especially for small $n_{12}$, but it is not clear whether they are worth the effort in the context of SIR relative to increasing the ratio $M/m$.

## ADDITIONAL REFERENCES

Cochran, W. G. (1977), *Survey Techniques*, New York: John Wiley.
Rubin, D. B. (1976), "Inference and Missing Data," *Biometrika*, 63, 581–592.

——(1983), "Progress Report on Project for Multiple Imputation of 1980 Codes," University of Chicago.
——(1985), "Bayesianly Justifiable and Relevant Frequency Calculations for the Applied Statistician," *The Annals of Statistics*, 12, 1151–1172.
——(1987) *Multiple Imputation for Nonresponse in Surveys*, New York: John Wiley.
Rubin, D. B., and Schenker, N. (1986), "Multiple Imputation for Interval Estimation From Simple Random Samples With Ignorable Nonresponse," *Journal of the American Statistical Association*, 81, 366–374.
Wilks, S. S. (1932), "Moments and Distributions of Estimates of Population Parameters From Fragmentary Samples," *Annals of Mathematical Statistics*, 2, 163–195.

# Comment

SHELBY J. HABERMAN*

Tanner and Wong have provided a fascinating illustration of how serious problems of Bayesian inference can be addressed. Their approach is of interest not only to committed Bayesians but also to other statisticians who need to evaluate posterior distributions without unrealistically simple models or unrealistically simple prior distributions. Their iterative use of sampling is of particular interest.

The techniques used in this article raise a question addressed briefly by the authors. Computational requirements are quite considerable even for small problems, and the techniques used do involve judgment in their use. These problems may decrease in significance as computers improve and as investigators have more practice with the authors' approach; nonetheless, the really important question is the frequency with which the approach of Tanner and Wong will be used effectively by regular users of statistical procedures.

This computational issue is particularly troublesome given that statistical analyses of the examples are available that involve little calculation. To begin, consider the linkage example. In this case, the maximum likelihood estimate of $\theta$ is obtained by solution of a simple quadratic equation, and there is a straightforward normal approximation to the distribution of the estimate. In samples of adequate size for serious study of linkage, the normal approximation is quite adequate. A disciple of Bayes can obtain a polynomial expression for the posterior density of $\theta$ given the data under the uniform prior of the authors or even under more realistic mixtures of beta priors. Thus this example serves to illustrate feasibility of the method rather than an efficient way to treat the problem at hand.

Normally, procedures such as the EM algorithm are readily used to obtain maximum likelihood estimates in the covariance example. To be sure, the sample is too small for reliance on the normal approximation for the distribution of maximum likelihood estimates, but it is necessary to ask why anyone would try to apply a five-parameter model to so little data.

The last example can be analyzed by use of published algorithms that perform maximum likelihood estimation and obtain asymptotic standard errors of parameter estimates with little expenditure of computer time. In contrast, the analysis in this article required almost 3 hours on a VAX 750.

Thus one issue that must be considered is under what circumstances will a typical consumer of statistics be willing to make the additional investment in the authors' techniques given their current cost and given the possibility that alternative methods of analysis may be satisfactory enough.

A second issue to consider involves Bayesian inference. The use of prior distributions derived from the Dirichlet family by conditioning seems both inefficient and unrealistic. For example, it seems more realistic in the latent-class example to use a joint multivariate normal prior on the log-odds ratios $\log(P_{abcdx}/P_{11111})$, for $a$, $b$, $c$, $d$, and $x$ equal to 1 or 2. Such a prior is easily constructed given that these log-odds satisfy a linear model under the traditional latent-class model.

Despite these reservations, Tanner and Wong are to be congratulated for their powerful new approach to Bayesian inference. I await with interest their views on the likely future of their method.

* Shelby J. Haberman is Professor, Department of Statistics, Northwestern University, Evanston, IL 60201.


Journal of the American Statistical Association
June 1987, Vol. 82, No. 398, Theory and Methods

# Comment

A. O'HAGAN*

The idea of data augmentation is certainly intriguing. I have tried to consider in what applications it would be valuable. My problem in this regard is that the authors have not provided such applications themselves. When one sees a new numerical technique proposed, one expects to be shown examples where the new method requires less computation than the old ones. The opposite is true of the present article. In all of the examples there is a single parameter whose unnormalized posterior density may be written down directly. Simple numerical integration suffices to normalize it and obtain any summaries of interest. The authors have, therefore, not demonstrated any improvement over existing, well-tried techniques.

It is true that in the genetic linkage model a normal approximation to the posterior density is poor, but who would use a normal approximation for a parameter in [0, 1]? A beta approximation, having the same mode and curvative at the mode, is easily fitted. Trivial computation shows it to be excellent in all of the cases shown in Figures 1, 3, and 4. Indeed, on the scale used in those diagrams it would also "superimpose" the true posterior. For the record, the beta approximation in the case of Figure 4 is Beta(8.163, .872).

So when would data augmentation be valuable? Presumably only when existing methods require very substantial computation. Furthermore, there must be a data augmentation available that dramatically improves the problem. I suggest, therefore, that the authors might look at problems with many parameters, for here we can obtain high-dimensional posterior densities that require laborious numerical integration to obtain moments or marginal densities. Then we require a data augmentation to simplify the posterior density so that it becomes very much more tractable.

It is hard to think of applications having both of these features. However, the authors could examine some problems of the following kind. A traditional designed experiment, such as a factorial experiment, is performed but has missing data. With the full data set the likelihood factorizes and, supposing that the corresponding parameters are independent a priori, they will be independent a posteriori. Therefore, data augmentation makes marginal densities immediately available, and other summaries are easily computed. With missing data, orthogonality is lost and the likelihood no longer factorizes in the same way. If we had normal prior distributions, we could still invoke general linear model theory and obtain marginal densities, et cetera. Nonnormal (but independent) priors, however, would induce a massive numerical integration problem. This is still not a good example, because it requires rather special kinds of prior belief. On the other hand, one often has prior knowledge about the sign of an effect, which would produce truncated priors. Alternatively, it is my opinion that prior beliefs often demand thick-tailed prior distributions. So, if one is happy to accept normality in the likelihood, data augmentation may offer a tractable way of using such priors.

I hope that the authors will justify their approach better in the future by presenting convincing examples. Until then I will remain doubtful of its practical value. Opportunities for data augmentation are not obviously common; *good* opportunities, where data augmentation produces computational advantages, seem likely to be rare.

* A. O'Hagan is Lecturer, Department of Statistics, University of Warwick, Coventry CV4 7AL, United Kingdom.


Journal of the American Statistical Association
June 1987, Vol. 82, No. 398, Theory and Methods

# Rejoinder

MARTIN A. TANNER and WING HUNG WONG

## 1. REPLY TO DEMPSTER, MORRIS, AND RUBIN

### 1.1 Hierarchical Models

Dempster and Morris point out the usefulness of data augmentation in the analysis of hierarchical models, and we agree with them completely. We are especially grateful to Morris for his derivation of the explicit calculations needed for the I and P steps in the multivariate normal means example. A type of educational data, recently brought to our attention by Michael Meltzer of the University of Chicago, can serve to illustrate the technique, as well as some possible extensions. In this application, one has student performance data $y$, which is modeled given school indicators $X_1$ and other student-level covariates $X_2$; that is, the observational model is

$$y = X_1 z_1 + X_2 z_2 + \varepsilon_1.$$

In addition, there is a latent model for the school effects given school characteristics ($W$) such as whether it is a Catholic or a public school; that is,

$$z_1 = W\gamma + \varepsilon_2.$$

Here $y$ is the observed data, $z$ is the latent data and $\theta = (\gamma, \sigma_1, \sigma_2)$ if the errors in the two models are assumed to be Gaussian with standard deviations $\sigma_1$ and $\sigma_2$, respectively. Typically, an EM-type algorithm is used to obtain the maximum likelihood estimate of $\theta$ (e.g., Raudenbush and Bryk 1986). As remarked by Dempster, however, it would be desirable to perform full Bayesian analyses with genuine priors, and data augmentation offers a natural method to accomplish this. Under standard conjugate priors, the computations needed for the I and P steps can be derived in the same way as in the example in Morris's discussion. To see the effect of changing priors, one can attach weights (ratio between the true prior and the conjugate prior) to the sampled values of $\theta$.

Very often it is desirable and prudent to perform the analysis using heavy-tailed errors, especially for the latent model, for which diagnostics are hard to come by and a Gaussian error may lead to unwarranted pooling of truly outlying schools. Thus a natural elaboration of the above model is to use a $t$ distribution with a small degree of freedom for $\varepsilon_2$. The data augmentation scheme can also be extended to handle this situation. This is done by using the normal/gamma mixture representation of the $t$ distribution; that is, $\varepsilon_2$ is distributed as normal/$\sqrt{q}$, where $q$ is a mean square variable. In such a problem, there are three sets of quantities whose posterior distributions may be of interest, namely, $z$, $q$, and $\theta$. The posterior distribution of either one given the other two is easy to obtain. The data augmentation algorithm will now be extended to sample iteratively from each of these conditional posterior distributions in turn (compare Li 1985). We hope that this example can illustrate the freedom and flexibility offered by the data augmentation scheme (and to a much greater extent, the Bayesian outlook) for the application to, and study of, more complex and realistic models—the need for which being succinctly summarized by Dempster in his opening paragraph.

### 1.2 Importance Sampling

Both Dempster and Rubin comment on the use of importance sampling, and Rubin proposes a highly innovative method (SIR algorithm) for creating a few imputations when the fraction of missing information is modest, which also exploits the simplicity of inference based on the augmented data. We agree with their opinion that importance sampling is a powerful tool that should be employed whenever appropriate. We regard data augmentation iteration and importance sampling as complementary, however, rather than rival concepts. To clarify this issue, we compare the efficiency of the two procedures in the canonical problem of estimating a posterior moment of the parameters. We do, however, wish to point out that in so doing we have distorted Rubin's intention, since the SIR algorithm was initially designed for a more limited task under more limited conditions.

Recall that the basic principle of importance sampling is to generate values from a trial density, and then weight the values thus generated according to

$$\text{weight} \propto \frac{\text{true density}}{\text{trial density}}.$$

Let $g_*(\theta)$ be the true posterior density, $g_i(\theta)$ be the estimate for $g_*(\theta)$ at the $i$th iteration of the data augmentation algorithm, $h_*(\theta, z)$ be the true joint posterior density for $\theta$ and $z$, and $h(\theta, z)$ be the trial density in Rubin's SIR algorithm. To develop a quantitative comparison between the data augmentation algorithm and the SIR algorithm, let us consider the estimation of a specific posterior moment

$$\rho = \int a(\theta) g_*(\theta)\, d\theta.$$

For the data augmentation algorithm, we can use the estimator

$$\hat{\rho}_i = \frac{1}{m} \sum_{j=1}^{m} a(\theta_j),$$

where $\{\theta_1, \ldots, \theta_m\}$ is a sample from $g_i(\theta)$ at the $i$th iteration. For the SIR algorithm, we can use the estimator

$$\hat{\rho}_h = \frac{\sum r_j a(\theta_j)}{\sum r_j},$$


Journal of the American Statistical Association
June 1987, Vol. 82, No. 398, Theory and Methods

where $\{(\theta_1, z_1), \ldots, (\theta_m, z_m)\}$ is a sample from $h(\theta, z)$ and $\{r_1, \ldots, r_m\}$ are the weights calculated as in Step 3 of Rubin's discussion, his key observation being that the true joint posterior

$$h_*(\theta, z) \propto p(\theta)p(y, z \mid \theta)$$

is easy to compute (up to proportionality constant) because of the simplicity of the augmented data likelihood $p(y, z \mid \theta)$.

It is easy to obtain expressions for the means and variances:

$$E(\hat{\rho}_i) = \int a(\theta)g_i(\theta)\, d\theta,$$

$$\operatorname{var}(\hat{\rho}_i) = \frac{1}{m}\left[\int a^2 g_i\, d\theta - (E\hat{\rho}_i)^2\right],$$

$$E(\hat{\rho}_h) = \int a(\theta)g_*(\theta)\, d\theta,$$

$$\operatorname{var}(\hat{\rho}_h) = \frac{1}{m}\left[\int\int a^2(\theta)\frac{h_*(\theta, z)^2}{h(\theta, z)}\, dz\, d\theta - \rho^2\right].$$

We can now make the following comparisons: (a) $\hat{\rho}_h$ is unbiased and $\hat{\rho}_i$ is not—the bias of the latter, however, decreases with increasing $i$ because $g_i$ converges in $L_1$ to $g_*$; (b) although both variances decrease with increasing $m$, for any fixed $m$, the variance of $\hat{\rho}_h$ can be arbitrarily large because of its dependence on the ratio $h_*/h$.

Thus, if the trial density is not chosen well, the SIR estimate $\hat{\rho}_h$ can be very inaccurate. In practice, this difficulty is indicated if the weights come out to be highly skewed. Diagnostics for this phenomena are hence possible in the manner suggested by Rubin. But what should one do when the weights are highly skewed? It is here that we see a principal strength of the data augmentation algorithm, namely that each iteration offers an improvement over the previous estimate.

If the streamlined version of SIR is applicable, that is, when the likelihood $l(\theta \mid y)$ is easy to compute, then it is possible and may be desirable to use a procedure that combines the strength of both data augmentation iteration and importance sampling. Such a procedure would use data augmentation iteration to improve the initial approximation until the weights for importance sampling are satisfactorily distributed. Further effort will be necessary to formulate good strategies for such a combination and to explore its full potential.

## 2. REPLY TO HABERMAN

Haberman raises two related concerns regarding the potential acceptance of the data augmentation algorithm as a tool for routine data analysis. His first point relates to the computational intensiveness of the algorithm, and the second point questions whether the algorithm will be adopted "given the possibility that alternative methods of analysis may be satisfactory enough" (p. 546).

Regarding computational issues, it is noted that the algorithm is conceptually a very simple algorithm. One generates $m$ augmented data sets, and then based on these data sets one repeatedly evaluates the augmented posterior or likelihood. Thus the computational burden is not due to the implementation of a complex algorithm, but rather to the repeated evaluation of the same set of operations. As we have noted in our article, the amount of work performed at each iteration by the IP algorithm is determined by the number of imputations. Thus it is a *highly parallel algorithm* in the sense that the implementation of the algorithm can be tailored to make full use of a multiprocessor machine by the careful selection of $m$. In contrast, regarding both the EM algorithm and Newton–Raphson algorithm, the user is constrained by the problem of how much of the resources of the machine can be used per iteration. In this way, on a multiprocessor machine, the IP algorithm may actually require a shorter execution time than either the EM or Newton–Raphson algorithms.

Haberman questions whether the data augmentation algorithm will be adopted for a particular problem given that "alternative methods of analysis may be satisfactory enough." Our opinion is that one does not know whether the normal approximation, for example, is *good enough* unless one is able to examine the likelihood (or posterior). Thus the data augmentation algorithm may be useful as a check on the validity of the normal approximation, as in the covariance example. Moreover, we do not believe that the plausibility of such departures from normality, as noted in the covariance example, can be ignored. Although fitting a five-parameter bivariate model to 12 observations may be unwise, what can one say regarding the analysis of 40 seven-dimensional observations? It is also important to note that an examination of the likelihood (or posterior) can sometimes allow one to check the *adequacy* of a given model. This is again exemplified in the covariance example. Certainly, under the bivariate normal model, we

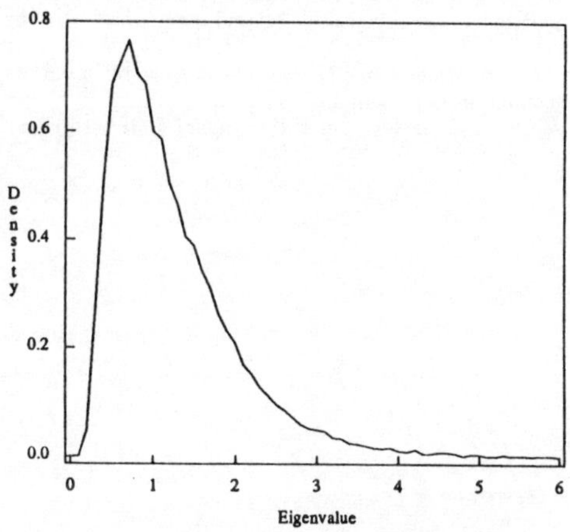


*Figure 1. Posterior Distribution of the Smallest Eigenvalue—Bivariate Case.*

would not expect 100 observations to yield the posterior encountered in our 12-observation data set. If such a posterior were to result from a large sample, however, we would be forced to question the validity of the bivariate normal assumption. In other words, even if we have a "large sample," we may find evidence against a model by noting some peculiarity in the likelihood. As such, by ignoring the likelihood (or posterior) and focusing on a point estimate, one may miss some feature of the data that relates to the adequacy of the model.

Regarding our analysis of the traditional latent-class model, Haberman seems to interpret our use of the Dirichlet distribution as a specification of the prior distribution for the conditional probabilities. We use the Dirichlet distribution as a vehicle for generating the probabilities, not to specify a prior. We can, however, incorporate a prior distribution by assigning weights to the probabilities.

Haberman queries our views on the future of the *method*. It is our opinion that the data augmentation algorithm is not only a useful method for calculating likelihoods (or posteriors) but it also provides a *paradigm* for handling missing data problems. For example in the context of grouped and censored data, we have used this paradigm to develop an algorithm for the nonparametric estimation of the hazard function (Tanner and Wong 1987).

## 3. REPLY TO O'HAGAN

We do not agree with the basic premise of O'Hagan that *the* measure of the value of a new methodological contribution is whether it minimizes numerical computation. We feel that data augmentation is valuable, not only as a numerical tool for computing posteriors, but as a way of thinking about certain types of problems (e.g., Morris's hierarchical model), as well as providing a diagnostic for both the normal approximation and the model. To focus, as does O'Hagan, on data augmentation as *only* a computational technique misses several major themes of the article.

It is unfortunate that O'Hagan misses several important features in our examples and as a consequence of this reaches the conclusion that the examples are unconvincing. He states: "In all of the examples there is a single parameter whose unnormalized posterior density may be written down directly. Simple numerical integration suffices to normalize it and obtain any summaries of interest" (p. 547). This is incorrect. In both the covariance and latent-class examples there are more than one parameter. In fact, the conditional independence model for the latent-class data has 12 parameters. We will be very surprised if in that example anyone can "write down" the unnormalized posterior density for an arbitrary parameter, for example, the quantity $\pi_{11}^{AX}$, upon which we focus. The point that we wish to make is that the data augmentation algorithm offers an approach of treating the nuisance parameters. This is because the algorithm enables one to readily obtain the marginal posterior distribution of any parameter or the joint posterior distribution of any two parameters. To illustrate, suppose that in the covariance example we are interested in the posterior distribution of the smallest eigenvalue. We can obtain this posterior distribution by finding the smallest eigenvalue for each matrix in the sample of covariance matrices drawn from the final estimate of $p(\Sigma \mid y)$. The histogram of these values yields the desired posterior (see Fig. 1 here).

O'Hagan suggests that we "look at problems with many parameters" (p. 547), as well as factorial experiments with missing data. In this regard, he seems to have missed the latent-class example. The associated log-linear model has 12 parameters and the associated $3 \times 2^4$ table, clearly, has a factorial design. O'Hagan is critical of the use of the normal approximation in the linkage example. We feel, however, that based on our analysis of the original data, the normal approximation is quite acceptable. Clearly, the adequacy of the normal approximation depends on the *nature of the data* and it is the use of data augmentation that can assist the statistician in assessing the adequacy of the approximation *for the particular data set*.

## ADDITIONAL REFERENCES

Raudenbush, S., and Bryk, A. S. (1986), "A Hierarchical Model for Studying School Effects," *Sociology of Education*, 59, 1–17.

Tanner, M. A., and Wong, W. H. (1987), "An Application of Imputation to an Estimation Problem in Grouped Lifetime Analysis," *Technometrics*, 29, 23–32.

# [11]

# Sampling-Based Approaches to Calculating Marginal Densities

ALAN E. GELFAND AND ADRIAN F. M. SMITH*

Stochastic substitution, the Gibbs sampler, and the sampling–importance–resampling algorithm can be viewed as three alternative sampling- (or Monte Carlo- ) based approaches to the calculation of numerical estimates of marginal probability distributions. The three approaches will be reviewed, compared, and contrasted in relation to various joint probability structures frequently encountered in applications. In particular, the relevance of the approaches to calculating Bayesian posterior densities for a variety of structured models will be discussed and illustrated.

KEY WORDS: Conditional probability structure; Data augmentation; Gibbs sampler; Hierarchical models; Importance sampling; Missing data; Monte Carlo sampling; Posterior distributions; Stochastic substitution; Variance components.

## 1. INTRODUCTION

In relation to a collection of random variables, $U_1, U_2, \ldots, U_k$, suppose that either (a) for $i = 1, \ldots, k$, the conditional distributions $U_i \mid U_j\ (j \neq i)$ are available, perhaps having for some $i$ reduced forms $U_i \mid U_j\ (j \in S_i \subset \{1, \ldots, k\})$, or (b) the functional form of the joint density of $U_1, U_2, \ldots, U_k$ is known, perhaps modulo the normalizing constant, and at least one $U_i \mid U_j\ (j \neq i)$ is available, where *available* means that samples of $U_i$ can be straightforwardly and efficiently generated, given specified values of the appropriate conditioning variables.

The problem addressed in this article is the exploitation of the kind of structural information given by either (a) or (b), to obtain numerical estimates of nonanalytically available marginal densities of some or all of the $U_i$ (when possible) simply by means of simulated samples from available conditional distributions, and without recourse to sophisticated numerical analytic methods. We do not claim that the sampling methods to be described are necessarily computationally efficient compared with expert use of the latter. Instead, the attraction of the sampling-based methods is their conceptual simplicity and ease of implementation for users with available computing resources but without numerical analytic expertise. All that the user requires is insight into the relevant conditional probability structure and techniques for the efficient generation of appropriate random variates (e.g., as described by Devroye 1986 and Ripley 1987).

In Section 2, we discuss and extend three alternative approaches put forward in the literature for calculating marginal densities via sampling algorithms. These are (variants of) the data-augmentation algorithm described by Tanner and Wong (1987), the Gibbs sampler algorithm introduced by Geman and Geman (1984), and the form of importance-sampling algorithm proposed by Rubin (1987, 1988). We note that the Gibbs sampler has been widely taken up in the image-processing literature and in other large-scale models such as neural networks and expert systems, but that its general potential for more conventional statistical problems seems to have been overlooked. As we show, there is a close relationship between the Gibbs sampler and the substitution or data-augmentation algorithm proposed by Tanner and Wong (1987). We generalize the latter and show that it is as least as efficient as the Gibbs sampler, and potentially more efficient, given the availability of distinct conditional distributions in addition to those in (a). We note that as a consequence of the relationship between the two algorithms, the convergence results established by Geman and Geman (1984) are applicable to the generalized substitution algorithm. The stronger convergence results established by Tanner and Wong (1987) require the availability of a particular set of conditional distributions, including those in (a). Both the substitution and Gibbs sampler algorithms are iterative Monte Carlo procedures, applicable when the kind of structural information given by (a) is available. When the structural information is of the kind described by (b), we see that an importance-sampling algorithm based on that of Rubin (1987, 1988) provides a noniterative Monte Carlo integration approach to calculating marginal densities.

In Section 3, we illustrate various model structures occurring frequently in applications, where one or more of these three approaches offers an easily implemented solution. In particular, we consider the calculation of Bayesian posterior distributions in incomplete-data problems, conjugate hierarchical models, and normal data models. In Section 4, we briefly summarize the results of some preliminary computational experience in two simple cases. (Detailed applications to complex, real-data problems will be presented in a subsequent paper.) Finally, in Section 5 we provide a summary discussion.

* Alan E. Gelfand is Professor, Department of Statistics, University of Connecticut, Storrs, CT 06269-3120. Adrian F. M. Smith is Professor, Department of Mathematics, Imperial College of Science, Technology, and Medicine, University of London, London SW7 2AZ, England. This work was carried out under the auspices of the U.K. Science and Engineering Research Council Complex Stochastic Systems Initiative, which provided travel support to the first author and supported Susan Hills, to whom the authors are indebted for carrying out extensive computer experimentation. The second author's collaboration with Amy Racine provided the initial stimulation for this work, and the authors have benefited from access to related independent work by David Clayton.


Journal of the American Statistical Association
June 1990, Vol. 85, No. 410, Theory and Methods

## 2. SAMPLING APPROACHES

In the sequel, we assume that we are dealing with real, possibly vector-valued random variables having a joint distribution whose density function is strictly positive over the (product) sample space. This ensures that knowledge of all full conditional specifications [such as in (a) of Sec. 1] uniquely defines the full joint density (e.g., see Besag 1974). Throughout, we assume the existence of densities with respect to either Lebesgue or counting measure, as appropriate, for all marginal and conditional distributions. The terms *distribution* and *density* are therefore used interchangeably.

Densities are denoted generically by brackets, so joint, conditional, and marginal forms, for example, appear as $[X, Y]$, $[X \mid Y]$, and $[X]$. Multiplication of densities is denoted by $*$; for example, $[X, Y] = [X \mid Y] * [Y]$. The process of marginalization (i.e., integration) is denoted by forms such as $[X \mid Y] = \int [X \mid Y, Z, W] * [Z \mid W, Y] * [W \mid Y]$, with the convention that all variables appearing in the integrand but not in the resulting density have been integrated out. Thus the integration is with respect to $Z$ and $W$. More generally, we use notation such as $\int h(Z, W) * [W]$ to denote, for given $Z$, the expectation of the function $h(Z, W)$ with respect to the marginal distribution for $W$.

### 2.1 Substitution or Data-Augmentation Algorithm

The substitution algorithm for finding fixed-point solutions to certain classes of integral equations is a standard mathematical tool that has received considerable attention in the literature (e.g., see Rall 1969). Its potential utility in statistical problems of the kind we are concerned with was observed by Tanner and Wong (1987) (who called it a data-augmentation algorithm) and the associated discussion. Briefly reviewing the essence of their development using the notation introduced previously, we have

$$[X] = \int [X \mid Y] * [Y] \tag{1}$$

and

$$[Y] = \int [Y \mid X] * [X], \tag{2}$$

so substituting (2) into (1) gives

$$[X] = \int [X \mid Y] * \int [Y \mid X'] * [X']$$

$$= \int h(X, X') * [X'], \tag{3}$$

where $h(X, X') = \int [X \mid Y] * [Y \mid X']$, with $X'$ appearing as a dummy argument in (3), and of course $[X] \equiv [X']$. Now, suppose that on the right side of (3), $[X']$ were replaced by $[X]_i$, to be thought of as an estimate of $[X] \equiv [X']$ arising at the $i$th stage of an iterative process. Then, (3) implies that for some $[X]_{i+1}$, $[X]_{i+1} = \int h(X, X') * [X']_i = I_h[X]_i$, in a notation making explicit that $I_h$ is the integral operator associated with $h$. Exploiting standard theory of such integral operators, Tanner and Wong (1987) showed that under mild regularity conditions this iterative process has the following properties (with obviously analogous results for $[Y]$).

*TW1* (*uniqueness*). The true marginal density, $[X]$, is the unique solution to (3).

*TW2* (*convergence*). For almost any $[X]_0$, the sequence $[X]_1, [X]_2, \ldots$ defined by $[X]_{i+1} = I_h[X]_i$ ($i = 0, 1, \ldots$) converges monotonically in $L_1$ to $[X]$.

*TW3* (*rate*). $\int \|[X]_i - [X]\| \to 0$ geometrically in $i$.

Extending the substitution algorithm to three random variables $X$, $Y$, and $Z$, we may write [analogous to (1) and (2)]

$$[X] = \int [X, Z \mid Y] * [Y], \tag{4}$$

$$[Y] = \int [Y, X \mid Z] * [Z], \tag{5}$$

and

$$[Z] = \int [Z, Y \mid X] * [X]. \tag{6}$$

Substitution of (6) into (5) and then (5) into (4) produces a fixed-point equation analogous to (3). A new $h$ function arises with associated integral operator $I_h$, and hence TW1, TW2, and TW3 continue to hold in this extended setting. Extension to $k$ variables is straightforward. A noteworthy by-product, using TW1, is a simple proof that under weak conditions specification of the conditional distributions $[U_{r, r \neq s} \mid U_s]$ $(s = 1, 2, \ldots, k)$ uniquely determines the joint density.

### 2.2 Substitution Sampling

Returning to (1) and (2), suppose that $[X \mid Y]$ and $[Y \mid X]$ are available in the sense defined at the beginning of Section 1. For an arbitrary (possibly degenerate) initial density $[X]_0$ draw a single $X^{(0)}$ from $[X]_0$. Given $X^{(0)}$, since $[Y \mid X]$ is available draw $Y^{(1)} \sim [Y \mid X^{(0)}]$, and hence from (2) the marginal distribution of $Y^{(1)}$ is $[Y]_1 = \int [Y \mid X] * [X]_0$. Now, complete a cycle by drawing $X^{(1)} \sim [X \mid Y^{(1)}]$. Using (1), we then have $X^{(1)} \sim [X]_1 = \int [X \mid Y] * [Y]_1 = \int h(X, X') * [X']_0 = I_h[X]_0$. Repetition of this cycle produces $Y^{(2)}$ and $X^{(2)}$, and eventually, after $i$ iterations, the pair $(X^{(i)}, Y^{(i)})$ such that $X^{(i)} \xrightarrow{d} X \sim [X]$, and $Y^{(i)} \xrightarrow{d} Y \sim [Y]$, by virtue of TW2. Repetition of this sequence $m$ times each to the $i$th iteration generates $m$ iid pairs $(X_j^{(i)}, Y_j^{(i)})$ $(j = 1, \ldots, m)$. We call this generation scheme *substitution sampling*. Note that though we have independence across $j$, we have dependence within a given $j$.

If we terminate all repetitions at the $i$th iteration, the proposed density estimate of $[X]$ (with an analogous expression for $[Y]$) is the Monte Carlo integration

$$[\hat{X}]_i = \frac{1}{m}\sum_{j=1}^{m}[X \mid Y_j^{(i)}]. \tag{7}$$

Note that the $X_j^{(i)}$ are not used in (7) (see Sec. 2.6).

We note that this version of the substitution-sampling algorithm differs slightly from the imputation–posterior algorithm of Tanner and Wong (1987). At each iteration $l$ ($l = 1, 2, \ldots, i$), they proposed creation of the mixture density estimate, $[\hat{X}]_l$, of the form in (7), with subsequent sampling from $[\hat{X}]_l$ to begin the next iteration. This mechanism introduces the additional randomness of equally likely selection from the $Y_j^{(l)}$ before obtaining an $X^{(l)}$. We suspect this sampling with replacement of the $Y^{(l)}$ was introduced to allow $m$ to vary across iterations, which may be useful in reducing computational effort.

The $L_1$ convergence of $[\hat{X}]_i$ to $[X]$ is most easily studied by writing $\int \|[\hat{X}]_i - [X]\| \leq \int \|[\hat{X}]_i - [X]_i\| + \int \|[X]_i - [X]\|$. The second term on the right side can be made arbitrarily small as $i \to \infty$, as a consequence of TW2. The first term on the right can be made arbitrarily small as $m \to \infty$, since $[\hat{X}]_i \xrightarrow{P} [X]_i$ for almost all $X$ (Glick 1974).

Extension of the substitution-sampling algorithm to more than two random variables is straightforward. We illustrate using the three-variable case, assuming the three conditional distributions in (4)–(6) are available. Taking an arbitrary starting marginal density for $X$, say $[X]_0$, we draw $X^{(0)} \sim [X]_0$, $(Z^{(0)\prime}, Y^{(0)\prime}) \sim [Z, Y \mid X^{(0)}]$, $(Y^{(1)}, X^{(0)\prime}) \sim [Y, X \mid Z^{(0)\prime}]$, and finally $(X^{(1)}, Z^{(1)}) \sim [X, Z \mid Y^{(1)}]$. A full cycle of the algorithm (i.e., to generate $X^{(1)}$ starting from $X^{(0)}$) thus requires six generated variates, rather than the two we saw earlier. Repeating such a cycle $i$ times produces $(X^{(i)}, Y^{(i)}, Z^{(i)})$. The aforementioned theory ensures that $X^{(i)} \xrightarrow{d} X \sim [X]$, $Y^{(i)} \xrightarrow{d} Y \sim [Y]$, and $Z^{(i)} \xrightarrow{d} Z \sim [Z]$. If we repeat the entire process $m$ times we obtain iid $(X_j^{(i)}, Y_j^{(i)}, Z_j^{(i)})$ $(j = 1, \ldots, m)$ (independent between, but not within, $j$'s). Note that implementation of the substitution-sampling algorithm does not require specification of the full joint distribution. Rather, what is needed is the availability of $[X, Z \mid Y]$, $[Y, X \mid Z]$, and $[Z, Y \mid Z]$. Of course, in many cases sampling from, say, $[X, Z \mid Y]$ requires, for example, $[X \mid Y, Z]$ and $[Y \mid Z]$, that is, the availability of a full conditional and a reduced conditional distribution. Paralleling (7), the density estimator of $[X]$ becomes

$$[\hat{X}]_i = \frac{1}{m}\sum_{j=1}^{m}[X \mid Y_j^{(i)}, Z_j^{(i)}], \tag{8}$$

with analogous expressions for estimating $[Y]$ and $[Z]$. $L_1$ convergence of (8) to $[X]$ again follows.

For $k$ variables, $U_1, \ldots, U_k$, the substitution-sampling algorithm requires $k(k - 1)$ random variate generations to complete a cycle. If we run $m$ sequences out to the $i$th iteration [$mik(k - 1)$ random generations] we obtain $m$ iid $k$ tuples $(U_{1j}^{(i)}, \ldots, U_{kj}^{(i)})$ $(j = 1, \ldots, m)$, with the density estimator for $[U_s]$ $(s = 1, \ldots, k)$ being

$$[\hat{U}_s]_i = \frac{1}{m}\sum_{j=1}^{m}[U_s \mid U_t = U_{tj}^{(i)};\, t \neq s]. \tag{9}$$

## 2.3 Gibbs Sampling

Suppose that we write (4)–(6) in the form

$$[X] = \int [X \mid Z, Y] * [Z \mid Y] * [Y]$$

$$[Y] = \int [Y \mid X, Z] * [X \mid Z] * [Z]$$

$$[Z] = \int [Z \mid Y, X] * [Y \mid X] * [X]. \tag{10}$$

Implementation of substitution sampling requires the availability of all six conditional distributions on the right side of (10), rarely the case in our applications. As noted at the beginning of Section 2, the full conditional distributions alone, $[X \mid Y, Z]$, $[Y \mid Z, X]$, and $[Z \mid X, Y]$, uniquely determine the joint distribution (and hence the marginal distributions) in the situations under study. An algorithm for extracting the marginal distributions from these full conditional distributions was formally introduced by Geman and Geman (1984) and is known as the Gibbs sampler. An earlier article by Hastings (1970) developed essentially the same idea and suggested its potential for numerical problems arising in statistics.

The Gibbs sampler was developed and has mainly been applied in the context of complex stochastic models involving very large numbers of variables, such as image reconstruction, neural networks, and expert systems. In these cases, direct specification of a joint distribution is typically not feasible. Instead, the set of full conditionals is specified, usually by assuming that an individual full conditional distribution only depends on some "neighborhood" subset of the variables [a reduced form, in the terminology of (a) in Sec. 1]. More precisely, for the set of variables $U_1, U_2, \ldots, U_k$,

$$[U_i \mid U_j;\, \neq i] \equiv [U_i \mid U_j;\, j \in S_i], \qquad i = 1, \ldots, k, \tag{11}$$

where $S_i$ is a small neighborhood subset of $\{1, 2, \ldots, k\}$. A crucial question is under what circumstances the specification (11) uniquely determines the joint distribution. The answer is taken up in great detail by Geman and Geman (1984), involving concepts such as graphs, neighborhood systems, cliques, Markov random fields, and Gibbs distributions. In all of the examples we consider, the joint distribution is uniquely defined. Our $k$'s will be small to moderate, and the available set of full conditional distributions will, in fact, be calculated from specification of the joint density.

Gibbs sampling is a Markovian updating scheme that proceeds as follows. Given an arbitrary starting set of values $U_1^{(0)}, U_2^{(0)}, \ldots, U_k^{(0)}$, we draw $U_1^{(1)} \sim [U_1 \mid U_2^{(0)}, \ldots, U_k^{(0)}]$, $U_2^{(1)} \sim [U_2 \mid U_1^{(1)}, U_3^{(0)}, \ldots, U_k^{(0)}]$, $U_3^{(1)} \sim [U_3 \mid U_1^{(1)} U_2^{(1)}, U_4^{(0)}, \ldots, U_k^{(0)}]$, and so on, up to $U_k^{(1)} \sim [U_k \mid U_1^{(1)}, \ldots, U_{k-1}^{(1)}]$. Thus each variable is visited in the natural order and a cycle in this scheme requires $K$ random variate generations. After $i$ such iterations we would arrive at $(U_1^{(i)}, \ldots, U_k^{(i)})$. Under mild conditions, Geman and Geman showed that the following results hold.

*GG1* (*convergence*). $(U_1^{(i)}, \ldots, U_k^{(i)}) \xrightarrow{d} [U_1, \ldots, U_k]$ and hence for each $s$, $U_s^{(i)} \xrightarrow{d} U_s \sim [U_s]$ as $i \to \infty$. In fact, a slightly stronger result is proven. Rather than requiring that each variable be visited in repetitions of the natural order, convergence still follows under any visiting scheme, provided that each variable is visited infinitely often (io).

*GG2* (*rate*). Using the sup norm, rather than the $L_1$ norm, the joint density of $(U_1^{(i)}, \ldots, U_k^{(i)})$ converges to the true joint density at a geometric rate in $i$, under visiting in the natural order. A minor adjustment to the rate is required for an arbitrary io visiting scheme.

*GGS* (*ergodic theorem*). For any measurable function $T$ of $U_1, \ldots, U_k$ whose expectation exists,

$$\lim_{i \to \infty} \frac{1}{i} \sum_{l=1}^{i} T(U_1^{(l)}, \ldots, U_k^{(l)}) \xrightarrow{\text{a.s.}} E(T(U_1, \ldots, U_k)).$$

As in Section 2.3, Gibbs sampling through $m$ replications of the aforementioned $i$ iterations ($mik$ random variate generations) produces $m$ iid $k$ tuples $(U_{1j}^{(i)}, \ldots, U_{kj}^{(i)})$ $(j = 1, \ldots, m)$, with the proposed density estimate for $[U_s]$ having form (9).

## 2.4 Relationship Between Gibbs Sampling and Substitution Sampling

It is apparent that in the case of two random variables Gibbs sampling and substitution sampling are identical. For more than two variables, using (10) and its obvious generalization to $k$ variables, we see that Gibbs sampling assumes the availability of the set of $k$ full conditional distributions (the minimal set needed to determine the joint density uniquely). The substitution-sampling algorithm requires the availability of $k(k - 1)$ conditional distributions, including all of the full conditionals.

Gibbs sampling is known to converge slowly in applications with $k$ very large. Regardless, fair comparison with substitution sampling, in the sense of the total amount of random variate generation, requires that we allow the Gibbs sampling algorithm $i(k - 1)$ iterations if the substitution-sampling algorithm is allowed $i$. Even so, there is clearly scope for accelerated convergence from the substitution-sampling algorithm, since it samples from the correct distribution each time, whereas Gibbs sampling only samples from the full conditional distributions. To amplify, we describe how the substitution-sampling algorithm might be carried out under availability of just the set of full conditional distributions. We see that it can be viewed as the Gibbs sampler, but under an io visiting scheme different from the natural one. We present the argument in the three-variable case for simplicity. Returning to (10), if $[Y \mid X]$ is unavailable we can create a sub-substitution loop to obtain it by means of

$$[Y \mid X] = \int [Y \mid X, Z] * [Z \mid X]$$

$$[Z \mid X] = \int [Z \mid X, Y] * [Y \mid X]. \tag{12}$$

Similar subloops are clearly available to create $[X \mid Z]$ and $[Z \mid Y]$. In fact, for $k$ variables this idea can be straightforwardly extended to the estimation of an arbitrary reduced conditional distribution, given the full conditionals. We omit the details.

The previous analysis suggests that we could view the reduced conditional densities such as $[Y \mid X]$ as available, and that we could thus carry out the substitution algorithm as if all needed conditional distributions were available; however, $[Y \mid X]$ is not available in our earlier sense. Under the subloop in (12), we can always obtain a density estimate for $[Y \mid X]$, given any specified $X$, say $X^{(0)}$. At the next cycle of the iteration, however, we would need a brand-new density estimate for $[Y \mid X]$ at $X = X^{(1)}$. Nonetheless, suppose we persevered in this manner, making our way through one cycle of (10). The reader may verify that the only distributions actually sampled from are, of course, the available full conditionals, that at the end of the cycle each full conditional will have been sampled from at least once, and thus that under repeated iterations each variable will be visited io. Therefore, this version of the substitution-sampling algorithm is merely Gibbs sampling with a different but still io visiting order. As a result, GG1, GG2, and GG3 still hold (TW1, TW2, and TW3 apply directly only when all required conditional distributions are available). Moreover, there is no gain in implementing the Gibbs sampler in this complicated order; the natural order is simpler and equally good.

This discussion may be readily extended to the case of $k$ variables. As a result, we conclude that when only the set of $k$ full conditionals is available the substitution-sampling algorithm and the Gibbs sampler are equivalent. Furthermore, we can now see when substitution sampling offers the possibility of acceleration relative to Gibbs sampling. This occurs when some reduced conditional distributions, distinct from the full conditional distributions, are available. Suppose that we write the substitution algorithm with appropriate conditioning to capture these available reduced conditionals. As we traverse a cycle, we would sample from these distributions as we come to them, otherwise sampling from the full conditional distributions.

An example will help clarify this idea. One way to carry out the Gibbs sampler in (10) is to follow the substitution order rather than the natural order. That is, given an initial $X^{(0)}$, $Y^{(0)}$, and $Z^{(0)}$, we start at the bottom line of (10), for example, drawing (a) $Y^{(0)\prime}$ from $[Y \mid X^{(0)}, Z^{(0)}]$, (b) $Z^{(0)\prime}$ from $[Z \mid Y^{(0)\prime}, X^{(0)}]$, (c) $X^{(0)\prime}$ from $[X \mid Z^{(0)\prime}, Y^{(0)\prime}]$, (d) $Y^{(1)}$ from $[Y \mid X^{(0)\prime}, Z^{(0)\prime}]$, (e) $Z^{(1)}$ from $[Z \mid Y^{(1)}, X^{(0)\prime}]$, and (f) $X^{(1)}$ from $[Z \mid Y^{(1)}, Z^{(1)}]$. Thus, in this case, one cycle using the substitution order corresponds to two cycles using the natural order. Suppose, however, that in addition to the full conditional distributions, $[Z \mid Y]$, say, is available and distinct from $[Z \mid X, Y]$. Following the substitution order, at step (e) we would instead draw $Z^{(1)}$ from the correct distribution, $[Z \mid Y^{(1)}]$.

In Section 3, we provide classes of examples where distinct reduced conditional distributions are available and classes where they generally are not. Our computational experience shows that the acceleration in convergence that arises from having available distributions in addition to the full conditionals is inconsequential (see Sec. 4).

## 2.5 The Rubin Importance-Sampling Algorithm

Rubin (1987) suggested a noniterative Monte Carlo method for generating marginal distributions using importance-sampling ideas. We first present the basic idea in the two-variable case. Suppose that we seek the marginal distribution of $X$, given only the functional form (modulo the normalizing constant) of the joint density $[X, Y]$ and the availability of the conditional distribution $[X \mid Y]$ [a special case of the conditions described in (b) of Sec. 1].

Suppose further (as is typically the case in applications) that the marginal distribution of $Y$ is not known. Choose an importance-sampling distribution for $Y$ that has positive support wherever $[Y]$ does and that has density $[Y]_s$, say. Then, $[X \mid Y] * [Y]_s$ provides an importance-sampling distribution for $(X, Y)$. Suppose that we draw iid pairs $(X_l, Y_l)$ $(l = 1, \ldots, N)$ from this joint distribution, for example, by drawing $Y_l$ from $[Y]_s$ and $X_l$ from $[X \mid Y_l]$. Rubin's idea is to calculate $r_l = [X_l, Y_l]/[X_l \mid Y_l] * [Y_l]_s$ $(l = 1, \ldots, N)$ and then estimate the marginal density for $[X]$ by

$$[\hat{X}] = \sum_{l=1}^{N} [X \mid Y_l] r_l \Big/ \sum_{l=1}^{N} r_l. \tag{13}$$

Note the important fact that $[X, Y]$ need only be specified up to a constant, since the latter cancels in (13). In other words, we do not need to evaluate the normalizing constant for $[X, Y]$. This feature is exploited in the examples of Section 3. By dividing the numerator and denominator of (13) by $N$ and using the law of large numbers, we immediately have the following.

*R1 (convergence).* $[\hat{X}] \rightarrow [X]$ with probability 1 as $N \rightarrow \infty$ for almost every $X$.

In addition, if $[Y \mid X]$ is available we immediately have an estimate for the marginal distribution of $Y$: $[\hat{Y}] = \sum_{l=1}^{N} [Y \mid X_l] r_l / \sum_{l=1}^{N} r_l$.

The successful performance of (13) typically depends strongly on the choice of $[Y]_s$ and its closeness to $[Y]$. Thus the suggestion of Tanner and Wong (1987) in their rejoinder to Rubin, to perhaps use for $[Y]_s$ the density estimate created after $i$ iterations of the substitution algorithm, merits further investigation. In fact, the whole problem of general strategies for synthesizing both the iterative and noniterative approaches under a fixed-budget (total number of random generations) criterion needs considerable further study.

The extension of the Rubin importance-sampling idea to the case of $k$ variables is clear. For instance, when $k = 3$, suppose that we seek the marginal distribution of $X$, given the functional form of $[X, Y, Z]$ up to a constant and the availability of the full conditional $[X \mid Y, Z]$. In this case, the pair $(Y, Z)$ plays the role of $Y$ in the two-variable case discussed before, and in general we need to specify an importance-sampling distribution $[Y, Z]_s$. Nevertheless, if $[Y \mid Z]$ is available, for example, we only need to specify $[Z]_s$. In any case, we draw iid triples $(X_l, Y_l, Z_l)$ $(l = 1, \ldots, N)$ and calculate $r_l = [X_l, Y_l, Z_l]/([X_l \mid Y_l, Z_l] * [Y_l, Z_l]_s)$. The marginal density estimate for $[X]$ then becomes [analogous to (13)]

$$[\hat{X}] = \sum_{l=1}^{N} [X \mid Y_l, Z_l] r_l \Big/ \sum_{l=1}^{N} r_l. \tag{14}$$

We note that in the $k$-variable case the Rubin importance-sampling algorithm requires $Nk$ random variate generations, whereas Gibbs sampling stopped at iteration $i$ requires $mik$ generations. For fair comparison of the two algorithms, we should therefore set $N = mi$. The relationship between the estimators (7) and (13) may be clarified if we resample $Y_1^*, Y_2^*, \ldots, Y_m^*$ from the distribution that places mass $r_l/\Sigma r_l$ at $Y_l$ $(l = 1, \ldots, N)$. We could then replace (13) with

$$[\hat{X}] = \frac{1}{m} \sum_{j=1}^{m} [X \mid Y_j^*], \tag{15}$$

so (7) and (15) are of the same form. Relative performance on average depends on whether the distribution of $Y^{(i)}$ or $Y^*$ is closer to $[Y]$. Empirical work described in Section 4 suggests that under fair comparison (7) performs better than (14) or (15). It seems preferable to iterate through a learning process with small samples rather than to draw a one-off large sample at the beginning [an idea that underlies much modern work in adaptive Monte Carlo; e.g., see Smith, Skene, Shaw, and Naylor (1987)].

## 2.6 Density Estimation

In this section, we consider the problem of calculating a final form of marginal density from the final sample produced by either the substitution- or Gibbs sampling algorithms. Since for any estimated marginal the corresponding full conditional has been assumed available, efficient inference about the marginal should clearly be based on using this full conditional distribution. In the simplest case of two variables, this implies that $[X \mid Y]$ and the $Y_j^{(i)}$ $(j = 1, \ldots, m)$ should be used to make inferences about $[X]$, rather than imputing $X_j^{(i)}$ $(j = 1, \ldots, m)$ and basing inference on these $X_j^{(i)}$'s. Intuitively, this follows, because to estimate $[X]$ using the $X_j^{(i)}$ requires a kernel density estimate. Such an estimate ignores the known form $[X \mid Y]$ that is mixed to obtain $[X]$. The formal argument is essentially based on the Rao–Blackwell theorem. We sketch a proof in the context of the density estimator itself. If $X$ is a continuous $p$-dimensional random variable, consider any kernel density estimator of $[X]$ based on the $X_j^{(i)}$ (e.g., see Devroye and Györfi 1985) evaluated at $X_0$: $\Delta_{X_0}^{(i)} = (1/mh_m^p) \sum_{j=1}^{m} K[(X_0 - X_j^{(i)})/h_m]$, say, where $K$ is a bounded density on $R^p$ and the sequence $\{h_m\}$ is such that as $m \rightarrow \infty$, $h_m \rightarrow 0$, whereas $mh_m^p \rightarrow \infty$. To simplify notation, set $Q_{m,X_0}(X) = (1/h_m^p)K[(X_0 - X)/h_m]$ so that $\Delta_{X_0}^{(i)} = (1/m) \sum_{j=1}^{m} Q_{m,X_0}(X_j^{(i)})$. Define $\gamma_{X_0}^{(i)} = (1/m) \sum_{j=1}^{m} E(Q_{m,X_0}(X) \mid Y_j^{(i)})$. By our earlier theory, both $\Delta_{X_0}^{(i)}$ and $\gamma_{X_0}^{(i)}$ have the same expectation. By the Rao–Blackwell theorem, var $E(Q_{m,X_0}(X \mid Y)) \leq$ var $Q_{m,X_0}(X)$, and hence MSE$(\gamma_{X_0}^{(i)}) \leq$ MSE$(\Delta_{X_0}^{(i)})$, where MSE denotes the mean squared error of the estimate of $[X_0]$.

Now, for fixed $Y$, as $m \rightarrow \infty$, $E(Q_{m,X_0}(X \mid Y)) \rightarrow [X_0 \mid Y]$ for almost every $X_0$, by the Lebesgue density theorem (see Devroye and Györfi 1985, p. 3). Thus in terms of random variables we have $E(Q_{m,X_0}(X \mid Y)) \xrightarrow{d} [X_0 \mid Y]$, so for large $m$, $\gamma_{X_0}^{(i)} \sim [\hat{X}_0]_i$ and $\text{MSE}(\gamma_{X_0}^{(i)}) \approx \text{MSE}([\hat{X}_0]_i)$, and hence $[\hat{X}_0]_i$ is preferred to $\Delta_{X_0}^{(i)}$.

The argument is simpler for estimation of $\eta = E(T(X)) = \int T(X) * [X]$, say. Here, $\hat{\eta}_1 = (1/m) \sum_{j=1}^{m} T(X_j^{(i)})$ is immediately seen to be dominated by $\hat{\eta}_2 = (1/m) \sum_{j=1}^{m} E(T(X) \mid Y_j^{(i)})$.

## 3. EXAMPLES

A major area of potential application of the methodology we have been discussing is in the calculation of marginal posterior densities within a Bayesian inference framework. In recent years, there have been many advances in numerical and analytic approximation techniques for such calculations (e.g., see Geweke 1988; Naylor and Smith 1982, 1988; Shaw 1988; Smith et al. 1987; Smith, Skene, Shaw, Naylor, and Dransfield 1985; Tierney and Kadane 1986), but implementation of these approaches typically requires sophisticated numerical analytic expertise, and possibly specialist software. By contrast, the sampling approaches we have discussed are straightforward to implement. For many practitioners, this feature will more than compensate for any relative computational inefficiency. To provide a flavor of the kinds of areas of application for which the methodology is suited, we present six illustrative examples.

### 3.1 A Class of Multinomial Models

We extend the one-parameter genetic-linkage example described by Tanner and Wong (1987, p. 530), which in its most general form involves multinomial sampling, where some observations are not assigned to individual cells but to aggregates of cells (see Dempster, Laird, and Rubin 1977; Hartley 1958). We give the model and distribution theory in detail for a two-parameter version, from which the extension to $k$ parameters should be clear. Let the vector $Y = (Y_1, \ldots, Y_5)$ have a multinomial distribution $\text{mult}(n, a_1\theta + b_1, a_2\theta + b_2, a_3\eta + b_3, a_4\eta + b_4, c(1 - \theta - \eta))$, where $a_i, b_i \geq 0$ are known and $0 < c = 1 - \sum_{i=1}^{4} b_i = a_1 + a_2 = a_3 + a_4 < 1$. Thus $\theta$ and $\eta$ range over $\theta \geq 0$, $\eta \geq 0$, and $\theta + \eta \leq 1$, so a three-parameter Dirichlet distribution, $\text{Dirichlet}(\alpha_1, \alpha_2, \alpha_3)$, may be a natural choice of prior density for $(\theta, \eta)$. From the form of $[Y \mid \theta, \eta] * [\theta, \eta]$, note that obtaining the exact marginals $[\theta \mid Y]$ and $[\eta \mid Y]$ is somewhat messy (involving a two-dimensional numerical integral). Nevertheless, all three sampling approaches we have described are readily applicable here by considering the unobservable nine-cell multinomial model for $X = (X_1, X_2, \ldots, X_9)$, given by $\text{mult}(n, a_1\theta, b_1, a_2\theta, b_2, a_3\eta, b_3, a_4\eta, b_4, c(1 - \theta - \eta))$. From the form of $[X \mid \theta, \eta] * [\theta, \eta]$ we see that $[\theta, \eta \mid X] \sim \text{Dirichlet}(X_1 + X_3 + \alpha_1, X_5 + X_7 + \alpha_2, X_9 + \alpha_3)$, and hence $[\theta \mid X]$ and $[\eta \mid X]$ are available as beta distributions for sampling. Furthermore, $[\theta \mid X, \eta]$ and $[\eta \mid X, \theta]$ are available as scaled beta distributions, scaled to the intervals $[0, 1 - \eta]$ and $[0, 1 - \theta]$, respectively. If we let $Y_1 = X_1 + X_2$, $Y_2 = X_3 + X_4$, $Y_3 = X_5 + X_6$, $Y_4 = X_7 + X_8$, and $Y_5 = X_9$ and define $Z = (X_1, X_3, X_5, X_7)$, we see that specification of $X$ is equivalent to specification of $(Y, Z)$. In addition, $[Z \mid Y, \theta, \eta]$ is the product of four independent binomials for $X_1$, $X_3$, $X_5$, and $X_7$, given by $[X_i \mid Y, \theta, \eta] = \text{binomial}(Y_i, a_i\theta/(a_i\theta + b_i))$ $(i = 1, 3, 5, 7)$, which are therefore readily available for sampling.

In the context of Section 2, we have a three-variable case, $(\theta, \eta, Z)$, with interest in the marginal distributions $[\theta \mid Y]$, $[\eta \mid Y]$, and $[Z \mid Y]$. Gibbs sampling requires $[\theta \mid Y, Z, \eta]$, $[\eta \mid Y, Z, \theta]$, and $[Z \mid Y, \theta, \eta]$, all of which are available. But in this case the reduced distributions $[\theta \mid Y, Z]$ and $[\eta \mid Y, Z]$ are available as well, enabling study of accelerated convergence. These reduced distributions substantially simplify the Rubin importance-sampling algorithm in obtaining $[\theta \mid Y]$ and $[\eta \mid Y]$; only an importance-sampling distribution $[Z \mid Y]_s$ need be specified (e.g., a default choice might be binomials with chance equal to $\frac{1}{2}$). Detailed comparison of the performance of the three algorithms for a specific case of this multinomial class is given in Section 4.

### 3.2 Hierarchical Models Under Conjugacy

Consider a general Bayesian hierarchical model having $k$ stages. In an obvious notation, we write the joint distribution of the data and parameters as

$$[Y \mid \theta_1] * [\theta_1 \mid \theta_2] * [\theta_2 \mid \theta_3] * \cdots * [\theta_{k-1} \mid \theta_k] * [\theta_k], \quad (16)$$

where we assume all components of prior specification to be available for sampling. Primary interest is usually in the marginal posterior $[\theta_1 \mid Y]$. The hierarchical structure implies that

$$\begin{aligned} [\theta_i \mid Y, \theta_{j \neq i}] &= [\theta_1 \mid Y, \theta_2], & i &= 1 \\ &= [\theta_i \mid \theta_{i-1}, \theta_{i+1}], & 1 &< i < k - 1 \\ &= [\theta_k \mid \theta_{k-1}], & i &= k. \end{aligned} \quad (17)$$

Suppose that we assume proper conjugate distributions at each stage. This is common practice in the formulation of such models, except perhaps for $[\theta_k]$, which is often assumed vague. Nevertheless, conjugate priors can generally be made arbitrarily diffuse by appropriate choices of hyperparameters, so this case is implicitly subsumed within the conjugate framework. In fact, $[\theta_k]$ can be vague, provided $[\theta_k \mid \theta_{k-1}]$ is still proper and available (see Secs. 3.4 and 3.5). Conjugacy implies that the densities in (17) will be available as updated versions of the respective priors (e.g., see Morris 1983a). Typically, no distinct reduced conditional distributions are available, and Gibbs sampling would be used to estimate the desired marginal posterior densities. To clarify this latter point, consider the case $k = 3$. The six conditional distributions in (10) would be $[\theta_1 \mid y, \theta_2, \theta_3]$, $[\theta_2 \mid y, \theta_1, \theta_3]$, $[\theta_3 \mid y, \theta_1, \theta_2]$, $[\theta_3 \mid y, \theta_2]$, $[\theta_1 \mid y, \theta_3]$, and $[\theta_2 \mid y, \theta_1]$. The first three are available as in (17), the fourth is available but not distinct from the third, and the last two are usually unavailable.

As a concrete illustration, consider an exchangeable Poisson model, which is illustrated further in Section 4 with the reanalysis of a published data set. Suppose that we observe independent counts, $s_i$, over differing lengths of time, $t_i$ (with resultant rate $\rho_i = s_i/t_i$) $(i = 1, \ldots, p)$. Assume $[s_i \mid \lambda_i] = P_0(\lambda_i t_i)$ and that the $\lambda_i$ are iid from $G(\alpha,$

$\beta$), with density $\lambda_i^{\alpha-1}e^{-\lambda_i/\beta}/\beta^\alpha\Gamma(\alpha)$. The parameter $\alpha$ is assumed known (in practice, we might treat $\alpha$ as a tuning parameter, or perhaps, in an empirical Bayes spirit, estimate it from the marginal distribution of the $s_i$'s), and $\beta$ is assumed to arise from an inverse gamma distribution $IG(\gamma, \delta)$ with density $\delta^\gamma e^{-\delta/\beta}/\beta^{\gamma+1}\Gamma(\gamma)$. (A diffuse version of this final-stage distribution is obtained by taking $\delta$ and $\gamma$ to be very small, perhaps 0.)

Letting $Y = (s_1, \ldots, s_p)$, the conditional distributions $[\lambda_j \mid Y]$ are sought. The full conditional distribution of $\lambda_j$ is given by

$$[\lambda_j \mid Y, \beta, \lambda_{i,i\neq j}] = G(\alpha + s_j, (t_j + 1/\beta)^{-1}), \qquad j = 1, \ldots, p, \quad (18)$$

whereas the full conditional distribution for $\beta$ is given by

$$[\beta \mid Y, \lambda_1, \ldots, \lambda_p] = IG(\gamma + p\alpha, \Sigma\lambda_i + \delta). \quad (19)$$

No distinct reduced conditional distributions are available. The conditional distribution of $\lambda_j$, given $Y$ and $\beta$, is (18), regardless of which or how many $\lambda_i$ ($i \neq j$) are given. The conditional distribution of $\beta$, given $Y$ and any subset of the $\lambda_j$'s, is unavailable. Given $(\lambda_1^{(0)}, \lambda_2^{(0)}, \ldots, \lambda_p^{(0)}, \beta^{(0)})$, the Gibbs sampler draws $\lambda_j^{(1)} \sim G(\alpha + s_j, (t_j + 1/\beta^{(0)})^{-1})$ ($j = 1, \ldots, p$) and $\beta^{(1)} \sim IG(\gamma + \alpha p, \Sigma_{j=1}^p \lambda_j^{(1)} + \delta)$ to complete one cycle. If we carry out $m$ repetitions each of $i$ iterations, generating $(\lambda_{1l}^{(i)}, \ldots, \lambda_{pl}^{(i)}, \beta_l^{(i)})$ ($l = 1, \ldots, m$), the marginal density estimate for $\lambda_j$ is

$$[\lambda_j \uparrow Y] = \frac{1}{m}\sum_{l=1}^{m} G\left(\alpha + s_j, \left(t_j + \frac{1}{\beta_l^{(i)}}\right)^{-1}\right), \qquad j = 1, \ldots, p, \quad (20)$$

whereas

$$[\beta \uparrow Y] = \frac{1}{m}\sum_{l=1}^{m} IG(\gamma + \alpha p, \Sigma\lambda_{jl}^{(i)} + \delta). \quad (21)$$

Rubin's importance-sampling algorithm is applicable in the setting (16) as well, taking a particularly simple form in the cases $k = 2, 3$. For $k = 3$, suppose that we seek $[\theta_1 \mid y]$. The joint density $[\theta_1, \theta_2, \theta_3 \mid Y] = [Y, \theta_1, \theta_2, \theta_3]/[Y]$, where the functional form of the numerator is given in (16). An importance-sampling density for $[\theta_1, \theta_2, \theta_3 \mid Y]$ could be sampled as $[\theta_1 \mid Y, \theta_2] * [\theta_3 \mid \theta_2] * [\theta_2 \mid Y]$, for some $[\theta_2 \mid Y]_s$. As remarked in Section 2.5, a good choice for $[\theta_2 \mid Y]_s$ might be obtained through a few iterations of the substitution-sampling algorithm. In any case, for $l = 1, \ldots, N$ we would generate $\theta_{2l}$ from $[\theta_2 \mid Y]_s$, $\theta_{3l}$ from $[\theta_3 \mid \theta_{2l}]$, and $\theta_{1l}$ from $[\theta_1 \mid Y, \theta_{2l}]$. Calculating

$$r_l = \frac{[Y, \theta_{1l}, \theta_{2l}, \theta_{3l}]}{[\theta_{1l} \mid Y, \theta_{2l}] * [\theta_{3l} \mid \theta_{2l}] * [\theta_{2l} \mid Y]_s},$$

we obtain the density estimator $[\theta_1 \uparrow Y] = \Sigma[\theta_1 \mid Y, \theta_{2l}]r_l/\Sigma r_l$. Note that (in the terminology of Rubin) the algorithm in this case can be streamlined by writing the joint density in the numerator of $r_l$ as $[\theta_{1l} \mid Y, \theta_{2l}] * [Y \mid \theta_{2l}] * [\theta_{2l} \mid \theta_{3l}] * [\theta_{3l}]$ and noting that $r_l$ does not involve $\theta_{1l}$, so we need not actually generate the $\theta_{1l}$.

Returning to the exchangeable Poisson model, the estimator of the marginal density of $\lambda_j$ under Rubin's importance-sampling algorithm is

$$[\lambda_j \uparrow Y] = \sum_{l=1}^{N} G\left(\alpha + s_j, \left(t_j + \frac{1}{\beta_l}\right)^{-1}\right) r_l \Big/ \sum_{l=1}^{N} r_l.$$

Here $r_l = [Y \mid \beta_l] * [\beta_l]/[\beta_l \mid Y]_s$, where $[Y \mid \beta_l]$ is the product of negative binomial densities; that is,

$$[Y \mid \beta_l] = \prod_{j=1}^{p}\left(\frac{\Gamma(s_j + \alpha)t_j^{s_j}\beta_l^{s_j}}{s_j!\Gamma(\alpha)(t_j + \beta_l)^{s_j+\alpha}}\right),$$

and $[\beta_l]$ is the IG prior evaluated at $\beta_l$. If $[\beta \mid Y]_s$ is not obtained from the substitution-sampling algorithm, as in (21), an alternative choice is $IG(\gamma + \alpha p, \Sigma_{i=1}^p \rho_i + 1)$. This arises because $[\beta \mid Y] = E_{[\lambda_1,\ldots,\lambda_p \mid Y]}[\beta \mid Y, \lambda_1, \ldots, \lambda_p] \approx [\beta \mid Y, \hat\lambda_1, \ldots, \hat\lambda_p]$, using $\hat\lambda_j = \rho_j$ in (19).

### 3.3 Multivariate Normal Sampling

A commonly occurring problem in combining continuous multivariate data is that often not all variables are observed for each experimental unit (e.g., see Dempster et al. 1977). If the data are sampled from multivariate normal populations with conjugate priors for the mean and covariance matrix, we have a general class of models where all full conditional distributions and at least some reduced conditional distributions are available. We illustrate in the simplest case, where we assume that $\binom{U_{1i}}{U_{2i}}$ ($i = 1, \ldots, n_1$), $\binom{V_{1j}}{V_{2j}}$ ($j = 1, \ldots, n_2$), and $\binom{W_{1k}}{W_{2k}}$ ($k = 1, \ldots, n_3$) are all iid $N(\theta, \Delta)$ with $\theta \sim N(\mu, \Sigma)$, where $\theta = \binom{\theta_1}{\theta_2}$ is not observable but $\mu = \binom{\mu_1}{\mu_2}$, $\Delta$, and $\Sigma$ are assumed known. Let $U = \binom{U_1}{U_2} = \binom{U_{11} \cdots U_{1n_1}}{U_{21} \cdots U_{2n_1}}$, with similar notation for $V$ and $W$. Finally, let $X = (U, V, W)$ and $2 \times N$, with $\bar{X} = N^{-1}X1$, where 1 is a column vector of $N$ 1s and $N = n_1 + n_2 + n_3$. Standard calculations show that $[\theta \mid X]$ is $N(\eta, \Omega)$, where $\eta = (N\Delta^{-1} + \Sigma^{-1})^{-1}(N\Delta^{-1}\bar{X} + \Sigma^{-1}\mu)$ and $\Omega = (N\Delta^{-1} + \Sigma^{-1})^{-1}$. With the obvious partitioning, $\eta = \binom{\eta_1}{\eta_2}$, $\Omega = \begin{pmatrix}\Omega_{11} & \Omega_{12}\\ \Omega_{21} & \Omega_{22}\end{pmatrix}$, the marginals $[\theta_1 \mid X] = N(\eta_1, \Omega_{11})$, and $[\theta_2 \mid X] = N(\eta_2, \Omega_{22})$ are available. Suppose, however, that $V_2$ and $W_1$, say, are unobserved. Let $Y = (U, V_1, W_2)$ and $Z = (V_2, W_1)$ so that $X \equiv (Y, Z)$. As in Section 3.1, we have a three-variable problem, here involving $\theta_1$, $\theta_2$, and $Z$. The full conditional distributions are all normal and hence available. For $\theta_1$ and $\theta_2$, $[\theta_1 \mid Y, Z, \theta_2] = N(\eta_1 + \Omega_{12}\Omega_{22}^{-1}(\theta_2 - \eta_2), \Omega_{11} - \Omega_{12}\Omega_{22}^{-1}\Omega_{21})$ and $[\theta_2 \mid Y, Z, \theta_1] = N(\eta_2 + \Omega_{21}\Omega_{11}^{-1}(\theta_1 - \eta_1), \Omega_{22} - \Omega_{21}\Omega_{11}^{-1}\Omega_{12})$. Letting $\bar{U}_1 = n_1^{-1}U_11$, with similar notation for $\bar{U}_2, \bar{V}_1, \bar{V}_2, \bar{W}_1$, and $\bar{W}_2$, we note by sufficiency that with regard to $Z$ we only need the full posterior $[\bar{Z} \mid \bar{Y}, \theta_1, \theta_2]$, where $\bar{Z}^T = (\bar{V}_2, \bar{W}_1)$ and $\bar{Y}^T = (\bar{U}_1, \bar{U}_2, \bar{V}_1, \bar{W}_2)$. Since

$$\bar{X}^T \equiv (\bar{U}_1, \bar{U}_2, \bar{V}_1, \bar{V}_2, \bar{W}_1, \bar{W}_2) \mid (\theta_1, \theta_2) \sim N\left(\begin{pmatrix}\theta\\ \theta\\ \theta\end{pmatrix}, \begin{pmatrix}n_1^{-1}\Delta & 0 & 0\\ 0 & n_2^{-1}\Delta & 0\\ 0 & 0 & n_3^{-1}\Delta\end{pmatrix}\right),$$

the conditional distribution $[\bar{Z} \mid \bar{Y}, \theta_1, \theta_2]$ is clearly normal. With the full conditionals and the reduced conditionals $[\theta_1 \mid Y, Z]$ and $[\theta_2 \mid Y, Z]$ available, the accelerated substitution algorithm can be used to obtain $[\theta_1 \mid Y]$ and $[\theta_2 \mid Y]$.

The Rubin importance-sampling algorithm is straightforward in this case. Simplifying notation by working with the sufficient statistic $(\bar{Y}, \bar{Z})$, suppose that we seek the density estimator of $[\theta_1 \mid Y]$ for instance. We have $[\theta_1 \mid Y] = \Sigma[\theta_1 \mid \bar{Y}, \bar{Z}_l, \theta_{2l}]r_l/\Sigma r_l$, where

$$r_l = \frac{[\bar{X}_l \mid \theta_{1l}, \theta_{2l}] * [\theta_{1l}, \theta_{2l}]}{[\theta_{1l} \mid \bar{Y}, \bar{Z}_l, \theta_{2l}] * [\theta_{2l} \mid \bar{Y}, \bar{Z}_l] * [\bar{Z}_l \mid \bar{Y}]_s},$$

with $\bar{X}_l \equiv (\bar{Y}, \bar{Z}_l)$ and $[\bar{Z} \mid \bar{Y}]_s$ a specified importance-sampling density. Thus for $l = 1, \ldots, N$ we generate $\bar{Z}_l \sim [\bar{Z} \mid \bar{Y}]$, $\theta_{2l} \sim [\theta_2 \mid \bar{Y}, \bar{Z}_l]$, and $\theta_{1l} \sim [\theta_1 \mid \bar{Y}, \bar{Z}_l, \theta_{2l}]$. Again, the choice of $[\bar{Z} \mid \bar{Y}]_s$ could be made using a few iterations of substitution sampling, or perhaps based on the intuitively appealing estimated conditional form, $[\bar{Z} \mid \bar{Y}, \hat{\theta}_1, \hat{\theta}_2]$, where $\hat{\theta}_1 = (n_1\bar{U}_1 + n_2\bar{V}_1)/(n_1 + n_2)$ and $\hat{\theta}_2 = (n_1\bar{U}_2 + n_3\bar{W}_3)/(n_1 + n_3)$.

### 3.4 Variance Component Models

Bayesian inference for variance components has typically required subtle numerical analysis or intricate analytic approximation (e.g., as evidenced by Box and Tiao 1973, chaps. 5 and 6). In marked contrast to such sophistication, marginal posterior densities for variance components are readily obtained through simple Gibbs sampling.

We illustrate this for the simplest variance components model defined by $Y_{ij} = \theta_i + \varepsilon_{ij}$ $(i = 1, \ldots, K, j = 1, \ldots, J)$, where, assuming conditional independence throughout, $[\theta_i \mid \mu, \sigma_\theta^2] = N(\mu, \sigma_\theta^2)$ and $[\varepsilon_{ij} \mid \sigma_e^2] = N(0, \sigma_e^2)$, so $[Y_{ij} \mid \theta_i, \sigma_e^2] = N(\theta_i, \sigma_e^2)$.

Let $\theta = (\theta_1, \ldots, \theta_K)$ and $Y = (Y_{11}, \ldots, Y_{KJ})$ and assume that $\mu$, $\sigma_\theta^2$, and $\sigma_e^2$ are independent, with priors specified by $[\mu] \sim N(\mu_0, \sigma_0^2)$, $[\sigma_\theta^2] \sim \mathrm{IG}(a_1, b_1)$, and $[\sigma_e^2] \sim \mathrm{IG}[a_2, b_2]$, where $\mu_0$, $\sigma_0^2$, $a_1$, $b_1$, $a_2$, and $b_2$ are assumed known (possibly chosen to correspond to diffuse priors).

The joint distribution $[Y, \theta, \mu, \sigma_\theta^2, \sigma_e^2]$ can be written as

$$[Y \mid \theta, \sigma_e^2] * [\theta \mid \mu, \sigma_\theta^2] * [\mu] * [\sigma_\theta^2] * [\sigma_e^2], \tag{22}$$

and we follow Box and Tiao (1973, chap. 5) in focusing interest on $[\sigma_\theta^2 \mid Y]$ and $[\sigma_e^2 \mid Y]$.

From the Gibbs sampling perspective, we have a four-variable system, $(\theta, \mu, \sigma_\theta^2, \sigma_e^2)$, with the following full conditional distributions:

$$[\sigma_\theta^2 \mid Y, \mu, \theta, \sigma_e^2] = [\sigma_\theta^2 \mid \mu, \theta] = \mathrm{IG}(a_1 + \tfrac{1}{2}K, b_1 + \tfrac{1}{2}\Sigma(\theta_i - \mu)^2),$$

$$[\sigma_e^2 \mid Y, \mu, \theta, \sigma_\theta^2] = [\sigma_e^2 \mid Y, \theta] = \mathrm{IG}(a_2 + \tfrac{1}{2}KJ, b_2 + \tfrac{1}{2}\Sigma\Sigma(Y_{ij} - \mu_i)^2),$$

$$[\mu \mid Y, \theta, \sigma_\theta^2, \sigma_e^2] = [\mu \mid \sigma_\theta^2, \theta] = N\left(\frac{\sigma_\theta^2\mu_0 + \sigma_0^2\Sigma\theta_i}{\sigma_\theta^2 + K\sigma_0^2}, \frac{\sigma_\theta^2\sigma_0^2}{\sigma_\theta^2 + K\sigma_0^2}\right),$$

and

$$[\theta \mid Y, \mu, \sigma_\theta^2, \sigma_e^2] = N\left(\frac{J\sigma_\theta^2}{J\sigma_\theta^2 + \sigma_e^2}\bar{Y} + \frac{\sigma_e^2}{J\sigma_\theta^2 + \sigma_e^2}\mu 1, \frac{\sigma_\theta^2\sigma_e^2}{J\sigma_\theta^2 + \sigma_e^2} I\right),$$

where $\bar{Y}^T = (\bar{Y}_1, \ldots, \bar{Y}_I)$, $\bar{Y}_{i.} = (1/J)\sum_{j=1}^{J} Y_{ij}$, 1 is a $K \times 1$ column vector of 1s, and $I$ is a $K \times K$ identity matrix.

Since all of these full conditionals are available, implementation of the Gibbs sampler is straightforward. Moreover, extensions to more elaborate variance component models follow precisely the same pattern, since the full conditional distributions for $\mu$ and $\theta$ continue to be normal, and those for the variance components continue to be IG.

### 3.5 Normal Means Model

The exchangeable $k$-group normal means model with different, unknown measurement variances in each group provides a simple example of an unbalanced class of models that has proved difficult to handle using empirical Bayes approaches to estimating posterior distributions (e.g., see Morris 1983b, 1987). Such models are straightforwardly handled by iterative sampling approaches, as we saw with the Poisson example of Section 3.2 and further illustrate here for this classical normal means example.

Suppose, then, assuming conditional independence throughout, that $Y_{ij} \sim N(\theta_i, \sigma_i^2)$, $\theta_i \sim N(\mu, \tau^2)$, $\sigma_i^2 \sim \mathrm{IG}(a_1, b_1)$ $(i = 1, \ldots, I, j = 1, \ldots, J_i)$, $\mu \sim N(\mu_0, \sigma_0^2)$, and $\tau^2 \sim \mathrm{IG}(a_2, b_2)$, where $\mu_0, \sigma_0^2, a_1, b_1, a_2$, and $b_2$ are assumed known (possibly chosen to reflect diffuse prior information). By sufficiency, we can confine attention to $Y = \{(\bar{Y}_{i.}, S_i^2); i = 1, \ldots, I\}$, where $\bar{Y}_{i.} = (1/J_i)\Sigma Y_{ij}$ and $S_i^2 = (1/J_i)\Sigma(Y_{ij} - \bar{Y}_{i.})^2$. Then, if we write $\theta = (\theta_1, \ldots, \theta_I)$ and $\sigma^2 = (\sigma_1^2, \ldots, \sigma_I^2)$, the joint distribution of $Y, \theta, \sigma^2$, $\mu$, and $\tau^2$ takes the form

$$[Y \mid \theta, \sigma^2] * [\theta \mid \mu, \tau^2] * [\sigma^2] * [\mu] * [\tau^2], \tag{23}$$

where

$$[Y \mid \theta, \sigma^2] * [\theta \mid \mu, \tau^2] * [\sigma^2] = \prod_{i=1}^{I} [\bar{Y}_i \mid \theta_i, \sigma_i^2] * [S_i^2 \mid \sigma_i^2] * [\theta_i \mid \mu, \tau^2] * [\sigma_i^2].$$

Of course, there is an obvious similarity between (22) and (23), but here we focus on $[\theta_i \mid Y]$ $(i = 1, \ldots, I)$. From the Gibbs sampling perspective, this is a $(2I + 2)$-variable problem: $(\theta_i, \sigma_i^2)$ $(i = 1, \ldots, I)$, together with $\mu$ and $\tau^2$. To identify the forms of the full conditionals, we first note that

$$[\theta \mid Y, \sigma^2, \mu, \tau^2] = N(\theta^*, D^*), \tag{24}$$

where $\theta_i^* = (J_i\bar{Y}_{i.}\tau^2 + \mu\sigma_i^2)/(J_i\tau^2 + \sigma_i^2)$, $D_{ii}^* = \sigma_i^2\tau^2/(J_i\tau^2 + \sigma_i^2)$, and $D_{ij}^* = 0$ $(i \neq j)$. Thus the full conditional distributions $[\theta_i \mid Y, \theta_j j \neq i, \sigma^2, \mu, \tau^2]$ $(i = 1, \ldots, I)$ are just the normal marginals of (24) and therefore available for sampling. From (23), we easily see that $[\sigma^2 \mid Y, \theta, \mu, \tau^2] = [\sigma^2 \mid Y, \theta] = \prod_{i=1}^{I} [\sigma_i^2 \mid \bar{Y}_{i.}, S_i^2, \theta_i]$, where $[\sigma_i^2 \mid \bar{Y}_{i.}, S_i^2, \theta_i] = \mathrm{IG}(a_1 + \tfrac{1}{2}J_i, b_1 + \tfrac{1}{2}\Sigma_j(Y_{ij} - \theta_i)^2)$. Finally, and closely resembling the forms obtained in Section 3.4,

$$[\mu \mid Y, \theta, \sigma^2, \tau^2] = [\mu \mid \theta, \tau^2] = N\left(\frac{\tau^2\mu_0 + \sigma_0^2\Sigma\theta_i}{\tau^2 + I\sigma_0^2}, \frac{\tau^2\sigma_0^2}{\tau^2 + I\sigma_0^2}\right),$$

and $[\tau^2 \mid Y, \theta, \sigma^2, \mu] = [\tau^2 \mid \theta, \mu] = \text{IG}(a_2 + \tfrac{1}{2}I, b_2 + \tfrac{1}{2}\Sigma(\theta_i - \mu)^2)$.

### 3.6 An Errors-in-Variables Model

Again, we consider a simple illustrative special case. Consider $Y$ to be a vector of responses assumed related to levels $X$ of a covariate according to the straight-line model

$$Y \sim N\left((1 X)\begin{pmatrix}\theta_1\\ \theta_2\end{pmatrix}, \sigma^2 I\right).$$

Responses are obtained at specified levels $X_0$ of the covariate, but suppose that these are not the actual levels $X_a$. Rather, given the former, beliefs about the latter are represented by $X_a \sim N(X_0, \tau^2 I)$. Interest centers on $\theta = (\theta_1, \theta_2)$, and to complete the distributional specification, suppose that we place independent conjugate priors on $\theta$, $\sigma^2$, and $\tau^2$. The joint distribution on $(Y, X_a, \theta, \sigma^2, \tau^2)$ then has the form

$$[Y \mid X_a, \theta, \sigma^2] * [X_a \mid \tau^2] * [\tau^2] * [\theta] * [\sigma^2], \quad (25)$$

where again there is obvious similarity to (22) and (23). The Gibbs sampler requires $[\theta \mid Y, X_a, \sigma^2, \tau^2] = [\theta \mid Y, X_a, \sigma^2]$, $[\sigma^2 \mid Y, X_a, \theta, \tau^2] = [\sigma^2 \mid Y, X_a, \theta]$, $[\tau^2 \mid Y, X_a, \theta, \sigma^2] = [\tau^2 \mid X_a]$, and $[X_a \mid Y, \theta, \sigma^2, \tau^2]$. If we assume a normal prior for $\theta$ and IG priors for $\sigma^2$ and $\tau^2$, we obtain normal full conditionals for $\theta$ and $X_a$ and IG full conditionals for $\sigma^2$ and $\tau^2$. We omit the details, which are somewhat similar to those in Sections 3.4 and 3.5.

## 4. NUMERICAL ILLUSTRATIONS

### 4.1 A Multinomial Model

We provide some preliminary insights into the relative performance and properties of the substitution-, Gibbs, and Rubin importance-sampling approaches by considering an artificial problem based on the class of multinomial models discussed in Section 3.1.

*Table 1. Comparison of Substitution (S) and Gibbs (G) Samplers*

| | | Estimate (SE) | | | | S closer than G | |
|---|---|---|---|---|---|---|---|
| | | $\theta$ | | $\eta$ | | | |
| Cycle | cdf value | G | S | G | S | $\theta$ | $\eta$ |
| 1 | .05 | .231 (.08) | .217 (.08) | .033 (.01) | .044 (.01) | 56% | 75% |
| | .25 | .504 (.10) | .492 (.09) | .177 (.04) | .225 (.04) | 55% | 78% |
| | .50 | .713 (.08) | .706 (.08) | .380 (.06) | .459 (.06) | 54% | 80% |
| | .75 | .873 (.05) | .871 (.05) | .620 (.06) | .706 (.06) | 51% | 80% |
| | .95 | .978 (.01) | .978 (.01) | .878 (.04) | .926 (.03) | 49% | 80% |
| 2 | .05 | .067 (.04) | .055 (.03) | .047 (.01) | .048 (.01) | 56% | 51% |
| | .25 | .286 (.07) | .266 (.07) | .236 (.04) | .241 (.04) | 56% | 52% |
| | .50 | .535 (.08) | .522 (.07) | .478 (.06) | .487 (.06) | 53% | 52% |
| | .75 | .773 (.06) | .768 (.05) | .728 (.05) | .737 (.05) | 51% | 52% |
| | .95 | .956 (.02) | .956 (.02) | .940 (.02) | .944 (.02) | 51% | 53% |
| 3 | .05 | .052 (.03) | .049 (.03) | .049 (.01) | .049 (.01) | 51% | 50% |
| | .25 | .254 (.06) | .252 (.06) | .247 (.04) | .247 (.04) | 51% | 50% |
| | .50 | .505 (.07) | .508 (.07) | .496 (.06) | .496 (.06) | 51% | 49% |
| | .75 | .754 (.06) | .760 (.05) | .746 (.05) | .747 (.05) | 51% | 50% |
| | .95 | .951 (.02) | .954 (.02) | .949 (.02) | .949 (.02) | 51% | 50% |
| 4 | .05 | .050 (.03) | .047 (.03) | .050 (.01) | .050 (.01) | 51% | 51% |
| | .25 | .250 (.06) | .249 (.06) | .250 (.04) | .249 (.04) | 50% | 51% |
| | .50 | .500 (.07) | .505 (.07) | .499 (.06) | .499 (.06) | 51% | 51% |
| | .75 | .751 (.06) | .757 (.05) | .750 (.05) | .751 (.05) | 51% | 51% |
| | .95 | .950 (.02) | .953 (.02) | .950 (.02) | .951 (.02) | 51% | 49% |

NOTE: Standard errors (SE's) are in parentheses

We suppose that data $Y = (Y_1, Y_2, Y_3, Y_4, Y_5) = (14, 1, 1, 1, 5)$ are available as a sample from the multinomial distribution $\text{mult}(22, \tfrac{1}{4}\theta + \tfrac{1}{8}, \tfrac{1}{4}\eta, \tfrac{1}{4}\eta + \tfrac{3}{8}, \tfrac{1}{2}(1 - \theta - \eta))$, and that the prior for $(\theta, \eta)$ is taken to be a Dirichlet(1, 1, 1) distribution. In the general notation of Section 3.1, we therefore have $a_1 = \tfrac{1}{4}$, $b_1 = \tfrac{1}{8}$, $a_2 = \tfrac{1}{4}$, $b_2 = 0$, $a_3 = \tfrac{1}{4}$, $b_3 = 0$, $a_4 = \tfrac{1}{4}$, $b_4 = \tfrac{3}{8}$, and $\alpha_1 = \alpha_2 = \alpha_3 = 1$, with interest centering on the calculation of the marginal posterior densities $[\theta \mid Y]$ and $[\eta \mid Y]$.

By considering instead a split-cell multinomial, which in this case takes the form

$$X = (X_1, X_2, \ldots, X_7) \sim \text{mult}(22, \tfrac{1}{4}\theta, \tfrac{1}{8}, \tfrac{1}{4}\theta, \tfrac{1}{4}\eta, \tfrac{1}{4}\eta, \tfrac{3}{8}, \tfrac{1}{2}(1 - \theta - \eta)),$$

we can use the analysis of Section 3.1 for this special case of a seven-cell multinomial to construct substitution- and Gibbs sampling algorithms involving $\theta$, $\eta$, and $Z = (X_1, X_5)$.

As noted in Section 3.1, we can compare the two forms of iterative sampling. To do so, we first obtained very accurate numerical estimates of $[\theta \mid Y]$ and $[\eta \mid Y]$ using techniques described by Smith et al. (1985, 1987), and from these obtained the true 5, 25, 50, 75, and 95 posterior percentile points for each parameter. Iterative cycles of the two samplers were then run, calibrated so that the total number of random variates generated was the same in both cases (as described in Sec. 2.4). The initialization was defined (for an arbitrary generating seed) in each case by taking independent samples from $\theta \sim U(0, 1)$ and $\eta \sim U(0, 1)$, subject to $0 \leq \theta + \eta \leq 1$. At each cycle, $m = 10$ drawings of the parameters were then made, and from estimates of the form (9) estimates of the cumulative posterior probabilities corresponding to each of the five true percentile points for each parameter were obtained. This process was replicated 5,000 times, enabling us to study the mean estimates of the cumulative probabilities, together with their standard errors, as well as the percentage of occasions on which each sampler was closest to the true value. A summary of the results following each of the first four cycles is given in Table 1.

We note from Table 1 that initially (cycles 1 and 2) the substitution sampler adapts more quickly than the Gibbs sampler, particularly for $\eta$. By the time we reach the third and fourth cycles, however, the two approaches are performing indistinguishably. What is astonishing, perhaps, is how remarkably good their performance is. By the fourth cycle, using only $m = 10$ drawings and starting from a default noninformative baseline, the marginal posterior density estimators based on (8) are providing on average extremely accurate estimates of cumulative probabilities. Our experiences with this and other examples (see Sec. 4.2) suggest that satisfactory convergence with iterative sampling requires only a small fraction of the levels of random variate generation reported by Tanner and Wong (1987).

The noniterative Rubin importance-sampling algorithm (Sec. 2.5) requires us to choose a sampling density, $[Z \mid Y]_s$, and then to proceed as follows, for $l = 1, \ldots, m$: Draw $Z_l$ from $[Z \mid Y]_s$, $\eta_l$ from $[\eta \mid Z, Y]$, and $\theta_l$ from $[\theta \mid \eta, Z, Y]$, with the latter two distributions as detailed previously, thus creating a triple $(\theta_l, \eta_l, Z_l)$. Then, calculate

$$r_l = \frac{[Y, Z_l \mid \theta_l, \eta_l] * [\theta_l, \eta_l]}{[\theta_l \mid \eta_l, Z_l, Y] * [\eta_l \mid Z_l, Y] * [Z_l \mid Y]_s},$$

and form estimates $[\theta \upharpoonright Y] = \sum_{l=1}^m [\theta \mid \eta_l, Z_l, Y] r_l / \sum_{l=1}^m r_l$ and $[\eta \upharpoonright Y] = \sum_{l=1}^m [\eta \mid \theta_l, Z_l, Y] r_l / \sum_{l=1}^m r_l$.

Table 2 shows the average cumulative posterior probability estimates from this approach, based on 2,500 replicates of $m = 40$ and $m = 200$ and taking $[Z \mid Y]_s$ to be the product of $X_1 \sim \text{binomial}(Y_1, \frac{1}{2})$ and $X_5 \sim \text{binomial}(Y_4, \frac{1}{2})$. Despite the much larger number of drawings compared with the iterative samplers, the estimation is rather poor. In general, experience suggests that the algorithm is highly sensitive to the choice of $[Z \mid Y]_s$ and that the larger one-off simulation is no match for iterative adaptation via small simulations.

### 4.2 A Conjugate Hierarchical Model

We apply the exchangeable Poisson model discussed in Section 3.2 to data on pump failures previously analyzed by Gaver and O'Muircheartaigh (1987) (reproduced here in Table 3), where $s_i$ is the number of failures and $t_i$ is the length of time in thousands of hours.

Recalling the model structure of Section 3.2 and the forms of conditional distribution given by (18) and (19), we illustrate the use of the Gibbs sampler for this data set, with $p = 10$, $\delta = 1$, $\gamma = 0.1$, and, for the purposes of illustration, $\alpha = \bar{\rho}^2/(S_\rho^2 - p^{-1}\bar{\rho}\sum_{i=1}^p t_i^{-1})$, with the latter derived by a method-of-moments empirical Bayes argument based on $E(\rho_i) = EE(\rho_i \mid \lambda_i) = \alpha/\beta \approx \bar{\rho}$:

$$V(\rho_i) = VE(\rho_i \mid \lambda) + EV(\rho_i \mid \lambda_i)$$
$$= (\alpha/\beta^2) + (\alpha/\beta t_i) \approx S_\rho^2 = p^{-1}\Sigma(\rho_i - \bar{\rho})^2.$$

Figure 1 shows a selection of four marginal posterior densities (for $\lambda_2, \lambda_4, \lambda_8, \lambda_9$) calculated from (20) following a run of 10 cycles of the algorithm. In fact, three densities are superposed: One corresponds to $m = 10$, one to $m = 100$, and the third is the exact density calculated using techniques described by Smith et al. (1985, 1987). Even in the cases of $\lambda_8$ and $\lambda_9$ (chosen as worst cases from $\lambda_1, \ldots, \lambda_{10}$), the densities are hardly distinguishable—a remarkable convergence from such a small number of drawings.

*Table 2. Estimates From the Rubin Importance-Sampling Algorithm*

| | Estimates: $m$ = 40 (200) | |
|---|---|---|
| cdf value | $\theta$ | $\eta$ |
| .05 | .105 (.150) | .049 (.049) |
| .25 | .311 (.351) | .244 (.241) |
| .50 | .521 (.537) | .485 (.477) |
| .75 | .739 (.734) | .729 (.714) |
| .95 | .939 (.932) | .934 (.921) |

*Table 3. Pump-Failure Data*

| Pump system | $s_i$ | $t_i$ | $\rho_i$ ($\times 10^2$) |
|---|---|---|---|
| 1 | 5 | 94.320 | 5.3 |
| 2 | 1 | 15.720 | 6.4 |
| 3 | 5 | 62.880 | 8.0 |
| 4 | 14 | 125.760 | 11.1 |
| 5 | 3 | 5.240 | 57.3 |
| 6 | 19 | 31.440 | 60.4 |
| 7 | 1 | 1.048 | 95.4 |
| 8 | 1 | 1.048 | 95.4 |
| 9 | 4 | 2.096 | 191.0 |
| 10 | 22 | 10.480 | 209.9 |

## 5. DISCUSSION

We have emphasized providing a comparative review and explication of three possible sampling approaches to the calculation of intractable marginal densities. The substitution-, Gibbs, and importance-sampling algorithms are all straightforward to implement in several frequently occurring practical situations, thus avoiding complicated numerical or analytic approximation exercises (often necessitating intricate attention to reparameterization and other subtleties requiring case-by-case consideration). For this latter reason if for no other the techniques deserve to be better known and experimented with for a wide range of problems. We hope that the unified exposition attempted here will provide a general, clarifying perspective within which to view the work of Geman and Geman (1984), Rubin (1987, 1988), and Tanner and Wong (1987), and to evaluate its potential for other structured problems. For example, in addition to the model structures given in Section 3, the methods find immediate and powerful application to problems involving ordered parameters or change points. In future work we shall provide detailed and extensive numerical illustration of many such problems.

The preliminary computational experience reported here illustrates the following points. Iterative, adaptive sampling (substitution or Gibbs) invariably provides better value, in terms of efficient use of generated variates, than an equivalent sample-size, noniterative, one-off approach (Rubin), provided a suitable structure for iterative sampling exists. In problems where certain reduced conditionals are available, there is scope for accelerating the substitution algorithm so that it becomes more efficient (particularly in early cycles) than the Gibbs algorithm; however, the gain in efficiency is only likely to be of consequence when the number of reduced conditionals is a relatively large fraction of the total number of conditionals involved in a cycle. There are important practical problems in tuning monitoring and stopping-rules procedures for iterative sampling in large-scale complex problems; we shall report on these in future work as well. Finally, we note that even in cases where ultimate convergence of the iterative sampling procedures proves slow, moment or other information provided by a few initial cycles can be used to provide highly effective starting values for more

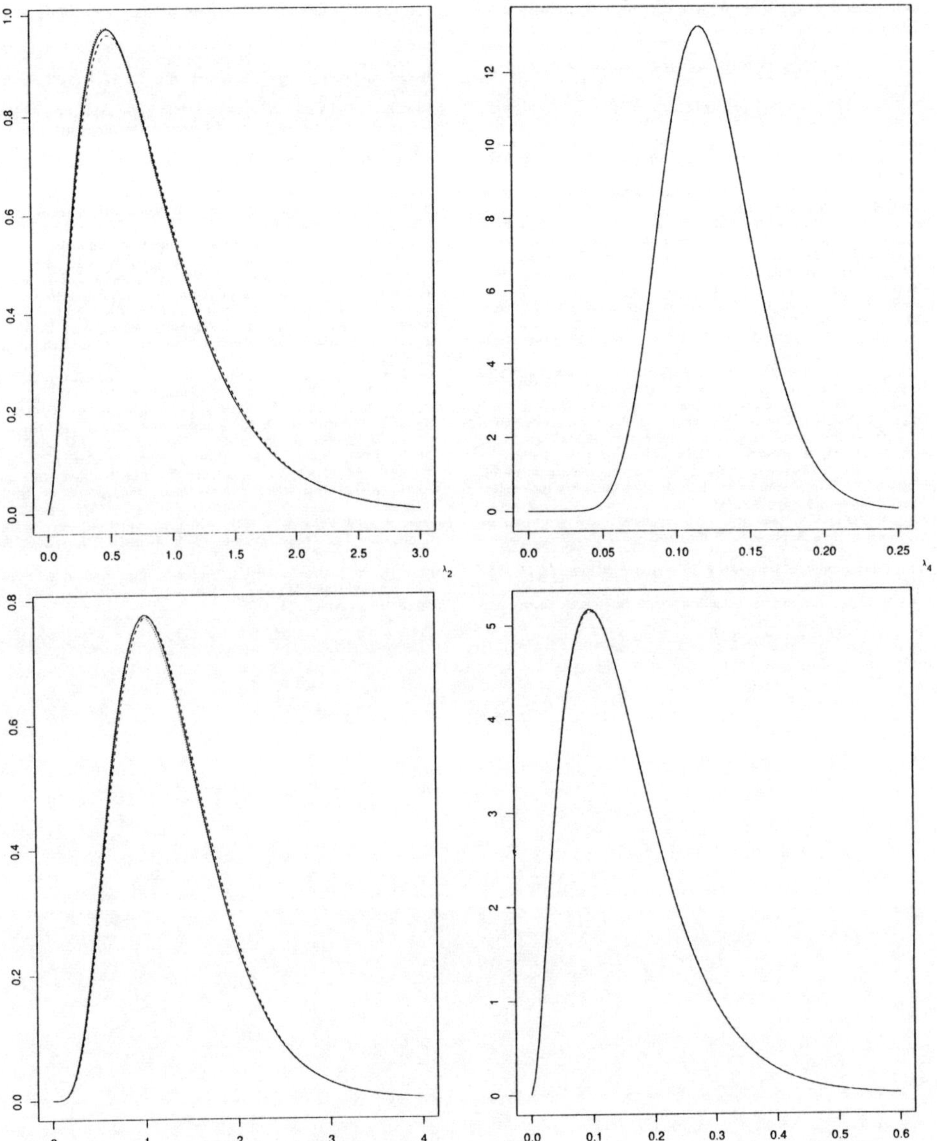


*Figure 1. Density Estimates for Pump-Failure Data: ···, m = 10; - - -, m = 100; ——, Exact.*

sophisticated numerical or analytic approximation techniques.

[*Received November 1988. Revised October 1989.*]

## REFERENCES


Besag, J. (1974), "Spatial Interaction and the Statistical Analysis of Lattice Systems" (with discussion), *Journal of the Royal Statistical Society*, Ser. B, 36, 192–326.

Box, G. E. P., and Tiao, G. C. (1973), *Bayesian Inference in Statistical Analysis*, Reading, MA: Addison-Wesley.

Dempster, A., Laird, N., and Rubin, D. B. (1977), "Maximum Likelihood From Incomplete Data Via the EM Algorithm" (with discussion), *Journal of the Royal Statistical Society*, Ser. B, 39, 1–38.

Devroye, L. (1986), *Non-uniform Random Variate Generation*, New York: Springer-Verlag.

Devroye, L., and Györfi, L. (1985), *Non-parametric Density Estimation: The $L_1$ View*, New York: John Wiley.

Gaver, D., and O'Muircheartaigh, I. (1987), "Robust Empirical Bayes Analysis of Event Rates," *Technometrics*, 29, 1–15.

Geman, S., and Geman, D. (1984), "Stochastic Relaxation, Gibbs Distributions and the Bayesian Restoration of Images," *IEEE Transactions on Pattern Analysis and Machine Intelligence*, 6, 721–741.

Geweke, J. (1988), "Antithetic Acceleration of Monte Carlo Integration in Bayesian Inference," *Journal of Econometrics*, 38, 73–90.

Glick, N. (1974), "Consistency Conditions for Probability Estimators and Integrals of Density Estimators," *Utilitas Mathematica*, 6, 61–74.

Hartley, H. (1958), "Maximum Likelihood Estimation From Incomplete Data," *Biometrics*, 14, 174–194.

Hastings, W. K. (1970), "Monte Carlo Sampling Methods Using Markov Chains and Their Applications," *Biometrika*, 87, 97–109.

Morris, C. (1983a), "Natural Exponential Families With Quadratic Variance Functions: Statistical Theory," *The Annals of Statistics*, 11, 515–529.

——— (1983b), "Parametric Empirical Bayes Inference: Theory and Applications," *Journal of the American Statistical Association*, 78, 47–59.

——— (1987), "Determining the Accuracy of Bayesian Empirical Bayes Estimates in Familiar Exponential Families," in *Statistical Decision Theory and Related Topics* (Vol. 4), eds. S. S. Gupta and J. O. Berger, New York: Springer-Verlag, pp. 251–264.

Naylor, J. C., and Smith, A. F. M. (1982), "Applications of a Method for the Efficient Computation of Posterior Distributions," *Applied Statistics*, 31, 214–225.

——— (1988), "Econometric Illustrations of Novel Numerical Integration Strategies for Bayesian Inferences," *Journal of Econometrics*, 38, 103–126.

Rall, L. (1969), *Computational Solution of Non-linear Operator Equations*, New York: John Wiley.

Ripley, B. (1987), *Stochastic Simulation*, New York: John Wiley.

Rubin, D. B. (1987), Comment on "The Calculation of Posterior Distributions by Data Augmentation," by M. A. Tanner and W. H. Wong, *Journal of the American Statistical Association*, 82, 543–546.

——— (1988), "Using the SIR Algorithm to Simulate Posterior Distributions," in *Bayesian Statistics 3*, eds. J. M. Bernardo, M. H. DeGroot, D. V. Lindley, and A. F. M. Smith, Oxford, U.K.: Oxford University Press, pp. 395–402.

Shaw, J. E. H. (1988), "A Quasirandom Approach to Integration in Bayesian Statistics," *The Annals of Statistics*, 16, 895–914.

Smith, A. F. M., Skene, A. M., Shaw, J. E. H., and Naylor, J. C. (1987), "Progress With Numerical and Graphical Methods for Bayesian Statistics," *The Statistician*, 36, 75–82.

Smith, A. F. M., Skene, A. M., Shaw, J. E. H., Naylor, J. C., and Dransfield, M. (1985), "The Implementation of the Bayesian Paradigm," *Communications in Statistics—Theory and Methods*, 14, 1079–1102.

Tanner, M., and Wong, W. (1987), "The Calculation of Posterior Distributions by Data Augmentation" (with discussion), *Journal of the American Statistical Association*, 82, 528–550.

Tierney, L., and Kadane, J. (1986), "Accurate Approximations for Posterior Moments and Marginal Densities," *Journal of the American Statistical Association*, 81, 82–86.

# Part IV
# Bayesian Econometrics

# [12]

*Econometrica*, Vol. 57, No. 6 (November, 1989), 1317–1339

## BAYESIAN INFERENCE IN ECONOMETRIC MODELS USING MONTE CARLO INTEGRATION

BY JOHN GEWEKE[1]

Methods for the systematic application of Monte Carlo integration with importance sampling to Bayesian inference in econometric models are developed. Conditions under which the numerical approximation of a posterior moment converges almost surely to the true value as the number of Monte Carlo replications increases, and the numerical accuracy of this approximation may be assessed reliably, are set forth. Methods for the analytical verification of these conditions are discussed. Importance sampling densities are derived from multivariate normal or Student $t$ approximations to local behavior of the posterior density at its mode. These densities are modified by automatic rescaling along each axis. The concept of relative numerical efficiency is introduced to evaluate the adequacy of a chosen importance sampling density. The practical procedures based on these innovations are illustrated in two different models.

KEYWORDS: Importance sampling, numerical integration, Markov chain model, ARCH linear model.

### 1. INTRODUCTION

ECONOMETRIC MODELS are usually expressed in terms of an unknown vector of parameters $\underline{\theta} \in \Theta \subseteq R^k$, which fully specifies the joint probability distribution of the observations $X = \{\underline{x}_1, \ldots, \underline{x}_T\}$. In most cases there exists a probability density function $f(X|\underline{\theta})$, and classical inference then often proceeds from the likelihood function $L(\underline{\theta}) = f(X|\underline{\theta})$. The asymptotic behavior of the likelihood function is well understood, and as a consequence there is a well developed set of tools with which problems of computation and inference can be approached; Quandt (1983) and Engle (1984) provide useful surveys. The analytical problems in a new model are often far from trivial, but there are typically several approaches that can be explored systematically with the realistic anticipation that one or more will lead to classical inference procedures with an asymptotic justification.

Bayesian inference proceeds from the likelihood function and prior information which is usually expressed as a probability density function over the parameters, $\pi(\underline{\theta})$, it being implicit that $\pi(\underline{\theta})$ depends on the conditioning set of prior information. The posterior distribution is proportional to $p(\underline{\theta}) = \pi(\underline{\theta})L(\underline{\theta})$. This formulation restricts the prior probability measure for $\Theta$ to be absolutely continuous with respect to Lebesgue measure. Extensions that allow concentrations of prior probability mass are generally straightforward but notationally cumbersome, and are postponed to the concluding section. Most Bayesian inference problems can be expressed as the evaluation of the expectation of a

[1]Financial support for this work was provided by NSF Grant SES-8605867, and by a grant from the Pew Foundation to Duke University. This work benefited substantially by suggestions from an editor, Jean-Francois Richard, Herman van Dijk, Robert Wolpert, and two anonymous referees, who bear no responsibility for any errors of commission or omission.

function of interest $g(\underline{\theta})$ under the posterior,

$$(1) \qquad E[g(\underline{\theta})] = \int_{\Theta} g(\underline{\theta}) p(\underline{\theta})\, d\underline{\theta} \Big/ \int_{\Theta} p(\underline{\theta})\, d\underline{\theta}.$$

Approaches to this problem are nowhere near as systematic, methodical, or general as are those to classical inference problems: classical inference is carried out routinely using likelihood functions for which the evaluation of (1) is at best a major and dubious undertaking and for most practical purposes impossible. If the integration in (1) is undertaken analytically then the range of likelihood functions that can be considered is small, and the class of priors and functions of interest that can be considered is severely restricted. Many numerical approaches, like quadrature methods, require special adaptation for each $g$, $\pi$, or $L$, and become unworkable if $k$ exceeds, say, three. (Good surveys of these methods may be found in Davis and Rabinowitz, 1975; Hammersley and Handscomb, 1979; and Rubinstein, 1981.)

With the advent of powerful and cheap computing, numerically intensive methods for the computation of (1) have become more attractive. Monte Carlo integration with importance sampling provides a systematic approach that—in principle—can be applied in any situation in which $E[g(\underline{\theta})]$ exists, and is practical for large values of $k$. It was discussed by Hammersley and Handscomb (1964, Section 5.4) and brought to the attention of econometricians by Kloek and van Dijk (1978). The main idea is simple. Let $\{\underline{\theta}_i\}$ be a sequence of $k$-dimensional random vectors, i.i.d. in theory and in practice generated synthetically and therefore a very good approximation to an i.i.d. sequence. If the probability distribution function of the $\underline{\theta}_i$ is proportional to $p(\underline{\theta})$, then $n^{-1}\Sigma_{i=1}^{n} g(\underline{\theta}_i) \to E[g(\underline{\theta})]$; if it is proportional to $L(\underline{\theta})$, then $n^{-1}\Sigma_{i=1}^{n} g(\underline{\theta}_i)\pi(\underline{\theta}_i) \to E[g(\underline{\theta})]$. (Except in Section 4, all convergence is in $n$, the number of Monte Carlo replications. Almost sure convergence is indicated " $\to$ ", convergence in distribution " $\Rightarrow$ ".) Only in a very limited set of simple cases is it feasible to generate synthetic variates whose p.d.f. is in proportion to the posterior or the likelihood function (but the set does include common and interesting cases in which classical and analytical Bayesian inference fail; see Geweke, 1986). More generally, suppose that the probability distribution function of the $\underline{\theta}_i$ is $I(\underline{\theta})$, termed the *importance sampling density*. Then, under very weak assumptions given in Section 2,

$$(2) \qquad \bar{g}_n \equiv \sum_{i=1}^{n} [g(\underline{\theta}_i) p(\underline{\theta}_i)/I(\underline{\theta}_i)] \Big/ \sum_{i=1}^{n} [p(\underline{\theta}_i)/I(\underline{\theta}_i)] \to E[g(\underline{\theta})].$$

The rate of almost sure convergence in (2) depends critically on the choice of the importance sampling density. Under stronger assumptions, developed in Section 3,

$$(3) \qquad n^{1/2}\{\bar{g}_n - E[\bar{g}(\underline{\theta})]\} \Rightarrow N(0, \sigma^2),$$

and $\sigma^2$ may be estimated consistently. This result was indicated by Kloek and van Dijk (1978), but does not follow from the result of Cramer (1946) they cite, and is repeated by Bauwens (1984) but without explicit discussion of the moments whose existence is assumed. Loosely speaking, the importance sampling density should mimic the posterior density, and it is especially important that the tails of $I(\underline{\theta})$ not decay more quickly than the tails of $p(\underline{\theta})$. This is keenly appreciated by those who have done empirical work with importance sampling, including Zellner and Rossi (1984), Gallant and Monahan (1985), Bauwens and Richard (1985), and van Dijk, Kloek, and Boender (1985). These and other investigators have experienced substantial difficulties in tailoring importance sampling densities to the problem at hand. This is an important failing, for the approach is neither attractive nor methodical if special arcane problems in numerical analysis have to be resolved in each application.

This paper approaches these problems analytically, and presents new results which should make the application of Monte Carlo integration by importance sampling much more routine. Section 3 provides sufficient and easily verified conditions for (3). It also introduces a measure of relative numerical efficiency which is natural to use in evaluating the effectiveness of any given importance sampling density. Based on these considerations, Section 4 outlines a systematic approach to the choice of $I(\underline{\theta})$. It utilizes the local behavior of the posterior density at its mode, and a new class of importance sampling densities, the multivariate split-normal and multivariate split-Student. These developments are illustrated in Section 5, with the homogeneous Markov chain model, and in Section 6, with the ARCH linear model. Some directions for future research are taken up in the last section.

## 2. BAYESIAN INFERENCE WITH IMPORTANCE SAMPLING

We begin with some additional notation, and a set of assumptions basic to what follows. Since expectations are taken with respect to a variety of distributions, it is useful to denote $E_f[h(\underline{\theta})] = [\int_\Theta f(\underline{\theta})\,d\theta]^{-1}\int_\Theta f(\underline{\theta})h(\underline{\theta})\,d\underline{\theta}$ where $f(\underline{\theta})$ is proportional to a probability density and $h(\underline{\theta})$ is integrable with respect to the probability measure that is induced by $f(\underline{\theta})$. Similarly denote by $\text{var}_f[h(\underline{\theta})]$ and $md_f[h(\underline{\theta})]$ the variance and mean deviation of $h(\underline{\theta})$ under this probability measure, when the respective moments exist.

ASSUMPTION 1: *The product of the prior density, $\pi(\underline{\theta})$, and the likelihood function, $L(\underline{\theta})$, is proportional to a proper probability density function defined on $\Theta$.*

ASSUMPTION 2: *$\{\underline{\theta}_i\}_{i=1}^{\infty}$ is a sequence of i.i.d. random vectors, the common distribution having a probability density function $I(\underline{\theta})$.*

ASSUMPTION 3: *The support of $I(\underline{\theta})$ includes $\Theta$.*

ASSUMPTION 4: *$E[g(\underline{\theta})]$ exists and is finite.*

Assumption 1 is generally satisfied and usually not difficult to verify. It explicitly allows the prior density to be improper, so long as $\int_{\Theta} p(\underline{\theta})\, d\underline{\theta} \equiv c < \infty$. For $h(\underline{\theta})$ integrable with respect to the posterior probability measure, the simpler notation $\bar{h} \equiv E[h(\underline{\theta})] \equiv E_p[h(\underline{\theta})]$ (and similarly var$[h(\underline{\theta})]$ and md$[h(\underline{\theta})]$) will be used. Assumptions 2 and 3 are specific to the method of Monte Carlo integration with importance sampling; their role is obvious from the discussion in the introduction. Most inference problems can be cast in the form of determining $E[g(\underline{\theta})]$, through the function of interest, $g(\underline{\theta})$.

The value of $\bar{g}_n$ in (2) is invariant with respect to arbitrary scaling of the posterior and importance sampling densities. In certain situations it is convenient not to work with normalizing constants for $I(\underline{\theta})$, so we shall re-express $\bar{g}_n$ with $I^*(\underline{\theta}) = d^{-1}I(\underline{\theta})$ in place of $I(\underline{\theta})$. Denote the *weight function* $w(\underline{\theta}) \equiv p(\underline{\theta})/I^*(\underline{\theta})$; then $\bar{g}_n = \sum_{i=1}^n g(\underline{\theta}_i)w(\underline{\theta}_i)/\sum_{i=1}^n w(\underline{\theta}_i)$. Consistency of $\bar{g}_n$ for $g$ is a direct consequence of the strong law of large numbers.

THEOREM 1: *Under Assumptions* 1–4, $\bar{g}_n \to E[g(\underline{\theta})]$.

This simple result provides a foundation for the numerical treatment of a very wide array of practical problems in Bayesian inference. A few classes of examples are worth mention.

(i) If the value of a Borel measurable function of the parameters, $h(\underline{\theta})$, whose expectation exists and is finite, is to be estimated with a quadratic loss function, then $g(\underline{\theta}) = h(\underline{\theta})$.

(ii) If a decision is to be made between actions $a_1$ and $a_2$, according to the expectation of loss functions $l(\underline{\theta}; a_1)$ and $l(\underline{\theta}; a_2)$, then $g(\underline{\theta}) = l(\underline{\theta}; a_1) - l(\underline{\theta}; a_2)$.

(iii) Posterior probabilities may be assessed through indicator functions of the form $\chi(\underline{\theta}; S) = 1$ if $\underline{\theta} \in S$, 0 if $\underline{\theta} \notin S$. To evaluate the posterior probability $P[h(\underline{\theta}) \in A]$, let $g(\underline{\theta}) = \chi(\underline{\theta}; \{\underline{\theta}: h(\underline{\theta}) \in A\})$. Assumption 4 is satisfied trivially, so given the other assumptions $\bar{g}_n \to P[h(\underline{\theta}) \in A]$. Construction of posterior c.d.f.'s is a special case that is particularly relevant if $h(\underline{\theta})$ itself has no posterior moments. (An example is provided subsequently in Section 5; see especially Table IV.)

(iv) The application to c.d.f.'s suggests how to construct quantiles. Suppose $P[h(\underline{\theta}) \leqslant x] = \alpha$, and for a given $\varepsilon > 0$ $P[h(\underline{\theta}) \leqslant x - \varepsilon] < \alpha$, $P[h(\underline{\theta}) < x + \varepsilon] > \alpha$. If $q_n$ is defined to be any number such that $\sum_{i: h(\underline{\theta}_i) \leqslant q_n} w(\underline{\theta}_i)/\sum_{i=1}^n w(\underline{\theta}_i) \geqslant \alpha$ and $\sum_{i: h(\underline{\theta}_i) \geqslant q_n} w(\underline{\theta}_i)/\sum_{i=1}^n w(\underline{\theta}_i) \geqslant 1 - \alpha$, then by Theorem 1 $\lim_{n\to\infty} P[q_n \in (x - \varepsilon, x + \varepsilon)] = 1$. Hence if the c.d.f. of $h(\underline{\theta})$ is strictly increasing at $x$, the $\alpha$th quantile can be computed routinely as a byproduct of Monte Carlo integration with importance sampling.

(v) In forecasting problems functions of interest are of the form $g(\underline{\theta}) = P(x_{T+s} \in A \,|\, \underline{\theta}, X)$ and in this case $g(\underline{\theta})$ itself is an integral. Monte Carlo integration with an importance sampling density for the parameters may be combined with Monte Carlo integration for future realizations of the random vector $\underline{x}_t$ to form predictive densities (Geweke (1988b)).

### 3. EVALUATING THE IMPORTANCE SAMPLING DENSITY

Standing alone these results are of little practical value, because nothing can be said about rates of convergence and there is no formal guidance for choosing $I(\underline{\theta})$ beyond the obvious requirement that subsets of the support of the posterior should not be systematically excluded. Moreover, as noted in the introduction, for seemingly sensible choices of $I(\underline{\theta})$, $\bar{g}_n$ can behave badly. Poor behavior is usually manifested by values of $\bar{g}_n$ that exhibit substantial fluctuations after thousands of replications, which in turn can be traced to extremely large values of $w(\underline{\theta}_i)$ that turn up occasionally. The key problem is the distribution of the ratio of two means in $\bar{g}_n$, and this may be characterized by a central limit theorem.

THEOREM 2: *In addition to Assumptions* 1–4, *suppose*

$$E[w(\underline{\theta})] = c^{-1}\int_{\Theta}[p(\underline{\theta})^2/I^*(\underline{\theta})]\,d\underline{\theta}$$

*and*

$$E\left[g(\underline{\theta})^2 w(\underline{\theta})\right] = c^{-1}\int_{\Theta}\left[g(\underline{\theta})^2 p(\underline{\theta})^2/I^*(\underline{\theta})\right]d\underline{\theta}$$

*are finite. Let*

$$\sigma^2 \equiv d^{-1}E\left\{[g(\underline{\theta})-\bar{g}]^2 w(\underline{\theta})\right\} = c^{-1}d^{-1}\int_{\Theta}[g(\underline{\theta})-\bar{g}]^2 w(\underline{\theta})p(\underline{\theta})\,d\underline{\theta},$$

$$\hat{\sigma}_n^2 \equiv \sum_{i=1}^{n}[g(\underline{\theta}_i)-\bar{g}_n]^2 w(\underline{\theta}_i)^2 \bigg/ \left[\sum_{i=1}^{n} w(\underline{\theta}_i)\right]^2.$$

*Then*

$$n^{1/2}(\bar{g}_n-\bar{g}) \Rightarrow N(0,\sigma^2),$$

$$n\hat{\sigma}_n^2 \to \sigma^2.$$

PROOF: Define $A(\underline{\theta}) \equiv g(\underline{\theta})w(\underline{\theta})$ and observe that $\bar{g}_n = [n^{-1}\Sigma_{i=1}^n A(\underline{\theta}_i)]/[n^{-1}\Sigma_{i=1}^n w(\underline{\theta}_i)]$. From Theorem 1, $n^{-1}\Sigma_{i=1}^n A(\underline{\theta}_i) \to cd\bar{g}$, $n^{-1}\Sigma_{i=1}^n w(\underline{\theta}_i) \to cd$. By a Central Limit Theorem (e.g. Billingsley (1979, Theorem 29.5)), $n^{1/2}(\bar{g}_n - \bar{g}) \Rightarrow N(0,\sigma^2)$,

$$\sigma^2 = c^{-2}d^{-2}\left\{\mathrm{var}[A(\underline{\theta})] + \bar{g}^2\,\mathrm{var}[w(\underline{\theta})] - 2\bar{g}\,\mathrm{cov}[A(\underline{\theta}), w(\underline{\theta})]\right\}$$

$$= c^{-2}d^{-1}\int_{\Theta}[g(\underline{\theta})-\bar{g}]^2 w(\underline{\theta})p(\underline{\theta})\,d\underline{\theta}. \qquad Q.E.D.$$

We shall refer to $\hat{\sigma}_n = (\hat{\sigma}_n^2)^{1/2}$ as the numerical standard error of $\bar{g}_n$.

The conditions of Theorem 2 must be verified, analytically, if that result is to be used to assess the accuracy of $\bar{g}_n$ as an approximation of $E[g(\underline{\theta})]$. Failure of these conditions does not vitiate the use of $\bar{g}_n$: there is no compelling reason why $n^{1/2}(\bar{g}_n - \bar{g})$ needs to have a limiting distribution. But it is essential to have some indication of $|\bar{g}_n - \bar{g}|$, and this task becomes substantially more difficult if a central limit theorem cannot be applied. Practicality suggests that $I(\underline{\theta})$ be chosen so that Theorem 2 applies. The additional assumptions in Theorem 2 typically can be verified by showing either

$$(4) \qquad w(\underline{\theta}) < \bar{w} < \infty \quad \forall \underline{\theta} \in \Theta, \quad \text{and} \quad \text{var}[g(\underline{\theta})] < \infty;$$

$$(5) \qquad \theta \text{ is compact}, \quad \text{and} \quad p(\theta) < \bar{p} < \infty, \quad I(\underline{\theta}) > \varepsilon > 0, \quad \forall \underline{\theta} \in \Theta.$$

Demonstration of (4) involves comparison of the tail behaviors of $p(\underline{\theta})$ and $I(\underline{\theta})$. Demonstration of (5) is generally simple, although $\Theta$ may be compact only after a nontrivial reparameterization of the prior and the likelihood. Meeting these conditions does not establish the reliability of $\bar{g}_n$ and $\hat{\sigma}_n^2$ in any practical sense: examples in Section 5 demonstrate uncontrived cases in which (5) applies but $\hat{\sigma}_n^2$ would be unreliable even for $n \approx 10^{20}$. The strength of these conditions is, rather, that they provide a starting point to use numerical methods in constructing $I(\underline{\theta})$. We shall return to this endeavor in Section 4.

The expression for $\sigma^2$ in Theorem 2 indicates that the numerical standard error is adversely affected by large var$[g(\underline{\theta})]$, and by large relative values of the weight function. The former is inherent in the function of interest and in the posterior density, but the latter can in principle be controlled through the choice of the importance sampling density. A simple benchmark for comparing the adequacy of importance sampling densities is the numerical standard error that would result if the importance sampling density were the posterior density itself, i.e., $I^*(\underline{\theta}) \propto p(\underline{\theta})$. In this case $\sigma^2 = \text{var}[g(\underline{\theta})]$ and the number of replications controls the numerical standard error relative to the posterior standard deviation of the function of interest—e.g., $n = 10{,}000$ implies the former will be one percent of the latter. This is a very appealing metric for numerical accuracy.

Only in special cases is it possible or practical to construct synthetic variates whose distribution is the same as the posterior. But since var$[g(\underline{\theta})]$ can be computed as a routine byproduct of Monte Carlo integration, it is possible to see what the numerical variance would have been, had it been possible to generate synthetic random vectors directly from the posterior distribution. Define the *relative numerical efficiency* of the importance sampling density for the function of interest $g(\underline{\theta})$,

$$RNE \equiv \text{var}[g(\underline{\theta})]/d^{-1}E\{[g(\underline{\theta}) - \bar{g}]^2 w(\underline{\theta})\} = \text{var}[g(\underline{\theta})]/\sigma^2.$$

The $RNE$ is the ratio of number of replications required to achieve any specified numerical standard error using the importance sampling density $I(\underline{\theta})$, to the number required using the posterior density as the importance sampling density. The numerical standard error for $\bar{g}_n$ is the fraction $(RNE \cdot n)^{-1/2}$ of the posterior standard deviation.

Low values of $RNE$ indicate that there exists an importance sampling density (namely, the posterior density itself) that does not have to be tailored specifically to the function of interest, and provides substantially greater numerical efficiency. Thus they alert the investigator to the possibility that more efficient, yet practical, importance sampling densities might be found. If one is willing to consider importance sampling densities that depend on the function of interest—which will be impractical if expected values of many functions of interest are to be computed—then even greater efficiencies can generally be achieved. A lower bound on $\sigma^2$ for all such functions can be expressed.

THEOREM 3: *In addition to Assumptions* 1–4 *suppose that* $md[g(\underline{\theta})]$ *is finite. Then the importance sampling density that minimizes* $\sigma^2$ *has kernel* $|g(\underline{\theta}) - \bar{g}|\, p(\theta)$, *and for this choice* $\sigma^2 = \{md[g(\underline{\theta})]\}^2$.

PROOF: From Theorem 2, $\sigma^2 = c^{-2}\int_{\Theta} K(\underline{\theta})^2/I(\underline{\theta})\, d\underline{\theta}$, when we impose the constraint $d^{-1} = \int_{\Theta} I(\underline{\theta})\, d\underline{\theta} = 1$ and define $K(\underline{\theta}) = |g(\underline{\theta}) - \bar{g}|\, p(\underline{\theta})$. Following Rubinstein (1981, Theorem 4.3.1; the restriction to bounded $\Theta$ is inessential), $\sigma^2$ is minimized by $I(\underline{\theta}) \propto |K(\underline{\theta})|$. *Q.E.D.*

Theorem 3 has some interesting direct applications. An important class of cases consists of model evaluation problems: $\underline{\theta} \in \Theta^*$ or $\underline{\theta} \in \bar{\Theta}^*$, $g(\underline{\theta})$ is an indicator function for $\Theta^*$, and so $E[g(\underline{\theta})] = P[\underline{\theta} \in \Theta^*] \equiv p^*$. Then $\sigma^2$ is minimized if $I(\underline{\theta}) \propto (1 - p^*)p(\underline{\theta})$ for $\underline{\theta} \in \Theta^*$ and $I(\underline{\theta}) \propto p^* p(\underline{\theta})$ for $\underline{\theta} \in \bar{\Theta}^*$: half the drawings should be made in $\Theta^*$ and half in $\bar{\Theta}^*$, and in proportion to the posterior in each case.

It appears impractical to construct the densities with kernel $|g(\underline{\theta}) - \bar{g}|\, p(\underline{\theta})$, in general: a different function would be required for each $g(\underline{\theta})$; a preliminary evaluation of $\bar{g}$ with a less efficient method would be a prerequisite; and methods for sampling from those importance sampling densities would need to be devised. However, Theorem 3 suggests that importance sampling densities with tails thicker than the posterior density (a characteristic of densities with kernel $|g(\underline{\theta}) - \bar{g}|\, p(\underline{\theta})$) might be more efficient than the posterior density itself (and hence have $RNE > 1$) although still suboptimal. This is found in practice with some frequency, and some examples are given in Section 5.

## 4. CHOOSING THE IMPORTANCE SAMPLING DENSITY

There are two logical steps in choosing the importance sampling density. The first is to determine a class of densities that satisfy the conditions of Theorem 2, usually by using (4) or (5). The second step is to find a density within that class that attains a satisfactory *RNE* for the functions of interest. The first objective can only be achieved analytically. The second is a well-defined problem in its own right, that can be solved numerically.

When $\Theta$ is compact the first step will usually be trivial. For example, if the classical regularity conditions for the asymptotic normal distribution of the maximum likelihood estimator $\hat{\underline{\theta}}$ obtain, with $\hat{\underline{\theta}} \stackrel{.}{\sim} N(\underline{\theta}^*, V)$, then the $N(\hat{\underline{\theta}}, V)$ importance sampling density satisfies (5). If $\Theta$ is not compact, it will generally be necessary to investigate the tail behavior of the likelihood function to verify (4). There are many instances in which $L(\underline{\theta})$ behaves like $k\theta_i^{-q}$ for large values of $\theta_i$ and it is not difficult to determine $q$. Given a multivariate normal prior on $\underline{\theta}$ the tail of the marginal posterior density in $\theta_i$ will behave like $\exp(-c\theta_i^2)$ for large values of $\theta_i$ and a multivariate normal importance sampling density may be appropriate. On the other hand, if the prior is improper, and flat for one or more parameters, then in these instances no multivariate normal importance sampling density will satisfy Theorem 2. An instructive leading example is the computation of the expected value of a function of the mean of a univariate normal population with unknown mean and variance, using the $N(\hat{\mu}, s^2/n)$ importance sampling density. It is not hard to see that (a) Theorem 2 does not apply; (b) if one computes $\hat{\sigma}_n^2$ and $\hat{\text{var}}(\bar{g}_n) \equiv \Sigma_{i=1}^n[g(\underline{\theta}_i) - \bar{g}_n]^2 w(\theta_i)/[\Sigma_{i=1}^n w(\theta_i)]^2$, then the computed $RNE$ approaches 0 almost surely as $n \to \infty$; (c) if $n$ is held fixed, computed $RNE \to 1$ as sample size approaches infinity. These results are typical of cases in which the classical asymptotic expansion of the log-likelihood function is valid, and $\Theta$ is not compact. They underscore the importance of investigating the tail behavior of the likelihood function analytically rather than numerically.

Once the tail behavior of the likelihood function is characterized a class of importance sampling densities is usually suggested. For example, if the tail is multivariate Student with variance matrix $\Sigma$ and degrees of freedom $\nu$, a multivariate $t$ importance sampling density with weakly larger variance and weakly smaller degrees of freedom will satisfy Theorem 2. In general let $I(\underline{\theta}; \underline{\alpha})$ be a family of importance sampling densities indexed by the $g \times 1$ vector $\underline{\alpha}$, and let $w(\underline{\theta}; \underline{\alpha}) \equiv p(\underline{\theta})/I(\underline{\theta}; \underline{\alpha})$ be the corresponding family of weight functions. It will not usually be practical to choose a distinct importance sampling density for each function of interest. A reasonable objective is the minimization of $E[w(\underline{\theta}, \underline{\alpha})] \propto \int_{\Theta}[p(\underline{\theta})^2/I(\underline{\theta}; \underline{\alpha})]\, d\underline{\theta}$. This function appears in the first condition of Theorem 2 and in the denominator of each $E[g(\underline{\theta})]$; it is precisely the function one would minimize if $g(\underline{\theta})$ were an indicator function for $\Theta^*$, $p(\Theta^*) = 1/2$.

Two related classes of importance sampling densities have turned out to be quite useful in our experience with this approach. They involve densities that have not, to our knowledge, been described in the literature. The heuristic idea is to begin with minus the inverse of the Hessian of the log posterior density evaluated at its mode as the variance matrix of a multivariate normal importance sampling density, and shift to the multivariate Student $t$, if warranted, with degrees of freedom indicated by the tail behavior of the posterior density. If the prior is diffuse, the asymptotic variance of the maximum likelihood estimator may be substituted for minus the inverse of the Hessian of the log posterior density evaluated at the mode. In practice either importance sampling density can be poor, because the Hessian poorly predicts (especially if it underpredicts) the value of the posterior density away from the mode, because the posterior density is substantially asymmetric, or both. Illustrations of these problems are provided in the next two sections. To cope with them we modify the multivariate normal or Student $t$, comparing the posterior density and importance sampling density in each direction along each axis, in each case adjusting the importance sampling density to reduce $E[w(\underline{\theta}, \underline{\alpha})]$. With $k$ parameters, there are $2k$ adjustments in the last step, so $\underline{\alpha}$ is a $2k \times 1$ vector.

A little more formally, let $\text{sgn}^+(\ )$ be the indicator function for nonnegative real numbers, and let $\text{sgn}^-(\ ) = 1 - \text{sgn}^+(\ )$. The $k$-variate split normal density $N^*(\underline{\mu}, T, \underline{q}, \underline{r})$ is readily described by construction of a member $\underline{x}$ of its population:

$$\underline{\varepsilon} \sim N(\underline{0}, I_k);$$

$$\eta_i = [q_i \,\text{sgn}^+(\varepsilon_i) + r_i \,\text{sgn}^-(\varepsilon_i)]\,\varepsilon_i \qquad (i = 1, \ldots, k);$$

$$\underline{x} = \underline{\mu} + T\underline{\eta}.$$

The log-p.d.f. at $\underline{x}$ is (up to an additive constant)

$$-\sum_{i=1}^{n} [\log(q_i)\text{sgn}^+(\varepsilon_i) + \log(r_i)\text{sgn}^-(\varepsilon_i)] - (1/2)\underline{\varepsilon}'\underline{\varepsilon}.$$

In the application here, $\underline{\mu}$ is the posterior mode, and $T$ is a factorization such that the inverse of $TT'$ is the negative of the Hessian of the log posterior density evaluated at the mode. A variate from the multivariate split Student density $t^*(\underline{\mu}, T, \underline{q}, \underline{r}, \nu)$ is constructed the same way, except that $\eta_i = [q_i \,\text{sgn}^+(\varepsilon_i) + r_i \,\text{sgn}^-(\varepsilon_i)]\varepsilon_i(\zeta/\nu)^{-1/2}$, with $\zeta \sim \chi^2(\nu)$. The log-p.d.f. is (up to an additive constant)

$$-\sum_{i=1}^{n} [\log(q_i)\text{sgn}^+(\varepsilon_i) + \log(r_i)\text{sgn}^-(\varepsilon_i)] - [(\nu + k)/2]\log(1 + \nu^{-1}\underline{\varepsilon}'\underline{\varepsilon}).$$

Choosing $\underline{\alpha}' = (\underline{q}', \underline{r}')$ to minimize $E[w(\underline{\theta}; \underline{\alpha})]$ requires numerical integration at each step of the usual algorithms for minimization of a nonlinear function. Solution of this optimization problem with any accuracy would, in most applications, likely be more time consuming than the computation of $E[g(\underline{\theta})]$ itself. A much more efficient procedure can be used to select values for $\underline{q}$ and $\underline{r}$ that, in many applications, produces high RNE's. The intuitive idea is to explore each axis in each direction, to find the slowest rate of decline in the posterior density relative to a univariate normal (or Student) density, and then choose the variance of the normal (or Student) density to match that slowest rate of decline. A little more formally, let $\underline{e}^{(i)}$ be a $k \times 1$ indicator vector, $e_i^{(i)} = \underline{e}^{(i)\prime}\underline{e}^{(i)} = 1$. For the split normal define $f_i(\delta)$ according to

$$\text{(6)} \quad \begin{aligned} &p(\hat{\underline{\theta}} + \delta T\underline{e}^{(i)})/p(\hat{\underline{\theta}}) = \exp[-\delta^2/2f_i(\delta)^2] \\ &\Rightarrow f_i(\delta) = |\delta|\{2[\log p(\hat{\underline{\theta}}) - \log p(\hat{\underline{\theta}} + \delta T\underline{e}^{(i)})]\}^{1/2} \end{aligned}$$

Then take $q_i = \sup_{\delta > 0} f_i(\delta)$ and $r_i = \sup_{\delta < 0} f_i(\delta)$. For the split Student the procedure is the same except that

$$f_i(\delta) = \nu^{-1/2}|\delta|\left\{\left[p(\hat{\theta})/p(\underline{\hat{\theta}} + \delta T\underline{e}^{(i)})\right]^{2/(\nu+k)} - 1\right\}^{-1/2}.$$

In practice, carrying out the evaluation for $\delta = 1/2, 1, \ldots, 6$, seems to be satisfactory.

The two examples that follow illustrate how these methods are applied.

## 5. EXAMPLES: THE BINOMIAL AND MARKOV CHAINS

In this section we take up a set of successively complex examples in which the parameter space $\Theta$ is compact. We begin with situations which can be studied analytically in some detail, and consequently there would be no need for a numerical approach at all. This is strictly for illustrative purposes. As will be seen, not much elaboration is required to generate realistic models in which there are no analytical solutions.

### 5.1. *A Simple Binomial Model*

Let a sample of size $N$ be drawn from a population in which $P[x=1]=\theta$, $P[x=0]=1-\theta$, and suppose there are $M$ occurrences of "1". If the prior is $\pi(\theta)=1$, $0 \leqslant \theta \leqslant 1$, then the posterior is

$$p(\theta) = [B(M+1, N-M+1)]^{-1}\theta^{M}(1-\theta)^{N-M} \quad (0 \leqslant \theta \leqslant 1). \tag{7}$$

Since $N^{1/2}(\hat{\theta}-\theta) \Rightarrow N(\theta, \theta(1-\theta))$, a normal approximation based on the asymptotic distribution of the m.l.e. $\hat{\theta}$ is

$$I(\theta) = \left[2\pi N^{-1}\hat{\theta}(1-\hat{\theta})\right]^{1/2}\exp\left\{-N(\hat{\theta}-\theta)^2/2\hat{\theta}(1-\hat{\theta})\right\}.$$

Since $0 \leqslant \theta \leqslant 1$ condition (5) is satisfied.

Consider two cases. In Case A, $N=69$, $M=6$, $\hat{\theta}=.087$, the asymptotic standard error for $\hat{\theta}$ is .034 and the asymptotic variance is .00115. In Case B, $N=71$, $M=54$, $\hat{\theta}=.761$, the asymptotic standard error for $\hat{\theta}$ is .051, and the asymptotic variance is .00256. For the corresponding normal importance sampling densities Figure 1 shows the weight functions $w(\theta)=p(\theta)/I(\theta)$ indicated by the heavy lines. In each case $w(\theta)$ appears badly behaved, attaining a value of over $10^{70}$ for $\theta$ around .9 in Case A, and attaining a value of over $10^4$ for $\theta$ between .2 and .3 in Case B. For our purposes it is $E[w(\theta)] = \int_\Theta w(\theta)p(\theta)\,d\theta$ and not $w(\theta)$ which matters. The light lines in Figure 1 show $w(\theta)p(\theta)$, which is much less than $w(\theta)$ when $w(\theta)$ is large. In Case A $E[w(\theta)] = 1.65 \times 10^{24}$, and in Case B $E[w(\theta)] = 1.05$. Clearly the importance sampling density $I(\theta)$ is adequate in the latter case but not the former. The difficulty in Case A is that the likelihood function declines more slowly than its asymptotic normal approximation, for $\theta > \hat{\theta}$. A split normal importance sampling density with variance .00115 for $\theta < \hat{\theta}$ and variance greater than .00115 for $\theta > \hat{\theta}$ performs better, as shown in Figure 2. As the latter variance increases, $E[w(\theta)]$ decreases exponentially, until the variance reaches .0016; $E[w(\theta)]$ attains its minimum at .0019, and then increases slowly.

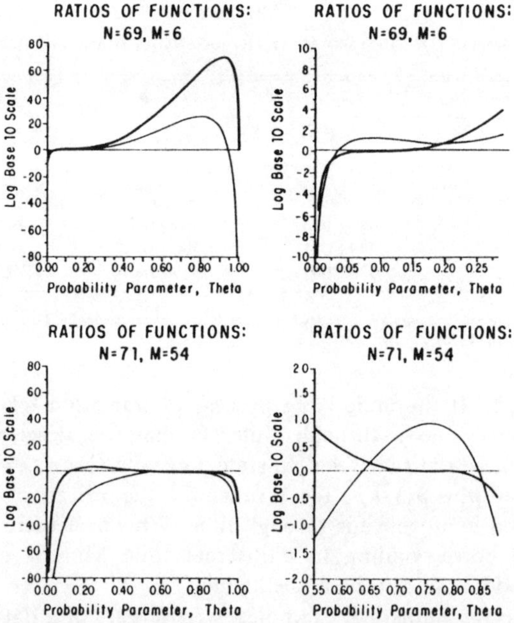


FIGURE 1.—Values of $\log(w(\theta))$ (heavy lines) and $\log(w(\theta)p(\theta))$ (light lines) for the two binomial models discussed in Section 5.1.

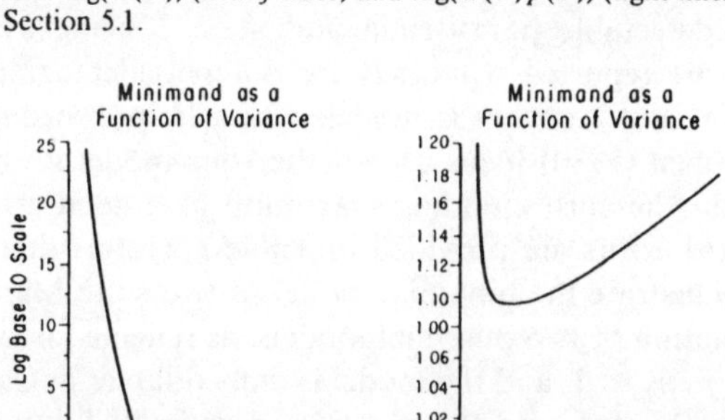


FIGURE 2.—Values of $E(w(\theta))$ as a function of the variance of a split normal importance sampling density, centered at $\hat{\theta}$, for Case A discussed in Section 5.1.

## 5.2. *The Two-State Homogeneous Markov Chain Model*[2]

Suppose an agent can occupy one of two states, with $P$[In state $i$ at time $t$ | In state $j$ at time $t-1$] $= p_j$, $i \neq j$. Let $N$ agents be observed at times $t-1$ and $t$, with $m_{ij}$ agents in state $i$ at time $t-1$ and state $j$ at time $t$. With a prior density $\pi(p_1, p_2) = 1$, $0 < p_i < 1$ $(i = 1, 2)$, the posterior p.d.f. is proportional to

$$p(p_1, p_2) = p_1^{m_{12}} p_2^{m_{21}} (1 - p_1)^{m_{11}} (1 - p_2)^{m_{22}}, \tag{8}$$

[2]For elaboration on the Markov chain model and the problem of embeddability, see Singer and Spilerman (1976) or Geweke, Marshall, and Zarkin (1986a, 1986b).

TABLE I

THREE EXAMPLES FOR THE TWO-STATE HOMOGENEOUS MARKOV CHAIN MODEL[a]

| | Case I | Case II | Case III |
|---|---|---|---|
| $m_{11}$ | 63 | 21 | 68 |
| $m_{12}$ | 6 | 66 | 28 |
| $m_{21}$ | 17 | 6 | 17 |
| $m_{22}$ | 54 | 24 | 4 |
| $\hat{p}_1$ | .087 (.034) | .759 (.046) | .292 (.046) |
| $\hat{p}_2$ | .239 (.051) | .200 (.073) | .810 (.086) |
| $\hat{p}_1 + \hat{p}_2$ | .326 (.061) | .959 (.086) | 1.102 (.097) |

[a] Carats denote mlc's: asymptotic standard errors are shown parenthetically. Data are taken from Singer and Cohen (1980).

$0 < p_i < 1$ ($i = 1, 2$). If the underlying process of transition takes place continuously and homogeneously through time, it may be described in terms of $\lim_{\delta \to 0} \delta^{-1} P[\text{In state } i \text{ at time } t + \delta | \text{In state } j \text{ at time } t] = r_j$, $i \neq j$. If $p_1 + p_2 < 1$, then $r_j = p_j \log(1 - p_1 - p_2)/(p_1 + p_2)$; otherwise the process of transition cannot be continuous and homogeneous through time. Whenever there exists a continuous time model corresponding to a discrete time Markov chain model, the discrete time model is said to be embeddable.

To provide specific numerical examples, we use part of a data set reported by Singer and Cohen (1980). The data pertain to the incidence of malaria in Nigeria, state 1 being no detectable parasitemia and state 2 being a positive test for parasitemia. Data are reported separately for different demographic groups. The process of infection and recovery is modeled as a homogeneous Markov chain, and Singer and Cohen (1980) focus on whether this model is embeddable. Data for three cases, and the corresponding maximum likelihood estimates and their asymptotic standard errors are provided in Table I. Notice that the Case I data are those used to illustrate the binomial model; a two-state Markov chain model is simply a combination of two binomial models, as reflected in comparison of (7) and (8). In Case I $\hat{p}_1 \hat{p}_2 \ll 1$, and the model is embeddable; in Case II $\hat{p}_1 + \hat{p}_2 < 1$ and in Case III $\hat{p}_1 + \hat{p}_2 > 1$, and the question of embeddability is open.

We conduct inference in this model with two priors and six functions of interest. The first prior is uninformative: $\pi(p_1, p_2) = 1$ for all $p_j \in (0, 1)$. The second prior imposes embeddability but otherwise is the same as the first: $\pi(p_1, p_2) = 2$ for all $(p_i, p_j)$: $p_i > 0$, $p_j > 0$, $p_1 + p_2 < 1$. The functions of interest are $p_1$, $p_2$, $p_1^{-1}$, $p_2^{-1}$, $r_1^{-1}$, and $r_2^{-1}$. The function $p_j^{-1}$ is the mean duration in state $j$ in the discrete time model, while $r_j^{-1}$ is the mean duration in state $j$ assuming a continuous time model. Since $r_j^{-1}$ is undefined if $p_1 + p_2 \geqslant 1$, $E(r_j^{-1})$ is defined only under the second prior; $E[r_j]$ is infinite for either prior, although the posterior distribution of $r_j$ may be obtained readily as suggested in examples (iii) and (iv) of Section 2.

Results obtained using the methods proposed in earlier sections are presented in Table II. Asterisks in column 1 denote the use of the second prior; otherwise the first is used. The first block of columns provides results with the normal

TABLE II

INFERENCE IN THE TWO-STATE HOMOGENEOUS MARKOV CHAIN MODEL[a]

| Importance Density $g(\ )$[b] | Normal $E[g]$ | Normal $sd[g]$ | Normal $100\hat{\sigma}_n$ | Normal $RNE$[c] | Split Normal $E[g]$ | Split Normal $sd[g]$ | Split Normal $100\hat{\sigma}_n$ | Split Normal $RNE$[c] | Likelihood Function $E[g]$ | Likelihood Function $sd[g]$ | Likelihood Function $100\hat{\sigma}_n$ | Likelihood Function $RNE$[c] | Actual $E[g]$ | Actual $sd[g]$ |
|---|---|---|---|---|---|---|---|---|---|---|---|---|---|---|
| Case I | | | | | | | | | | | | | | |
| $p_1$ | .098 | .034 | .051 | .441 | .099 | .035 | .033 | 1.13 | .099 | .035 | .035 | 1.00 | .099 | .035 |
| $p_2$ | .246 | .049 | .056 | .769 | .247 | .051 | .050 | 1.01 | .248 | .050 | .050 | 1.00 | .247 | .050 |
| $p_1^{-1}$ | 11.7 | 4.98 | 5.09 | .960 | 11.7 | 4.94 | 4.15 | 1.41 | 11.6 | 5.37 | 5.37 | 1.00 | 11.7 | 4.98 |
| $p_2^{-1}$ | 4.25 | .926 | .970 | .911 | 4.24 | .934 | .891 | 1.10 | 4.21 | .910 | .910 | 1.00 | 4.24 | .925 |
| $r_1^{-1*}$ | 9.60 | 4.35 | 4.44 | .961 | 9.58 | 4.32 | 3.62 | 1.42 | 9.55 | 4.68 | 4.68 | 1.00 | — | — |
| $r_2^{-1*}$ | 3.49 | .887 | .932 | .905 | 3.48 | .897 | .854 | 1.10 | 3.45 | .874 | .874 | 1.00 | — | — |
| Case II | | | | | | | | | | | | | | |
| $p_1$ | .753 | .046 | .051 | .808 | .753 | .045 | .044 | 1.05 | .752 | .046 | .046 | 1.00 | .753 | .046 |
| $p_2$ | .218 | .070 | .083 | .720 | .219 | .072 | .070 | 1.05 | .220 | .072 | .072 | 1.00 | .219 | .072 |
| $p_1^{-1}$ | 1.33 | .083 | .096 | .759 | 1.33 | .083 | .081 | 1.05 | 1.33 | .084 | .084 | 1.00 | 1.33 | .083 |
| $p_2^{-1}$ | 5.18 | 2.09 | 1.92 | 1.19 | 5.16 | 2.07 | 1.77 | 1.37 | 5.15 | 2.06 | 2.06 | 1.00 | 5.17 | 2.08 |
| $p_1^*$ | .739 | .043 | .060 | .526 | .740 | .043 | .052 | .686 | .739 | .044 | .054 | .647 | — | — |
| $p_2^*$ | .183 | .050 | .064 | .629 | .182 | .049 | .060 | .672 | .183 | .050 | .063 | .647 | — | — |
| $p_1^{-1*}$ | 1.36 | .082 | .117 | .491 | 1.36 | .081 | .098 | .681 | 1.36 | .082 | .103 | .647 | — | — |
| $p_2^{-1*}$ | 6.00 | 2.13 | 2.21 | .928 | 6.01 | 2.09 | 2.18 | .921 | 5.97 | 2.07 | 2.58 | .647 | — | — |
| $r_1^{-1*}$ | .472 | .129 | .170 | .576 | .472 | .130 | .163 | .642 | .471 | .132 | .164 | .647 | — | — |
| $r_2^{-1*}$ | 2.18 | 1.23 | 1.30 | .891 | 2.18 | 1.22 | 1.30 | .880 | 2.17 | 1.22 | 1.51 | .647 | — | — |
| Case III | | | | | | | | | | | | | | |
| $p_1$ | .296 | .045 | .051 | .810 | .296 | .046 | .046 | 1.01 | .297 | .046 | .046 | 1.00 | .296 | .046 |
| $p_2$ | .782 | .084 | .118 | .507 | .784 | .083 | .082 | 1.03 | .781 | .085 | .085 | 1.00 | .783 | .084 |
| $p_1^{-1}$ | 3.46 | .558 | .583 | .915 | 3.46 | .565 | .547 | 1.06 | 3.46 | .556 | .556 | 1.00 | 3.46 | .562 |
| $p_2^{-1}$ | 1.29 | .153 | .246 | .387 | 1.29 | .152 | .152 | .998 | 1.30 | .156 | .156 | 1.00 | 1.29 | .154 |
| $p_1^*$ | .269 | .041 | .145 | .080 | .268 | .041 | .092 | .201 | .270 | .041 | .089 | .208 | | |
| $p_2^*$ | .669 | .062 | .257 | .057 | .670 | .060 | .138 | .189 | .668 | .062 | .136 | .208 | — | — |
| $p_1^{-1*}$ | 3.80 | .600 | 1.82 | .108 | 3.82 | .615 | 1.32 | .217 | 3.80 | .595 | 1.30 | .208 | — | — |
| $p_2^{-1*}$ | 1.51 | .150 | .686 | .048 | 1.51 | .147 | .340 | .185 | 1.51 | .152 | .333 | .208 | — | — |
| $r_1^{-1*}$ | 1.21 | .394 | .116 | .115 | 1.22 | .419 | 0.93 | .206 | 1.21 | .408 | .893 | .208 | — | — |
| $r_2^{-1*}$ | .488 | .170 | .698 | .060 | .488 | .173 | .401 | .185 | .490 | .179 | .391 | .208 | — | — |

[a] Expected values of functions of interest $g(\ )$ were computed by Monte Carlo integration with importance sampling, using 10,000 replications.

[b] Prior is $\pi(p_1, p_2) = 1$ if no asterisk; with asterisk, prior is $\pi(p_1, p_2) = 2$ if $p_1 + p_2 < 1$. In Case I, essentially all the mass of the posterior satisfies $p_1 + p_2 < 1$.

[c] Relative numerical efficiency.

importance sampling density motivated by the asymptotic distribution: $p_1$ and $p_2$ are independent, $p_j \sim N(\hat{p}_j, \hat{p}_j(1 - \hat{p}_j)/(m_{j1} + m_{j2}))$. The second block of columns provides results with the split normal importance sampling density constructed as described in Section 4. Since the posterior density is the product of a binomial in $p_1$ and a binomial in $p_2$, each $p_j$ may be expressed as a ratio involving two independent chi-square random variables (Johnson and Kotz (1970, Section 24.2)). Hence it is straightforward to use the likelihood function itself as the importance sampling density; the third block of columns provides results obtained this way. Finally, the posterior moments of $p_j$ and $p_j^{-1}$ have simple analytical expressions under the first prior, and these are provided in the last block of columns. (Under the second prior the posterior moments of $p_j$ and

$p_j^{-1}$ could probably be computed more efficiently by numerical evaluation of the incomplete beta, but this has not been pursued because it does not extend to multi-state Markov chain models.)

In Case I, $P[p_1+p_2<1]\doteq 1$ under the first prior; none of the 10,000 replications with any importance sampling density turned up $p_1+p_2>1$. In Case II, $P[p_1+p_2<1]\doteq .65$ and in Case III $P[p_1+p_2<1]\doteq .21$, under the first prior. The computed expected values of all functions of interest with each importance sampling density are quite similar to each other and (where available) to the actual values. There are no surprises given the computed numerical standard errors. There are greater relative differences among the computed standard deviations of the functions of interest, and it may be verified analytically that this must be the case. One could, of course, compute numerical standard errors for the posterior standard deviations as well. The relative numerical efficiencies indicate that, with a few small exceptions in Case II, the split normal is a better importance sampling density than the normal. The superiority of the split normal increases systematically as $E(p_j)$ approaches zero and the asymptotic normal becomes a poorer approximation to the likelihood function. Many examples of relative numerical efficiencies in excess of 1.0 turn up with the split normal importance sampling density, precisely where we would expect: $E(p_j)$ closer to zero, and moments of the $p_j$ showing greater dispersion, like $p_j^{-1}$. By construction RNE is 1.0 for importance sampling from the posterior density, and for importance sampling from the likelihood function it is the posterior probability of an imposed prior, evaluated assuming a prior that is uniform in the parameter space of the likelihood function.

TABLE III

SOME DIAGNOSTICS FOR COMPUTATIONAL ACCURACY IN THE TWO-STATE HOMOGENEOUS MARKOV CHAIN MODEL

| | Normal Sampling Density | | Split Normal Sampling Density | |
|---|---|---|---|---|
| | $n = 10{,}000$ | $n = 50{,}000$ | $n = 10{,}000$ | $n = 50{,}000$ |
| Case I | | | | |
| $RNE, p_1$ | .441 | .269 | 1.137 | 1.139 |
| $RNE, p_2$ | .769 | .657 | 1.014 | 1.012 |
| $\omega_1$ | 186.2 | 1,774.7 | 2.5 | 2.5 |
| $\omega_{10}$ | 72.9 | 497.0 | 2.5 | 2.5 |
| Case II | | | | |
| $RNE, p_1$ | .808 | .774 | 1.054 | 1.047 |
| $RNE, p_2$ | .720 | .564 | 1.054 | 1.053 |
| $\omega_1$ | 31.2 | 277.9 | 1.9 | 1.9 |
| $\omega_{10}$ | 17.5 | 106.3 | 1.9 | 1.9 |
| Case III | | | | |
| $RNE, p_1$ | .810 | .790 | 1.011 | 1.009 |
| $RNE, p_2$ | .507 | .462 | 1.032 | 1.030 |
| $\omega_1$ | 101.0 | 348.7 | 1.8 | 1.8 |
| $\omega_{10}$ | 52.8 | 161.3 | 1.8 | 1.8 |

From the example of Section 5.1, we know that the population *RNE* for the normal importance sampling density in Case I should be on the order of $10^{-24}$. The normal is such a bad importance sampling density that, even with $10^4$ replications, it will almost certainly not produce the values of the $p_j$ that lead to difficulties, and computed *RNE* will bear no resemblance to population *RNE*. As the number of replications increases computed *RNE*'s tend to decline in situations like this, as is clear from consideration of Figure 1. Table III contrasts some computed *RNE*'s for $n = 10{,}000$ and $n = 50{,}000$ and bears out the difficulties with Case I. In these situations the largest weights $w(\underline{\theta}_i)$ increase rapidly as the number of replications, $n$, increases, and order statistics pertaining to the $w(\underline{\theta}_i)$ are often useful in identifying very poorly conditioned importance sampling densities. Since the largest $w(\underline{\theta}_i)$'s may be retained systematically as $n$ increases and $\Sigma_{i=1}^n w(\underline{\theta}_i)$ and $\Sigma_{i=1}^n w(\underline{\theta}_i)^2$ are computed in any event, the diagnostic of the form $\omega_m = (n/m)\Sigma_{i=1}^m w^{(n+1-i)}(\underline{\theta})^2/\Sigma_{i=1}^n w(\underline{\theta}_i)^2$ (where superscripts denote order statistics and $m$ is small) can be quite useful as an indication of thin tails in the importance sampling density relative to the posterior. The values of these diagnostics for our examples are provided in Table III, again for $n = 10{,}000$ and $n = 50{,}000$; they indicate—correctly—that the split normal is a much better importance sampling density than is the normal.

### 5.3. *The Quartile Homogeneous Markov Chain Model*[3]

Consider last a four-state homogeneous Markov chain model, state $j$ denoting the $j$th quartile of income. Let $p_{ij} \equiv P[\text{In state } j \text{ at time } t \,|\, \text{In state } i \text{ at time } t-1]$, and arrange the $p_{ij}$ in a $4 \times 4$ matrix $P$. Since $\Sigma_{i=1}^4 p_{ij} = \Sigma_{i=1}^4 p_{ji} = 1$ for $j = 1, \ldots, 4$, the likelihood function has 9 free parameters and is not proportional to a product of multinomial densities. We suppose that the process of transition between income quartiles takes place in continuous time; define

$$r_{ij} = \lim_{\delta \to 0} \delta^{-1} P[\text{In state } j \text{ at time } t+\delta \,|\, \text{In state } i \text{ at time } t]$$

for all $i \neq j$, and let $r_{ii} = -\Sigma_{j \neq i} r_{ij}$. If the discrete time model is embeddable, then $R = Q \log[\Lambda] Q^{-1}$, where $Q$ is the matrix whose columns are right eigenvectors of $P$ and $\Lambda$ is the diagonal matrix of corresponding eigenvalues. A given discrete time model is embeddable if and only if the off-diagonal elements of $Q \log[\Lambda] Q^{-1}$ are all real and nonnegative and the diagonal elements are all real and nonpositive for some combination of the branches of the logarithms of the eigenvalues. The problems of inference subject to the constraint of embeddability is ill suited to classical treatment, and a purely analytical Bayesian approach is precluded by the complexity of the indicator function for embeddability.

[3]Details on the construction of the normal importance sampling density for this example may be found in Geweke, Marshall, and Zarkin (1986a). Interest in this problem in general and in measures of mobility in particular is motivated by Geweke, Marshall, and Zarkin (1986b).

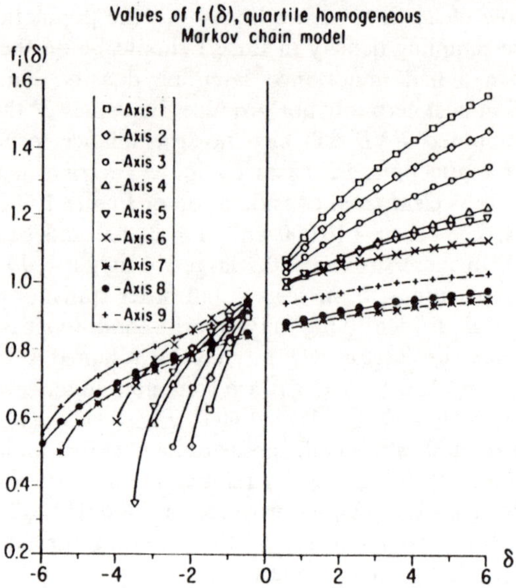


FIGURE 3.—Values of $f_i(\delta)$ along nine orthogonal axes of the likelihood function, constructed as described in Section 5.3.

The $9 \times 1$ vector of parameters for the likelihood function consists of those $p_{ij}$ for which $j \neq i$ and $j \neq i+1$, ordered so that the $\hat{p}_{ij}$ appear in ascending order.[4] The split normal importance sampling density was constructed as described in Section 4, with $T$ being the lower triangular Choleski factorization of the aymptotic variance matrix of $\underline{\theta}$. Some idea of the behavior of the likelihood surface is conveyed in Figure 3, which shows values of the $f_i(\delta)$ defined in (6). For the lower-numbered axes, the boundary of the parameter space is only a few standard deviations in the negative direction, and so $f_i(\delta)$ is truncated on the left. For these axes $f_i'(\delta)$ is greater for $\delta > 0$ than it is for the higher-numbered axes, indicating that the likelihood function tapers most slowly relative to the asymptotic normal approximation in these directions. This is not surprising in view of the example in Section 5.1.

The functions of interest are $-r_{jj}^{-1}$, mean duration in each state, $r_{jj}$, rate of transition to and from each state, and three summary measures of mobility that have been proposed in the literature: $M^*(R) = -.25\,\mathrm{tr}(R)$, $M_B^*(R) = .25\Sigma_{i=1}^4 \Sigma_{j=1}^4 |i-j| r_{ij}$, and $M_2^*(R) = -\log(\lambda_2)$ where $\lambda_2$ is the second largest eigenvalue of $P$. We impose the prior $\pi(P) \propto 1$ if $P$ is embeddable and $\pi(P) = 0$ otherwise. Under this prior posterior means of the $r_{jj}^{-1}$, but not the $r_{jj}$ or the mobility measures, exist. The data set was extracted from the National Longitudinal Survey data file, consisting of those 460 white males who were not enrolled

[4] For details, see the Appendix of Geweke, Marshall, and Zarkin (1986a).

in school, employed full time, and married with spouse present during 1968, 1969, and 1970; and whose family incomes were coded for 1969, 1970, and 1971.

Monte Carlo integration with 10,000 replications, using the normal and split normal importance sampling densities was carried out. The normal importance sampling density performed very poorly: $\omega_1 = 2{,}829.6$. The corresponding value for the split normal was 22.1. The posterior probability of embeddability is .246 (with a numerical standard error of .0009, computed using the split normal importance sampling density). This probability, rather than 1.0, provides the norm against which *RNE*'s should be compared, for any importance sampling density based on the likelihood function rather than the posterior.

The computed values in Table IV are essentially the same, for expected values and medians of functions of interest, with the two importance sampling densities employed. Higher moments, and quantile points farther from the median, show greater differences. The computed *RNE*'s suggest that the split normal is substantially more efficient than the normal. However, for the normal importance sampling density computed *RNE* tends to vary widely with each set of 10,000 replications, and deteriorates as the number of replications increases. In this application, and—we conjecture—in most applications with more than a few dimensions, careful modification of asymptotic normal and other symmetric distributions is required if computed numerical standard errors are to be interpreted reliably using Theorem 2.

TABLE IV

INFERENCE IN THE QUARTILE HOMOGENEOUS MARKOV CHAIN MODEL[a]

| Importance Density $g(\;)$ | Split Normal $E[g]$ | $sd[g]$ | $100\hat{\sigma}_n$ | $RNE$[b] | Normal $E[g]$ | $sd[g]$ | $100\hat{\sigma}_n$ | $RNE$[b] |
|---|---|---|---|---|---|---|---|---|
| $-r_{11}^{-1}$ | 2.35 | .338 | .786 | .184 | 2.36 | .341 | .919 | .138 |
| $-r_{22}^{-1}$ | 1.33 | .170 | .393 | .187 | 1.34 | .170 | .470 | .131 |
| $-r_{33}^{-1}$ | 1.33 | .162 | .380 | .182 | 1.33 | .169 | .546 | .095 |
| $-r_{44}^{-1}$ | 2.28 | .324 | .786 | .169 | 2.28 | .323 | 1.056 | .094 |

| Importance Density $g(\;)$ | Split Normal Quantile .01 | .25 | .50 | .75 | .99 | Normal Quantile .01 | .25 | .50 | .75 | .99 |
|---|---|---|---|---|---|---|---|---|---|---|
| $-r_{11}^{-1}$ | 1.68 | 2.10 | 2.33 | 2.56 | 3.28 | 1.68 | 2.12 | 2.32 | 2.56 | 3.21 |
| $-r_{22}^{-1}$ | .987 | 1.22 | 1.32 | 1.45 | 1.76 | .965 | 1.21 | 1.32 | 1.44 | 1.76 |
| $-r_{33}^{-1}$ | .979 | 1.21 | 1.32 | 1.43 | 1.74 | .973 | 1.21 | 1.32 | 1.44 | 1.75 |
| $-r_{44}^{-1}$ | 1.60 | 2.05 | 2.26 | 2.48 | 3.13 | 1.65 | 2.05 | 2.25 | 2.48 | 3.14 |
| $-r_{11}$ | .304 | .391 | .428 | .477 | .594 | .311 | .390 | .430 | .472 | .589 |
| $-r_{22}$ | .566 | .692 | .759 | .821 | 1.01 | .566 | .691 | .755 | .825 | 1.03 |
| $-r_{33}$ | .572 | .698 | .760 | .823 | 1.02 | .570 | .694 | .757 | .823 | 1.02 |
| $-r_{44}$ | .319 | .403 | .443 | .488 | .621 | .318 | .402 | .444 | .488 | .598 |
| $M^*(R)$ | .485 | .565 | .598 | .637 | .741 | .487 | .563 | .597 | .638 | .745 |
| $M_B^*(R)$ | .586 | .675 | .718 | .760 | .881 | .586 | .676 | .716 | .759 | .864 |
| $M_2^*(R)$ | .317 | .435 | .810 | 1.03 | 1.48 | .316 | .418 | .795 | 1.00 | 1.49 |

[a] Expected values, standard deviations, and fractiles of functions of interest $g(\;)$ were computed by Monte Carlo integration with importance sampling from a split normal density, using 10,000 replications.
[b] Relative numerical efficiency.

## 6. EXAMPLES: ARCH LINEAR MODELS

In this section we take up an example in which the parameter space $\Theta$ is not compact. It involves constraints which cannot readily be imposed in classical inference, as discussed in detail elsewhere (Geweke (1988a)). For this, among other reasons, exact Bayesian inference is appealing in this model.

The ARCH linear model was proposed by Engle (1982) to allow persistence in conditional variance in economic time series. Let $\underline{x}_t$: $k \times 1$ and $y_t$ be time series for which the distribution of $y_t$ conditional on $\psi_{t-1} = \{\underline{x}_{t-s}, y_{t-s}, s \geqslant 1\}$ is

$$(9) \qquad y_t | \psi_{t-1} \sim N(\underline{x}_t'\underline{\beta}, h_t).$$

Defining $\varepsilon_t = y_t - \underline{x}_t'\underline{\beta}$, take

$$h_t = \alpha_0 + \sum_{j=1}^{p} \alpha_j \varepsilon_{t-j}^2 = h(\varepsilon_{t-1}, \ldots, \varepsilon_{t-p}; \underline{\gamma}_{q \times 1}).$$

The parameterization in terms of $\underline{\gamma}$ allows restrictions like $\alpha_0 = \gamma_0$, $\alpha_j = \gamma_1(p+1-j)$, the linearly declining weights employed by Engle (1982, 1983). For (9) to be plausible it is necessary that $\alpha_0 > 0$ and $\alpha_j \geqslant 0$ $(j = 1, \ldots, p)$. For $\{\varepsilon_t\}$ to be stationary it is necessary and sufficient that the roots of $1 - \sum_{j=1}^{p} \alpha_j z^j$ all be outside the unit circle.

Given the sample $(\underline{x}_t, y_t; t = 1, \ldots, T)$ the log-likelihood function is (up to an additive constant)

$$(10) \qquad l = -(1/2) \sum_{t=p+1}^{T} \log h_t - (1/2) \sum_{t=p+1}^{T} h_t^{-1} \varepsilon_t^2.$$

Engle (1982) has shown that for large samples $\partial^2 l / \partial \underline{\beta} \, \partial \underline{\gamma}' \doteq 0$, and the maximum likelihood estimates $\hat{\underline{\beta}}$ of $\underline{\beta}$ and $\hat{\underline{\gamma}}$ of $\underline{\gamma}$ are asymptotically independent. A scoring algorithm may be used to maximize the likelihood function, and revision of the estimate $\underline{\gamma}^{(i)}$ of $\underline{\gamma}$ and $\underline{\beta}^{(i)}$ of $\underline{\beta}$ may proceed separately at each step. We programmed the procedure described in Engle (1982) and then replicated other results (Engle (1983)) using data furnished by the *Journal of Money, Credit, and Banking*.

Since $\hat{\underline{\beta}}$ and $\hat{\underline{\gamma}}$ are asymptotically independent we construct the importance sampling density as the product of a density for $\underline{\beta}$ and a density for $\underline{\gamma}$. To employ condition (4) and the procedures of Section 4 first compare the tail behavior of the multivariate $t$ density in $m$ dimensions,

$$f(\underline{x}; \underline{\mu}, \Sigma, \nu) \propto \left[1 + \nu^{-1}(\underline{x} - \underline{\mu})'\Sigma^{-1}(\underline{x} - \underline{\mu})\right]^{-(\nu+m)/2},$$

with that of the likelihood function. Allowing $|x_j| \to \infty$ while fixing all other $x_i$, the log of $f(\underline{x}; \underline{\mu}, \Sigma, \nu)$ behaves like $-(\nu + m)\log|x_j|$. The log likelihood function (10) behaves like $-[(T-p)/2]\log(\gamma_j)$ as $\gamma_j \to \infty$ with all other parameters fixed, and like $-(T-p)\log|\beta_j|$ as $|\beta_j| \to \infty$ with all other parameters fixed. Hence for $\underline{\beta}$ we shall use a split Student importance sampling density with $T - p - k$ degrees

TABLE V

INFERENCE IN THE ARCH LINEAR MODEL[a]

| $g(\ )$ | $E[g]$ | $sd[g]$ | $100\hat{\sigma}_n$ | $RNE$[b] |
|---|---|---|---|---|
| $\beta_1$ | 1.096 | .092 | .110 | .700 |
| $\beta_2$ | .954 | .135 | .155 | .757 |
| $\gamma_0$ | 1.001 | .291 | .373 | .609 |
| $\gamma_1$ | .281 | .066 | .079 | .693 |
| $P$[Stable] | .795 | — | .475 | .722 |

[a] Expected values and standard deviations of functions of interest $g(\ )$ were computed by Monte Carlo integration with importance sampling from a split Student density, using 10,000 replications.
[b] Relative numerical efficiency.

of freedom, and for $\gamma$ we shall use a split Student importance sampling density with $(T-p)/2-q$ degrees of freedom. The variance matrices are given by expressions (28) and (32), respectively, in Engle (1982).

An artificial sample of size 200 provides a concrete example:

$$y_t = \sum_{j=1}^{2} \beta_j x_{tj} + \varepsilon_t, \qquad \beta_1 = \beta_2 = 1;$$

$$x_{t1} = 1, \qquad x_{t2} = \cos(2\pi t/200);$$

$$\varepsilon_t \sim N(0, h_t), \qquad h_t = 1 + .5\varepsilon_{t-1}^2 + .25\varepsilon_{t-2}^2.$$

The parameterization of the conditional variance process in the model is $h_t = \gamma_0 + \gamma_1(2\varepsilon_{t-1}^2 + \varepsilon_{t-2}^2)$; $\alpha_0 = \gamma_0$, $\alpha_1 = 2\gamma_1$, $\alpha_2 = \gamma_1$. We assume a flat prior on $\beta$ and $\gamma$, and estimate these four parameters and assess the posterior probability of stability $P[|\gamma_1| < 1/3]$. Results using the split Student importance sampling density constructed as described in Section 4 are presented in Table V.

The local behavior of the log-likelihood function is portrayed in Figure 4. For each of the four axes this figure indicates the actual log-likelihood function (solid thick line), the multivariate "$t$" approximation to the likelihood function using the information matrix (Engle (1982, (28) and (32))) (dotted line), and the split Student approximation to the likelihood function constructed as described in Section 4 (thin line). (In the case of $\gamma$, the curves terminate at $\gamma_0 = 0$ or $\gamma_1 = 0$.) For $\beta$ the classical asymptotic theory provides a good approximation to the likelihood function, but for $\gamma$ the approximation is poor.

In Table VI we compare diagnostics for computational accuracy using the unmodified and split normal densities (which are unjustified theoretically since $E[w(\theta)]$ is infinite in these cases) and the unmodified and split Student densities. The unmodified densities perform quite poorly, for reasons made clear in Figure 4. The split normal and split Student sampling densities lead to acceptable performance, in the sense that $RNE$ is apparently of the same order of magnitude as would be achieved if one could sample directly from the likelihood function. In the case of the split normal we know this is an illustration, but from

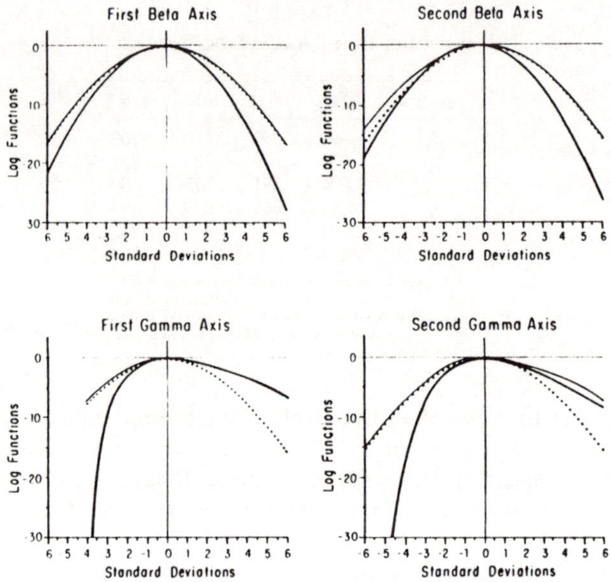


FIGURE 4.—All functions are normalized to have the value 1 at 0 standard deviations, which corresponds to the asymptotic maximum likelihood estimator. The solid thick line is the log likelihood function; the dotted line is the log multivariate $t$ importance sampling density; and the solid thin line is the log split Student importance sampling density.

TABLE VI

SOME DIAGNOSTICS FOR COMPUTATIONAL ACCURACY ARCH LINEAR MODEL

| | Multivariate Normal Sampling Density | | Split Normal Sampling Density | |
|---|---|---|---|---|
| | $n = 10{,}000$ | $n = 50{,}000$ | $n = 10{,}000$ | $n = 50{,}000$ |
| $RNE, \beta_1$ | .006 | .016 | .686 | .666 |
| $RNE, \beta_2$ | .005 | .014 | .682 | .708 |
| $RNE, \gamma_0$ | .001 | .002 | .564 | .529 |
| $RNE, \gamma_1$ | .006 | .012 | .748 | .714 |
| $RNE, p$ | .018 | .044 | .744 | .738 |
| $\omega_1$ | 9,872.8 | 39,893.6 | 51.0 | 225.4 |
| $\omega_{10}$ | 994.4 | 4,857.2 | 22.3 | 66.0 |
| | **Multivariate $t$ (99) Sampling Density** | | **Split Student Sampling Density** | |
| | $n = 10{,}000$ | $n = 50{,}000$ | $n = 10{,}000$ | $n = 50{,}000$ |
| $RNE, \beta_1$ | .036 | .068 | .700 | .707 |
| $RNE, \beta_2$ | .027 | .065 | .757 | .742 |
| $RNE, \gamma_0$ | .002 | .006 | .609 | .501 |
| $RNE, \gamma_1$ | .017 | .041 | .692 | .690 |
| $RNE, p$ | .049 | .118 | .722 | .741 |
| $\omega_1$ | 9,726.9 | 37,067.8 | 19.9 | 232.3 |
| $\omega_{10}$ | 979.6 | 4,476.4 | 16.0 | 70.9 |

TABLE VII

SOME DIAGNOSTICS FOR COMPUTATIONAL ACCURACY
ARCH LINEAR MODEL

| | Multivariate Sampling Density with Indicated Degrees of Freedom $n = 10{,}000$ | | | |
|---|---|---|---|---|
| | $\nu = 0.5$ | $\nu = 1.0$ | $\nu = 1.5$ | $\nu = 2.0$ |
| $RNE, \beta_1$ | .144 | .302 | .402 | .451 |
| $RNE, \beta_2$ | .150 | .307 | .410 | .455 |
| $RNE, \gamma_0$ | .127 | .243 | .303 | .313 |
| $RNE, \gamma_1$ | .129 | .262 | .339 | .365 |
| $RNE, p$ | .127 | .282 | .368 | .417 |
| $\omega_1$ | 52.2 | 42.2 | 41.2 | 40.2 |
| $\omega_{10}$ | 47.2 | 32.3 | 29.8 | 34.2 |
| | $\nu = 3.0$ | $\nu = 4.0$ | $\nu = 6.0$ | $\nu = 12.0$ |
| $RNE, \beta_1$ | .536 | .529 | .487 | .356 |
| $RNE, \beta_2$ | .540 | .525 | .537 | .430 |
| $RNE, \gamma_0$ | .336 | .317 | .271 | .197 |
| $RNE, \gamma_1$ | .410 | .399 | .366 | .292 |
| $RNE, p$ | .493 | .504 | .522 | .495 |
| $\omega_1$ | 53.4 | 69.0 | 136.3 | 321.4 |
| $\omega_{10}$ | 34.2 | 42.7 | 71.2 | 130.0 |

Figure 4 astronomical values of $n$ are required before $RNE$ is mainly determined by the relative tail behavior of the likelihood function and importance sampling density. Diagnostics for the split Student and split normal importance sampling densities are essentially the same.

The alternative approach to Monte Carlo integration with importance sampling for posterior densities with tails thick relative to the normal has been to employ multivariate $t$ distributions with small degrees of freedom (Zellner and Rossi (1984), van Dijk, Hop, and Louter (1986)). We conclude this example with an examination of diagnostics for computational accuracy for this procedure. Beginning with the asymptotic variance matrices for $\underline{\beta}$ and $\underline{\gamma}$, multivariate importance sampling densities with $\nu = .5$, 1, 1.5, 2, 3, 4, 6, and 12 degrees of freedom were used. The resulting diagnostics for numerical accuracy are given in Table VII. $RNE$ for the four parameters attains a maximum at $\nu = 3$, that for $p$ at $\nu = 6$. For $\nu < 3$ the sampling density is too diffuse, and as $\nu$ becomes larger it approaches the multivariate normal which is much too compact. That $\omega$-diagnostics are sensitive to sampling densities with tails that are too thin rather than too thick is clearly indicated in Table VII. The $RNE$ of the multivariate $t$ with low degrees of freedom is decidedly less than that of the split Student $t$, but of the same order of magnitude for proper choice of $\nu$. In the absence of any systematic or theoretical basis for choosing $\nu$, however, costly numerical experimentation is required to find the appropriate value.

## 7. CONCLUSION

Integration by Monte Carlo with importance sampling provides a paradigm for Bayesian inference in econometric models. Thanks to increasingly cheap computing it is practical to obtain the exact posterior distributions of any function of interest of the parameters, subject to any diffuse prior. This is done through systematic exploration of the likelihood function, and the asymptotic sampling distribution theory for maximum likelihood estimators provides the organization for this exploration. Diffuse priors may incorporate inequality restrictions which arise frequently in applied work but are impractical if not impossible to handle in a classical setting. Concentrations of prior probability mass may be treated by applying the methods in this paper to both the original and lower dimensional problems, and combining results through the appropriate posterior odds ratios. By choosing functions of interest appropriately, formal and direct answers to the questions that motivate empirical work can be obtained. The investigator can routinely handle problems in inference that would be analytical nightmares if attacked from a sampling theoretic point of view, and is left free to craft his models and functions more according to substantive questions and less subject to the restrictions of what asymptotic distribution theory can provide.

Integration by Monte Carlo is an attractive research tool because it makes numerical problems much more routine than do other numerical integration methods. The principle analytical task left to the econometrician is the characterization of the tail behavior of the likelihood function. This characterization leads to a family of importance sampling densities from which a good choice can be made automatically. Prior distributions and functions of interest can then be specified at will, taking care that computed moments actually exist. There is no *guarantee* that these methods will produce computed moments of reasonable accuracy in a reasonable number (say, 10,000) of Monte Carlo replications. Pathological likelihood functions will demand more analytical work: for example analytical marginalization may be possible, and transformation of parameters to obtain more regular posterior densities can be quite helpful, but these tasks must be approached on a case-by-case basis. It is in precisely such cases that sampling theoretic asymptotic theory and normal or other approximations to the posterior are also likely to be inadequate: the difficulty is endemic to the model and not the approach. Clearly there is much to be learned about more complicated likelihood functions. In approaching these problems, the emphasis should be—as it has been in this paper—on generic rather than specific solutions.

*Institute of Statistics and Decision Sciences, Duke University, Durham, NC 27706, U.S.A.*

*Final manuscript received March, 1989.*

## REFERENCES

BAUWENS, W. (1984): *Bayesian Full Information Analysis of Simultaneous Equation Models Using Integration by Monte Carlo.* Berlin: Springer-Verlag.

BAUWENS, W., AND J. F. RICHARD (1985): "A 1-1 Poly-*t* Random Variable Generator with Application to Monte Carlo Integration," *Journal of Econometrics*, 29, 19–46.

BILLINGSLEY, P. (1979): *Probability and Measure*. New York: Wiley.

CRAMER, H. (1946): *Mathematical Methods of Statistics*. Princeton: Princeton University Press.
DAVIS, P. J., AND P. RABINOWITZ (1975): *Methods of Numerical Integration*. New York: Academic Press.
ENGLE, R. F. (1982): "Autoregressive Conditional Heteroscedasticity with Estimates of the Variance of United Kingdom Inflation," *Econometrica*, 50, 987–1008.
——— (1983): "Estimates of the Variance of U.S. Inflation Based on the ARCH Model," *Journal of Money, Credit, and Banking*, 15, 286–301.
——— (1984): "Wald, Likelihood Ratio and Lagrange Multiplier Tests in Econometrics," Chapter 13 of *Handbook of Econometrics*, Vol. II, ed. by Z. Griliches and M. D. Intriligator. Amsterdam: North-Holland.
GALLANT, A. R., AND J. F. MONAHAN (1985): "Explicitly Infinite-Dimensional Bayesian Analysis of Production Technologies," *Journal of Econometrics*, 30, 171–202.
GEWEKE, J. (1986): "Exact Inference in the Inequality Constrained Normal Linear Regression Model," *Journal of Applied Econometrics*, 1, 127–141.
——— (1988a): "Exact Inference in Models with Autoregressive Conditional Heteroscedasticity," in *Dynamic Econometric Modeling*, ed. by E. Berndt, H. White, and W. Barnett. Cambridge: Cambridge University Press, 73–104.
——— (1988b): "Antithetic Acceleration of Monte Carlo Integration in Bayesian Inference," *Journal of Econometrics*, 38, 73–90.
GEWEKE, J., R. C. MARSHALL, AND G. ZARKIN (1986a): "Exact Inference for Continuous Time Markov Chain Models," *Review of Economic Studies*, 53, 653–699.
——— (1986b): "Mobility Indices in Continuous Time Markov Chains," *Econometrica*, 54, 1407–1423.
HAMMERSLEY, J. M., AND D. C. HANDSCOMB (1964): *Monte Carlo Methods*. London: Methuen (First Edition).
——— (1979): *Monte Carlo Methods*. London: Chapman and Hall (Second Edition).
JOHNSON, N. C., AND S. KOTZ (1970): *Continuous Univariate Distributions-2*. New York: Wiley.
KLOEK, T., AND H. K. VAN DIJK (1978): "Bayesian Estimates of Equation System Parameters: An Application of Integration by Monte Carlo," *Econometrica*, 46, 1–20.
QUANDT, R. E. (1983): "Computational Problems and Methods," Chapter 12 of *Handbook of Econometrics*, Vol. I, ed. by Z. Griliches and M. D. Intriligator. Amsterdam: North-Holland.
RUBINSTEIN, R. Y. (1981): *Simulation and the Monte Carlo Method*. New York: Wiley.
SINGER, B., AND J. COHEN (1980): "Estimating Malaria Incidence and Recovery Rates from Panel Surveys," *Mathematic Biosciences*, 49, 273–305.
SINGER, B., AND S. SPILERMAN (1976): "The Representation of Social Processes by Markov Models," *American Journal of Sociology*, 82, 1–54.
VAN DIJK, H. K., J. P. HOP, AND A. S. LOUTER (1986): "An Algorithm for the Computation of Posterior Moments and Densities Using Simple Importance Sampling," Erasmus University Econometric Institute Report 8625/E.
VAN DIJK, H. K., T. KLOEK, AND C. G. E. BOENDER (1985): "Posterior Moments Computed by Mixed Integration," *Journal of Econometrics*, 29, 3–18.
ZELLNER, A., AND P. ROSSI (1984): "Bayesian Analysis of Dichotomous Quantal Response Models," *Journal of Econometrics*, 25, 365–394.

# [13]

S. E. Fienberg and A. Zellner (eds.)
Studies in Bayesian Econometrics and Statistics.


## 5.1 BAYESIAN THEORY OF IDENTIFICATION IN SIMULTANEOUS EQUATIONS MODELS*

Jacques H. DRÈZE
*University of Louvain*

## *1. Introduction*

### *1.1. Contents*

This paper treats, from a Bayesian viewpoint, the identification problem for parameters of the simultaneous equations model.

That model is introduced in section 1.2. The theory of identification for *parameters* of that model is summarized in section 2. The results in that section are proved, for instance, in [9]. A more detailed treatment can be found in [3].

Section 3 first points out the scope of a Bayesian approach, which permits replacing some of the deterministic constraints of a simultaneous equations model by probabilistic constraints—that is, by a prior density on parameters. This approach may lead the econometrician to analyze, with an informative prior density, models that are not identified. A simple theorem captures the implications of underidentification for the posterior density. The classical conditions for identification are then extended into necessary conditions for the existence of a proper posterior density.

* An earlier version of this paper was presented to the Third NBER-NSF Seminar on Bayesian Inference Econometrics, Harvard Univ., October 1971. I have greatly benefitted from the discussion during this meeting, and in particular from the remarks of L. J. Savage, to whose memory this paper is dedicated. I have also been influenced by the work of Joseph Kadane [5] and Dennis V. Lindley [6]. I am grateful to Anton Barten, Michel Mouchart, Jean-François Richard, Thomas Rothenberg, Tran Van Hoa, Léon Wegge and Arnold Zellner for their comments on an earlier version of this paper.

## 1.2. *The simultaneous equations model*

In its simplest form, the simultaneous equations model consists of a system of $m$ linear stochastic equations

$$By' + \Gamma z' = u' \tag{1}$$

where

$y'$ is an $m \times 1$ vector of endogenous variables;

$z'$ is an $n \times 1$ vector of exogenous variables;

$B$ is an $m \times m$ nonsingular matrix of coefficients, with $B_{ii} = 1$, $i = 1 \ldots m$ (normalization);

$\Gamma$ is an $m \times n$ matrix of coefficients;

$u'$ is an $m \times 1$ vector of unobserved disturbances.

Let $(Y, Z)$ denote a $T \times (m+n)$ matrix of observations, and $U = YB' + Z\Gamma'$ be the corresponding $T \times m$ matrix of unobserved disturbances. The data-generating process may be described by the joint density of the disturbances

$$\psi(U|Z)\,\mathrm{d}U = \psi(YB' + Z\Gamma')\,\mathrm{d}U. \tag{2}$$

Let $\mathring{\Sigma}$ denote the parameters of the density $\psi$—an arbitrary set. The joint density of the observations, or data density, is obtained from (2) by performing the integrand transformation (in $Tm$ space) from $U$ to $Y$, with Jacobian $\|B\|^T$. Thus:

$$p(Y|Z, B, \Gamma, \mathring{\Sigma})\,\mathrm{d}Y = \psi(YB' + Z\Gamma')\|B\|^T\,\mathrm{d}Y. \tag{3}$$

The set of parameters of this density is $(B, \Gamma, \mathring{\Sigma})$.

Since $B$ is nonsingular, we may solve (1) explicitly for $y$, obtaining:

$$y' = -B^{-1}\Gamma z' + B^{-1}u' =_{\text{def}} \Pi z' + v'. \tag{4}$$

Equation (4) is called the "reduced form" of the "structural system" (1). The $T \times m$ matrix of unobserved reduced form disturbances is

$$V = Y - Z\Pi' = UB'^{-1}. \tag{5}$$

The joint density of the reduced form disturbances is obtained from (2) by performing the integrand transformation from $U$ to $V$, with Jacobian $\|B\|^T$. Thus:

$$\phi(V|Z)\,\mathrm{d}V = \psi(VB')\|B\|^T\,\mathrm{d}V. \tag{6}$$

Let $\mathring{\Omega}$ denote the parameters of the density $\phi$—a set of functions of the parameters $(B, \mathring{\Sigma})$. From (5) and (6), we obtain an alternative expression for the data density, namely:

$$p(Y|Z, \Pi, \mathring{\Omega})\, \mathrm{d}Y = \phi(Y - Z\Pi')\, \mathrm{d}Y. \tag{7}$$

**Notation.** If $A$ is a matrix, vec $A$ denotes the row-vector formed from the elements of $A$, taken one row at a time.

## 2. *Observational equivalence and identification*

### 2.1. *Observational equivalence*

A *structure* $S$ is a complete specification of the probability distribution function $P(y)$ of a set of observations $y$. A set of structures $\mathscr{S}$ is called a *model*. A typical problem of statistical inference consists in making judgments (either deterministic or probabilistic) about structures, given the model $\mathscr{S}$ and the observations $y$. In most applications, $S$ is a point, and $\mathscr{S}$ is a set of points, in parameter space.

**Definition 2.1.** *Two structures $S$ and $\bar{S}$ in $\mathscr{S}$ are* "observationally equivalent" *if and only if $P(y|S) = P(y|\bar{S})$ identically in $y$.*

Given an arbitrary structure $S \in \mathscr{S}$, the set of all structures in $\mathscr{S}$ which are observationally equivalent to $S$ is an *equivalence class*; it will be denoted $\mathscr{E}(S)$.

**Application 2.1.** Consider the system (1) under the specification that $\psi\ (U|Z)$ is $N(\text{vec } U'|0, \Sigma \otimes I_T)$. The subset $\mathscr{S}$ of $R^{m(2m+n)}$, $\mathscr{S} = \{(B, \Gamma, \Sigma) \in R^{m(2m+n)} \,|\, |B| \neq 0, B_{ii} = 1, i = 1 \ldots m, \Sigma \text{ P.D.S.}\}$ defines a model. A point $S \in \mathscr{S}$ is a structure. Let

$$\Pi =_{\text{def}} -B^{-1}\Gamma, \qquad B\Pi + \Gamma = 0;$$
$$\Omega =_{\text{def}} B^{-1}\Sigma B'^{-1}, \quad B\Omega B' - \Sigma = 0. \tag{8}$$

It follows from (6) that $\phi(V|Z)$ is $N(\text{vec } V'|0, \Omega \otimes I_T)$, and from (7) that $p(Y|Z, \Pi, \Omega)$ is $N(\text{vec } Y'|\text{vec } \Pi Z', \Omega \otimes I_T)$. Consequently, the probability density of the data is completely specified by the reduced form parameters $(\Pi, \Omega)$. If $S$ and $\bar{S}$ are two points in $\mathscr{S}$ such that

$$B^{-1}\Gamma = \bar{B}^{-1}\bar{\Gamma}, \quad B^{-1}\Sigma B'^{-1} = \bar{B}^{-1}\bar{\Sigma}\bar{B}'^{-1}, \tag{9}$$

then $S$ and $\bar{S}$ are observationally equivalent.

### 2.2. *Identification*

**Definition 2.2.** (i) *The structure $S$ in $\mathscr{S}$ is* "identified" *if and only if $\mathscr{S}$ does not own any $\bar{S} \neq S$ such that $S$ and $\bar{S}$ are observationally equivalent; i.e. if and only if $\mathscr{E}(S) = \{S\}$.*

(ii) *The model $\mathscr{S}$ is* "identified" *if and only if every $S \in \mathscr{S}$ is identified; i.e., if and only if the partition of $\mathscr{S}$ into equivalence classes is a partition into singletons.*

**Application 2.2.** Under the normality assumption of application 2.1, the structural model $\mathscr{S} = \{(B, \Gamma, \Sigma) \in R^{m(2m+n)} \mid |B| \neq 0,\ B_{ii} = 1,\ i = 1 \ldots m, \Sigma$ P.D.S.$\}$ is not identified; but the reduced form model $\mathscr{S} = \{(\Pi, \Omega) \in R^{m(m+n)} | \Omega$ P.D.S.$\}$ is identified. Conditionally on non-singular $B$, the transformation from the parameter space $(B, \Gamma, \Sigma)$ to the parameter space $(B, \Pi, \Omega)$ is one-to-one. In the parameter space $(B, \Pi, \Omega)$, the subspace defined by "$B = \bar{B}, \Omega$ P.D.S." is an identified model; and the intersection of $\mathscr{S}$ with the subspace defined by "$\Pi = \bar{\Pi}, \Omega = \bar{\Omega}$" is an equivalence class.

### 2.3. *Local identification*

When $\mathscr{S}$ is a subset of a metric space, one can define the distance between two structures. It is then possible to introduce a weaker, very useful concept.

**Definition 2.3.** *The structure $S$ is* "locally identified" *if and only if, for some open neighborhood $\mathscr{B}(S)$, $\mathscr{S} \cap \mathscr{B}(S)$ does not own any $\bar{S} \neq S$ such that $S$ and $\bar{S}$ are observationally equivalent, i.e., if and only if $\mathscr{E}(S) \cap \mathscr{B}(S) = \{S\}$.*

**Application 2.3.** Under the normality assumption of application 2.1, the structural model

$$\mathscr{S} = \{(B, \Gamma, \Sigma) \in R^{m(2m+n)} \mid |B| \neq 0, \quad \Psi_k(B, \Gamma, \Sigma) = 0, \quad k = 1 \ldots K\} \tag{10}$$

is identified whenever the set of relations

$$B\Pi + \Gamma = 0, \quad B\Omega B' - \Sigma = 0, \quad \Psi_k(B, \Gamma, \Sigma) = 0, \quad k = 1 \ldots K, \tag{11}$$

can be solved *uniquely* for $(B, \Gamma, \Sigma)$ as functions of $(\Pi, \Omega)$.

In some cases ("just identification"), a solution exists for all $(\Pi, \Omega) \in R^{m(m+n)}$, $\Omega$ P.D.S.; in other cases ("overidentification"), a solution exists for some but not all $(\Pi, \Omega) \in R^{m(m+n)}$, $\Omega$ P.D.S. The uniqueness problem is raised for values of $(\Pi, \Omega)$ such that a solution exists.

(i) Let $(\beta, \gamma, \sigma)$ be the triplet of $1 \times m^2$, $1 \times mn$ and $1 \times m^2$ vectors (vec $B$, vec $\Gamma$, vec $\Sigma$). Let also

$$\Psi_\beta = \left[\frac{\partial \Psi_i}{\partial \beta_j}\right], \quad \Psi_\gamma = \left[\frac{\partial \Psi_i}{\partial \gamma_j}\right], \quad \Psi_\sigma = \left[\frac{\partial \Psi_i}{\partial \sigma_j}\right]$$

be the $K \times m^2$, $K \times mn$ and $K \times m^2$ matrices of partial derivatives of the constraints $\Psi_k$, $k = 1 \ldots K$, with respect to the structural parameters $(\beta, \gamma, \sigma)$. At points where the $K \times m^2$ matrix

$$[\Psi_\beta \ \Psi_\gamma \ \Psi_\sigma][I \otimes B \ \ I \otimes \Gamma \ \ I \otimes 2\Sigma]' = \Psi^*$$

is of rank $m^2$, solutions to (11) are locally unique. This sufficient condition is necessary for local uniqueness, at regular points of the matrix $\Psi^*$. When the constraints $\Psi_k$ are linear in $(B, \Gamma)$ and independent of $\Sigma$, local uniqueness implies global uniqueness.

(ii) A simple case is obtained when the conditions $\Psi_k = 0$, $k = 1 \ldots K \geqq m^2$, consist in assigning predetermined values to a subset of the parameters $(B, \Gamma)$.

An arbitrary equation from (1), say the first, may then be written as:

$$\bar{B}_{11} y_1' + B_{12} y_2' + \bar{\Gamma}_{11} z_1' + \Gamma_{12} z_2' = u_1 \tag{12}$$

where $(\bar{B}_{11}, \bar{\Gamma}_{11})$ is the subvector of predetermined parameters. The first row of the first matrix equation in (11) may be written as:

$$(\bar{B}_{11} \quad B_{12}) \begin{bmatrix} \Pi_{11} & \Pi_{12} \\ \Pi_{21} & \Pi_{22} \end{bmatrix} + (\bar{\Gamma}_{11} \quad \Gamma_{12}) = 0. \tag{13}$$

Equation (13) may be solved uniquely for $B_{12}$ and $\Gamma_{12}$ (as functions of $\Pi$) if and only if $\Pi_{21}$ has full row-rank (rank condition). This requires that the dimension of $\bar{\Gamma}_{11}$ should be at least equal to that of $B_{12}$ (order condition); i.e., that $(\bar{B}_{11}, \bar{\Gamma}_{11})$ be of dimension $m$ at least.

*Remarks* 2.3. (i) Definition 2.2 gives a precise content to the *concept* of identification, for structures and models. The *theory* of identification, for a specific class of parametric models, is concerned with alternative statements of necessary and/or sufficient conditions for a model to be identified. The restatements aim at providing easy checks of the fact that a particular model is, or is not identified. These checks can be applied directly to the (a priori) specifications of the model, without explicit reference to the likelihood. (Of course, analysis of the likelihood underlies the theory.) The order conditions for identification of a simultaneous equations model by means of exclusion restrictions provide a good example.

(ii) The ultimate purpose of a study of identification has been aptly linked to the design of experiments (see the remark by L. J. Savage, quoted by Kadane [5, p. 9]). The statistician who wishes to make inferences about certain parameters will attempt to design an experiment such that these parameters are identified. Because econometric research proceeds usually from nonexperimental data, that remark should be extended. The econometrician who is concerned with inference about parameters that are not identified may try to overcome this difficulty by collecting richer data, or by resorting to a more restrictive theory. For instance, he may switch from annual to quarterly data, and introduce a specific lag structure. A study of identification may then prove useful to guide his strategy for collecting data and developing theoretical specifications.

## 3. *A Bayesian extension*

### 3.1. *Scope of a Bayesian analysis*

Whether it be based upon theory or past observation, economic knowledge is typically subject to uncertainties of various kinds. Probabilistic statements would seem more adequate than deterministic statements to reflect such knowledge as we have. In general, model specification should thus be partly stochastic. The prior information would then be described by means of exact constraints on some of the parameters, as well as by means of a probability distribution function defined on some or all of the remaining parameters.

Classical methods are not very well suited to handle such situations. The econometrician relying on these methods will be tempted to disregard probabilistic prior information, or else to translate it into deterministic constraints.[1] It would be more natural to carry out the analysis

[1] I do not mean to deny the possibility of analyzing from a classical viewpoint models with random parameters, or pooled data. However, little use has been made of this possibility, in the econometric literature.

within the Bayesian framework. Statistical methods have been developed to that effect, for the simultaneous equations model [2, 4, 7, 8, 10].[2]

This approach may well lead to models that are not identified, according to the definitions given above, if one considers only the exact prior restrictions. The Bayesian econometrician should thus be prepared to carry out inference within unidentified models—a topic largely ignored (and understandably so) by classical econometric theory.

### 3.2. *Implications of underidentification*

To say that a model is unidentified is to say that some equivalence classes contain more than one element. These elements are observationally equivalent structures, the respective likelihoods of which are identical. Whatever judgements an econometrician might make about the relative probabilities of these structures must remain unaltered by observation. This common-sense remark is reflected in the theorem below about simultaneous equations models. Its counterpart, from a classical viewpoint, is that no tests can discriminate among alternative specifications of unidentified, or just identified, models.

Let, as in section 1.2, the set of structural parameters be $(B, \Gamma, \mathring{\Sigma})$, where $\mathring{\Sigma}$ is the set of parameters of the density of the disturbances, $\Psi(V|Z)$. The set of parameters of the density $\phi(V|Z)$ of reduced form disturbances is $(\mathring{\Omega})$, where the elements of $(\mathring{\Omega})$ are functions of $(B, \mathring{\Sigma})$. Let $f(B, \Gamma, \mathring{\Sigma})$ be the prior density on the structural parameters, and assume that, for given $B$, the transformation from $(\mathring{\Sigma})$ to $(\mathring{\Omega})$ is non-singular. One may then perform an integrand transformation from the parameter space $(B, \Gamma, \mathring{\Sigma})$ to the parameter space $(B, \Pi, \mathring{\Omega})$ to define the alternative prior density $h(B, \Pi, \mathring{\Omega})$. To simplify notation, it is assumed here that the normalization and other exact constraints are implicit in $f(B, \Gamma, \mathring{\Sigma})$; that is, the domain of $f$ is $\mathscr{S}$.

**Example 3.2.** When $\Psi(V|Z)$ is $N(\text{vec}\, U'|0, \Sigma \otimes I_T)$, then $(\mathring{\Sigma}) = \Sigma$,

[2] "This remark should bring comfort to those who regard underidentification as the natural state of an econometric model: indeed, stochastic prior information, even if imprecise, should enable them to proceed with structural estimation. Turn now to overidentification. In principle, deterministic restrictions are maintained, even if they conflict with the observations: the prior information dominates the sample information. In contrast, stochastic prior information gets combined with sample information through Bayes' theorem; neither type of information dominates the other, but both count in proportion to their respective precisions" [1, p. 9].

$(\mathring{\Omega}) = \Omega = B^{-1}\Sigma B'^{-1}$. The Jacobian of the transformation from $(B, \Gamma, \Sigma)$ to $(B, \Pi, \Omega)$ is $\|B\|^{n+m+1}$ and $h(B, \Pi, \Omega)\,\mathrm{d}B\,\mathrm{d}\Pi\,\mathrm{d}\Omega = f(B, -B\Pi, B\Omega B')\cdot\|B\|^{n+m+1}\cdot\mathrm{d}B\,\mathrm{d}\Pi\,\mathrm{d}\Omega$.

The prior density $h(B, \Pi, \mathring{\Omega})$ may always be factorized into a conditional density on $B$, given $(\Pi, \mathring{\Omega})$, and a marginal density on $(\Pi, \mathring{\Omega})$, that is:

$$h(B, \Pi, \mathring{\Omega})\,\mathrm{d}B\,\mathrm{d}\Pi\,\mathrm{d}\mathring{\Omega} \equiv h_c(B|\Pi, \mathring{\Omega})\,\mathrm{d}B\,h_m(\Pi, \mathring{\Omega})\,\mathrm{d}\Pi\,\mathrm{d}\mathring{\Omega}. \tag{14}$$

Let $h(B, \Pi, \mathring{\Omega}|\,Y, Z)$ be the posterior density, in parameter space $(B, \Pi, \mathring{\Omega})$, which obtains (from $h$) after the sample $(Y, Z)$ has been observed.

**Theorem 3.2.**

$$p(Y|Z, B, \Gamma, \mathring{\Sigma}) \underset{Y}{\equiv} p(Y|Z, \Pi, \mathring{\Omega})$$

*implies*

$$h(B, \Pi, \mathring{\Omega}\,|\,Y, Z)\,\mathrm{d}B\,\mathrm{d}\Pi\,\mathrm{d}\mathring{\Omega} = h_c(B|\Pi, \mathring{\Omega})\,\mathrm{d}B\,h_m(\Pi, \mathring{\Omega}|\,Y, Z)\,\mathrm{d}\Pi\,\mathrm{d}\mathring{\Omega},$$

*whenever the posterior density exists.*

*Remark.* The conclusion in theorem 3.2 could equivalently be written as: $h_c(B|\,Y, Z, \Pi, \mathring{\Omega}) \underset{Y}{\equiv} h_c(B|\Pi, \mathring{\Omega})$. Stated in this way, the conclusion remains valid even when $h_m(\Pi, \mathring{\Omega}|\,Y, Z)$ is not a proper density.

*Proof*

$$h(B, \Pi, \mathring{\Omega}|\,Y, Z) =_{\text{def}} \frac{h(B, \Pi, \mathring{\Omega})\cdot p(Y|Z, B, \Pi, \mathring{\Omega})}{\int h(B, \Pi, \mathring{\Omega})p(Y|Z, B, \Pi, \mathring{\Omega})\,\mathrm{d}B\,\mathrm{d}\Pi\,\mathrm{d}\mathring{\Omega}}$$

$$= \frac{h_c(B|\Pi, \mathring{\Omega})h_m(\Pi, \mathring{\Omega})p(Y|Z, \Pi, \mathring{\Omega})}{\int h_m(\Pi, \mathring{\Omega})p(Y|Z, \Pi, \mathring{\Omega})\cdot[\int h_c(B|\Pi, \mathring{\Omega})\,\mathrm{d}B]\,\mathrm{d}\Pi\,\mathrm{d}\mathring{\Omega}}$$

$$=_{\text{def}} h_c(B|\Pi, \mathring{\Omega})\cdot h_m(\Pi, \mathring{\Omega}|\,Y, Z),$$

where the definitions are given by Bayes' theorem.

*Remarks* 3.2. (i) A general theorem, of which the foregoing may be viewed as a special case, has been proved by Kadane [5, theorem 5].

Kadane comments upon my result as follows: "Because the map from the structure to the reduced form is identifying, the entire Bayesian analysis can be done on the reduced form" [5, p. 191].

(ii) In the parameter space $(B, \Pi, \mathring{\Omega})$ equivalence classes are obtained by fixing $\Pi$ and $\mathring{\Omega}$ (see application 2.2). Thus, the conditional density $h_c(B|\Pi, \mathring{\Omega})$ defines, for each equivalence class, a conditional probability measure over the elements of the class.

In lecturing on this subject, I have used the term "identification in probability" to characterize a situation where there exists a proper conditional prior density over the elements of every equivalence class. This choice was motivated by analogy with the use of the term "identification" to characterize a situation where every equivalence class is a singleton.

Reactions to this presentation have led me to recognize that it was misleading to use the word "identification" in defining a property of the prior density for the parameters of unidentified models. I agree with Kadane's view "that identification is a property of the likelihood function, and is the same whether considered classically or from the Bayesian approach" [5, p. 175].[3]

### 3.3. *Analysis of prior density*

Whenever the prior density function is well defined, and the likelihood is dominated, the posterior density will be well defined as well, whether the model is identified or not. This is the rationale for Lindley's remark [6, p. 46]: "In passing it might be noted that unidentifiability causes no real difficulty in the Bayesian approach." Indeed, the difficulty has been shifted to the level of defining a prior density on *all* the parameters of the model; in the context of simultaneous equations models, which always involve many parameters, elicitation of a proper prior density may raise serious difficulties. In particular, natural-conjugate prior densities will prove inadequate, in the case of unidentified models. Indeed, a natural-conjugate prior density may be expressed in terms of reduced form parameters only, because the likelihood may be so expressed.

[3] The "property" under consideration is that equivalence classes should be singletons. The property should be verified *after* all the relevant constraints have been embodied in the likelihood function. The property can sometimes be verified without full specification of the functional form of the likelihood function—see, e.g., [9, theorem 4].

When the prior density is expressed in terms of the parameters $(B, \Pi, \mathring{\Omega})$, the decomposition appearing in theorem 3.2 is suggestive, because it brings out explicitly a conditional density that will not be altered by the observations, namely $h_c(B|\Pi, \mathring{\Omega})$. Analysis of that conditional density reveals immediately certain characteristics of the posterior density. By contrast, the prior $h_m(\Pi, \mathring{\Omega})$ will be altered by the observations, to an extent that increases (loosely speaking) with the precision of the sample.

Typically, however, the prior information will be expressed more naturally in terms of the structural parameters $(B, \Gamma, \mathring{\Sigma})$. And one may expect the prior density to be truly informative about a subset of these parameters only. One may then wish to know whether the prior density on such a proper subset of structural parameters implies a proper prior density on $B$, given the reduced form parameters $(\Pi, \mathring{\Omega})$. In that case, one may also wish to evaluate the Jacobian of the transformation. A theory relating the prior information on an arbitrary subset of structural parameters to the implied prior information on $B$, given $(\Pi, \mathring{\Omega})$, plays, in the Bayesian analysis, a role exactly similar to the classical theory of identification, which relates an arbitrary set of constraints on the structural parameters to identification of the model. The application below shows that the similarity exists at the technical as well as the logical level.

**Application 3.3.** Consider once more the structural model

$$\mathscr{S} = \{(\beta, \gamma, \sigma) \in R^{m(2m+n)} \big| \, |B| \neq 0, \quad \Psi_k(\beta, \gamma, \sigma) = 0, \quad k = 1 \ldots K\}.$$

Let $(\tilde{\beta}, \tilde{\gamma}, \tilde{\sigma})$ be a *subvector* of $(\beta, \gamma, \sigma)$, with $\tilde{\beta}$, $\tilde{\gamma}$ and $\tilde{\sigma}$ of dimension $1 \times \mu$, $1 \times \nu$ and $1 \times \rho$ respectively, and define $\lambda = \mu + \nu + \rho$.

We want to investigate the relationship between a prior density on $(\tilde{\beta}, \tilde{\gamma}, \tilde{\sigma})$ and the prior density on $B$, conditionally on $(\Pi, \Omega)$; and we want to investigate that relationship under the constraints $\Psi_k(\beta, \gamma, \sigma) = 0$, $k = 1 \ldots K$.

A major technical simplification in the evaluation of Jacobians is obtained by treating $\Psi_k$, $k = 1 \ldots K$, as a random variable, with the implicit proviso that inference should proceed conditionally on $\Psi_k = 0$. This approach would not be well suited to calculate the posterior density but it is very convenient for our present purposes.

Let then $\psi = (\Psi_1 \ldots \Psi_k \ldots \Psi_K)$, a $1 \times K$ vector, and choose $(\tilde{\beta}, \tilde{\gamma}, \tilde{\sigma})$ of dimensionality $\lambda = m^2 - K$, by restricting attention to an appropriate subvector of $(\beta, \gamma, \sigma)$. One may then consider the integrand transformation between the $m^2$-dimensional vector $(\psi, \tilde{\beta}, \tilde{\gamma}, \sigma)$ and the $m^2$-dimensional vector $\beta$. Conditionally on $(\Pi, \Omega)$, that transformation is well defined by means of (8), provided the Jacobian of the transformation has full rank $m^2$. The transformation relates two alternative parametrizations of the prior density, namely $(B, \Pi, \Omega)$ and $(\psi, \tilde{\beta}, \tilde{\gamma}, \tilde{\sigma}, \Pi, \Omega)$. If the transformation is well defined, a proper prior density on $(\tilde{\beta}, \tilde{\gamma}, \tilde{\sigma})$, conditionally on $\psi(=0)$ and $(\Pi, \Omega)$, implies a proper prior density on $B$, conditionally on $\psi(=0)$ and $(\Pi, \Omega)$.

(i) *Conditionally on* $(\Pi, \Omega)$, we have from (8):

$$\left[\frac{\partial\gamma'}{\partial\beta}\right] = [I_m \otimes -\Pi'], \qquad \left[\frac{\partial\sigma'}{\partial\beta}\right] = [I_m \otimes 2B\Omega]. \tag{15}$$

Remember (from application 2.3) that

$$\Psi_\beta = \left[\frac{\partial\psi'}{\partial\beta}\right], \qquad \Psi_\gamma = \left[\frac{\partial\psi'}{\partial\gamma}\right], \qquad \Psi_\sigma = \left[\frac{\partial\psi'}{\partial\sigma}\right],$$

where these matrices have dimension $K \times m^2$, $K \times mn$ and $K \times m^2$ respectively.

Define the $\lambda \times m^2$, $\lambda \times mn$ and $\lambda \times m^2$ matrices $\Phi_\beta$, $\Phi_\gamma$ and $\Phi_\sigma$ as follows:

Let $\tilde{\beta}_i$, the $i$th element of $\tilde{\beta}$, be the $k$th element of $\beta (i \leqq \mu, k \leqq m^2)$; row $i$ of $\Phi_\beta$ has an entry equal to 1 in the $k$th position, entries equal to 0 everywhere else, $i = 1 \ldots \mu$; rows $\mu+1$ to $\lambda$ of $\Phi_\beta$ have all their entries equal to 0.

Similarly, let $\tilde{\gamma}_i = \gamma_k$; row $\mu+i$ of $\Phi_\gamma$ has an entry equal to 1 in the $k$th position, entries equal to 0 everywhere else, $i = 1 \ldots \nu$; rows 1 to $\mu$ and $\mu+\nu+1$ to $\lambda$ of $\phi_\gamma$ have all their entries equal to 0.

Let finally $\tilde{\sigma}_i = \sigma_k$; row $\mu+\nu+i$ of $\Phi_\sigma$ has an entry equal to 1 in the $k$th position, entries equal to 0 everywhere else; rows 1 to $\mu+\nu$ of $\Phi_\sigma$ have all their entries equal to 0.

*Conditionally on* $(\Pi, \Omega)$, the $(K+\lambda) \times m^2$ Jacobian matrix of the transformation from $(\psi, \tilde{\beta}, \tilde{\gamma}, \sigma)$ to $\beta$ is defined by:

$$\begin{bmatrix} \Psi_\beta + \Phi_\gamma \left[\dfrac{\partial\gamma'}{\partial\beta}\right] + \Psi_\sigma \left[\dfrac{\partial\sigma'}{\partial\beta}\right] \\[2ex] \Phi_\beta + \Phi_\gamma \left[\dfrac{\partial\gamma'}{\partial\beta}\right] + \Phi_\sigma \left[\dfrac{\partial\sigma'}{\partial\beta}\right] \end{bmatrix} =$$

$$= \begin{bmatrix} \Psi_\beta & \Psi_\gamma & \Psi_\sigma \\ & & \\ \Phi_\beta & \Phi_\gamma & \Phi_\sigma \end{bmatrix} \begin{bmatrix} I_m \otimes I_m \\ I_m \otimes -\Pi' \\ I_m \otimes 2B\Omega \end{bmatrix}. \tag{16}$$

That Jacobian must be of full rank $m^2$ in order for a one-to-one transformation to exist, conditionally on $(\Pi, \Omega)$, between $\beta$ and

$(\psi, \tilde{\beta}, \tilde{\gamma}, \sigma)$.

Under the assumption $|B| \neq 0$, the rank of the Jacobian matrix in (16) is the same as the rank of the matrix

$$\begin{bmatrix} \Psi_\beta & \Psi_\gamma & \Psi_\sigma \\ & & \\ \Phi_\beta & \Phi_\gamma & \Phi_\sigma \end{bmatrix} \begin{bmatrix} I_m \otimes B' \\ I_m \otimes \Gamma' \\ I_m \otimes 2\Sigma \end{bmatrix} = \begin{bmatrix} \Psi^* \\ \\ \Phi^* \end{bmatrix}. \tag{17}$$

The matrix

$$\begin{bmatrix} \Psi^* \\ \Phi^* \end{bmatrix}$$

is a natural extension of the matrix $\Psi^*$, whose rank was found relevant (in application 2.3) to ascertain the identification of the model.

(ii) Let us now consider case (ii) of application 2.3, under a mixture of exact and stochastic prior information. More specifically, let us assume that the conditions $\Psi_k = 0$, $k = 1 \ldots K$, consist in assigning predetermined values to a subset of the parameters $(B, \Gamma)$; and that the prior density factors into $h(\Pi, \Omega)$ and

$$f(\tilde{\beta}, \tilde{\gamma}, \sigma) = \prod_{i=1}^{m} f_i(\tilde{B}_i, \tilde{\Gamma}_i),$$

where $(\tilde{B}_i, \tilde{\Gamma}_i)$ denotes a subvector from the $i$th row of $(B, \Gamma)$.

Consider then an arbitrary equation from (1), say the first, and write it as follows (under appropriate relabeling, if necessary):

$$\bar{B}_{11}y_1' + B_{12}y_2' + B_{13}y_3' + B_{14}y_4' + \bar{\Gamma}_{11}z_1' + \Gamma_{12}z_2' + \Gamma_{13}z_3' + \Gamma_{14}z_4' = u_1 \tag{18}$$

where

$(\bar{B}_{11} \quad \bar{\Gamma}_{11})$ is the subvector of predetermined parameters;

$(B_{12} \quad \Gamma_{12} \quad \Gamma_{13})$ is the subvector of parameters appearing as arguments of the prior density $f_1(\tilde{B}_1, \tilde{\Gamma}_1)$;

the dimension of $B_{13}$ does not exceed that of $\bar{\Gamma}_{11}$, and $B_{14}$ has positive dimension only when $B_{13}$ has maximal dimension;

the dimension of $\Gamma_{12}$ does not exceed that of $B_{14}$, and $\Gamma_{13}$ has positive dimension only when $\Gamma_{12}$ has maximal dimension.

The first row of the matrix equation $B\Pi + \Gamma = 0$ may be written as:

$$(\bar{B}_{11} \quad B_{12} \quad B_{13} \quad B_{14}) \begin{bmatrix} \Pi_{11} & \Pi_{12} & \Pi_{13} & \Pi_{14} \\ \Pi_{21} & \Pi_{22} & \Pi_{23} & \Pi_{24} \\ \Pi_{31} & \Pi_{32} & \Pi_{33} & \Pi_{34} \\ \Pi_{41} & \Pi_{42} & \Pi_{43} & \Pi_{44} \end{bmatrix}$$

$$+(\bar{\Gamma}_{11} \quad \Gamma_{12} \quad \Gamma_{13} \quad \Gamma_{14}) = 0, \tag{19}$$

with $\Pi_{31}$ and $\Pi_{42}$ square. Solving explicitly for $B_{13}$ and $B_{14}$ in terms of $\bar{B}_{11}$, $B_{12}$, $\bar{\Gamma}_{11}$, $\Gamma_{12}$ and $\Pi$, we find

$$B_{13} = -[(\bar{B}_{11}\Pi_{11} + \bar{\Gamma}_{11}) + (B_{12}\Pi_{21} + B_{14}\Pi_{41})]\Pi_{31}^{-1} \tag{20}$$

$$\begin{aligned} B_{14} = {} & -[\bar{B}_{11}\Pi_{12} - (\bar{B}_{11}\Pi_{11} + \bar{\Gamma}_{11})\Pi_{31}^{-1}\Pi_{32} + B_{12} \\ & \times (\Pi_{22} - \Pi_{21}\Pi_{31}^{-1}\Pi_{32}) + \Gamma_{12}](\Pi_{42} - \Pi_{41}\Pi_{31}^{-1}\Pi_{32})^{-1}. \end{aligned} \tag{21}$$

The explicit solution requires that the matrix

$$\begin{pmatrix} \Pi_{31} & \Pi_{32} \\ \Pi_{41} & \Pi_{42} \end{pmatrix}$$

be nonsingular. This condition is equivalent to the rank and order conditions mentioned under application 2.3 (ii).

$(\bar{\Gamma}_{11} \quad \Gamma_{12})$ must have the dimension of $(B_{13} \quad B_{14})$, i.e.. the subvector $(\bar{B}_{11} \quad B_{12} \quad \bar{\Gamma}_{11} \quad \Gamma_{12} \quad \Gamma_{13})$ from $(B_1 \quad \Gamma_1)$ on which prior information (exact or stochastic) is available must have dimension at least $m$ (order condition);

the matrix

$$\begin{bmatrix} \Pi_{31} & \Pi_{32} & \Pi_{33} \\ \Pi_{41} & \Pi_{42} & \Pi_{43} \end{bmatrix}$$

must have full row rank (rank condition). This matrix is the counterpart of $\Pi_{21}$ in (13).

Provided these conditions are fulfilled, one may perform an integrand transformation from $\Gamma_{12}$ to $B_{14}$, with Jacobian $\|\Pi_{42}-\Pi_{41}\Pi_{31}^{-1}\Pi_{32}\|$, and write the prior density $f_1(\bar{B}_1, \bar{\Gamma}_1)$ as:

$$\begin{aligned} &f(B_{12}, B_{14}(\Pi_{42}-\Pi_{41}\Pi_{31}^{-1}\Pi_{32})+\bar{B}_{11}\Pi_{12} \\ &-(\bar{B}_{11}\Pi_{11}+\bar{\Gamma}_{11})\Pi_{31}^{-1}\Pi_{32}+B_{12}(\Pi_{22}-\Pi_{21}\Pi_{31}^{-1}\Pi_{32}),\Gamma_{13}) \\ &\cdot\|\Pi_{42}-\Pi_{41}\Pi_{31}^{-1}\Pi_{32}\|\cdot \mathrm{d}B_{12}\mathrm{d}B_{14}\mathrm{d}\Gamma_{13}. \end{aligned} \quad (22)$$

Conditionally on $\Pi$, the elements of $B_1$ are then subject to a prior density, with mass concentrated in a subspace having the dimensionality of $(B_{12}, B_{14})$; that density is defined by (20), (22) and $B_{11} = \bar{B}_{11}$. If similar conditions are satisfied for all other equations as well, prior information (either exact or stochastic) is available on all the elements of $B$, conditionally on $\Pi$ and $\Omega$, and $h_c(B|\Pi, \Omega)$ is a well-defined density.

In view of theorem 3.2, the necessary conditions for the existence of a well-defined conditional density $h_c(B|\Pi, \Omega)$ are also necessary for the existence of a well-defined posterior density. The conditions obtained in application 3.3 take the form of generalizations of the conditions for local identification mentioned in application 2.3.

*Remark* 3.3. The rank of the Jacobians in (16) and (22) depends upon the rank of the matrix $\Pi$, or of a submatrix from $\Pi$. The true value of $\Pi$ may be such that the rank condition is not satisfied. However, in a Bayesian analysis, the posterior density of $\Pi$ will typically be continuous. Because the set of values for $\Pi$ where the rank condition is *not* met has Lebesgue measure zero in $R^{mn}$, the posterior probability that the rank condition is fulfilled will then be equal to one, after observing an arbitrary finite sample. G. S. Maddala has recognized this problem, in a recent unpublished note,[4] and shown that it may have serious implications for inference. An adequate test of rank conditions should be incorporated into the existing Bayesian techniques.

[4] Weak priors and sharp posteriors in simultaneous equation models, mimeographed, 1972.

### 3.4. *Concluding remarks*

By way of conclusion, it may be appropriate to comment again about the relevance of identification, in a Bayesian approach. What is gained, from knowledge that a model is identified? Kadane [5, theorems 3 and 4] presents an interesting characterization of identification in the context of modification of opinion, and in the context of decision-making. I have elsewhere [1, p. 9] noted that identification (resp. local identification) is a necessary condition for uniqueness (resp. local uniqueness) of maximum likelihood estimators.[5] In a Bayesian approach, maximum likelihood estimators correspond (numerically) to posterior modes obtained from a locally uniform prior density.[6] From the viewpoint of making decisions, uniqueness of posterior modes is seldom relevant. Identification is neither necessary nor sufficient for other, more relevant characteristics of the posterior density to obtain. The existence of posterior moments is a good example. In this respect, it should be noted that the "order of overidentification" has been found relevant for the existence of sampling moments, in a classical approach, and of posterior moments, in a Bayesian approach.[7] Again, the "order of overidentification" is a by-product of the theory of identification in simultaneous equations models; it is not part of the general concept of identification.

[5] *Proof*: Let $S^*$ be a maximum likelihood estimator of the true structure. If there exists $\bar{S}$ observationally equivalent to $S^*$, then $\bar{S}$ is also a maximum likelihood estimator.

[6] In the case of an unidentified model, uniqueness (resp. local uniqueness) of the posterior mode $S^*$ requires uniqueness (resp. local uniqueness) of the prior conditional mode over the set of structures $\mathscr{E}(S^*)$ which are observationally equivalent to $S^*$.

[7] See [1, p. 11] or [2, theorems 3.1 and 3.2].

## *References*

[1] J. H. Drèze, Econometrics and decision theory. Econometrica 40 (1972), 1–17. (Reprinted in the present volume.)

[2] J. H. Drèze, Bayesian limited information analysis of the simultaneous equations model. CORE DP 7111.

[3] F. M. Fisher, The identification problem in econometrics. New York, McGraw-Hill, 1966.

[4] R. Harkema, Simultaneous equations, a Bayesian approach. Rotterdam, Rotterdam University Press, 1971.

[5] J. Kadane, The role of identification in Bayesian theory. Mimeographed, 1972. (A revised version of this paper appears in the present volume.)
[6] D. V. Lindley, Bayesian statistics: a review. Philadelphia, SIAM, 1971.
[7] J. A. Morales, Bayesian full information structural analysis, Berlin, Springer, 1971.
[8] J. F. Richard, Posterior and predictive densities for simultaneous equations models. Berlin, Springer, 1973.
[9] T. J. Rothenberg, Identification in parametric models. Econometrica 39 (1971), 577–591.
[10] A. Zellner, An introduction to Bayesian inference in econometrics, New York, Wiley, 1971.

# Bayesian Decision Theory and the Simplification of Models

*JOSEPH B. KADANE*

DEPARTMENT OF STATISTICS
CARNEGIE-MELLON UNIVERSITY
PITTSBURGH, PENNSYLVANIA

*and*

*JAMES M. DICKEY*

DEPARTMENT OF MATHEMATICS AND STATISTICS
STATE UNIVERSITY OF NEW YORK AT ALBANY
ALBANY, NEW YORK

## 1. Introduction

The problem of evaluating econometric models is here viewed as a particular case of a general class of problems called decision problems. Since the authors are attracted to a particular approach to decision problems, the Bayesian approach, we have written this paper in order to see what light can be shed by Bayesian decision theory on the evaluation of econometric models.

Briefly, the decisions $d$ that might be chosen lie in some decision space $\mathscr{D}$. There is some space $\Omega$ of possible alternative circumstances $\theta$, and your opinion over this space is given by the probability density with element $p(\theta)\,d\theta$. If one's utility function is $U(d, \theta)$, representing the relative desirability of the decision $d$ if the alternative $\theta$ were true, Bayesian decision theory recommends that you choose that decision $d$ which maximizes the integral

$$\int_{\Omega} U(d, \theta) p(\theta)\, d\theta \tag{1.1}$$

(See F. Ramsey, 1931, Von Neumann & Morgenstern, 1947, and Savage,




ISBN 0-12-416550-8

1954, for the theoretical background of this choice, called the principle of maximum expected utility.)

The view that this simple paradigm might play a fundamental role in statistics by providing a theory analogous to price theory in microeconomics has been met with considerable controversy among statisticians. Some statisticians maintain that the inputs, particularly the opinion $p(\theta)\,d\theta$ and the utility function $U(d, \theta)$, are very difficult to obtain in applied contexts, while others resist the notion of any kind of organizing principle for statistics. The burden of proof is on Bayesians to show that the principle suggested by (1.1) is useful as a way of thinking about statistical problems and that the inputs are worth the trouble to discover. In this light, the problem of choice between econometric models is a fine test case.

While we will speak of econometric models in general, the specific model we will refer to and have in the back of our minds most often is the choice between linear regressors, especially the choice whether or not to add some specified variable to a model. However, because our viewpoint is very general (and this is a distinct advantage of Bayesian theory), much of what we say has much wider applicability.

An enormous literature already exists on these and related questions, some of which is reviewed in Gaver & Geisel (1973), Harvey & Collier (1977), Hocking (1976), and J. Ramsey (1974). We cannot hope to take up each of the methods reviewed in these papers, but rather will deal with strengths and weaknesses we find in classes of such methods.

Some methods, typified by the use of tests of hypotheses, ask whether or not the submodel is a "true" restriction of the larger model. The level of the test is the probability under the submodel that the statistic in question would be as large as observed or larger. However, practicing statisticians have long been aware that the question of the exact "truth" of a sharp statistical hypothesis is hardly ever appropriate (Berkson, 1938). Nearly every model used in econometrics is known *a priori* to be literally false. Because of this, although a typical statistical test with a small sample will rarely reject, for large enough samples it will nearly always reject. Consequently, whether or not a test rejects a null hypothesis appears to have more to do with the sample size than with how true the null hypothesis is. For these reasons we find methods based on classical tests of hypotheses to be most unsatisfactory. A purist may object, saying that the level of a test should be set in consideration of the power that can be achieved with that sample size, but after all these years there is no satisfactory theory of how to decide what level of test to set, and most practitioners use .05 or .01 in an automatic way, without justification.

The fundamental problem with the test of a hypothesis, we believe, lies not in the failure to jiggle the level satisfactorily, but rather with the underlying fact of having asked a poor question. No method geared to the question

Is the null hypothesis true? will be statisfactory in the most common situation where we know the null hypothesis to be false. Even if one attempts to wriggle out of this argument by claiming that what is really meant by $x \sim N(10.0, 1.0)$ is that the mean $\mu$ satisfies the inequality $9.950 \leq \mu \leq 10.049$ and the variance $\sigma^2$ satisfies the inequality $0.950 \leq \sigma^2 \leq 1.049$, and one modifies the test accordingly, the above argument still holds with "false" changed to "extremely unlikely". Of course, if the intervals are widened, their joint probability content grows, but then they become that much less a representation of $X \sim N(10.0, 1.0)$.

The important question in practice is not whether a true effect is nonzero, for it is already known not to do exactly zero, but rather, How large is the effect? But then this question is only relevant in terms of How large is how important? This question in turn depends on the use to which the inference will be put, namely, on the utility function of the concerned scientist. Approaches which attempt to explain model simplification from the viewpoint of the inappropriate question, Is it true that . . . ? have a common thread in that they all proceed without reference to the utility function of the scientist. And therefore, from the decision theory view, they all impose normative conditions on the utility function which are seldom explicit and often far from the case in practice.

We explore in this paper the consequences of asking the more relevant question Is it useful to assume that . . . ?, which requires us to be more explicit about the investigator's utility function. We find the utility function in Bayesian theory to be especially useful to explain the behavior of a scientist who, with the same data (and hence opinion) studying the same phenomenon, sometimes uses a simpler model and sometimes a more complex one, depending on his purpose. His view of the truth has not changed, but his purpose, and hence his utility function, has.

In response to the popular interest in questions of "truth" and the natural interpretation of the posterior distribution as a measure of knowledge on such questions, we begin by examining the role of the posterior odds for a statistical hypothesis within a general two-decision setting (Section 2). We identify the extent to which the posterior odds can be a useful summary of data for decision making independent of the detailed utility structure. Our finding is essentially negative, that the odds are typically not useful as a data summary in this sense.

We turn then to the specific problem of how to choose predictors in the normal multiple regression model (Section 3). Without nontrivial prior belief in any null hypothesis, we consider the question of how to decide whether or not to accept a restriction or "simplification" on the class of available predictors. In other words, attention is redirected from the simplification of sampling models to the simplification of predictors. Results are summarized and compared for predictive utilities in the familiar squared-error-loss form (Lindley, 1968) and a new predictive utility in the form of a probability density function (Dickey & Kadane, 1977). Comparisons are made to various truth-test criteria and traditional predictor-selection criteria. A general montonicity is noted for the predictive expected utility as a function of the predictor class, whereby cost-free predictor variables tend to be included in the optimal predictor on a foregone basis. We discuss the consequent paradox of Bayesian overfitting (Section 4).

Alternative Bayesian discussions of selecting regression models are given by Leamer (1978a,b) and by Davis & DeGroot (1978).

## 2. Bayesian Response to Data; Odds

When one obtains new data $D$, Bayesian theory suggests how one should take this data into account in changing one's opinion about $\theta$. Suppose that the probability of observing $D$, if $\theta$ were the true state of nature, is, under one's opinion, given by the function $l(D|\theta)$. Then one's new opinion should be represented by your conditional distribution of $\theta$, given $D$, given by

$$p(\theta|D)\,d\theta = l(D|\theta)p(\theta)\,d\theta \Big/ \int_{\Omega} l(D|\theta)p(\theta)\,d\theta. \tag{2.1}$$

Technically, Eq. (2.1) has used a rather simple result from probability theory called Bayes' theorem, from which the name Bayesian statistics derives. Of course, the opinion used in the earlier Eq. (1.1) must be your own current opinion at the time of the decision, and hence must take into account all of the information available to you, that is, one's opinion must be changed by Eq. (2.1) every time new data appear.

We now undertake to model the simplest sort of decision that might be made. We suppose that there are two hypotheses, $H$ and $H^c$. You are sure both cannot be true ($H \cap H^c = \varnothing$), but that one of them is true ($P(H \cup H^c) = 1$). We denote the *odds for* $H$, $O(H)$, as

$$O(H) = P(H)/P(H^c) = P(H)/(1 - P(H)) \tag{2.2}$$

When new data $D$ are observed, the prior odds $O(H)$ change to posterior odds $O(H|D)$ in the following way, using Bayes Theorem:

$$O(H|D) = B_D(H)O(H), \tag{2.3}$$

where

$$B_D(H) = p(D|H)/p(D|H^c)$$

and

$$p(D|J) = \int_J l(D|\theta)p(\theta)\,d\theta / \int_J p(\theta)\,d\theta \qquad \text{for} \quad J = H, H^c.$$

The quantity $B_D(H)$ is called the Bayes factor and is constant among persons having the same likelihood and the same conditional priors, given $H$ and $H^c$. The usefulness of the Bayes factor as an expression for how strongly your current beliefs support $H$ has been explored in a series of papers by Jeffreys (1961), Edwards, Lindman, & Savage (1963), and Dickey (1971, 1973a, 1974, 1976, 1977, 1979); see also Lindley (1957, 1961).

The goal of this section is to show that Bayes factors have a very limited usefulness in the two-decision context. We note that Bayes factors do appear in other contexts (see, for example, Zellner & Vandaele (1975, pp. 640–641)) not discussed here.

## 2.1. Evidence for a Dichotomous Decision

Suppose that the relevant decision space D has only two desicions, $d_1$ and $d_2$, which we can think of as deciding in favor of or against $H$, respectively. With respect to the utility function $U(d, \theta)$, the optimal decision is to choose that decision $d$ at which the maximum of $\{\bar{U}_1, \bar{U}_2\}$ occurs, where

$$\bar{U}_i = E[U(d_i, \tilde{\theta})] = \int_\Omega U(d_i, \theta)p(\theta)\,d\theta, \qquad i = 1, 2. \tag{2.4}$$

(Here the tilde over a symbol emphasizes that the symbol denotes a random variable.) The expectation is conditional on all information, including any sample data, as explained in Eq. (2.1).

We need a notation for the conditional expectation of a utility function under the different hypotheses. Thus we write

$$\bar{U}_i(J) = E[U(d_i, \tilde{\theta})|J] = \int_J U(d_i, \theta)p(\theta)\,d\theta \Big/ \int_J p(\theta)\,d\theta, \tag{2.5}$$

for $i = 1, 2$ and $J = H, H^c$. Using this notation, we can write

$$\bar{U}_i = \bar{U}_i(H)P(H) + \bar{U}_i(H^c)P(H^c), \qquad i = 1, 2. \tag{2.6}$$

Finally, we introduce the notation $V(\theta) = U(d_1, \theta) - U(d_2, \theta)$, and write

$$\bar{V}(J) = E(V(\tilde{\theta})|J) = \bar{U}_1(J) - \bar{U}_2(J), \qquad J = H, H^c. \tag{2.7}$$

Then

$$\begin{aligned} \bar{U}_1 - \bar{U}_2 &= \bar{U}_1(H)P(H) + \bar{U}_1(H^c)P(H^c) - \bar{U}_2(H)P(H) - \bar{U}_2(H^c)P(H^c) \\ &= \bar{V}(H)P(H) + \bar{V}(H^c)P(H^c). \end{aligned} \tag{2.8}$$

Without loss of generality, we assume $\bar{V}(H) > 0$, so that if we knew $H$ were true, with our present distribution of probability over $H$, we would prefer to make decision $d_1$ ($H$ is true), rather than decision $d_2$ ($H^c$ is true). Then it is easy to see from (2.8) that $\bar{U}_1 \geq \bar{U}_2$ if and only if

$$O(H) = P(H)/P(H^c) \geq -\bar{V}(H^c)/\bar{V}(H). \tag{2.9}$$

It is tempting to consider (2.9) as a threshold criterion on the odds. To do so, we are particularly interested in the situation when only $O(H)$, and not the right-hand side of (2.9), depends on the distribution over $\Omega$.

TABLE 1

| | | True | |
|---|---|---|---|
| | | $H$ | $H^c$ |
| Accept | $H$ | 0 | $-L_{12}$ |
| | $H^c$ | $-L_{21}$ | 0 |

An important special case of this model is where the utilities only depend on whether $\theta \in H$ or $\theta \in H^c$. Thus we might have the utility structure in Table 1 (see, for example, Edwards *et al.*, 1963; Zellner, 1971; and Gaver & Geisel, 1973), where $L_{12}$ and $L_{21}$ are both typically positive. The analysis above specializes to the criterion "accept" $H$ (decide $d_1$) iff

$$O(H) \geq L_{12}/L_{21}. \tag{2.10}$$

After data $D$ are observed, (2.10) can be expressed equivalently as

$$O(H|D) \geq L_{12}/L_{21}, \tag{2.11}$$

or, by using (2.3) with (2.11),

$$B_D(H) \geq L_{12}/[L_{21}O(H)]. \tag{2.12}$$

Thus the Bayes factor is the appropriate function of the data on which to have a threshold (and $L_{12}/[L_{21}O(H)]$ is the appropriate threshold), above which it is optimal to choose $d_1$, and below which it is optimal to choose $d_2$. This optimistic special case suggests that perhaps there is something special about the Bayes factor in general that makes it a canonical summary of the data. That this is not so is the burden of the next result.

For the Bayes factor to be a sufficient summary of the data, the right-hand side of (2.9) must not depend on the data. However, when data are available, the expectations on the right-hand side of (2.9) are *posterior* expectations and in general depend on the data. A sufficient condition for their independence of the data is that $V(\theta)$ be constant within $H$ and within $H^c$. In general, the condition is also necessary in a sense now to be defined. The result applies separately to the numerator and to the denominator of the right-hand side of (2.9), though the statement is made in terms of the denominator.

**Theorem.** *Let $\mathscr{P}_H$ be the class of all continuous-type distributions with densities $p_H(\theta)$ over an analytic segment $H$ of a finite dimensional Euclidean space. Assume $V(\theta)$ is bounded and continuous. Then the expectation $E[V(\theta)|H] = \int V(\theta)p_H(\theta)\,d\theta$ is constant in $p_H$ over $\mathscr{P}_H$ if and only if $V$ is constant in $\theta$ over $H$.*

*Proof.* The sufficiency of constant $V$ is immediate. To see that this condition is also necessary, assume that it does not hold. Suppose $\theta_1,\theta_2$ are two values in $H$ for which $V(\theta_1) > V(\theta_2)$. Choose neighborhoods $N_1$ and $N_2$ of $\theta_1$ and $\theta_2$, respectively, for which the two image sets are disjoint, $V(N_1) \cap V(N_2) = \varnothing$. If two densities $p_H^1$ and $p_H^2$ have their support sets contained in $N_1$ and $N_2$, respectively, then $\int V(\theta)p_H^1(\theta)\,d\theta > \int V(\theta)p_H^2(\theta)\,d\theta$ (since the two integrands satisfy this same inequality), which then contradicts the constancy of the expection. ■

The proof extends easily to any subclass of $\mathscr{P}_H$ which contains distributions having arbitrarily small probability for the complements of arbitrary neighborhoods. This typically true for the set of conditional posterior distributions from all possible data in a given statistical decision problem. Thus we see that when utility takes more than two values within $H$ and/or $H^c$, the threshold on the odds, and hence the preferred decision, will depend on the conditional uncertainties given $H$ and given $H^c$.

Consequently, the adequacy of the Bayes factor as a summary of the data depends on the acceptability of a utility structure similar to Table 1. Such a structure cannot be close to correct for the most common problems because if I wrongly accept $H$ (make decision $d_1$), it matters for almost every practical purpose whether, although wrong, $H$ is close to correct or very far from correct. Thus in the simple versus composite case (say, testing whether or not a normal mean is 10.0), the Bayes factor cannot summarize the data. Of course, this holds as well for the yet more common composite versus composite cases.

For this reason it is our judgement that the Bayes factor theory, originally justified in the simple versus simple case as the Bayesian version of hypothesis testing, does not fill the need of econometricians for a useable, theoretically well-based way of choosing models. Note that we arrive at this conclusion even when positive probability is put on the truth of the null hypothesis, although, as explained in Section 1, we doubt that this is appropriate in general.

## 2.2. Hypothesis Defined by the Decision Problem

In Section 2.1 we started with a hypothesis $H$ (for example, $\theta = 10.0$) and an alternative (for example, $\theta \neq 10.0$), and found (restrictive) conditions under which the Bayes factor is sufficient as a representation of the impact of new data. Our method was essentially to take the hypothesis $H$ (and therefore also its complement $H^c$) as fixed and to try to modify the utility functions in ways that would lead to the Bayes factor. In this section, by contrast, we try the reverse, that is, we begin with a reasonable utility function for a two-decision problem and work backward to see what sort of hypothesis $H$ a procedure based on the Bayes factor would imply. We propose to choose the hypothesis $H$ by the requirement

$$\begin{aligned} U(d_1, \theta) &\geq U(d_2, \theta) \quad \text{for all} \quad \theta \in H, \\ U(d_1, \theta) &< U(d_2, \theta) \quad \text{for all} \quad \theta \in H^c. \end{aligned} \tag{2.13}$$

Then $d_1$ would be the preferred decision for any known $\theta$ in $H$, and $d_2$ would be preferred for any known $\theta$ in $H^c$. Consequently, we define a hypothesis $H$ for inference relative to a given two-decision problem by

$$H = \{\theta : U(d_1, \theta) \geq U(d_2, \theta)\}. \tag{2.14}$$

To connect this theory with the kind of theory discussed in Section 2.1, consider, for example, what happens if our objective is to give an accurate estimate of $\theta$. Let us suppose that decision $d_1$ simplifies the model by declaring that $\theta = \theta^*$, when $\theta^*$ is some particular value (like $\theta^* = 10$, for example). In this case, suppose that our reward is some function of $\theta$, say $S_1(\theta)$. Alternatively, if $\theta$ is not declared to have the value 10, then it must be estimated, say by some $\hat{\theta}$. Then the reward that we receive is a function of both $\hat{\theta}$ and $\theta$, say $S_2(\hat{\theta}, \theta)$. Anscombe (1963) considers an estimation loss function which is constant if the parameter is freely estimated ($S_2(\hat{\theta}, \theta) = U_2$) but proportional to the square of the parameter if the estimate is taken to be zero ($S_1(\theta) = U_1 \theta^2$, where $U_1$ is negative).

Now of course, $\hat{\theta}$ must be chosen as best it can be if the second decision is made, that is, $\hat{\theta} = \hat{\theta}_D$, where $\hat{\theta}_D$ maximizes $E[S_2(\hat{\theta}, \tilde{\theta}) | D]$.

Of particular interest are utility pairs differing by a constant "reward for simplicity" $U^* > 0$, where

$$S_1(\theta) = S_2(\theta^*, \theta) + U^*. \tag{2.15}$$

In this case, we obtain by (2.14) the "hypothesis"

$$H = \{\theta : S_2(\hat{\theta}_D, \theta) - S_2(\theta^*, \theta) \leq U^*\}. \tag{2.16}$$

Consider, for example, the familiar negative squared error, $S_2(\hat{\theta}, \theta) = -(\hat{\theta} - \theta)^2$. Then, if $\theta$ is to be freely estimated, the optimal estimate will be the posterior mean $\hat{\theta}_D = E[\tilde{\theta}|D]$, and $H$ takes the form

$$H = \{\theta: \theta \lessgtr \tfrac{1}{2}(\theta^* + \hat{\theta}_D) - \tfrac{1}{2}U^*/(\theta^* - \hat{\theta}_D)\}, \tag{2.17}$$

where the sense of the inequality is taken to match that of $\theta^* \lessgtr \hat{\theta}_D$. Thus $H$ turns out to be an infinite half-line, and worse, one that depends on the data through $\hat{\theta}_D$.

Our attempts to find a link decision theory and the problem of choosing a probability model have failed, and we make no further such attempts here. In both cases we sought to retain some kind of formal decision theoretic meaning for classical hypothesis-testing ideas. A similar such failure is reported in Edwards *et al.*(1963).

The consequence of this failure is that something must go, either our Bayesian approach or classical hypothesis testing. Having declared our purpose in this paper to explore the consequences of Bayesian decision theory for choosing models, with particular reference to econometrics, it will come as no surprise to our readers that our choice is to keep our Bayesian decision theory and see what its consequences are. In the next section, then, we begin anew to see where Bayesian theory will lead us. We introduce the normal linear model and the briefly review the behavior of classical and Bayesian "truth-testing" procedures for later comparisons before turning to our preferred Bayesian methods. Our purpose is constructive and is designed to lead to theoretically based but useful procedures, rather than retrospective, that is, to see if we can find some explanation for old procedures derived from other perspectives.

## 3. Prediction

### 3.1. The Distributions

To obtain a context in which one can judge the suitability of a decision model, we specify the usual normal linear sampling model in which the response variable $y$ has a normal distribution conditional on the concomitant $r$-dimensional vector variable $\mathbf{x}$ with linear mean $\mathbf{x}^T\boldsymbol{\beta}$ and variance $\sigma^2$. Given a sequence of values for $\mathbf{x}$, conditional independence is assumed under the sampling model for the corresponding values $y$. We shall consider the family of natural conjugate prior distributions for $\boldsymbol{\beta}$ and $\sigma$ and make direct use of the posterior predictive distribution for future $y$ given $\mathbf{x}$. This distribution is the mixture of sampling distributions for $y$ given $\mathbf{x}$, mixed over the posterior distribution of $\boldsymbol{\beta}$ and $\sigma$. It models one's coherent postdata personal uncertainty about future $y$ taking account of the fact that $\boldsymbol{\beta}$ and $\sigma$ are not known. See Raiffa & Schlaifer (1961) for details.

For much of our development the distribution of future $\mathbf{x}$ is arbitrary. But special interest attaches to the case when the vectors $\mathbf{x}_i$ are "random," more precisely are independently sampled from a multivariate normal distribution, say with unknown mean and variance $\xi$, and $\Sigma$. Given that the prior uncertainty is independent between $(\xi, \Sigma)$ and $(\boldsymbol{\beta}, \sigma)$, then by an obvious factorization of the likelihood function, so is the posterior uncertainty, and future $\mathbf{x}$ will be posterior independent of $(\boldsymbol{\beta}, \sigma)$. Geisser (1965), under "ignorance" priors on $(\xi, \Sigma)$, and Ando and Kaufman (1965), under conjugate priors, obtain multivariate-$t$ predictive distributions for $\mathbf{x}$ with moments approximately matching the empirical moments of observed $\mathbf{x}_1, \ldots, \mathbf{x}_n$

$$E(\mathbf{x} \mid D) = \bar{\mathbf{x}}, \qquad V(\mathbf{x} \mid D) = n^{-1} \sum_{i=1}^{n} (\mathbf{x}_i - \bar{\mathbf{x}})(\mathbf{x}_i - \bar{\mathbf{x}})^T. \tag{3.1}$$

We shall use (3.1) as an approximation.

### 3.2. Truth-Test Criteria

A variety of rather arbitrary criteria for selecting a linear model appear in practice, such as the modified multiple correlation statistic $\bar{R}^2 = 1 - s^2/\{\sum(y_i - \bar{y})^2/(n-1)\}$, where $vs^2$ is the sample error sum of squares, $n$ is the sample size, and $v = n - r$, where $r$ is the number of concomitant variables included in the regression. This is typically preferred to the unmodified statistic, $R^2 = 1 - vs^2/\sum(y_i - \bar{y})^2$, merely because the latter is monotonically increasing for any nested sequence of models, whereas $\bar{R}^2$ may sometimes decrease when a new variable is added. On the other hand, the modified statistic need not lie in the unit interval $[0, 1]$. (See Gaver & Geisel, 1973, for a sense in which $R^2$ is a Bayesian criterion for choice when comparing two models of equal dimensionality.)

Some criteria are intended to answer the question, Is the model true? The traditional $F$ test asks this question of the smaller model $H$ in a nested pair of models $H \subset \Omega$. Denote by $r, \mathbf{x}, \boldsymbol{\beta}$, and $\sigma$ the parameters in the larger, unquestioned model $\Omega$ where $r$ is the dimensionality of $\boldsymbol{\beta}$. Then if $H$ is defined by linear constraints on $\boldsymbol{\beta}$, we can, without loss of generality, define the variables in such a way that $H$ takes the nested discarded-variables form

$$H: \quad \boldsymbol{\beta}_K = \mathbf{0}, \tag{3.2a}$$

where

$$\mathbf{x} = (\mathbf{x}_H^{\mathrm{T}}, \mathbf{x}_K^{\mathrm{T}})^{\mathrm{T}}, \qquad \boldsymbol{\beta} = (\boldsymbol{\beta}_H^{\mathrm{T}}, \boldsymbol{\beta}_K^{\mathrm{T}})^{\mathrm{T}}. \tag{3.2b}$$

The parameters of $H$ as a normal linear model will then be denoted by $r_H$, $\mathbf{x}_H, \boldsymbol{\beta}_H, \sigma_H \equiv \sigma$.

Tests for inference on the hypothesis $H$ are available based on a sample of $n$ realizations of $(y, \mathbf{x})$. For example, the traditional $F$- test is based on the error sums of squares (SSE) for the two models, SSE($H$) and SSE($\Omega$), and the sum of squares between (SSB),

$$\mathrm{SSB}(H,\Omega) = \mathrm{SSE}(H) - \mathrm{SSE}(\Omega). \tag{3.3}$$

The test rejects $H$ if

$$\mathrm{SSB}(H,\Omega) > (r - r_H)s^2 F_{r-r_H,\,\nu}(.95), \tag{3.4}$$

where $s^2 = \mathrm{SSE}(\Omega)/\nu$, the usual unbiased estimate of $\sigma^2$. Note that this threshold on the between sum of squares is approximately constant in the sample size $n$ for large $n$, as was noted by Lindley (1968) in the $\sigma^2$-known case.

Under this same nesting, the odds-ratio test for "truth" has been developed in Dickey (1971). Consider a mixed-type prior distribution with positive probabilities $P(H), P(H^c)$, and conditional prior densities $P(\boldsymbol{\beta},\sigma|H^c)$ and $P(\boldsymbol{\beta}_H,\sigma|H)$ of the natural conjugate form.

Under the natural assumption of the *continuity condition*, that the conditional prior distributions $P(\boldsymbol{\beta}_H,\sigma|\boldsymbol{\beta}_K)$ are continuous-in-distribution in $\boldsymbol{\beta}_K$ at $\boldsymbol{\beta}_K = \mathbf{0}$, one obtains in the Bayes factor $B_D(H)$ a useful approximation to the Bayes factor for a more realistic hypothesis in the form of a neighborhood set closely surrounding $H$. The neighborhood Bayes factor would result from various integrable continuous-type distributions with a mound of high density over the neighbourhood set. Of course, the quality of the approximation depends on the data outcome $D$. (As always, the conditional distribution $P(\boldsymbol{\beta}_H,\sigma|H)$, and hence also the Bayes factor, is not unique to the conditioning event $H$, but depends on the choice of conditioning parameter $\boldsymbol{\beta}_K$ (Gunel & Dickey, 1974).)

The resulting approximate Bayes factor $B_D(H)$ is proportional to the *density ordinate* of the usual $F$ statistic (as opposed to the traditional tail area). The decision threshold on the between sum of squares, following from a fixed threshold on the Bayes factor, is proportional to the logarithm of sample size in the known-variance case

$$\mathrm{SSB}(H,\Omega) > C\log(n). \tag{3.5}$$

Contrast this with the threshold in (3.4). Even for small sample sizes, the tendency is for a Bayes factor to require more extreme data for rejection than the tail area test requires (Dickey, 1977).

For large sample sizes, either criterion, (3.4) or (3.5), is nearly certain to reject $H$ in statistical practice, since $H$ is nearly always known not to hold exactly, and the "power" of either test increases to unity in $n$. (This analysis for the Bayes factor assumes, of course, that the full model $\Omega = H \cup H^c$ holds exactly. However, we believe that the same phenomenon occurs in greater generality.) Of course, this phenomenon could be adjusted for by arbitrarily changing the level of the test (3.4) with sample size $n$ or some similar *ad hoc* response in the case of the Bayes factor (3.5). Rather than pursue such a line, we develop below a decision threshold on $\mathrm{SSB}(H, \Omega)$ that is proportional to the sample size $n$—whereby it will no longer hold that one nearly always decides against simplicity when $n$ is large.

### 3.3. Predictors and Utility

We now neglect the concept of a smaller sampling model $H$ and the artificial question of the "truth" of $H$. As a subset of the parameter space, $H$ will receive no special positive probability. The family of natural conjugate prior distributions will apply to $\boldsymbol{\beta}$, $\sigma$ over $\Omega$.

We begin with a continuous action space and derive a discreteaction space in a natural way. Consider the problem of predicting a future value of $y$ from a future concomitant value $\mathbf{x}$. That is, the decision maker must choose after sampling, based on the information he has gained, a *predictor*, or predictor function, $\hat{y}(\mathbf{x})$. He is assumed then to receive a utility $W$ depending on his choice and on the future outcome. Typically, $W$ is large for $\hat{y}(\mathbf{x})$ near $y$.

Since the decision maker must choose a whole function $\hat{y}(\cdot)$, it seems natural that the utility $W$ should depend on the chosen function, as well as on the realized values $\hat{y}(\mathbf{x})$, $y$. $W$ might, for example, impose penalties for choosing a complicated predictor function or a predictor depending on coordinate variables in $\mathbf{x}$ which are costly to observe. As an illustration, consider the utility

$$W\{\hat{y}(\mathbf{x}), y, \hat{y}(\cdot)\} = \begin{cases} c - [\hat{y} - y]^2 & \text{if } \hat{y}(\cdot) \text{ is constant in } \mathbf{x}, \\ -[\hat{y}(\mathbf{x}) - y]^2 & \text{otherwise.} \end{cases} \tag{3.6}$$

The utility function (3.6) offers a reward $c$ if you are willing to have the same prediction $\hat{y}$ regardless of $\mathbf{x}$. This means that $\mathbf{x}$ need not be measured for the prediction. Thus agreeing in advance to this restriction save some effort, whose cost is $c$. Intuitively, if $\mathbf{x}$ does not vary much anyway, and/or if the connection between $y$ and $\mathbf{x}$ is weak, and/or if $c$ is large, the simpler predictor would be the more favored.

We shall assume in general that there are classes of predictor functions defined such that within each class $I$, $W$ is constant in $\hat{y}(\cdot) \in I$ for given values $\hat{y}(\mathbf{x})$ and $y$. In the example shown in (3.6), there are two such classes: $I_1$ consisting of all constants $\hat{y}$ and $I_2$ of the general functions $\hat{y}(\mathbf{x})$. It will be natural in such predictor-choice problems to identify the derived problem of how to choose between classes $I$.

Thus we seek to compute

$$\max_{I} \max_{\hat{y}_I(\cdot)} \iint W(\hat{y}_I(\mathbf{x}), y, y_I(\cdot)) p(y \mid \mathbf{x})\, dy\, p(\mathbf{x})\, d\mathbf{x}. \tag{3.7}$$

This formula looks more formidable than it really is. The maximization with respect to $I$ and $\hat{y}_I(\cdot)$ occurs after the integration with respect to $y$ and $\mathbf{x}$ because they are chosen before $y$ and $\mathbf{x}$ are known.

According to the principle of maximal expected utility, a single posterior-predictive distribution based on a single parametrized sampling model should be used to calculate the expected utilities of all the competing predictors. Strictly speaking, we are no longer concerned with deciding between probability models, but rather with choosing a class of predictors $I$.

Since the class of utilities of particular interest to us are those for which the only feature of $\hat{y}_I(\cdot)$ to affect $W$ is its membership in the class $I$, we will write, with only slight abuse of notation, $W(\hat{y}_I(\mathbf{x}), y, I)$ instead of $W(\hat{y}(\mathbf{x}), y, \hat{y}(\cdot))$.

We now specialize our utility function to the form

$$W(\hat{y}, y, I) = U_I w(\hat{y}, y) + U_I^*, \tag{3.8}$$

where $w(\hat{y}, y)$ measures the utility of predicting $\hat{y}$ when $y$ turns out to be the realization of the process and $U_I$ and $U_I^*$ are constants depending on $I$ that reflect, for example, the cost of measuring and computing from the variables permitted by the class $I$.

To distinguish $w(\hat{y}, y)$ from $W(\hat{y}, y, I)$, we call the former a purely predictive utility component since it is, strictly speaking, not a utility function, but rather one aspect of the utility function (3.8). The illustration (3.6) is a special case of (3.8) in which

$$U_{I_1} = U_{I_2} = 1, \qquad U_{I_1}^* = c, \qquad U_{I_2}^* = 0, \qquad \text{and} \qquad w(\hat{y}, y) = -(\hat{y} - y)^2.$$

The form (3.8) is especially convenient because the associated expected utility preserves the linearity of $W$ in $w$ as

$$\bar{W}(I) = U_I \bar{w}(I) + U_I^*, \tag{3.9}$$

where

$$\bar{w}(I) = \max_{\hat{y}_I(\cdot)} \iint w(\hat{y}_I(\mathbf{x}), y) p(y \mid \mathbf{x})\, dy\, p(\mathbf{x})\, d\mathbf{x}$$

and

$$\bar{W}(I) = \max_{\hat{y}_I(\cdot)} \iint W(\hat{y}_I(\mathbf{x}), y, I) p(y \mid \mathbf{x})\, dy\, p(\mathbf{x})\, d\mathbf{x}.$$

Hence the optimal choice between two classes of predictors $I_1$ and $I_2$ will proceed by a comparison between the expected purely predictive performance of the two models ($\bar{w}(I_1)$ and $\bar{w}(I_2)$) to the multiplicative ($U_{I_1}$ and $U_{I_2}$) and/or additive ($U_{I_1}^*$ and $U_{I_2}^*$) rewards for simplicity, as summarized by (3.9).

Two important special cases suggest themselves. If $U_{I_1} = U_{I_2} = U$, so that the rewards for simplicity are purely additive, maximum expected utility reduces to the criterion: prefer $I_1$ over $I_2$ if and only if

$$\bar{w}(I_1) - \bar{w}(I_2) \geq (U^*_{I_1} - U^*_{I_2})/U. \tag{3.10}$$

Similarly, if $U^*_{I_1} = U^*_{I_2}$, so that the rewards for simplicity are purely multiplicative, maximum expected utility reduces to: prefer $I_1$ over $I_2$ if and only if

$$\bar{w}(I_1)/\bar{w}(I_2) \geq U_{I_2}/U_{I_1}. \tag{3.11}$$

The predictor classes considered here are of the discarded-variable type. The class of all predictor functions is restricted to dependence on only a specified subvector $\mathbf{x}_I$,

$$\hat{y}(\mathbf{x}) = \hat{y}_I(\mathbf{x}_I), \qquad \mathbf{x} = (\mathbf{x}_I^{\mathrm{T}}, \mathbf{x}_K^{\mathrm{T}})^{\mathrm{T}}, \tag{3.12}$$

where for ease of notation $\mathbf{x}_I$ consists of the first few coordinates of $\mathbf{x}$. The subscripts $I$ and $K$ are used analogously to $H$ and $K$ in (3.2). Note that if a specific independent variable, say $x_q$, is included in $\mathbf{x}_I$, then all functions of that variable are contained in the class $I$ of available predictor functions. For example, higher powers of $x_q$ cannot be excluded without excluding $x_q$ itself. Another limitation of this form is that one cannot consider whether or not to have a zero intercept as a choice between classes $I$. An advantage of this form is that if one considers that the main cost of having $x_q$ included in the set of variables on the basis of which a prediction $\hat{y}$ is to be made is the cost of measuring $x_q$, then it is quite reasonable to allow the most flexible use of $x_q$ for prediction.

An alternative class of predictors not pursued here is the class of constrained linear predictors. This is the class of all linear forms satisfying a specified linear constraint on the coefficients, or without loss of generality, the class of all linear forms in a specified subvector $\mathbf{x}_I$,

$$\hat{y}(\mathbf{x}) = \mathbf{x}_I^{\mathrm{T}}\hat{\boldsymbol{\beta}}_I. \tag{3.13}$$

Constrained linear predictors are studied in papers of Goldstein (1975a,b, 1976).

Having stated that we will study the discarded-variables class of predictors under the general form of utility (3.8), it remains only to specify the purely predictive utility component in order to have a closed-form decision problem ready for analysis. In this paper we report results for two such choices: the usual negative squared error

$$w(\hat{y}, y) = -(\hat{y} - y)^2, \tag{3.14a}$$

which is taken up in Section 3.4, and a less familiar function is probability density form

$$w(\hat{y}, y) = f_\alpha(\hat{y} - y), \tag{3.14b}$$

where $f_\alpha$ is the probability density of a random variable $\alpha$ as developed in Section 3.5.

### 3.4. Negative Squared Error as Purely Predictive Utility Component

Having outlined a decision model for consideration, we now summarize the resulting decision criterion in the discarded-variable case for comparison to other Bayesian procedures and to classical procedures. The derivations are not dealt with at great length since they are given in a line of studies including Anscombe (1963), Dickey (1967), Lindley (1968), Harrison & Stevens (1976) and Dickey & Kadane (1977).

We use the negative squared error form for the purely predictive utility (3.8), (3.14a), and the optimization procedure outlined in (3.7). Lindley (1968) derived the corresponding criterion in the case of known $\sigma^2$. Starting with the expectations over $y$ and $\mathbf{x}$, we have, given $I$ and $\hat{y}_I(\cdot)$,

$$\begin{aligned} E[\{\tilde{y} - y_I(\tilde{\mathbf{x}})\}^2] &= E[E[\{\tilde{y} - E(y|\mathbf{x}_I) + E(y|x_I) - \hat{y}_I(\mathbf{x}_I)\}^2|\tilde{\mathbf{x}}_I]] \\ &= E[E[\{y - E(y|\mathbf{x}_I)\}^2|\mathbf{x}_I]] \\ &\quad + E[E[\{E(y|\mathbf{x}_I) - y_I(\mathbf{x}_I)\}^2|\mathbf{x}_I]]. \end{aligned} \tag{3.15a}$$

This expectation is minimized by the choice

$$\hat{y}_I(\mathbf{x}_I) = E(y|\mathbf{x}_I), \tag{3.16}$$

which makes the second summand zero. Using this choice and resuming the calculation in (3.15a), we have

$$\begin{aligned} -w(I) &= E[\mathrm{Var}(y|\mathbf{x}_I)] \\ &= \mathrm{Var}\, y - \mathrm{Var}[E(y|\mathbf{x}_I)] \\ &= E(\sigma^2) + \mathrm{Var}(\tilde{\mathbf{x}}^{\mathrm{T}}\tilde{\boldsymbol{\beta}}) - \mathrm{Var}[E(\boldsymbol{\beta})^{\mathrm{T}}E(\mathbf{x}|\mathbf{x}_I)] \\ &= E(\sigma^2) + E(\tilde{\boldsymbol{\beta}}^{\mathrm{T}}\tilde{\mathbf{x}}\tilde{\mathbf{x}}^{\mathrm{T}}\tilde{\boldsymbol{\beta}}) - E(\tilde{\boldsymbol{\beta}}^{\mathrm{T}}\tilde{\mathbf{x}})E(\tilde{\mathbf{x}}^{\mathrm{T}}\tilde{\boldsymbol{\beta}}) - \mathrm{Var}[E(\boldsymbol{\beta})^{\mathrm{T}}E(\mathbf{x}|\mathbf{x}_I)] \\ &= E(\sigma^2) + \mathrm{tr}[E(\tilde{\mathbf{x}}\tilde{\mathbf{x}}^{\mathrm{T}})E(\tilde{\boldsymbol{\beta}}\tilde{\boldsymbol{\beta}}^{\mathrm{T}})] - E(\tilde{\boldsymbol{\beta}})^{\mathrm{T}}E(\tilde{\mathbf{x}}\tilde{\mathbf{x}}^{\mathrm{T}})E(\boldsymbol{\beta}) \\ &\quad + E(\tilde{\boldsymbol{\beta}})^{\mathrm{T}}E(\tilde{\mathbf{x}}\tilde{\mathbf{x}}^{\mathrm{T}})E(\boldsymbol{\beta}) - E(\tilde{\boldsymbol{\beta}})^{\mathrm{T}}E(\tilde{\mathbf{x}})E(\tilde{\mathbf{x}}^{\mathrm{T}})E(\tilde{\boldsymbol{\beta}}) - \mathrm{Var}(E(\boldsymbol{\beta})^{\mathrm{T}}E(\mathbf{x}|\mathbf{x}_I)] \\ &= E(\sigma^2) + \mathrm{tr}[E(\mathbf{x}\mathbf{x}^{\mathrm{T}})\,\mathrm{Var}(\boldsymbol{\beta})] + E(\boldsymbol{\beta})^{\mathrm{T}}\,\mathrm{Var}(\mathbf{x})E(\boldsymbol{\beta}) \\ &\quad - E(\boldsymbol{\beta})^{\mathrm{T}}\,\mathrm{Var}[E(\mathbf{x}|\mathbf{x}_I)]E(\boldsymbol{\beta}) \\ &= E(\sigma^2) + \mathrm{tr}[E(\mathbf{x}\mathbf{x}^{\mathrm{T}})\,\mathrm{Var}(\boldsymbol{\beta})] + E(\boldsymbol{\beta})^{\mathrm{T}}E\{\mathrm{Var}[\mathbf{x}|\mathbf{x}_I]\}E(\boldsymbol{\beta}), \end{aligned} \tag{3.15b}$$

which is equivalent to Lindley's (1968) Eq. (10).

To simplify the discussion of these findings, we specify the popular "ignorance" prior density within the natural conjugate family

$$p(\beta, \sigma) = c/\sigma. \tag{3.17}$$

The results for this case also hold as approximations in large samples for the more general case (Dickey, 1976).

For sample data $D = (\mathbf{y}, X)$, where $\mathbf{y} = (y_1, \ldots, y_n)$ and $X = (\mathbf{x}_1, \ldots, \mathbf{x}_n)^{\mathrm{T}}$, denote the usual sufficient statistics by

$$\begin{aligned} N &= X^{\mathrm{T}}X, & \mathbf{b} &= N^{-1}X^{\mathrm{T}}\mathbf{y}, \\ \nu &= n - r, & s^2 &= \nu^{-1}(\mathbf{y} - X\mathbf{b})^{\mathrm{T}}(\mathbf{y} - X\mathbf{b}) = \nu^{-1}\mathrm{SSE}(\Omega), \end{aligned} \tag{3.18}$$

where $r$ is the dimension of $\mathbf{x}$, that is, the number of independent variables in the regression model $\Omega$, and $\nu$ is the degrees of freedom. Then the resulting posterior distribution is given by

$$(\boldsymbol{\beta} \mid \sigma, D) \sim N^{(r)}(\mathbf{b}, \sigma^2 N^{-1}), \tag{3.19a}$$

an $r$-dimensional normal distribution with mean $\mathbf{b}$ and variance matrix $\sigma^2 N^{-1}$, and by

$$(\sigma^2 \mid D) \sim s^2/(\chi^2_\nu/\nu), \tag{3.19b}$$

$s^2$ divided by a chi-square random variable with $\nu$ degrees of freedom, divided by $\nu$. The optimal predictor in class $I$ (3.15) is then

$$\hat{y}_I(\mathbf{x}_I) = \mathbf{x}_I^{\mathrm{T}}\mathbf{c}_I, \tag{3.20}$$

where $\mathbf{c}_I$ denotes the usual subset-regression least-squares coefficient vector based on $D_I = (\mathbf{y}, X_I)$ where the matrix $X = (X_I, X_K)$ is partitioned according to $\mathbf{x}^{\mathrm{T}} = (\mathbf{x}_I^{\mathrm{T}}, \mathbf{x}_K^{\mathrm{T}})^{\mathrm{T}}$. The corresponding predictive utility (3.15b) becomes

$$-\bar{w}(I) = (1 + r/n)s^2\nu/(\nu - 2) + \mathrm{SSB}(I, \Omega)/n. \tag{3.21}$$

Note that in both (3.15b) and (3.21) one term alone depends on the class $I$. In the case of additive rewards ($U_I \equiv 1$), one prefers class $I_2$ over $I_1$ if and only if $n(U^*_{I_1} - U^*_{I_2})$ is exceeded by the difference

$$\mathrm{SSB}(I_1, \Omega) - \mathrm{SSB}(I_2, \Omega). \tag{3.22}$$

This difference will itself be a between sum of squares $\mathrm{SSB}(I_1, I_2)$ in the nested case $I_1 \subset I_2$. Hence as stated previously, one obtains a threshold on SSB proportional to sample size, in distinction to the previous thresholds (3.4) and (3.5).

Note the lack of dependence of the criterion on any feature of the model $\Omega$ other than the property that the predictor variables of $I_1$ and $I_2$ are both subsets of those of some model to which a prior distribution in the "ignorance" form applies. In particular, this lack of dependence holds whether or not $I_1$ and $I_2$ are nested.

The first term of (3.21), $(1 + r/n)s^2\nu/(\nu - 2)$, is the Bayesian posterior predictive squared error of the optimal predictor $\hat{y}_\Omega(\mathbf{x})$, based on the full set of $r$ variables. This term refers literally to the context in which inference and prediction are *both* made from the full model $\Omega$. Thus this term is the "cost of doing business," the cost we must pay for the prediction regardless of other costs. If we infer from the full model to obtain the posterior expected squared error for optimal prediction by a smaller class $I$, there is then the additional term $\mathrm{SSB}(I, \Omega)/n$. Hence a statistician obeying the principle of maximum expected utility and evaluating predictor classes from the standpoint of the full inference model would prefer the larger predictor class of any nested pair if no special rewards are given for simplicity (that is, if $U_I \equiv U$ and $U_I^* = U^*$). We shall return to this aspect in Section 4.

For the next few paragraphs we discuss expected squared errors calculated from a sampling theory viewpoint. Lest this change in point of view cause confusion, we pause here to remind the reader about the distinction. In Bayesian theory, the random variables are what is unknown, namely the values of parameters. The data, once observed, are no longer random (we know the outcome), and hence calculations like (2.1) are done conditional on the observed data outcomes. By contrast, in traditional sampling theory, the random variables are the "data that might have been," distributed conditionally on the values of the parameters. Each side in the discussion accepts the mathematics of the other side; the discussion is rather about which computations are relevant for inference.

Taking expectations now under the *sampling model*, holding the parameters fixed, we find

$$E(y - \hat{y}_I)^2 = (1 + r_I/n)\sigma_I^2, \tag{3.23}$$

where $\sigma_I^2$ is the sampling-model conditional variance of $y$ given $\mathbf{x}_I$. This expectation is commonly estimated by the statistic,

$$(1 + r_I/n)\mathrm{SSE}(I)/v_I. \tag{3.24}$$

Traditionally, one compares values of (3.23) for various subsets $I$. Such a criterion is less than appropriate for several reasons, including that the expectation is not conditional on the observed data $D$. Instead, it is an average over values of the coefficient estimate $\mathbf{c}_I$ which did *not* occur, and therefore violates the likelihood principle, which is a consequence of the Bayesian axioms.

A statistician who uses the estimate (3.23) ignores the data $X_K$. The similarity in form of (3.24) to the first term of (3.21) yields an interpretation of the traditional statistic (3.24) as an approximate posterior predictive squared error of $\hat{y}_I$ for a Bayesian statistician whose prior uncertainty implies conditional independence between $(y, \mathbf{x}_I)$ and $X_K$ given $\mathbf{y}$ and $X_I$. To take this position for various sets $\mathbf{x}_I$ would be to place very special conditions on the joint distribution of future data $(y, \mathbf{x})$ and known data $D = (\mathbf{y}, X_I, X_K)$.

It can be shown that a Bayesian who believes $H: \boldsymbol{\beta}_K = \mathbf{0}$ but otherwise has prior "ignorance" would have the same predictor as someone who was "ignorant" within $\Omega$ but was forced to discard $\mathbf{x}_K$, namely,

$$\begin{aligned}\hat{y}(\mathbf{x}) &= E(y \mid \mathbf{x}, \boldsymbol{\beta}_K = \mathbf{0}, D) = \mathbf{x}_I^{\mathrm{T}} E(\boldsymbol{\beta}_I \mid \boldsymbol{\beta}_K = \mathbf{0}, \mathrm{D}) \\ &= \mathbf{x}_I^{\mathrm{T}} \mathbf{c}_I, = \hat{y}_I(\mathbf{x}),\end{aligned} \tag{3.25}$$

using (3.20). The person believing $H$ would interpret (3.24) as his optimal posterior predictive squared error (as opposed to (3.21) for the believer in $\Omega$). To compare (3.24) for two difference choices $I_1$ and $I_2$ under this interpretation would be to first state a prior opinion that $\boldsymbol{\beta}_{J_1} = 0$ with "ignorance" on the remaining coefficients $\boldsymbol{\beta}_{I_1}$ and then an analogous prior with respect to $I_2$. To hold both opinions simultaneously would be a contradiction.

## 3.5. Probability Density Form of Purely Predictive Utility Component

We now consider the second kind of purely predictive utility component, that in the form of a probability density function (pdf) of $f_\alpha(\hat{y} - y)$; see (3.14). It will help the reader's intuition to think of the density $f_\alpha$ of the random variable $\alpha$ as symmetric and unimodal with mode at zero. Thus we most prefer (have highest utility for) a zero predictive error $(\hat{y} - y)$. Our preference decreases as the error increases in absolute value, and we are equally displeased by an over-prediction ($\hat{y} - y$ positive) and an under-prediction ($\hat{y} - y$ negative) of equal magnitude. Such a utility function has the advantage of being bounded, and hence stable (Kadane & Chuang, 1978); by comparison, negative square error has neither property. Tiao and Afonja (1976) discuss the special case in which $f$ is the normal pdf; some of these densities $f_\alpha$ are in the class of conjugate utilities proposed by Lindley (1976). Our analysis follows Dickey & Kadane (1977).

Suppose now that $\alpha$ has two moments, so that we may write

$$E(\alpha) = 0, \qquad \mathrm{Var}(\alpha) = a^2. \tag{3.26}$$

The variable $a$ has an interesting interpretation. If "a miss is as good as a mile," that is, if any but the most accurate estimates are considered nearly equally worthless, then $a$ would be taken to be quite small. On the other hand, if the idea is to be roughly correct, then $a$ could be taken to be larger. Thus $a$ measures an important aspect of an investigator's utility function, one that might change from one occasion to the next, even with the same data and the same formal decision problem, depending on the investigator's purpose.

Conditional on $I$, the expected purely predictive utility component achieved is

$$\bar{w}(I) = \max_{\hat{y}_I(\cdot)} \iint f_\alpha(\hat{y}_I(\mathbf{x}) - y)p(y|\mathbf{x}_I)\,dy\,p(\mathbf{x}_I)\,d x_I, \tag{3.27}$$

where we have already used the fact that $\hat{y}_I(\mathbf{x})$ depends on $\mathbf{x}$ only through $\mathbf{x}_I$. We notice that the inner integral is a familiar form. In fact, it is the formula for the convolution of two random variables $\alpha$ and $y$, with the resulting density evaluated at $\hat{y}_I(\mathbf{x})$. Thus we may write

$$\int f_\alpha(\hat{y}_I(\mathbf{x}) - y)p(y|\mathbf{x}_I)\,dy = f_{\alpha+y|\mathbf{x}_I}(\hat{y}_I(\mathbf{x}_I)). \tag{3.28}$$

Substituting this expression in (3.27), we have

$$\bar{w}(I) = \max_{\hat{y}_I(\cdot)} \int f_{\alpha+y|\mathbf{x}_I}(\hat{y}_I(\mathbf{x}_I)p(\mathbf{x}_I)\,d\mathbf{x}_I. \tag{3.29}$$

Since the class of predictors $\hat{y}_I(\cdot)$ are all functions of $\mathbf{x}_I$, the maximization in (3.29) can be performed pointwise in $\mathbf{x}_I$ before integration. Hence

$$\bar{w}(I) = \int \left[\max_{\hat{y}_I(\mathbf{x}_I)} f_{\alpha+y|\mathbf{x}_I}(\hat{y}_I(\mathbf{x}_I))\right] p(\mathbf{x}_I)\,d\mathbf{x}_I. \tag{3.30}$$

Now $\alpha$ is unimodal and symmetric, with center of symmetry 0. Similarly, conditionally on $\mathbf{x}_I$, $y - E(y|\mathbf{x}_I)$ is assumed unimodal and symmetric with center of symmetry 0. Since these two random variables are independent, their convolution is also unimodal and symmetric with mode at 0 (Wintner, 1938, p. 30). Because the mode of this convolution is at 0, the mode of $f_{\alpha+y|\mathbf{x}_I}$ is at $E(y|\mathbf{x}_I)$, and hence the best choice of $\hat{y}_I(\mathbf{x}_I)$ is

$$\hat{y}_I(\mathbf{x}_I) = E(y|\mathbf{x}_I). \tag{3.31}$$

The resulting purely predictive utility component is

$$\bar{w}(I) = \int f^{(0)}_{\alpha+y|\mathbf{x}_I - E(y|\mathbf{x}_I)} p(\mathbf{x}_I)\,d\mathbf{x}_I. \tag{3.32}$$

This is in the form of a mixture of densities (here evaluated at zero). The corresponding mixture random variable has zero mean and variance $V$, where

$$V = a^2 + E[\mathrm{Var}(y|\mathbf{x}_I)]. \tag{3.33}$$

Suppose $g$ is the density of some location-scale family (for specificity, one can think of $g$ as $\varphi$, the standardized normal density). Then we assume the approximation

$$\int f^{(z)}_{\alpha+y|\mathbf{x}_I - E(y|\mathbf{x}_I)} p(\mathbf{x}_I)\,d\mathbf{x}_I \doteq g(z/V^{1/2})/V^{1/2}, \tag{3.34}$$

so that, in particular,

$$\int f^{(0)}_{\alpha+y|\mathbf{x}_I - E(y|\mathbf{x}_I)} p(\mathbf{x}_I)\, d\mathbf{x}_I \doteq g(0)/V^{1/2}. \tag{3.35}$$

Hence we have the approximation

$$\bar{w}(I) \doteq g(0)/\{a^2 + E[\mathrm{Var}(y|x_I)]\}^{1/2} \tag{3.36}$$

For a normal approximation, $g(0) = (2\pi)^{-1/2}$. Note that $E[\mathrm{Var}(y|\mathbf{x}_I)]$ is evaluated in (3.15b). In the "ignorance" prior case, by (3.21),

$$V = (1 + r/n)s^2 v/(v - 2) + \mathrm{SSB}(I, \Omega)/n. \tag{3.37}$$

It is interesting to compare the decision criteria that result from (3.36) with those we have seen in (3.16) and (3.21) in the squared error loss case. In the case that all "rewards" are multiplicative ($U_I^* \equiv U$), the criterion for two nested predictor models, $I_1 \subset I_2$, prefers $I_2$ for

$$\mathrm{SSB}(I_1, I_2) > n[a^2 + V][(U_1/U_2)^2 - 1].$$

Compare this with the criterion (3.22) which also has the form of a threshold on the between sum of squares. Again, note the proportionality of the threshold to sample size.

## 4. Discussion

We have seen that the Bayesian odds ratio is not useful as a complete data summary for decision making. Indeed, the very concept of inference about the "truth" of a model has come under question. Also, in the specific decision problem of choice of predictor in a normal linear model, we have found that the predictive criterion behaved as a function of sample size in such a fashion that even if the $F$ test tail area and the Bayesian odds for a submodel were very small (3.4), (3.5), the optimal predictor could still be the predictor based only on the submodel, (3.22), (3.36). In other words, one can have coherent preference for the use of a simple model even if it is known to be "false."

According to Occam's razor, one should prefer scientific models that are simple. In our decision criteria for prediction it has turned out that positive rewards for simplicity are necessary to imply preferences for simple nested predictor classes. For our utilities $W(\hat{y}, y, I)$ (3.8), the obvious monotonicity of the purely predictive expected utility component

$$I_1 \subset I_2 \quad \Rightarrow \quad \bar{w}(I_1) \le \bar{w}(I_2),$$

implies that positive rewards, $(U_{I_1}^* - U_{I_2}^*)/U$ (3.10) or $\log(U_{I_1}/U_{I_2})$ (3.11), will be needed. This result is quite general since to maximize a functional $\bar{w}[\hat{y}(\cdot)]$ over a larger class of functions $\hat{y}(\cdot)$ can never lead to a smaller maximum value $\bar{w}$. Hence, to the extent that real decision makers are coherent, the fact that simple predictor models are used could mean that positive rewards do exist.

The principle is clear that one should not deliberately exclude a cost-free variable from influence on one's predictor, provided that one maintains the freedom to use it in an optimal way. It could turn out, though, that the optimal way itself implies a predictor which is a function of the other variables only. In our treatment of the normal linear model, this happens with probability zero for a prior distribution of continuous type. However, a mixed-type prior distribution could lead to such an event with nontrivial probability as follows.

Using a probability density form of predictive utility $w$, one may consider linear predictors based on a parameter estimate $\hat{\boldsymbol{\beta}}$. If the probability is positive for a submodel of the form $H:\boldsymbol{\beta}_K = \mathbf{0}$, then the predictive expected utility $\bar{w}(\hat{\boldsymbol{\beta}}) = E(w|H)P(H) + E(w|H^c)P(H^c)$ will typically have two local maxima in $\hat{\boldsymbol{\beta}}$, one for which have $\hat{\boldsymbol{\beta}}_J = \mathbf{0}$ and the other with $\hat{\boldsymbol{\beta}}_J \neq \mathbf{0}$. A high value of $P(H)$ will tend to imply that the overall maximum is achieved at the first of these points. Hence, even without positive rewards for simplicity, $\bar{W}$ will be maximized by the simpler predictor. (For details, see Dickey & Kadane, 1977.) In our view, however, mixed-type priors that put positive probability on low dimensional subspaces are not usually realistic in the context of most social sciences. Consequently, positive rewards for simplicity seem to be required to explain why good economists work with simple models.

Common sense and statistical folklore warn against using too many variables in a regression relative to the sample size. Least squares estimates are notoriously prone to outliers and other problems and become even more suspect in high dimensions. For example, the traditionally estimated expected squared error of prediction (3.24) increases for large values of $r_I/n$. Yet the Bayesian methods developed in Section 3 require the use of all available cost-free variables. This phenomenon has been called by Lindley (1978) the paradox of Bayesian overfitting.

We offer two explanations of this difficulty. The first, also cited by Dickey (1973b) and by Lindley (1978) in his discussion of Young (1977), is that the model chosen for illustration in Section 3 takes as its prior the ignorance form (3.17), which is justified as an approximation to a proper subjective prior distribution by the theory of stable estimation (Edwards *et al.*, 1963). This approximation yields the least squares estimates. It will be valid when sample size $n$ is large relative to the dimensionality of the regressors ($r_I$), so the approximation tends not to be valid precisely where the problem occurs ($r_I$ large relative to $n$).

Our second explanation, a basic premise of this paper, is that previous efforts in this area have not taken explicit account of the utility function of the economist. Just as in the past many efforts have been made to avoid proper subjective prior distributions on the parameters, many economists and statisticians try to avoid the statement of an explicit utility function for the problem of model choice. In problems of estimation this is not such a severe difficulty because, if the posterior distribution is symmetric, every symmetric utility function having its maximum at zero will result in the same estimate, namely the posterior center of symmetry. Thus it is not necessary to think about whether your utility is negative squared error, negative absolute error, or normal pdf form since they all result in the same estimate. However, as we have seen, the choice of models problem is essentially different in this respect. For choosing models, there seems to be no alternative to an honest assessment of your utility function and, in particular, of your personal weighting of accuracy against parsimony.

## ACKNOWLEDGMENTS

The authors are pleased to acknowledge the financial support of the National Science Foundation and the useful and helpful comments of Jan Kmenta, James B. Ramsey, and Michel Mouchart. Reproduction for any purpose of the U.S. Government is permitted.

## REFERENCES

Ando, A., & Kaufman, G. W. Bayesian analysis of the independent multinormal process—neither mean nor precision known. *Journal of the American Statistical Association*, 1965, **60**, 347–358.

Anscombe, F. J. Bayesian inference concerning many parameters with reference to supersaturated designs. *Bulletin of the International Statistical Institute*, 1963, **40**, 721–733.

Berkson, J. Some difficulties of interpretation encountered in the application of the chi-squared test. *Journal of the American Statistical Association*, 1938, **33**, 526–542.

Davis, W. W., & DeGroot, M. H. A new look at selecting regression models. Unpublished technical report, Carnegie-Mellon University, 1978.

Dickey, J. M. A Bayesian hypothesis-testing procedure. *Annals of the Institute of Statistical Mathematics*, 1967, **19**, 367–369.

Dickey, J. M. The weighted likelihood ratio, linear hypotheses on normal location parameters. *Annals of Mathematical Statistics*, 1971, **42**, 204–223.

Dickey, J. M. Scientific reporting and personal probabilities: Student's hypothesis. *Journal of the Royal Statistical Society Series B*, 1973, **35**(2), 285–305.(a)

Dickey, J. M. Discussion to "Combining possibly related estimation problems" by B. Efron and C. Morris. *Journal of the Royal Statistical Society*, *Series B*, 1973, **35**(3), 379–421.(b)

Dickey, J. M. Bayesian alternatives to the *F* test and the least squares estimate in the normal linear model. In Stephen E. Fienberg & Arnold Zellner (Eds.), *Studies in Bayesian econometrics and statistics* (dedicated to Leonard J. Savage). Amsterdam: North-Holland Publ., 1974. Pp. 515–554.

Dickey, J. M. Approximate posterior distributions. *Journal of the American Statistical Association*, 1976, **71**(September), 680–689.

Dickey, J. M. Is the tail area useful as an approximate Bayes factor? *Journal of the American Statistical Association*, 1977, **72**(March), 138–142.

Dickey, J. M. Approximate coherence for regression model inference-with a new analysis of Fisher's broadback wheatfield data. In A. Zellner (Ed.) *Studies in Bayesian econometrics and statistics* (in honor of Sir Harold Jeffreys). Amsterdam: North-Holland Publ., 1979, in press.

Dickey, J. M. & Kadane, Joseph B. Simplification of predictors when the utility has probability density form. Unpublished technical report, Department of Statistics, Carnegie-Mellon University, 1977.

Edwards, W., Lindman, H., & Savage, L. J. Bayesian statistical inference for psychological research. *Psychological Review*, 1963, **70**, 193–242.

Gaver, K. M., & Geisel, M. S. Discriminating among alternative models: Bayesian and non-Bayesian methods. In Paul Zarembka (Ed.) *Frontiers of econometrics*. New York: Academic Press, 1973.

Geisser, S. Bayesian estimation in multivariate analysis. *Annals of Mathematical Statistics*, 1965, **36**, 150–159.

Goldstein, M. Approximate Bayes solutions to some non-parametric problems. *Annals of Statistics*, 1975, **3**, 512–517.(a)

Goldstein, M. A note on some Bayesian non-parametric estimates. *Annals of Statistics*, 1975, **3**, 736–740.(b)

Goldstein, M. Bayesian analysis of regression problems. *Biometrika*, 1976, **63**(1), 51–58.

Gunel, E. & Dickey, J. M. Bayes factors for independence in contingency tables. *Biometrika*, 1974, **61**, 545–571.

Harrison, P. J., & Stevens, C. F. Bayesian forecasting (with discussion). *Journal of the Royal Statistical Society, Series B*, 1976, **38**(3), 205–247.

Harvey, A. C., & Collier, P. Testing for functional misspecifications in regression analysis. *Journal of Econometrics*, 1977, **6**, 103–119.

Hocking, R. R. The analysis and selection of variables in linear regression. *Biometrics*, 1976, **32**(1), 1–49.

Jeffreys, H. *Theory of probability*. (3rd ed.) London and New York: Oxford Univ. Press (Clarendon), 1961.

Kadane, J. B., & Chuang, D. Stable decision problems. *Annals of Statistics*, 1978, **6**, 1095–1110.

Leamer, E. *Specification searches*. New York: Wiley, 1978.(a)

Leamer, E. Regression selection strategies and revealed priors. *Journal of the American Statistical Association*, 1978, **73**, 580–587.(b)

Lindley, D. V. A statistical paradox. *Biometrika*, 1957, **44**, 187–192.

Lindley, D. V. The use of prior probability distributions in statistical inference and decision. *Proceedings of the 4th Berkeley symposium on statistics and probability*, 1961, **1**, 453–468.

Lindley, D. V. The choice of variables in multiple regression (with discussion). *Journal of the Royal Statistical Society, Series B*, 1968, **30**(1), 31–66.

Lindley, D. V. A class of utility functions. *Annals of Statistics*, 1976, **4**, 1–10.

Lindley, D. V. The Bayesian approach (with discussion). *Scandinavian Journal of Statistics*, 1978, **5**, 1–26.

Raiffa, H., & Schlaifer, R. *Applied statistical decision theory*. Boston: Division of Research, Harvard Business School. Republished in paperback: Cambridge, Mass.: MIT Press, 1968.

Ramsey, F. P. *The foundations of mathematics and other essays*. London: Kegan, Paul, Trench, Trubner & Co., Ltd., 1931.

Ramsey, J. B. Classical model selection through specification error tests. In Paul Zarembka (Eds.), *Frontiers of econometrics.* New York: Academic Press, 1974. Chapter 1.

Savage, L. J. *The foundations of statistics.* New York: Wiley, 1954.

Tiao, G. C., & Afonja, B. Some Bayesian considerations of the choice of design for ranking, selection, and estimation. *Annals of the Institute of Statistical Mathematics*, 1976, **28**, 167–185.

Von Neumann, J., & Morgenstern, O. *Theory of games and economic behavior.* Princeton, N.J.: Princeton Univ. Press, 1947.

Wintner, A. *Asymptotic distributions and infinite convolutions.* Ann Arbor, Mich.: Edwards, 1938.

Young, A. S. A Bayesian approach to prediction using polynomials. *Biometrika*, 1977, **64**, 309–317.

Zellner, A. *An introduction to Bayesian inference in econometrics.* New York.: Wiley, 1971.

Zellner, A., & Vandaele, W. Bayes–Stein estimators for R-means, regression and simultaneous equation models. In S. E. Fienberg & A. Zellner (Eds.), Studies in Bayesian Econometrics and Statistics. Amsterdam: North-Holland Publ., 1974. Pp. 627–653.

[15]

ECONOMETRICA

VOLUME 53 MARCH 1985 NUMBER 2

# BAYESIAN ECONOMETRICS

BY ARNOLD ZELLNER[1]

The widespread use of prior information in formulating, estimating, and using econometric models is reviewed. Attempts to avoid the use of prior information by formulating multivariate statistical VAR and ARMA time series models for economic time series data have resulted in heavily over-parametrized models. A simple demand, supply, and entry model is presented to contrast models utilizing prior information provided by economic theory and other sources with multivariate statistical time series models. Formal Bayesian methods for incorporating prior information in econometric estimation, testing, and prediction are presented. A number of published applied Bayesian studies are cited in which Bayesian methods have proved to be effective. It is concluded that wise use of the Bayesian approach will produce improved econometric results.

## 1. INTRODUCTION

I AM GRATEFUL to have this opportunity to share some of my thoughts on Bayesian econometrics with you. Before doing this, I would like to say that I have great admiration and respect for the work of Irving Fisher and Henry Schultz. Henry Schultz, who spent many years at the University of Chicago, made many significant research contributions. Similarly, Irving Fisher's research has had a profound effect on economics and econometrics. While I could spend the entire lecture attempting to summarize their research, I shall just emphasize that both of them produced key results relating to relatively simple models that have endured over the years. For example, Fisher put forward the famous Fisher equation that relates the nominal interest rate to the anticipated real rate and the anticipated rate of inflation. Often, when I am asked, "Are there any laws in economics?" I point to the Fisher equation as an example. Schultz worked on the laws of supply and demand, relatively simple relationships that are additional examples of sophisticatedly simple laws in economics. I shall discuss the role of "simplicity" in model-building later in my lecture.

Since the early 1960's, Bayesian econometrics has developed rapidly with considerable interaction among econometricians and statisticians. After the publication of Bayes' essay in 1763, many statisticians and scientists including Laplace, Edgeworth, Jeffreys, de Finetti, Box, Savage, Lindley, Raiffa, Jaynes, Schlaifer, Pratt, Good, Geisser, DeGroot, Tiao, and others have made important contributions to the development of Bayesian statistics. As regards modern Bayesian econometrics,[2] it commenced in the early 1960's with papers by Jacques Drèze, Tom Rothenberg, Walter Fisher, Albert Ando, Gordon Kaufman, myself, and a few others. Since then, there has been an upswell of work in Bayesian econometrics

[1] This is a written version of my Fischer-Schultz Lecture, presented at the European Econometric Society Meetings, Pisa, Italy, August 29-September 2, 1983. My research was financed in part by the National Science Foundation and by income from the H. G. B. Alexander Endowment Fund, Graduate School of Business, University of Chicago.

[2] See Zellner [61] for a review of the development of modern Bayesian econometrics.

with a good deal of interaction among econometricians and statisticians.[3] Now we have many published Bayesian monographs and papers. Also, recent and current econometrics texts include varying amounts of Bayesian material and Bayesian computer programs are available. Then too, the Bayesian learning model has been utilized in many works in economic theory.[4]

During the last twenty years, I have found econometricians and economists to be very interested in Bayesian ideas and methods. Contrary to some theories of how science progresses, it was the case that many were quite interested in learning about Bayesian ideas and methods as long as they were given evidence that the new methods work well in practice. This pragmatic attitude on the part of the profession is very commendable. I always like to learn about new ideas and approaches but the bottom line is how well they work in practice. This pragmatic dictum gets a lot of weight in the thinking of many econometricians and economists.

In view of the importance of the pragmatic dictum, mentioned above, I shall not describe in detail the various axiom systems that have been put forward to justify the Bayesian approach. Harold Jeffreys [26], a leading natural scientist, has an axiom system in his book, *Theory of Probability*, that leads him to conclude that the only sound approach to inference in science is the Bayesian approach based on a "reasonable degree of belief" definition of probability. In addition, Savage's [49] axiom system, put forward in his *Foundations of Statistics*, implies that procedures for acting rationally in making decisions under uncertainty are Bayesian procedures. While these axiom systems are not perfect, they do exist and may be of interest to some of you. As an aside, Savage and Milton Friedman were very good friends and both admirers of Bruno de Finetti's work. Savage and Friedman co-authored two well known papers on choice under uncertainty.[5] These papers brought to the fore the importance of economic utility theory in statistics by emphasizing the central role of maximization of expected utility or equivalently minimization of expected loss in making choices under uncertainty, for example in choosing estimates, predictions, and models. Their work helped to introduce economists and statisticians to the intimate connection between utility theory and statistical theory. There is thus considerable axiomatic support for Bayesian inference and decision procedures that have been applied in analyzing many problems in econometrics, for example estimation, prediction, testing, control, and other decision problems with very good results.

[3] Since 1970, semi-annual meetings of the NBER-NSF Seminar on Bayesian Inference in Econometrics have been held at which Bayesian research papers have been discussed by econometricians and statisticians. Some of these papers have been published in Fienberg and Zellner [13] and Zellner [62].

[4] See, e.g., papers and references in Fienberg and Zellner [13] and Zellner [62]. Also see the book edited by M. Boyer and R. E. Kihlstrom, *Bayesian Models in Economic Theory*, published in the North-Holland Publishing Company's *Series on Bayesian Econometrics and Statistics*, 1983.

[5] These papers, "The Utility Analysis of Choices Involving Risk" and "The Expected-Utility Hypothesis and the Measurability of Utility" have been reprinted in *The Writings of Leonard Jimmie Savage*, American Statistical Association: Washington, D.C., 1981.

Rather than dwell on axiomatic matters, I shall consider more down-to-earth matters. Why are we interested in the Bayesian approach? My point of view is that in econometrics and in other areas of science a good deal of prior information is employed, usually informally. The Bayesian approach provides formal methods for handling background or prior information. Some of you may say that this is subjective and not objective. I believe that the words, "subjective" and "objective" are loaded words fraught with difficulty. I refer to the philosopher Bishop Berkeley who was a philosophical idealist. He claimed, rightly in my opinion, that it is impossible to prove that the "real" world exists. On the other hand, there are philosophical realists who believe, without proof, that the "real" world exists and that we as scientists can learn about its properties. Finally, there are solipsists who claim that only I exist and nothing else exists. Since it's impossible to settle these issues, who is "objective" and who is "subjective"? My position is that the best we can do at present is to build models and determine how well they perform in explaining past data, predicting as yet unobserved data, and analyzing policy problems. The point that I make to you, very strongly, is that non-Bayesians, as well as Bayesians, use a good deal of prior information in building and using models. Further, since the use of prior information is unavoidable, it should be used carefully and more formally than has been done in the past. Finally, it appears to me that most econometricians usually do not use a frequency concept of probability in model-building, estimation, testing, and prediction, but rather a degree of reasonable belief concept of probability.[6]

To illustrate some of these points, let us consider modeling of time series data as, for example, in demand and supply modeling or macroeconometric modeling. How do we generally approach the modeling problem? Some take the Jeffreys-Wrinch Simplicity Postulate, Ockham's Razor, and the Principle of Parsimony very seriously and formulate relatively simple models with the *belief* that they will probably work well. If there is a need to capture much detail, these workers would formulate a large, relatively simple model and investigate the extent to which it is adequate and accomplishes the objectives of a modeling project. Those who advocate sophisticated simplicity in model-building include Harold Jeffreys, Ronald Fisher, Milton Friedman, George Box, myself, and many others. Note that there are many examples of sophisticatedly simple models that work reasonably well, for example, Newton's laws, the gas laws, Einstein's laws, Schrödinger's equation, laws of supply and demand, the Fisher equation, Friedman's consumption model, the quantity theory of money, and so on. On the other hand, there are others who *believe* that complicated models will probably work well in practice. Witness the large number of macroeconometric models that have hundreds of nonlinear stochastic difference equations containing thousands of parameters. It is not clear whether such models have unique or many solutions. I have challenged many audiences to give me examples of large, complicated models that work well in any area of science and have not yet heard of any. A main point to be appreciated in this discussion is that *prior beliefs* about the

[6] This point has also been made with respect to physical scientists in Jeffreys [26].

simplicity/complexity issue have a vital impact on the way in which econometric models are formulated.

To pursue this last point further and more explicitly, let me review several types of linear time series models that are currently being studied and used. The first is the multiple time series model or multivariate autoregressive moving average (MVARMA) model put forward by Quenouille **[48]** and others. The MVARMA model for a mean-corrected $p \times 1$ vector of variables, $z_t$, is given by

$$(1) \qquad H(L)z_t = F(L)e_t \qquad (t = 1, 2, \ldots, T),$$

where $H(L)$ and $F(L)$ are matrix lag operators given by

$$H(L) = H_0 + H_1 L + H_2 L^2 + \cdots + H_r L^r,$$

$$F(L) = I + F_1 L + F_2 L^2 + \cdots + F_q L^q,$$

with $L$ the lag operator ($L^i z_t = z_{t-i}$) and $e_t$ is a $p \times 1$ non-serially correlated error vector with zero mean and $p \times p$ covariance matrix $\Sigma$.

The issue is how successful will we probably be in using a MVARMA model given the types of samples that we often have in econometrics, say 20 years of quarterly data? One relevant consideration is the number of parameters in the model. This number, denoted by $N$ is given by

$$N = (r+1)p^2 + qp^2 + p + p(p+1)/2,$$

that is, $(r+1)p^2$ parameters in the $H_i$ matrices, $qp^2$ in the $F_i$ matrices, $p$ means, and $p(p+1)/2$ distinct parameters in the error covariance matrix. Clearly, the value of $N$, the number of parameters will depend on the values assigned to $p$, $r$, and $q$. Some may consider a model with 30 variables ($p = 30$) to be a small model but for illustration I shall assume that just six variables ($p = 6$) appear in the model, $r = 3$, and $q = 4$. With these assumed values for $p$, $q$, and $r$, there are 395 parameters in the model. Suppose that we have 20 years of quarterly data on the six variables, that is a total of 480 observations. The observation/parameter ratio is 480/395, not even two observations per parameter which is hardly satisfactory because with so few observations, parameter estimates and predictions will generally be imprecise. Also, if tests are performed to determine whether elements of the coefficient matrices are equal to zero, they probably will not be very powerful and will have an impact on subsequent inferences, the problem of pre-testing. These then are some of the consequences of entertaining a relatively complicated model. With nonlinear versions of the above model, there would be even greater difficulties. There is a need for restrictions or prior information, perhaps derived from economic theory, that can effectively reduce the number of parameters.

Some in economics and statistics have not employed MVARMA processes but have utilized vector autoregressions. Of course, if the matrix lag operator, $F(L)$ in the MVARMA model is invertible, the MVARMA process has an infinite autoregressive representation, $F^{-1}(L)H(L)z_t = e_t$, or

$$(I - G_1 L - G_2 L^2 - \cdots - G_n L^n - \cdots)z_t = e_t.$$

If we cut off the autoregressive operator at the term involving $L^n$, that is we consider an $n$th order vector autoregression to represent our data, note that each of the $G$ matrices is $p \times p$ and we have $n$ of them. Thus there are $np^2$ parameters in the $G$ matrices, $p(p+1)/2$ in the error term covariance matrix, and $p$ means. If again we assume that we have 6 variables and that $n = 10$, perhaps a reasonable value for quarterly, seasonally unadjusted data, then the system contains 387 parameters. If we have 20 years of quarterly data on the 6 variables, there are 480 observations and the observation/parameter ratio, 480/387 is abysmally low. Further, consider a single equation of the above vector autoregression, say that for $z_{it}$. There are $np = 10 \times 6 = 60$ coefficients in this relation. Fitting by least squares is like fitting a regression with 60 independent variables that are usually highly correlated. The precision of estimation and prediction is not going to be very high. Indeed Litterman [**35**] has fit such relations with U.S. quarterly macroeconomic data and found that estimates and predictions are not very precise, the latter relative to predictions yielded by Box-Jenkins models, univariate autoregressions, and several macroeconometric models. Thus an attempt to determine relations solely from the data, this time by use of unrestricted vector autoregressions is foiled by the appearance of an inordinate number of parameters in the system. Again there is a need for subject matter information, or prior information, that will help to reduce the number of free parameters.

Sometimes economic theory or prior information suggests that some of the variables in a MVARMA process are exogenously determined. If we partition the $z_t$ vector as follows, $z_t' = (y_t', x_t')$, we can write the MVARMA model in the following form:

$$(2) \qquad \begin{pmatrix} H_{11}(L) & H_{12}(L) \\ H_{21}(L) & H_{22}(L) \end{pmatrix} \begin{pmatrix} y_t \\ x_t \end{pmatrix} = \begin{pmatrix} F_{11}(L) & F_{12}(L) \\ F_{21}(L) & F_{22}(L) \end{pmatrix} \begin{pmatrix} e_{1t} \\ e_{2t} \end{pmatrix}.$$

The assumption that the variables in $x_t$ are exogenous implies, as pointed out in Zellner and Palm [**64**], that

$$H_{21}(L) \equiv 0, \quad F_{12}(L) \equiv 0, \quad \text{and} \quad F_{21}(L) \equiv 0.$$

Thus 3 matrix lag polynomials are identically zero, given that $x_t$ is exogenous. With these restrictions imposed, the above system becomes

$$(3) \qquad H_{11}(L)y_t + H_{12}(L)x_t = F_{11}(L)e_{1t},$$

$$(4) \qquad H_{22}(L)x_t = F_{22}(L)e_{2t}.$$

The first set of relations in (3) is in the form of a linear dynamic structural econometric model with $y_t$ endogenous and $x_t$ exogenous. The second set of relations in (4) is a model for the random exogenous variables that is in the MVARMA form. The question has arisen as to whether a model for $x_t$ is needed. For certain purposes it is not needed. For example in estimation of the structural equations, provided that they are identified, one can proceed conditional upon the observed values of the $x_t$'s. However, if the system is to be used in forecasting, the future values of $x_t$ are needed. Currently in forecasting with structural econometric models, it is often not clear how the future values of exogenous variables are obtained. In many cases workers *guess* the future values of the exogenous variables, a very important, informal use of prior information that usually has a vital impact on forecasts and forecast intervals. If the exogenous variables are generated by a MVARMA process, as in (4), it can be used to forecast their future values.

If in the above system, 2 of the 6 variables are assumed exogenous and if we use the same assumptions as above, namely $r = 3$ and $q = 4$, the total number of free parameters is 183. While this is an improvement relative to the unrestricted MVARMA process, the observation/parameter ratio, 480/183 is still quite low. However, we generally go further by assuming that particular variables do not appear in some of the equations of (3), assumptions that lead to zero restrictions which are important in achieving identification and representing our economic knowledge about the forms of structural relations. Such restrictions, as well as the assumption that some variables are exogenous, have come under vigorous attack in the recent literature as being "unrealistic." I do not believe that this is a reasonable criticism of these systems because in many applications these restrictions are reasonable and if there is some doubt about them, many of them can be tested.

To make this last point clearly, let's come down from these abstract algebraic systems to a very simple demand and supply system, the kind that Henry Schultz and many others have used. In this connection, I shall emphasize another area, that of model formulation in which prior information is very important. In formulating an initial variant of a model, say a demand and supply model, we all use much prior information in sorting through combinations of ideas. In this process, the role of unusual facts has been emphasized by Jacques Hadamard [21] in his book, *The Psychology of Invention in the Mathematical Field*. From a survey of his fellow mathematicians, he found that most of them made their major research breakthroughs by discovering unusual facts and providing explanations of them. In economics, econometrics and other sciences there are a number of examples of this process. Given that unusual facts are so important, in a recent paper (Zellner [63]) I suggested 8 ways to produce unusual facts, one of which is to look at the equations of current macroeconometric models. Almost every equation contains one or more unusual facts. I have had several students consider such unusual facts and investigate them using micro and industry data with very illuminating results—see Chau [8], Laub [29, 30], Levedahl [33], Montmarquette [38, 39], and Peck [44, 45]. Another way to produce unusual facts is to push current theories to extremes. For example, with respect to the proportionality hypothesis of Friedman's [16] consumption theory, consider what happens as permanent income approaches zero. The ratio of permanent consumption to permanent income probably rises toward one as I suggested years ago (Zellner [57]) and recently work with low income Indian consumers has supported this view—see Bhalla [4]. Also, at very high permanent incomes, the proportion of permanent income consumed probably falls. Thus, while proportionality is a pretty good approximation over the mid-range of permanent incomes, at very low and very high levels of permanent incomes, it probably breaks down. Discovering such unusual facts and rationalizing them can lead to significant new theories of consumption and saving—see, e.g., Gersovitch [**18**].

Recently, with respect to theoretical and applied demand and supply studies, I noticed some very unusual facts. In Muth [**41**] and some other papers on rational expectations models of competitive industries, the number of firms is assumed to be a constant. On the other hand, in most applied econometric demand-supply analyses, the number of firms variable does not appear explicitly in estimated models. As is well known, Alfred Marshall and many economists have emphasized the important role of entry and exit of firms in producing long run equilibrium in a competitive industry. Also, simple aggregation considerations indicate that the number of firms in operation is a variable that should appear in a competitive industry's supply function. With these somewhat unusual facts in mind, William Veloce, a current student of mine, and I have formulated a three equation demand and supply model (Veloce and Zellner [**55**]) containing demand, supply, and net entry equations for the following endogenous variables:

$$y_{1t} = (1-L)\log Q_t, \qquad y_{2t} = (1-L)\log P_t, \qquad \text{and}$$

$$y_{3t} = (1-L)\log N_t,$$

where $Q_t$, $P_t$, and $N_t$, are quantity, real price, and the number of firms in operation, respectively in period $t$. The three equations of our demand, supply, and entry (DSE) model of a competitive industry are:

$$\text{(5)} \qquad \text{Demand:} \quad y_{1t} = a_1 y_{2t} + \boldsymbol{b}_1' \boldsymbol{x}_{1t} + u_{1t},$$

$$\text{(6)} \qquad \text{Supply:} \quad y_{1t} = a_2 y_{2t} + y_{3t} + \boldsymbol{b}_2' \boldsymbol{x}_{2t} + u_{2t},$$

$$\text{(7)} \qquad \text{Net Entry:} \quad (1-cL)y_{3t} = \boldsymbol{b}_3' \boldsymbol{x}_{3t} + u_{3t}, \quad 0 < |c| < 1,$$

where $\boldsymbol{x}_{1t}$ and $\boldsymbol{x}_{2t}$ are vectors of exogenous variables appearing in the demand and supply equations, respectively, $\boldsymbol{x}_{3t}$ is a vector of predetermined variables appearing in the net entry equation, the $a$'s, $b$'s, and $c$ are structural parameters, and $u_{1t}$, $u_{2t}$, and $u_{3t}$ are zero-mean, serially uncorrelated disturbance terms.

From the net entry relation, we can solve for $y_{3t}$, that is

$$\text{(8)} \qquad y_{3t} = (\boldsymbol{b}_3' \boldsymbol{x}_{3t} + u_{3t})/(1-cL)$$

and substitute this expression in the supply equation (6) to obtain the following "reduced" two equation system:

$$\text{(9)} \qquad y_{1t} = a_1 y_{2t} + \boldsymbol{b}_1' \boldsymbol{x}_{1t} + u_{1t},$$

$$\text{(10)} \qquad (1-cL)y_{1t} = a_2(1-cL)y_{2t} + \boldsymbol{b}_3' \boldsymbol{x}_{3t} + \boldsymbol{b}_2'(1-cL)\boldsymbol{x}_{2t} + u_{3t} + (1-cL)u_{2t}.$$

This reduced two equation system is similar to those that have been estimated in applied econometric studies. Note that equation (9) is exactly the demand equation (5) of the three equation DSE system. However, the second equation (10) of the two equation reduced system is a mixture of the supply and net entry equations, (6) and (7) of the three equation DSE model. It contains lagged endogenous variables, autocorrelated disturbances, and predetermined variables and coefficients from the supply and net entry equations. In empirical work, lags and autocorrelated disturbances in an equation like (10) have been rationalized in terms of partial supply adjustment and/or adaptive expectations effects—see, e.g., Nerlove [**42**]. Here the lags and autocorrelation appear as a consequence of considering the three equation DSE model to be an appropriate model. In current work with annual data for a Canadian industry, Veloce has found that a three equation DSE model, like the one above, provides a good interpretation of the data.

Let us consider the number of variables in the three equation system, (5)-(7). There are three endogenous variables and the variables in three $x$ vectors. If each of these vectors contains three exogenous variables, then there are 12 variables in the system. Consider modeling these 12 variables in a vector autoregressive approach. Each matrix in the autoregressive system is $12 \times 12$, that is contains 144 parameters. Further, the $12 \times 12$ disturbance covariance matrix contains 78 distinct elements. It is clear that with about 30 annual observations for each variable that a vector autoregressive approach, as well as an unrestricted MVARMA approach, will probably be very unsatisfactory. If we omit some of the relevant variables, we shall be analyzing marginal processes, like the two equation system (9)-(10), and it will be very difficult to interpret estimation results in behavioral terms. These then are some of the practical consequences of employing approaches that fail to take account of relevant, available economic information, that is prior information.

What is the economic information that has been employed to produce the restrictions on the three equation model above? First, for example, it is assumed that the number of firms variable does not appear in the demand equation, certainly a reasonable restriction. Also, variables such as the real wage rate and the price of capital services that might be included as exogenous variables in the supply equation probably do not belong in the demand equation. These restrictions and others like them have been embedded in the three equation model which contains about 20 parameters. If some of these restrictions seem doubtful, they can be tested. In these testing problems as well as in formulating the three equation model, it is apparent that much prior information is used. It is clearly important to use such prior information carefully and to have good methods for incorporating it flexibly in estimation, testing, forecasting, and policy analysis procedures.

To provide another example of a situation in which much prior information is used and has a vital impact on the results of analyses, consider the topic of seasonal adjustment. In much work on seasonal adjustment (see, e.g., Zellner [**59**]) an observed variable, $y_t$, is assumed to be represented as the sum of unobservable random trend-cycle, seasonal, and noise components, all assumed mutually uncorrelated. There does not seem to be any economic theory that suggests that the random trend-cycle component is uncorrelated with the random seasonal component. Yet this a priori assumption is employed to produce seasonally adjusted variables that are used in econometric analyses. See

Havenner and Swamy [24] for an analysis that casts doubt on the adequacy of this assumption.

The discussion presented above indicates that much prior information is needed and used in work with econometric models. Thus good procedures for incorporating prior information in analyses are needed. Fortunately, the Bayesian approach is one in which prior information, as little or as much as an investigator desires, can be flexibly and formally incorporated in estimation, prediction, testing, model selection, and policy analysis procedures. Let me briefly review these procedures and comment on some of their salient features.[7]

First, with respect to estimation, Bayes' Theorem yields the general result that a posterior probability density function (pdf) for the parameters of a model is proportional to a prior pdf times the likelihood function; that is,

$$\text{Posterior pdf} \propto (\text{Prior pdf}) \times (\text{Likelihood Function})$$

or

$$(11) \qquad p(\boldsymbol{\theta}|D) = cp(\boldsymbol{\theta}|I_0)p(y|\boldsymbol{\theta}), \quad \boldsymbol{\theta} \subset \Theta, \quad y \subset R_y,$$

where $\boldsymbol{\theta}$ is a vector of parameters, $\Theta$ the parameter space, $y$ a vector of observations, $R_y$ the sample space, $I_0$ the prior information, $D = (y, I_0)$, $c$ a normalizing constant given by $c^{-1} = \int_\Theta p(\boldsymbol{\theta}|I_0)p(y|\boldsymbol{\theta})\, d\boldsymbol{\theta}$, $p(\boldsymbol{\theta}|I_0)$ the prior pdf, and $p(y|\boldsymbol{\theta})$ the likelihood function. Thus for any estimation problem, given a prior pdf and a likelihood function, Bayes' Theorem provides an *exact, finite sample* posterior pdf for the parameters.[8]

The prior pdf, $p(\boldsymbol{\theta}|I_0)$, in (11) can be chosen to represent available prior information about possible values of $\boldsymbol{\theta}$, a formal, probabilistic representation. Note that the ranges of the elements of $\boldsymbol{\theta}$ can be restricted to particular intervals, say zero to one if this is suggested by economic theory or other considerations. Further, (11) can be viewed as a transformation of the prior pdf into a posterior pdf that incorporates both prior and sample information. In addition, assuming a pdf for $\boldsymbol{\theta}$ is closely related to assuming a pdf for parameters in non-Bayesian analyses of random parameter models such as random regression coefficient models, the structural form of the errors in the variables model, etc. The adequacy of the prior pdf for $\boldsymbol{\theta}$ can be assessed by use of various diagnostic checks including posterior odds, described below, for alternative priors, and most importantly by determining the quality of predictions yielded by a model incorporating the information in the prior pdf as was done in Litterman [35]. Of course, in structural econometric models, prior information must be introduced to identify structural parameters. On many occasions, Drèze [11, 12] and others have emphasized that this needed prior information can be flexibly and satisfactorily represented by use of a prior pdf.

[7] Some works that provide detailed presentations of Bayesian methods are Box and Tiao [6], DeGroot [10], Jeffreys [26], Leamer [31], Lindley [34], and Zellner [58].

[8] In Baxter [3], a posterior pdf for the mean rate of growth of the money supply is computed and employed in an imaginative way to provide a measure of the public's degree of confidence in a monetary reform.

Analytical or numerical integration procedures are available for analyzing properties of the posterior pdf in (11). For example, if $\boldsymbol{\theta}' = (\boldsymbol{\theta}_1', \boldsymbol{\theta}_2')$, the marginal posterior pdf for $\boldsymbol{\theta}_1$ is

$$(12) \qquad p(\boldsymbol{\theta}_1|D) = \int_{\Theta_2} p(\boldsymbol{\theta}_1, \boldsymbol{\theta}_2|D)\, d\boldsymbol{\theta}_2 = \int_{\Theta_2} p(\boldsymbol{\theta}_1|\boldsymbol{\theta}_2)p(\boldsymbol{\theta}_2|D)\, d\boldsymbol{\theta}_2$$

where $\boldsymbol{\theta}_2 \subset \Theta_2$.[9] The integration in (12) is a very convenient means of getting rid of nuisance parameters, for example autocorrelation coefficients, disturbance variances and covariances, etc., a problem that is often difficult to solve in non-Bayesian approaches. The marginal posterior pdf for $\boldsymbol{\theta}_1$ in (12) can be used to make posterior probability statements about the elements of $\boldsymbol{\theta}_1$, to compute moments of the elements of $\boldsymbol{\theta}_1$, etc.

If a point estimate for $\boldsymbol{\theta}$ is desired, given that we have a convex loss function, $L(\boldsymbol{\theta}, \hat{\boldsymbol{\theta}})$, where $\hat{\boldsymbol{\theta}}$ denotes an estimate for $\boldsymbol{\theta}$, the optimal estimate is obtained by minimizing the posterior expectation of the loss function with respect to $\hat{\boldsymbol{\theta}}$, that is by solving the following problem:

$$(13) \qquad \min_{\hat{\boldsymbol{\theta}}} \int_{\Theta} L(\boldsymbol{\theta}, \hat{\boldsymbol{\theta}})p(\boldsymbol{\theta}|D)\, d\boldsymbol{\theta}.$$

The solution to (13), say $\hat{\boldsymbol{\theta}}^*$ is the Bayesian estimate that is tailored to be optimal relative to the particular loss function that is deemed reasonable. For example, for the quadratic loss function, $L(\boldsymbol{\theta}, \hat{\boldsymbol{\theta}}) = (\boldsymbol{\theta} - \hat{\boldsymbol{\theta}})'A(\boldsymbol{\theta} - \hat{\boldsymbol{\theta}})$, with $A$ a given, positive definite symmetric matrix, $\hat{\boldsymbol{\theta}}^*$ = the posterior mean of $\boldsymbol{\theta}$, i.e., $\hat{\boldsymbol{\theta}}^* = \int_{\Theta} \boldsymbol{\theta} p(\boldsymbol{\theta}|D)\, d\boldsymbol{\theta}$. For an absolute error loss function, the median of the posterior pdf is optimal. If other loss functions are employed, the minimization problem in (13) can be solved analytically or numerically to obtain an optimal point estimate. Generally, Bayesian estimates have been shown to have very good sampling properties in both small and large samples. A sufficient condition for Bayesian estimators to be admissible is

$$(14) \qquad \text{Average Risk} = \int_{\Theta} r(\boldsymbol{\theta})p(\boldsymbol{\theta}|I_0)\, d\boldsymbol{\theta} < \infty$$

where $r(\boldsymbol{\theta}) = \int_{R_y} L(\boldsymbol{\theta}, \hat{\boldsymbol{\theta}})p(y|\boldsymbol{\theta})\, dy$, the risk function for $\hat{\boldsymbol{\theta}}$ relative to the loss function $L(\boldsymbol{\theta}, \hat{\boldsymbol{\theta}})$. When (14) is satisfied, as it usually is when $p(\boldsymbol{\theta}|I_0)$ is a proper prior pdf, the Bayesian estimator, $\hat{\boldsymbol{\theta}}^*$, minimizes average risk given in (14). Thus Bayesian estimators have rather good sampling properties as shown analytically and in Monte Carlo studies.[10]

[9] The conditional posterior pdf, $p(\boldsymbol{\theta}_1|\boldsymbol{\theta}_2, D)$ has been used in many analyses to explore the effects of various assumptions about $\boldsymbol{\theta}_2$'s value on inferences about $\boldsymbol{\theta}_1$. See, e.g., Box and Tiao [6] and Zellner [58] for examples.

[10] See, e.g., Chetty [9], Fomby and Guilkey [15], Gibbons [19], Griffiths and Dao [20], Lee, Judge, and Zellner [32], Park [43], Surekha and Griffiths [50], Thornber [53], and Zellner [58].

As regards prediction procedures, let $y_f$ denote a vector of as yet unobserved variables with pdf, $p(y_f|\theta)$. Since $\theta$'s value is unknown, it is hard to use this pdf in practice. However, the joint pdf for $y_f$ and $\theta$ is $p(y_f|\theta)\times p(\theta|D)$, where $p(\theta|D)$ is the posterior pdf for $\theta$ given in (11). Thus the marginal or predictive pdf for $y_f$, $p(y_f|D)$, is readily obtained by integration, that is,

$$(15)\qquad p(y_f|D)=\int_{\Theta} p(y_f|\theta)p(\theta|D)\,d\theta$$

which can be used to make probability statements about the elements of $y_f$, to compute their moments, etc. Further, if $L(y_f, \hat{y}_f)$ is a convex loss function, where $\hat{y}_f$ is a point prediction, an optimal value of $\hat{y}_f$ can be obtained by minimizing expected loss with respect to $\hat{y}_f$, that is by solving the following problem:

$$(16)\qquad \min_{\hat{y}_f}\int_{R_y} L(y_f, \hat{y}_f)p(y_f|D)\,dy_f.$$

The solution to (16), say $\hat{y}_f^*$, is the Bayesian point prediction.[11] If, for example, the loss function is quadratic, then $\hat{y}_f^*$ is the mean of the predictive pdf, $p(y_f|D)$. As with Bayesian estimates, the Bayesian predictor has very good sampling properties in small and large samples.

With respect to large sample properties of Bayesian estimation procedures, it has been shown in the literature that the posterior pdf $p(\theta|D)$ in (11) approaches a multivariate normal pdf as the sample size grows with mean equal to the maximum likelihood estimate, $\hat{\theta}_{MLE}$, and covariance matrix equal to the inverse of the estimated Fisher information matrix, $\hat{Inf}^{-1}$, that is $p(\theta|D)\rightarrow N(\hat{\theta}_{MLE}, \hat{Inf}^{-1})$ as the sample size gets large. Heyde and Johnstone **[25]** and Hartigan [23] show that when the observations are i.i.d., the conditions needed for this result are identical to those needed for proving the asymptotic normality of maximum likelihood estimators. However, when the observations are stochastically dependent, as in time series analyses, Heyde and Johnstone **[25]** show that the conditions needed for the asymptotic normality of the posterior pdf are simpler and more robust than those needed to establish the asymptotic normality of the maximum likelihood estimator. While these results are interesting, it should be appreciated that Bayes' Theorem in (11) provides operational, exact finite sample results and thus there is no need in general to rely on approximate, asymptotic results that are often poor approximations in small sample cases.[12]

Bayesian posterior odds ratios are available for comparing and testing hypotheses and for solving model selection problems. For example, consider two mutually exclusive hypotheses, $H_1$ and $H_2$ and assume that the prior odds for these two hypotheses is $\Pi_1/\Pi_2$, where $\Pi_i$, $i=1, 2$, is the prior probability associated with $H_i$. Then, from Bayes' Theorem, the posterior odds associated with $H_1$ versus

[11] See Varian [54] for a very interesting computation of $\hat{y}_f^*$ for an asymmetric LINEX loss function. Also Litterman [35] computes $\hat{y}_f^*$ in his work with Bayesian VAR models.

[12] Zellner and Rossi [66] provide analysis to illustrate this point with respect to logit models.

$H_2$, denoted by $K_{12}$ is given by

$$(17) \qquad K_{12} = (\text{prior odds}) \times (\text{Bayes' factor})$$

$$= (\Pi_1/\Pi_2) p_1(y|H_1)/p_2(y|H_2)$$

where $y$ is an observation vector and $p_i(y|H_i)$ is the pdf for $y$ given $H_i$, $i = 1, 2$. If $H_1$ and $H_2$ are both simple hypotheses, the Bayes' factor, $p_1(y|H_1)/p_2(y|H_2)$ is identical to the likelihood ratio. On the other hand, if $H_1$ and $H_2$ are composite hypotheses involving free parameters, $\boldsymbol{\theta}_1$ and $\boldsymbol{\theta}_2$, respectively, then the Bayes' factor is given by

$$(18) \qquad p_1(y|H_1)/p_2(y|H_2)$$

$$= \int_{\Theta_1} p_1(y|\boldsymbol{\theta}_1, H_1) p_1(\boldsymbol{\theta}_1|I_1)\, d\boldsymbol{\theta}_1 \Big/ \int_{\Theta_2} p_1(y|\boldsymbol{\theta}_2, H_2) p_2(\boldsymbol{\theta}_2|I_2)\, d\boldsymbol{\theta}_2 ,$$

where $p_i(y|\boldsymbol{\theta}_i, H_i)$ is the pdf for $y$ given $\boldsymbol{\theta}_i$ and $H_i$ and $p_1(\boldsymbol{\theta}_i|I_i)$ is the prior pdf for $\boldsymbol{\theta}_i \subset \Theta_i$ given prior information $I_i$, $i = 1, 2$. Since $p_i(y|H_i)$ can be computed before observing $y$, the Bayes' factor in (18) is in the form of a ratio of predictive pdfs. Bayes' factors have been computed for many hypotheses relating to means, variances, regression coefficients, coefficients of ARIMA models, etc.,[13] and in a number of cases related to sampling theory test statistics and $P$-values. Also, analyses of more than two hypotheses can be performed and asymptotic properties of posterior odds have been derived. While these asymptotic results are interesting, it should be noted that $K_{12}$ in (17) is an exact finite sample measure of the relative degree of confidence associated with $H_1$ and $H_2$. Since non-Bayesian testing procedures do not involve the concept of the probability of a hypothesis, there is no non-Bayesian analogue of the posterior odds, $K_{12}$.

If the two hypotheses $H_1$ and $H_2$ are mutually exclusive and exhaustive, then $K_{12} = P/(1-P)$, where $P$ is the posterior probability associated with $H_1$. Suppose, for example, that $H_1$ is $\boldsymbol{\theta} = \boldsymbol{\theta}_0$, with $\boldsymbol{\theta}_0$ a given value of $\boldsymbol{\theta}$, and $H_2$ is $\boldsymbol{\theta} \neq \boldsymbol{\theta}_0$. From (17), the value of $K_{12} = P/(1-P)$ can be computed and from it a value of $P$ can be obtained. If our loss function is $L(\boldsymbol{\theta}, \hat{\boldsymbol{\theta}}) = (\boldsymbol{\theta} - \hat{\boldsymbol{\theta}})'A(\boldsymbol{\theta} - \hat{\boldsymbol{\theta}})$, with $A$ a given pds matrix, then expected loss is given by

$$EL(\boldsymbol{\theta}, \hat{\boldsymbol{\theta}}) = PL(\boldsymbol{\theta}_0, \hat{\boldsymbol{\theta}}) + (1-P) EL(\boldsymbol{\theta}, \hat{\boldsymbol{\theta}})|H_2$$

where $E$ on the right-hand side denotes the expectation based on the posterior pdf for $\boldsymbol{\theta}$ under $H_2$. On inserting the quadratic loss function and minimizing expected loss with respect to $\hat{\boldsymbol{\theta}}$, the optimal point estimate, $\hat{\boldsymbol{\theta}}^*$ is

$$(19) \qquad \hat{\boldsymbol{\theta}}^* = P\boldsymbol{\theta}_0 + (1-P)\bar{\boldsymbol{\theta}}_2$$

$$= \boldsymbol{\theta}_0 + \frac{1}{1+K_{12}}(\bar{\boldsymbol{\theta}}_2 - \boldsymbol{\theta}_0)$$

[13] See e.g., Jeffreys [26], Geisel [17], Leamer [31], Monahan [37], and Zellner [58].

where $\bar{\theta}_2$ is the posterior mean of $\theta$ under $H_2$. $\hat{\theta}^*$ is a Bayesian "pre-test" estimate.[14] The second line of (19) involves the "shrinkage" factor $1/(1+K_{12})$ and is in Stein-like form. The analysis can be easily modified to apply to other types of hypotheses and to combining predictions from alternative models.[15] That these procedures are relatively simple and permit flexible incorporation of prior information through the prior odds and prior pdfs for parameters is noteworthy.

With respect to control and other policy analysis problems, the Bayesian approach yields solutions that are operational, reflect uncertainty about parameters' values, and incorporate allowance for sequential learning. To illustrate, let $y_f = x\beta + u_f$ where $y_f$ is a future value of a policy variable, $x$ is a control variable, $\beta$ is a regression coefficient, and $u_f$ is a future error term with zero mean and variance $\sigma^2$. If the loss function is $L(y_f, y_*) = (y_f - y_*)^2$, where $y_*$ is a given target value for $y_f$, expected loss is given by $E(x\beta + u_f - y_*)^2$, or

$$E(y_f - y_*)^2 = x^2 E\beta^2 - 2y_* x E\beta + Eu_f^2 + y_*^2$$

under the assumption that $u_f$ and $\beta$ are independent. The value of $x$ that minimizes expected loss is

$$(20) \qquad x^* = y_* E\beta / E\beta^2 = (y_*/E\beta)\frac{1}{1+\operatorname{var}\beta/(E\beta)^2}$$

where $E\beta$ and $\operatorname{var}\beta$ are the posterior mean and variance of $\beta$, respectively.[16] It is seen that the expression for $x^*$ involves not only $E\beta$ but also $\operatorname{var}\beta$, the posterior variance of $\beta$, a measure of uncertainty about the value of $\beta$. In the literature, more elaborate one-period and multi-period control problems have been analyzed using Bayesian methods. In multiperiod problems, learning and design considerations enter. Comparisons of Bayesian and non-Bayesian control solutions have generally indicated that Bayesian solutions are better.[17]

Above, some Bayesian procedures and methods have been briefly reviewed. It is relevant to question whether these methods work well in applications. In this regard, it is important to realize that many, if not all, non-Bayesian results can be produced by Bayesian methods under special assumptions. For example, with a large sample assumption, it was mentioned earlier that a posterior mean that is optimal relative to quadratic loss is approximately equal to the maximum likelihood estimate. Also many non-Bayesian estimates such as SUR regression coefficient estimates, 2SLS and 3SLS estimates, Cochrane-Orcutt estimates, etc., have been shown to be means of *conditional* posterior pdfs, based on diffuse priors, that is conditional posterior pdfs in which nuisance parameters are set equal to sample estimates. In *large samples* such approximate results may be

[14] This estimate was presented in Zellner and Vandaele [**65**] and a similar approach was employed in Griffiths and Dao [**20**].

[15] Geisel [**17**] used this approach in an application to combine predictions from two non-nested models.

[16] See Fisher [**14**], Harkema [**22**], Prescott [**46**], and Zellner [**58**] for further analysis of this and other control problems.

[17] See, e.g., Harkema [**22**], Prescott [**46**], and Zellner [**58**].

satisfactory; however in small to moderate sized samples these conditional posterior pdfs are poor approximations to the relevant marginal posterior pdfs and thus such approximations lead to poor estimates and other inferences.[18]

Bayesian methods have been used successfully in a number of applied studies. I shall just mention a few examples. Litterman [35] has used Bayesian vector autoregressions in forecasting seven important U.S. quarterly macroeconomic variables including real GNP, the implicit deflator for GNP, the unemployment rate, etc. In his work, Litterman used and is using a cleverly formulated informative prior pdf for the many parameters of his seven variable VAR which effectively reduces the number of free parameters. His forecasts compare very favorably with those obtained from unrestricted VARs, univariate AR models, Box-Jenkins' models, and two large structural econometric models. Varian [54] has applied Bayesian regression techniques to the problem of real estate assessment employing an asymmetric LINEX loss function. Akaike and Ishiguro [1] and their colleagues have developed a Bayesian approach and computer program, BAYSEA, for seasonal adjustment. Comparative study of the output of BAYSEA with the results of other seasonal adjustment procedures indicate that the former results compare favorably with other results. Morris [40] has discussed and referenced many applications of empirical Bayes' procedures. Prescott [46], Bowman and Laporte [5], Harkema [22], and Zellner [58, Ch. 11] have compared Bayesian and non-Bayesian solutions to policy control problems with results indicating that the former are superior in general. Swamy [51] and Swamy and Mehta [52] have compared Bayesian and non-Bayesian solutions to the undersized sample problem with the finding that the former are as good or better than non-Bayesian solutions. Monahan [37] has developed and applied Bayesian procedures for Box-Jenkins' models that include very good model selection, estimation, and prediction methods. Wright [56] reports a very interesting application of Bayesian methods in solving an overhead cost allocation problem. Miller [36] reviews Bayesian applications in actuarial statistics. Many applications of Bayesian analysis are reported in Kanjii [28]. Fomby and Guilkey [15] and Griffiths and Dao [20] have shown in Monte Carlo experiments that Bayesian estimators have lower mean-squared error (MSE) than non-Bayesian estimators for a regression coefficient in a simple regression model with autocorrelated disturbances. Surekha's and Griffiths' [50] Monte Carlo results for estimation of regression coefficients when disturbance terms are heteroscedastic indicate that Bayesian estimators outperform leading non-Bayesian estimators according to a MSE criterion. Park's [43] Monte Carlo experimental results show that a Bayesian minimum expected loss (MELO) estimator performs somewhat better than OLS and 2SLS estimators according to several criteria including MSE and mean absolute deviation. Brown's [7], Bawa, Brown, and Klein's [2], and Jorion's [27] Bayesian procedures for portfolio choice perform better than certainty equivalence procedures.[19]

[18] For examples, see Box and Tiao [6], Leamer [31], and Zellner [58, 60].

[19] For further references, see the many papers listed under the key words "Bayes" and "Bayesian" in annual issues of the ASA/IMS *Current Index to Statistics*, Washington, DC: American Statistical Association.

In summary, I have emphasized that substantial prior information is being used in econometric analyses, usually informally and sometimes incoherently.[20] The Bayesian approach permits investigators to use prior information formally in a flexible, coherent framework to solve estimation, testing, prediction, model selection, design, and policy analysis problems. Some studies have been cited to demonstrate that the Bayesian approach not only has good theoretical properties but has also been successfully applied in practice. It is to be expected that many more Bayesian applications will be reported in the future given that a growing number of Bayesian computer programs are becoming available—see, e.g., Press [47], and that many more econometricians are becoming familiar with Bayesian methods. Since wise use of Bayesian techniques can lead to better solutions to econometric problems, Bayesian methods can contribute importantly to the progress of econometric science.

*University of Chicago*

*Manuscript received January, 1984; revision received May, 1984.*

REFERENCES

[1] AKAIKE, H., AND M. ISHIGURO: "Comparative Study of the X-11 and BAYSEA Procedures of Seasonal Adjustment," in *Applied Time Series Analysis of Economic Data*, ed. by A. Zellner. U.S. Government Printing Office: Washington, DC., 1983 (presented at the ASA-CENSUS-NBER Conference on Applied Time Series Analysis of Economic Data, Washington, DC, 1981).

[2] BAWA, V. S., S. J. BROWN, AND R. W. KLEIN: *Estimation Risk and Optimal Portfolio Choice*. Amsterdam: North-Holland, 1979.

[3] BAXTER, M.: "The Role of Expectations in Stabilization Policy," presented at the Econometric Society meeting, Pisa, Italy, 1983.

[4] BHALLA, S. S.: "Measurement Errors and the Permanent Income Hypothesis: Evidence from Rural India," *American Economic Review*, 69(1979), 295-307.

[5] BOWMAN, H. W., AND A. M. LAPORTE: "Stochastic Optimization in Recursive Equation Systems with Random Parameters with an Application to Control of the Money Supply," in *Studies in Bayesian Econometrics and Statistics in Honor of Leonard J. Savage*, ed. by S. E. Fienberg and A. Zellner. Amsterdam: North-Holland, 1975, pp. 427-440.

[6] BOX, G. E. P., AND G. C. TIAO: *Bayesian Inference in Statistical Analysis*. Reading, Massachusetts: Addison-Wesley, 1973.

[7] BROWN, S. J.: "Optimal Portfolio Choice Under Uncertainty," Ph.D. Thesis, Graduate School of Business, U. of Chicago, 1976.

[8] CHAU, L. C.: "Stock Adjustment, Liquidity, and Normal Income in the Determination of Household Expenditures on Durable Goods," Ph.D. Thesis, Department of Economics, University of Wisconsin at Madison, 1967.

[9] CHETTY, V. K.: "*Bayesian Analysis of Some Simultaneous Equation Econometric Models*," Ph.D. Thesis, Department of Economics, University of Wisconsin at Madison, 1966.

[10] DEGROOT, M. H.: *Optimal Statistical Decisions*. New York: McGraw Hill Book Co., 1970.

[11] DRÈZE, J. H.: "Econometrics and Decision Theory," *Econometrica*, 40(1972), 1-17.

[12] ——— "Bayesian Theory of Identification in Simultaneous Equation Models," in *Studies in Bayesian Econometrics and Statistics in Honor of Leonard J. Savage*, ed. by S. E. Fienberg and A. Zellner. Amsterdam: North-Holland, 1975, pp. 159-174.

[13] FIENBERG, S. E., AND A. ZELLNER (EDS.): *Studies in Bayesian Econometrics and Statistics in Honor of Leonard J. Savage*. Amsterdam: North-Holland, 1975.

[14] FISHER, W. D.: "Estimation in the Linear Decision Model," *International Economic Review*, 3(1962), 1-29.

[20] See Swamy [51] for some striking examples of incoherent use of prior information in some leading non-Bayesian "solutions" to the "undersized sample" problem.

[15] Fomby, T. B., and D. K. Guilkey: "On Choosing the Optimal Level of Significance for the Durbin-Watson Test and the Bayesian Alternatives," *Journal of Econometrics*, 8(1978), 203-213.
[16] Friedman, M.: *A Theory of the Consumption Function*. Princeton, New Jersey: Princeton University Press, 1957.
[17] Geisel, M. S.: "Bayesian Comparisons of Simple Macroeconomic Models," in *Studies in Bayesian Econometrics and Statistics in Honor of Leonard J. Savage*, ed. by S. E. Fienberg and A. Zellner. Amsterdam: North-Holland, 1975, pp. 227-256.
[18] Gersovitch, M.: "Savings and Nutrition at Low Incomes," *Journal of Political Economy*, 91(1983), 841-855.
[19] Gibbons, M. R.: "Estimating the Parameters of the Capital Asset Pricing Model (CAPM): A Minimum Expected Loss (MELO) Approach," Technical Report, Graduate School of Business, Stanford University, 1980.
[20] Griffiths, W. E., and D. Dao: "A Note on the Bayesian Estimator in an Autocorrelated Error Model," *Journal of Econometrics*, 12(1980), 389-392.
[21] Hadamard, J.: *The Psychology of Invention in the Mathematical Field*. New York: Dover Publications, Inc., 1945.
[22] Harkema, R.: "An Analytical Comparison of Certainty Equivalence and Sequential Updating," *Journal of the American Statistical Association*, 70(1975), 348-350.
[23] Hartigan, J. A.: *Bayes Theory*. New York: Springer-Verlag, 1983.
[24] Havenner, A., and P. A. V. B. Swamy: "A Random Coefficient Approach to Seasonal Adjustment of Economic Time Series," *Journal of Econometrics*, 15(1981), 177-209.
[25] Heyde, C. C., and I. M. Johnstone: "On Asymptotic Posterior Normality for Stochastic Processes," *Journal of the Royal Statistical Society, Series B*, 41(1979), 184-189.
[26] Jeffreys, H.: *Theory of Probability*. London: Oxford University Press, 1967.
[27] Jorion, P.: "Portfolio Analysis of International Equity Investments: Bayes-Stein Estimators for the Mean and Implications for Risk Reduction," Ph.D. Thesis, Graduate School of Business, University of Chicago, 1983.
[28] Kanji, G. K. (Ed.): "Proceedings of the Annual Conference on Practical Bayesian Statistics" Cambridge University, 1982, published in *Journal of the Institute of Statisticians*, 32(1983), 1-278.
[29] Laub, P. M.: "The Dividend-Earning Relationship: A Study of Corporate Quarterly Panel Data, 1945-65," Ph.D. Thesis, Graduate School of Business, University of Chicago, 1971.
[30] Laub, P. M.: "Some Aspects of the Aggregation Problem in the Dividend-Earning Relationship," *Journal of the American Statistical Association*, 67 (1972), 552-559.
[31] Leamer, E. E.: *Specification Searches*. New York: John Wiley, 1978.
[32] Lee, T. C., G. G. Judge, and A. Zellner: *Estimating the Parameters of the Markov Probability Model from Aggregate Time Series Data*. Amsterdam: North-Holland, 1977.
[33] Levedahl, J. W.: "The Effects of Households' Permanent and Transitory Income on Both Their Timing of Automobile Purchases and Their Net Outlay for Automobile Services," Ph.D. Thesis, Department of Economics, University of Chicago, 1976.
[34] Lindley, D. V.: *Introduction to Probability and Statistics from a Bayesian Viewpoint* (2 vols.). Cambridge: Cambridge University Press, 1965.
[35] Litterman, R.: "A Bayesian Procedure for Forecasting with Vector Autoregressions," unpublished manuscript, Department of Economics, Massachusetts Institute of Technology, 1980.
[36] Miller, R.: "Actuarial Applications of Bayesian Statistics," in *Bayesian Analysis in Econometrics and Statistics: Essays in Honor of Harold Jeffreys*, ed. by A. Zellner. Amsterdam: North-Holland, 1980, pp. 197-212.
[37] Monahan, J. F.: "Fully Bayesian Analysis of ARIMA Time Series Models," *Journal of Econometrics*, 21(1983), 307-332.
[38] Montmarquette, C.: "A Model of Inventory Holdings with Empirical Application to Canadian Manufacturing Industries" Ph.D. Thesis, Department of Economics, University of Chicago, 1973.
[39] ———: "Random Walk Behavior of Finished Goods Inventory Investment," *Révue Economique*, 28(1977), 352-375.
[40] Morris, C. N.: "Parametric Empirical Bayes Inference: Theory and Applications," *Journal of the American Statistical Association*, 78(1983), 47-54.
[41] Muth, J.: "Rational Expectations and the Theory of Price Movements," *Econometrica*, 29(1961), 315-335.
[42] Nerlove, M.: "Lags in Economic Behavior," *Econometrica*, 40(1972), 221-251.
[43] Park, S. B.: "Some Sampling Properties of Minimum Expected Loss (MELO) Estimators of Structural Coefficients;" *Journal of Econometrics*, 18 (1982), 295-312.

[44] PECK, S. C.: "A Test of Alternative Theories of Investment Using Data from the Electric Utilities Industry," Ph.D. Thesis, Graduate School of Business, University of Chicago, 1973.

[45] PECK, S. C.: "Alternative Investment Models for Firms in the Electric Utilities Industry," *Bell Journal of Economics and Management Science*, 5(1974), 420-458.

[46] PRESCOTT, E. C.: "Adaptive Decision Rules of Macroeconomic Planning," in *Studies in Bayesian Econometrics and Statistics in Honor of Leonard J. Savage*, ed. by S. E. Fienberg and A. Zellner. Amsterdam: North-Holland, 1975, pp. 427-440.

[47] PRESS, S. J.: "Bayesian Computer Programs," in *Bayesian Analysis in Econometrics and Statistics: Essays in Honor of Harold Jeffreys*, ed. by A. Zellner. Amsterdam: North-Holland, 1980, pp. 429-442.

[48] QUENOUILLE, M. H.: *The Analysis of Multiple Time-Series*. New York: Hafner Publishing Co., 1957.

[49] SAVAGE, L. J.: *The Foundations of Statistics*. New York: John Wiley, 1954.

[50] SUREKHA, K., AND W. E. GRIFFITHS: "A Monte Carlo Comparison of Some Bayesian and Sampling Theory Estimators in Two Heteroscedastic Error Models," manuscript, Indian Statistical Institute, New Delhi, and University of New England, Armidale, Australia, 1983.

[51] SWAMY, P. A. V. B.: "A Comparison of Estimators for Undersized Samples," *Journal of Econometrics*, 14(1980), 161-181.

[52] SWAMY, P. A. V. B., AND J. S. MEHTA: "Further Results on Zellner's Minimum Expected Loss and Full Information Maximum Likelihood Estimators for Undersized Samples," *Journal of Business and Economic Statistics*, 1(1983), 154-162.

[53] THORNBER, H.: "Finite Sample Monte Carlo Studies: An Autoregressive Illustration," *Journal of the American Statistical Association*, 62(1967), 801-808.

[54] VARIAN, H. R.: "A Bayesian Approach to Real Estate Assessment," in *Studies in Bayesian Econometrics and Statistics in Honor of Leonard J. Savage*, ed. by S.E. Fienberg and A. Zellner. Amsterdam: North-Holland, 1975, pp. 195-208.

[55] VELOCE, W., AND A. ZELLNER: "Modelling a Competitive Industry with Entry: Implications for Demand and Supply Analysis," *Economics Letters*, 16 (1984), 71-75.

[56] WRIGHT, R. L.: "Measuring the Precision of Statistical Cost Allocations," *Journal of Business and Economic Statistics*, 1(1983), 93-100.

[57] ZELLNER, A.: "Tests of Some Basic Propositions in the Theory of Consumption," *American Economic Review*, 50(1960), 565-573.

[58] ———: *An Introduction to Bayesian Inference in Econometrics*. New York: John Wiley, 1971.

[59] ———: *Seasonal Analysis of Economic Time Series*, U.S. Bureau of the Census, Washington, DC: U.S. Government Printing Office, 1978.

[60] ———: "Statistical Analysis of Econometric Models," *Journal of the American Statistical Association*, 74(1979), 628-641.

[61] ———: "The Current State of Bayesian Econometrics," *Basic Issues in Econometrics*, ed. by A. Bellner, Chicago: University of Chicago Press, 1984, 306-321.

[62] ———: *Bayesian Analysis in Econometrics and Statistics: Essays in Honor of Harold Jeffreys*. Amsterdam: North-Holland, 1980.

[63] ———: "Philosophy and Objectives of Econometrics," in *Macroeconomic Analysis: Essays in Macroeconomics and Econometrics*, ed. by D. Currie, R. Nobay and D. Peel. England: Croom Helm, 1981.

[64] ZELLNER, A., AND F. PALM: "Time Series Analysis and Simultaneous Equation Econometric Models," *Journal of Econometrics*, 2(1974), 17-54.

[65] ZELLNER, A., AND W. VANDAELE: "Bayes-Stein Estimators for $k$-Means, Regression and Simultaneous Equation Models," in *Studies in Bayesian Econometrics and Statistics in Honor of Leonard J. Savage*, ed. by S.E. Fienberg and A. Zellner. Amsterdam: North-Holland, 1975, pp. 627-653.

[66] ZELLNER, A., AND P. E. ROSSI: "Bayesian Analysis of Dichotomous Quantal Response Models," *Journal of Econometrics*, 25 (1984), 365-393.

# Name Index